Advances in Carbon Management Technologies

Volume 1

Carbon Removal, Renewable and Nuclear Energy

Editors

Subhas K Sikdar

Retired, Cincinnati, OH, USA
formerly Associate Director for Science
National Risk Management Research Laboratory
US Environment Protection Agency, Cincinnati, Ohio, USA

Frank Princiotta

Retired, Chapel Hill, North Carolina, USA
formerly Director, Air Pollution Prevention and Control Division
National Risk Management Research Laboratory
US Environment Protection Agency, Research Triangle Park, NC, USA

T0225326

CRC Press
Taylor & Francis Group
Boca Raton London New York

CRC Press is an imprint of the
Taylor & Francis Group, an **informa** business

A SCIENCE PUBLISHERS BOOK

CRC Press
Taylor & Francis Group
6000 Broken Sound Parkway NW, Suite 300
Boca Raton, FL 33487-2742

First issued in paperback 2022

© 2020 by Taylor & Francis Group, LLC
CRC Press is an imprint of Taylor & Francis Group, an Informa business

No claim to original U.S. Government works

Version Date: 20200210

ISBN-13 978-0-367-19842-8 (hbk)
ISBN-13: 978-0-367-53364-9 (pbk)

DOI: 10.1201/9780429243608

**Visit the Taylor & Francis Web site at
http://www.taylorandfrancis.com**

**and the CRC Press Web site at
http://www.crcpress.com**

Dedication

To the fond memories of my mother, Biva; to my elder sister, Ratna; and my daughter, Manjorie; all of whom have deeply affected my professional and social attitude.

—Subhas Sikdar

To my children Thomas, Elizabeth and John. They, their generation and generations to follow, will reap from seeds we sow. If humanity doesn't get its act together in a hurry, climate change will drastically degrade the habitability of the home planet for them and their fellow species.

—Frank Princiotta

Preface

Average global temperatures, computed from measurements on land, ocean, and via satellites, have steadily increased since the Industrial Revolution, the time we started extracting and burning fossil fuels. Keeping pace with this, greenhouse gases in the atmosphere, principally CO_2, have been accumulating monotonically. Experts have developed climate models which suggest that in the absence of near-term dramatic emission reductions, warming over time could yield impacts jeopardizing life on Earth as we know it. To avoid such potential catastrophic impacts, experts, including the U.S. National Academy of Sciences, have prescribed replacement of fossil fuels, such as coal, petroleum, and natural gas with renewable and non-carbon fuels, principally solar, wind, and nuclear. This aspiration has been called the principal grand challenge of this century. This book, Advances in Carbon Management Technologies, was designed to take a stock of the state-of-the-science development along this journey.

Advances in carbon Management Technologies comprises 43 chapters, in 2 volumes, contributed by experts from all over the world. Volume 1 of the book, containing 22 chapters, discusses the status of technologies capable of yielding substantial reduction of carbon dioxide emissions from major combustion sources. Such technologies include renewable energy sources that can replace fossil fuels, and technologies to capture CO_2 after fossil fuel combustion or directly from the atmosphere, with subsequent permanent long-term storage. The introductory chapter emphasizes the gravity of the issues related to greenhouse gas emission-global temperature correlation, the state of the art of key technologies and the necessary emission reductions needed to meet international warming targets. Section 1 deals with global challenges associated with key fossil fuel mitigation technologies, including removing CO_2 from the atmosphere, and emission measurements. Section 2 presents technological choices for coal, petroleum, and natural gas for the purpose of reducing carbon footprints associated with the utilization of such fuels. Section 3 deals with promising contributions of alternatives to fossil fuels, such as hydropower, nuclear, solar photovoltaics, and wind.

Volume 2 of Advances in Carbon Management Technologies has 21 chapters. It presents the introductory chapter again, for framing the challenges that confront the proposed solutions discussed in this volume. Section 1 presents various ways biomass and biomass wastes can be manipulated to provide a low-carbon footprint of the generation of power, heat and co-products, and of recovery and reuse of biomass wastes for beneficial purposes. Section 2 provides potential carbon management solutions in urban and manufacturing environments. This section also provides state-of-the-art of battery technologies for the transportation sector. The chapters in section 3 deals with electricity and the grid, and how decarbonization can be practiced in the electricity sector.

The overall topic of advances in carbon management is too broad to be covered in a book of this size. It was not intended to cover every possible aspect that is relevant to the topic. Attempts were made, however, to highlight the most important issues of decarbonization from technological viewpoints. Over the years carbon intensity of products and processes has decreased, but the proportion of energy derived from fossil fuels has been stubbornly stuck at about 80%. This has occurred despite very rapid development of renewable fuels, because at the same time the use of fossil fuels has also increased. Thus, the challenges are truly daunting. It is hoped that the technology choices provided here will show the myriad ways that solutions will evolve. While policy decisions are the driving forces for technology development, the book was not designed to cover policy solutions.

As editors, we are thankful to the contributing authors for their great efforts in delivering the chapters in a timely fashion. We commend the chapter reviewers for great engagement with the topics and for providing constructive comments in each case. Without the utmost cooperation of the authors and reviewers we would not be able to meet the deadline to produce this timely book. Throughout the formative stages of this development, publisher's representative, Mr. Vijay Primlani, has provided encouragement and assisted us in every way possible to complete the project. We owe our gratitude to him.

Subhas K Sikdar
Cincinnati, Ohio

Frank Princiotta
Chapel Hill, North Carolina

Contents

Section 3. Wind/Solar/Hydro/Nuclear

INTRODUCTION

What Key Low-Carbon Technologies are Needed to Meet Serious Climate Mitigation Targets and What is their Status?

Frank Princiotta

1. Introduction

Since the industrial revolution, humanity has emitted Gigaton quantities of carbon dioxide (CO_2) and other greenhouse gases. Figure 1 (NASA GISS, 2019) shows that the warming that has occurred since 1880 has been in the order of 1.1 degrees centigrade higher than pre-industrial levels. As a result of manmade emissions, carbon dioxide concentrations have dramatically spiked to unprecedented levels when viewed from an 800,000-year perspective. Current concentrations of carbon dioxide are now approximately 410 PPM, relative to the 280-ppm level just before the industrial revolution. Note that in the absence of a serious global emission reduction program, CO_2 concentrations are projected to rise as high as 1000 ppm later this century.

Figure 2 compares actual warming (NASA GISS, 2019) to model projections. The model used was the Model for the Assessment of Greenhouse Gas Induced Climate Change (MAGICC), using middle of the road model assumptions and assuming a fossil fuel intensive emission trajectory (A1FI). It is important to note the close correlation of the actual warming relative to the model projections. As can be seen, if we continue on this fossil fuel intensive emission path and the model continues to accurately predict warming, the planet would be 1.5 °C warmer by 2035 and 2 °C warmer by 2045. These warming levels are particularly relevant since the international community has set a warming target of no greater than 2 °C and optimally below 1.5 °C by the end of this century. If we were to continue on our current fossil fuel intensive trajectory, 2100 warming is projected to be greater than 4 °C and rising.

To put the significance of such warming in perspective, Figure 3 was generated based on data from a reconstruction of global temperature for the last 11,300 years (Marcot, 2013), with more recent warming data and model projections included. When current and projected warming is viewed from this long-term perspective, it becomes clear that humankind has, in just 240 years, fundamentally changed the heat transfer characteristics of the planet, with even more dramatic change projected. Note that, as of 2017, warming was about 0.2 °C warmer than any time in the last 11,300 years. If we continue on our fossil fuel intensive emission trajectory, warming is projected to be in the order of 3.5 °C greater than any time over this period.

Retired Research Director, USEPA, 100 Longwood Drive, Chapel Hill, NC 27514.
Email: fprinciotta@msn.com

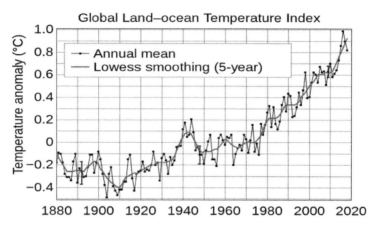

Figure 1. Warming that has occurred since industrial revolution.

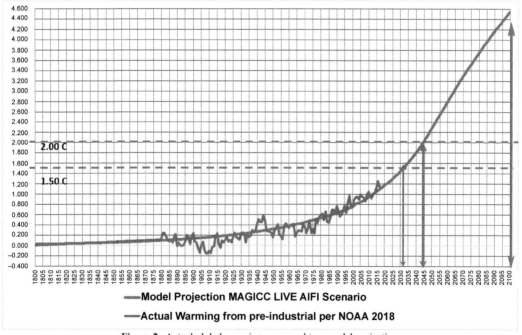

Figure 2. Actual global warming compared to a model projection.

2. Only Emissions of Greenhouse Gases (GHGs) Can Account for the Warming Experienced Since the Industrial Revolution

As discussed, it is clear that the planet has warmed considerably since the industrial revolution. A legitimate question Is whether such warming is a result of human emissions of greenhouse gases or the result of natural factors, such as solar variations and volcanic eruptions. Such eruptions can cool the planet after the reflective particles are driven into the stratosphere, while the planet could warm back up after the particles have settled out of the atmosphere. Figure 4 (USEPA, 2017) illustrates that when

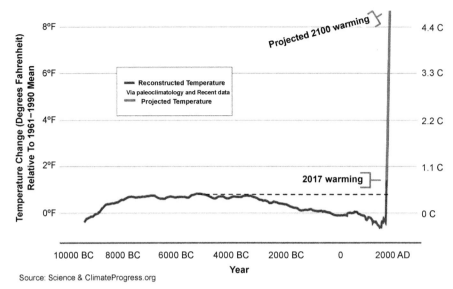

Source: Science & ClimateProgress.org

Figure 3. Temperature change (relative to the 1961–1990 mean) over past 11,300 years (in blue) plus projected warming this century on humanity's current emissions path (in red).

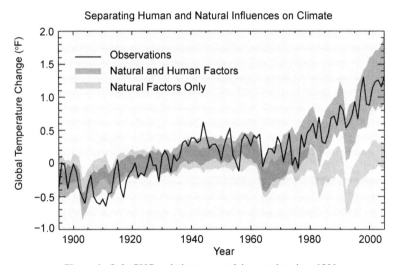

Figure 4. Only GHG emissions can explain warming since 1900.

comparing actual warming to warming predicted by models when only natural factors are considered versus when one accounts for the greenhouse gas impacts, it is clear that human GHG emissions have provided the driving force for the observed warming. Note that Figure 2 reinforces this conclusion, since model warming projections, assuming only GHG emission impacts, yield results consistent with the actual warming.

3. The Heat Added by Anthropogenic Emissions of GHGs is Already Yielding Major Impacts, with More to Come

It is not possible to change the heat balance of the earth so substantially, as humanity has done by adding large quantities of GHGs, without major impacts. It has been calculated (Skeptical Science, 2019) that

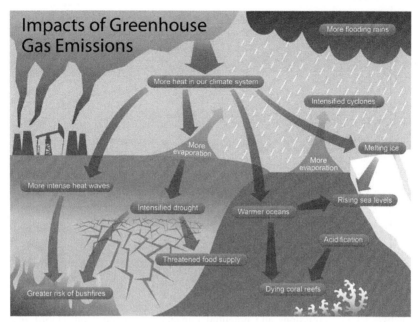

Figure 5. The impacts of greenhouse gas emissions.

the added heat associated with elevated levels of GHGs in the atmosphere since 1998 is equivalent to the detonation of 2.74 billion Hiroshima size atomic bombs. Currently, heat equivalent to four such bomb detonations is being added each second. Ten bombs per second of heat is projected to be added later this century, in the absence of serious global mitigation efforts. Figure 5 (Cook, 2017) illustrates the impacts of all this heat being added to the atmosphere. More heat means both higher temperatures in the atmosphere as well as greater rates of evaporation, yielding more flooding rains. The impacts of higher temperatures and greater evaporation rates are depicted in Figure 5. They include: More intensive heat waves and drought, potentially threatened food supplies, greater risk of wild fires, melting ice yielding seawater rise and more intense weather events, such as more dangerous cyclones. The ocean's ecosystems are also at risk due to a combination of ocean warming and acidification, since about 90% of all the heat and 25% of the CO_2 ends up in the oceans. CO_2 absorbed in the ocean generates carbonic acid which increases the ocean's acidity.

As Figure 2 illustrated, we face the prospects of warming at the 4 °C level later this century. Warren (2010) summarized the implications of a 4 °C warmer world as follows: "Enormous adaptation challenges in the agricultural sector, with large areas of cropland becoming unsuitable for cultivation. … large losses in biodiversity, forests, coastal wetlands… supported by an acidified and potentially dysfunctional marine ecosystem. Drought and desertification would be widespread, with large numbers of people experiencing increased water stress. … Human and natural systems would be subject to increasing levels of agricultural pests and diseases, and increases in the frequency and intensity of extreme weather events."

4. There is the Danger of a Runaway Situation if Warming occurs Too Rapidly and Activates Tipping Points Associated with Amplifying Feedbacks

It is important to note, that current projection models do not account for the possibility that there could be accelerating warming due to "tipping points" associated with driving forces that could yield a point in time when the global climate changes from one stable state to another, a threshold which reaches a point of "no return" that can change the planet irreversibly. Such points could cascade, yielding a "hothouse Earth". Figure 6 (an updated/upgraded figure from Climate Change Knowledge (2014)) illustrates the

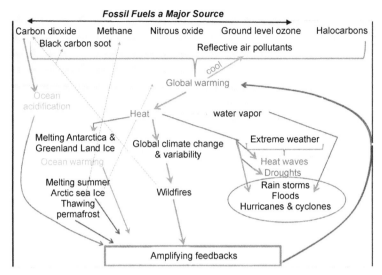

Figure 6. Amplifying feedbacks could yield "runaway" warming.

relationship between GHG emissions and potential impacts with a focus on Amplifying Feedbacks. Examples of such feedbacks, that if cascaded could contribute to such a runaway state include:

- Accelerated Melting of Arctic sea ice and Antarctica/Greenland land ice (such melting would decrease Earth's reflectivity, allowing more heat to be absorbed by the atmosphere).
- Melting of permafrost in Siberia, Canada and Alaska (large additional source of CO_2 & CH_4 not accounted for in models).
- Ocean warming and acidification along with increasing number and intensity of wildfires (these would have the effect of weakening CO_2 sinks, absorption on land and in the ocean, yielding an acceleration of growth of GHG concentrations in the atmosphere).

A recent study (Steffen, 2018) examined this issue and concluded that potential planetary thresholds yielding accelerating and potentially irreversible warming could occur at a temperature rise as low as 2.0 °C above preindustrial levels. They concluded that limiting warming to a maximum of 1.5 °C would dramatically lower the risk of this potentially catastrophic instability.

Although not discussed in the study, it follows that warming in the vicinity of 3–4 °C would substantially raise the probability of such tipping points, yielding a "hothouse earth".

5. Growing Global Emissions, the Result of a Growing Population Demanding an Expanding Array of Resource Intensive Goods and Services

The dramatic growth in GHG emissions since the industrial revolution are driven by two key drivers. First, world population has been growing relentlessly. World population is now at 7.5 billion, has tripled since 1950, and is expected to grow to over 9 billion by 2050. Second, in developed nations, people have expanded their list of "needs" to include personal transportation, residences with energy-intensive heating, cooling, and lighting, a diet heavily oriented toward meat consumption, and an ever-growing array of consumer goods and services. Developing countries are moving in the same direction, albeit at earlier stages. World population has been growing annually at about 1.2% and CO_2 emissions at 2.3% over the last 17 years.

Figure 7 illustrates the factors responsible for the challenges to long term sustainability with a focus on climate change, the most serious sustainability threat. The middle of the figure indicates that these human needs are met by means of a large array of industrial, agricultural, and energy technologies and

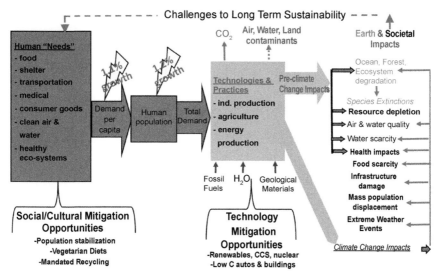

Figure 7. Drivers yielding GHG emissions and the two key mitigation approaches.

practices. Although, there are a multitude of sustainability impacts associated with these "technologies and practices", independent of climate change. The major threats are shown color coded in two categories: Earth and Societal impacts. These include, degradation of air and water quality, depletion of minerals and fresh water supplies and ecosystem damage. Unique climate change impacts are listed on the right side of the figure and include: Potential food scarcity, infrastructure damage, mass population displacement and extreme and damaging weather events. As indicated by the red return arrows, in addition to such unique impacts, climate change has the potential to exacerbate impacts associated with other human activities, such as ocean and forest degradation. The bottom of the figure indicates that there are two classes of mitigation opportunities. The most commonly considered approach is replacing/upgrading current technologies and practices. Another, less discussed, but potentially important if technology modifications alone are insufficient to avoid serious climate impacts, would be to modify social and cultural behavior toward energy-efficient and resource-intensive lifestyles.

6. Each Country Has a Unique GHG Emission Trajectory and Mitigation Challenge

When one examines GHG emissions on a country by country basis, fundamental differences in emission characteristics and mitigation challenges are observed. Table 1 (generated based on databases from Global Carbon Atlas (2018)) summarizes CO_2 emission data for the 14 largest emitters in 2017. They are positioned in the order of the magnitude of their emissions. China, the EU and the U.S. are by far the largest emitters. The developed countries are identified by the normal font, while those in various stages of economic transition are in the bold font. Let us briefly discuss the situation in key developed countries and then in developing countries. At this point it should be noted that the IPCC (2013) has concluded that, in order to have a chance of limiting warming to no greater than 2 °C, global per capita emissions should be between 1.1 and 2.2 t/person in 2050 and zero in 2100.

China is by far the largest emitter, passing the U.S. in 2006. It is considered somewhere between a developing and a developed country. Their 17-year emission growth rate at 6.6% is unmatched by any other country. They have rapidly transformed from a low-end developing country to a country with unprecedented economic growth via rapid urbanization, industrialization (supported by an unprecedented power generation expansion, primarily based on coal), and major growth of their on-road transportation fleet. Population growth at 0.6% has been only a minor factor in influencing their rapid emission growth. Their per capita emission has grown dramatically to 7 t/p and still growing.

Table 1. CO_2 emission data for countries with the greatest emissions in 2017.

Country	2017 Emissions GT CO_2	2017 per Capita Emissions tonnes/person	2017 Population millions	2000 to 2017 Annual Emission Growth Rate	2000 to 2017 Annual Population Growth Rate
China	9863	7.0	*1409*	6.6%	0.6%
USA	5184	16.0	324	−0.9%	0.8%
EU	3544	7.0	507	−1.0%	0.2%
India	*2446*	*1.8*	*1359*	*5.2%*	*1.5%*
Russia	1716	12.0	143	0.8%	−0.1%
Japan	1207	9.5	127	−0.3%	0.0%
Iran	*672*	*8.3*	*81*	*3.6%*	*1.2%*
Saudi Arabia	627	19.0	33	4.5%	2.7%
S. Korea	616	12.0	51	1.9%	0.5%
Canada	592	16.0	37	0.2%	1.0%
Brazil	*481*	*2.3*	*209*	*2.3%*	*1.0%*
Indonesia	*475*	*1.8*	*264*	*3.5%*	*1.3%*
S. Africa	456	8.0	57	1.1%	1.3%
Australia	408	17.0	24	0.9%	1.4%
Rest of World	*7866*	*2.6*	*2995*	*3.7%*	*2.2%*
Total	36153	4.7	7620	2.3%	1.3%

IPCC: Per Capita target for 2 °C maximum warming = 1.1 to 2.2 in 2050 and near Zero in 2100.

The U.S. and the EU are both highly developed with similar standards of living, yet the EU's per capita emissions are less than half those of the U.S. The EU has been the most conscientious regarding minimizing GHG emissions, has had a culture of treating energy conservation seriously, and uses less electricity per capita, in part because of very high electric rates. They have also has been leaders in utilizing wind and solar power. As a more population dense area with less dependence on large low-efficiency cars and trucks, their transportation emissions are much lower as well. Finally, the very low population growth has also been a factor. However, for both the EU and especially the US, reducing per capita emissions to the 1.1 to 2.2 level by 2050 will be a monumental challenge.

The developing world is in a fundamentally different situation. Per Table 1, countries in this category would be include India, Iran, Brazil, Indonesia and many African, South American and Asian countries in the "Rest of the world" designation. It is estimated that over 4 billion people fall in this category. They generally have similar characteristics: Low standards of living, modest per capita incomes, high birth rates and relatively low per capita emissions. The challenge for developing countries is to improve their economies while at the same time lowering or at least not raising their per capita emissions. India is a particularly important case since its population is close to that of China, with a fast-growing economy and a major increase in emissions (and per capita emissions) in recent years. This is not the direction this sector of the world's economy should be heading if we are serious about avoiding unacceptable climate change.

7. Greenhouse Gas Emissions are Associated with All Energy, Industrial and Agricultural/Land Use Sectors

In order to appreciate the scope of the mitigation challenge, it is important to understand the relationships between: The energy, industrial and agricultural/land use sectors, the related activities that are needed for the desired societal end uses ("needs") and the resulting GHG emissions. Figure 8 (WRI, 2007) quantifies these relationships for the world as of 2005, the last year such a chart was published. An example of these relationships follows: If the end use/activity is heating, cooling and lighting residential buildings, the relevant sectors are Electricity and Other Fuel Combustion. This is the case since some residences are heated with electricity (heat pumps or resistance heating) and others via direct combustion of a fuel, such as natural gas. So, both electricity and fuel combustion sectors contribute to CO_2 emissions associated with residential buildings, as indicated in Figure 8.

As can be seen, key end uses that drive energy-related emissions include road travel, residential and commercial building cooling, heating and lighting and the production of chemicals, cement, and iron & steel needed for production of goods. The net result are huge emissions of CO_2, most of which are associated with the combustion of coal, oil and natural gas. Also, the agricultural sector is responsible for the majority of emitted methane, the second most important greenhouse gas. Land use change was another important contributor to raising CO_2 concentrations, however less active deforestation in recent years has decreased the importance of this sector. Note that in 2005, 77% of the anthropogenic warming was associated with CO_2, with methane and N_2O contributing 15% and 7%, respectively. Also note, that the term CO_2 equivalent [$CO_2(e)$] emissions, is the amount of CO_2 which would have the equivalent global warming impact when accounting for CO_2 plus the other GHG gases. For 2005 that number globally was 44 Gt(e).

8. Major Emission Mitigation from All Sectors and GHGs is Required Immediately in Order to Have a Chance of Meeting International Targets

In order to have a chance of limiting warming to between 1.5 and 2.0 °C per the international community's stated target, it will be necessary to drastically reduce emissions as soon as possible. Figures 9 and 10 have been generated in order to help quantify this monumental challenge. The previously mentioned MAGICC model was used. Warming projected should be considered as "best guess" values, considering that there are uncertainties in such values, especially for long term projections. Both figures assess the

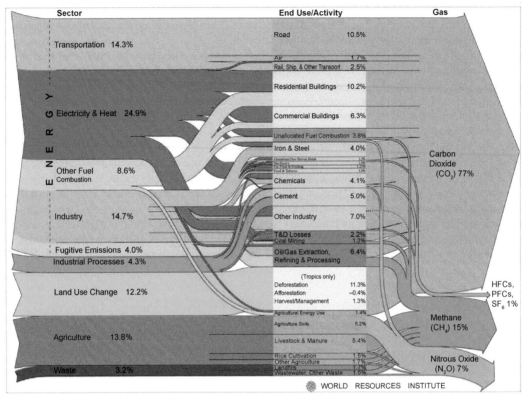

Figure 8. Global Greenhouse Gas emissions in 2005 by sector, end use and gas.

warming implications of six emission scenarios; Figure 9 shows the assumed emission scenarios in Gt CO_2 and Figure 10 shows the projected warming all the way to 2100, associated with these emission scenarios:

1) A business as usual case that assumes continued reliance on fossil fuels with no serious mitigation reductions for methane and nitrous oxide, the two other key GHGs.
2) A scenario consistent with the 2015 Paris Agreement where most countries "promised" significant but relatively modest emission reductions.
3) A serious CO_2 global emission reduction program with near term emission stabilization and 3% annual emission reductions, starting in **2035** and continuing for 65 years.
4) Scenario 3 above, with the addition of a complimentary CH_4 and N_2O emission reduction program, starting in 2035 as well.
5) Scenario 4 above, but with the addition of a major Direct Atmospheric Capture (DAC) program to remove CO_2 from the atmosphere, starting in **2030** and continuing for 70 years.
6) Scenario 5 above, but with CO_2, methane and N_2O mitigation starting ten years earlier, in **2025,** and maintained for 75 years.

As can be seen, with 3% annual CO_2 mitigation starting in 2035, it will be difficult to limit warming to below 2 °C. The addition of CH_4 and N_2O mitigation significantly reduces the warming, but limiting warming to 2 °C still appears unlikely. When one adds a major Direct Air Capture (DAC) component to the mitigation strategy it appears that limiting warming to 2 °C may be achievable. Finally, if we start CO_2, CH_4 and N_2O emission reductions ten years earlier, in 2025, and again supplement with a major DAC complimentary program, further warming reduction is achieved, raising the probability that warming would be limited to 2 °C. It is worth noting that none of these options appear to be able to limit warming to 1.5 °C, a likely unattainable target. DAC technology involves the construction of massive chemical plants designed to remove CO_2 directly from the atmosphere.

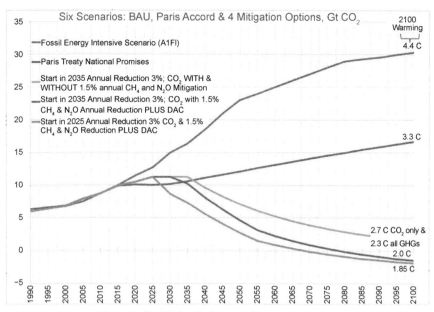

Figure 9. Six GHG emission scenarios, Gt CO_2 per year.

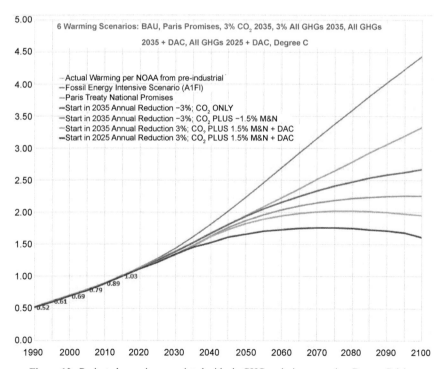

Figure 10. Projected warming associated with six GHG emission scenarios, Degree Celsius.

It should be noted that DAC technology is at a very early stage of development and implementation costs are likely to be very high per ton of carbon captured. Keith (2018) has analyzed a chemical sorbent process, aqueous sorbent KOH coupled to a calcium caustic sorbent recovery loop, and estimated capture costs at $94 to $232 per ton. Options 5 and 6 described above, assume DAC capture would start in 2030 with progressively greater annual removal quantities for a total of 37 Gt for the 70-year mitigation period.

Given this quantity, the estimated cost of such DAC capture would be between $1.3 to $2.5 trillion per year or $72 and $178 trillion over the 70-year period! This study did not include the required costs for permanent storage of the CO_2, probably in deep underground saline aquifers, which will likely add trillions of US$ to the cost of such an enterprise.

9. Emerging Technologies Will need to be Available and Extensively Utilized if Global Warming Target Levels Stand a Chance of Being Achieved

If the international community ultimately sets sufficiently aggressive emission requirements in order to minimize the worst consequences of climate change, affordable, practical low-carbon technologies would need to be commercially available within the next ten years. Of particular importance would be Carbon Capture and Storage (CCS) technologies, advanced nuclear generators, low-cost renewable generation with energy storage capability, efficient buildings and low-emission vehicles. Also, given the importance of methane and N_2O emissions (see Figures 10 and 11), the international community must agree on emission reductions for these pollutants as soon as possible as well. For methane, leakage from oil, gas and coal operations are particularly important. Agriculture operations are important sources for CO_2, methane and N_2O.

Figure 11, derived from IEA (2016), quantifies the amount of CO_2 that would have to be mitigated by technology, by sector, between now and 2050 in order to have a chance of limiting warming to 2 °C. As can be seen, all sectors require major CO_2 reductions.

Tables 2 to 5 have been generated with the aim of summarizing for key technologies by sector: The current state of the art, issues that could limit near term utilization and research, development and demonstration priorities.

Note that for Table 2 a column has been added which quantifies what IEA (2016) has determined would be the potential mitigation impact of each power generation technology, in terms of Gt of CO_2 mitigated between now and 2050. It is worth noting that, in recent years, solar and wind technologies have seen their capital costs reduced substantially along with cost reductions in storage technologies needed if these technologies can continue to generate power when the sun doesn't shine and the wind doesn't blow. On the other hand, CCS technology development has stumbled. Princiotta (2017) summarized the state of the art of CCS technology as of 2016 and concluded that, despite an active demonstration program initiated 10 years ago, 43 projects have been shut down due to overruns yielding unacceptably high costs, degradation of power plant efficiency and serious capture and storage technical difficulties. Such technology is particularly important for relatively new coal and natural gas-fired power generators,

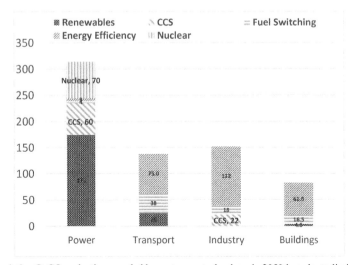

Figure 11. Cumulative Gt CO_2 reductions needed by sector per technology in 2050 in order to limit warming to 2 °C.

Table 2. Power generation low-carbon technologies.

Technology	IEA 2050 carbon emission reduction, Gt; 2 °C goal	Current state of the art	Issues	Technology RD&D needs
Carbon Capture and Storage	3.7	Early commercialization for coal with many demos having cost overrun and operating issues	High capital costs, 20–30% conversion efficiency degradation, complexity and potential reliability concerns; Underground Storage: Cost, safety, efficacy and permanency issues	**High:** Demos on next generation technology on a variety of units burning coals & natural gas; enhanced Underground Storage program with long term demos evaluating large number of geological formations
Solar-Photovoltaic and Concentrating (renewable)	3.7	First generation commercial	Solar resource intermittent and variable, although costs have been reduced further efficiency/cost reductions needed	**High:** Research needed to develop and demo cells with higher efficiency, and lower capital costs; develop/commercialize affordable storage technology
Wind Power (renewable)	2.6	Commercial (on-shore)	Costs very dependent on strength of wind source, large turbines visually obtrusive, intermittent power source	**Medium:** Higher efficiencies, off-shore demonstrations. Affordable storage technology
Nuclear Power-advanced & next generation	2.6	Commercial BWR, PWR; Developmental: Generation III+ and IV: e.g., Pebble Bed Modular Reactor	Deployment targeted by 2030 with a focus on lower cost, minimal waste, enhanced safety and resistance to proliferation	**High:** Demonstrations of key advanced technologies with complimentary research on important issues; commercialization of fusion technology could be transformational, might be possible, mid to late century
Biomass as fuel gasified or co-fired with coal (renewable)	1.0	Early Commercial	Important to assess true renewability of biomass source, limited to 20% when co-fired with coal	**Medium:** Biomass/IGCC would enhance efficiency and CO_2 benefit; also, genetic engineering to enhance biomass plantations
Fuel Switching coal to gas	0.2	Commercial (w/o CCS)	Effectiveness of CCS on natural gas generators; CH4 emissions during hydro fracturing could reduce GHG reduction benefit	**High:** Hydro fracturing environmental mgt., CCS demos needed, especially in the U.S.
Smart Grids	Not calculated, but supports renewables	Early Commercial, with active research focused on next generation technologies	Telecommunications cost high, security concerns and questions regarding consumer acceptance/ participation	**High:** Enhanced smart grid modeling, reduce telecommunication cost component, demonstrate effectiveness in maximizing solar and wind power production in overall mix

Table 3. Mobile source technologies.

Technology	IEA number of light duty vehicles in 2050, millions	Current state of the art	Issues	Technology RD&D needs
Electric & Hybrid Gasoline and Diesel	91	Early commercial	For electric plugs-in, mileage (battery) limitations; charging durations and high purchase prices; benefits greater if power from low-carbon sources	**High:** Battery improvements in storage capability, cost and lifetimes important
Fuel Cell Electric Vehicle	27	Developmental	Fuel cell costs and fuel cell stack life; also, H_2 production and need for fueling infrastructure	**High:** Breakthrough R,D&D needed to develop competitive, long lived fuel cell stack; viable H_2 production and storage, with a focus on safety, needed
Ethanol from cellulosic biomass sources, e.g., wood	0	Developmental	Important to assess true renewability of biomass source; inability to convert wide range of biomass sources with competitive production costs	**High:** Breakthrough RD&D needed to develop economical technology capable of generating large quantities; especially critical for the aircraft industry
Biodiesel & other fuels from biomass; thermo chemical processes	0	Developmental	Important to assess true renewability of biomass source; inability to convert wide range of biomass sources with competitive production costs	**High:** Breakthrough RD&D needed to develop economical technology capable of generating large quantities; especially critical for the aircraft industry

Table 4. Industrial technologies.

Technology	Current state of the art	Issues	RD&D needs
CO_2 Capture and Storage (IEA ETP 2015 projects 1.6 Gt mitigation in 2050)	Early development	Applicability limited to large energy-intensive industries, including fuel transformation processes; key questions: Cost, safety, efficacy	**High:** Major program with long term demos evaluating large number of geological formations to evaluate efficacy, cost and safety
Motor Systems	Commercial	For most industries not a major cost; lack of expertise for some industries	**Medium:** Lower costs and higher efficiencies desirable
Enhanced energy efficiency: Existing basic material processes	Commercial	Developing countries can have low energy efficiency due to lack of incentive and/or expertise	**Low**
Steam systems (required for many industries)	Commercial	For most industries not a major cost; lack of expertise for some industries	**Low**
Materials/Product Efficiency	First generation: commercial	Little incentive to minimize the CO_2 "content" of materials and products; life cycle analyses required	**Medium:** Conduct life cycle analyses of key materials and products with the aim of minimizing CO_2 "content"
Cogeneration (combined heat and power)	Commercial	Limited by electric grid access that would allow the ability to feed electricity back to grid; also high capital costs	**Low**
Enhanced energy efficiency: New basic material processes	Developmental to Near-commercial depending on industry	New, innovative production processes require major RD&D and would need reasonable payback to replace more C intensive processes	**Medium/High:** Develop and demonstrate less carbon intensive production processes for key industries
Fuel Substitution in Basic Materials Production	Commercial	Natural gas substitution for oil and coal can be expensive	**Low**
Feedstock Substitution in key industries	Commercial	Biomass and bioplastics can substitute for petroleum feedstocks and products; however, cost high and availability low	**Medium:** Develop affordable substitute feedstocks and products based on biomass

Table 5. Building technologies.

Technology	IEA 2050 carbon emission reduction, Gt; 2 °C goal	Current state of the art	Issues	Technology RD&D priority and needs
Enhanced energy mgt. and high efficiency building envelope: Insulation, sealants, windows, etc.	2.5	Commercial	Lack of incentive, high initial costs, long building lifetime	**Low/medium** priority: Incremental improvements to lower cost and enhance performance
High efficiency building heating and cooling, including heat pumps	0.8	Commercial	Lack of incentive, high initial costs	**Low/medium** priority: Incremental improvements to lower cost and enhance performance
Solar heating and cooling	0.5	First generation commercial	High initial costs, availability of low-cost efficient biomass heating systems	**Medium:** Focus on development of advanced biomass stoves and solar heating technology in developing countries

since it is unlikely that such plants would be prematurely retired in favor of renewable or nuclear plants. The current generation of nuclear power generation is commercially available but is burdened with high capital costs, waste disposal issues and serious safety concerns. The March 2011 Fukushima Daichi nuclear disaster, caused by a severe tsunami, has been responsible for several countries reconsidering their commitment to future nuclear reactor construction.

Table 3 lists the key mobile source technologies projected to play an important emission reduction role between now and 2050. Included is a column based on IEA (2016) which projects the number of light duty vehicles on the road in 2050 for key low-carbon technologies. As can be seen, electric and hybrid electric gasoline and diesel vehicles are projected to be the most important. Their cost and, therefore, their market penetration will be heavily dependent on the costs and performance characteristics of the vehicle storage batteries. Also projected to be important are fuel cell vehicles. These vehicles are in the early stages of commercialization. Fuel cells generate electricity to power the motor, generally using oxygen from the air and compressed hydrogen. They are generally classified as zero emission vehicles since the only effluent is water, However, it is important to account for any carbon emissions that would be associated with the production of hydrogen. Biomass fueled vehicles are at early stages of development, but have the potential to be significant low-carbon mobile source options.

Table 4 summarizes key industrial low-carbon technologies. The technology with the greatest change chance of making the greatest impact in this sector is carbon capture and storage. This technology would be applicable to major industrial sources of CO_2, including cement, iron and steel, oil refining and pulp and paper operations. Unfortunately, as discussed in the power generation discussion above, serious cost, reliability and permanent storage issues. Clearly, an enhanced research, development and demonstration program utilizing the next generation CCS technologies is required.

Table 5 lists key low-carbon building technologies. Emission reduction in this sector depends less on the development of new technology and more on providing incentives to promote the use of state-of-the-art technology for retrofitting existing buildings and incorporating in new buildings. For this reason, priority for research, development and demonstration programs is lower than in the power, mobile source and industrial sectors.

10. Technology RD&D is Woefully Inadequate

Given the need for dramatic emission reductions required in all economic sectors, it is essential that global RD&D expenditures are adequate to ensure the availability of high performing low-cost technologies in the near term. Figure 12 summarizes IEA (2011) analysis of actual versus needed global energy technology

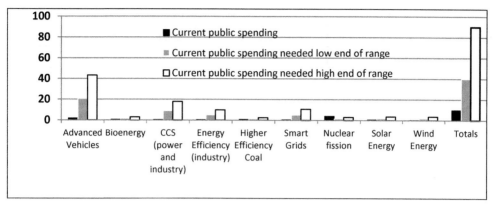

Figure 12. IEA analysis comparing actual low-carbon technology RD&D versus required spending, $ billions.

RD&D annual funding for key technologies consistent with reducing global emissions 50% by 2050. This analysis concludes that actual expenditures are a small fraction of what is required. The total required funding is estimated at $40–$90 billion per year, whereas actual spending for these key technologies is estimated to be only $10 billion annually. Although substantial, the $30–$80 billion annual funding gap is minuscule compared to the $1.7 trillion global military spending in 2017 (CNBC, 2018).

11. Geoengineering Options should be Studied

As previously discussed, meeting the global community's goal of limiting warming to 1.5–2 °C will be a monumental challenge. Given the massive and radical infrastructure change that will be needed in the near term, meeting such a target may not be possible. It has been suggested that geoengineering options could, at least in theory, buy us time to make the necessary infrastructure/technology changes to dramatically reduce global GHG emissions. They have been described as both a delaying tactic or as a possible "last resort" action to limit catastrophic climate change. Geoengineering measures attempt to compensate for GHG emissions via two fundamentally distinct approaches: (1) changing Earth's solar radiation balance by increasing reflectivity and (2) removing CO_2 from the atmosphere. Figure 13 mentions several of the most discussed geoengineering concepts in both categories. Table 6 compares the characteristics of these two concepts.

Note that two scenarios in Figures 9 and 10 assumed availability of commercial Direct Air Capture Technologies to compliment emission reduction efforts. DAC is sometimes referred to as the negative CO_2 emissions option. However, this approach, as well as all the options mentioned in Figure 13, are only

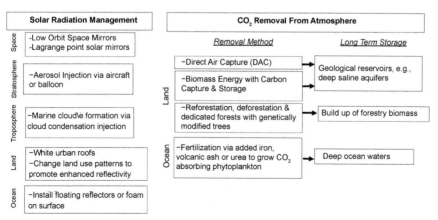

Figure 13. Solar radiation management and atmospheric CO_2 removal geoengineering concepts.

Table 6. Comparing solar radiation and atmospheric CO_2 removal concepts.

Characteristic	Carbon dioxide removal proposals	Solar radiation management proposals
Do they directly address the cause of GHG-induced climate change?	Yes: Such techniques act as negative CO_2 emitters	No: They utilize reflection of incoming solar radiation to compensate for heat added by GHGs; would not mitigate ocean acidification by CO_2
Can they intoduce novel risks?	No	Yes: Effects on stratspheric ozone levels and deleterious meterological impacts possible
How expensive?	Generally at least as expensive as emission reduction techniques	Although costs are very preliminary they appear relatively low re. emission control
How fast would results be realized?	Would take decades to realize significant results	Although still at conceptual stage, results could be realized within several years after installation
What level of international cooperation would be required?	Major cooperation involving multi-trillion US$ financial commitments needed	Less cooperation needed re. financial commitments but potential societal risks suggest international agreement before implementation needed
Key feasibility questions	Are the processes effective and affordable?	Are the processes effective with acceptable risks?

at a conceptual stage with serious performance, environmental impact and economic issues. Nevertheless, it is the author's view that, given the magnitude of the mitigation challenge, such approaches warrant serious feasibility evaluations, as soon as possible.

14. Conclusions

Humanity has dug itself a very deep hole. To limit the damage, it will take a concerted international effort, building on the 2015 Paris Accord, to aggressively reduce emissions in all sectors as soon as possible. Low-carbon technologies will of necessity play a crucial role in this process. It is the goal of this book to provide an assessment of the status and prospects of key low-carbon technologies.

References

CNBC. https://www.cnbc.com/2018/05/02/global-military-spend-rose-to-1-point-7-trillion-in-2017-arms-watchdog-says.

Cook, J. 2017. Impacts of Greenhouse Gas Emissions, Skeptical Science, https://www.beforetheflood.com/explore/the-deniers/fact-climate-change-is-very-very-dangerous/.

Climate Change Knowledge. 2014. http://www.climate-change-knowledge.org/uploads/GHG_pollution_schema.png.

Global Carbon Atlas. 2018. http://www.globalcarbonatlas.org/en/ CO_2-emissions.

International Energy Agency: Energy Technology Perspectives 2010 (IEA Publications, Paris, France, 2011).

International Energy Agency: Energy Technology Perspectives 2015 (IEA Publications, Paris, France, 2016).

Keith, D. 2018. A Process for Capturing CO_2 from the Atmosphere, Joule, Volume 2, August 2018.

Marcot, S. 2013. A reconstruction of regional and global temperature for the past 11,300 years, science. Science 08 Mar 2013: 339(6124): 1198–1201. DOI: 10.1126/science.1228026.

NASA GISS. 2019. Https://data.giss.nasa.gov/gistemp/graphs_v3/.

Princiotta, F. 2016. We are losing the climate change mitigation challenge, can we recover? MRS Energy & Sustainability, June, 2016, https://www.cambridge.org/core/journals/mrs-energy-and-sustainability/article/we-are-losing-the-climate-change-mitigation-challenge-is-it-too-late-to-recover/7EABEBD608FC3C651FE75931B19E7157/core-reader.

Skeptical Science. 2019. Global Warming at 4 Hiroshima Atomic Bombs Per Second https://4hiroshimas.com/.

Steffen, W. et al. 2018. Trajectories of the Earth System in the Anthropocene, 2018, Proceedings of the National Academy of Sciences of the USA.

USEPA. 2017. https://www.env-econ.net/2017/01/epa-separating-human-and-natural-influences-on-climate.html.

Warren, R. 2010. The Royal Society, The role of interactions in a world implementing adaptation and mitigation solutions to climate change. Philos. Trans. R. Soc. A 2011(369): 233. doi:10.1098/rsta.2010.0271 (published November 29, 2010).

World Resources Institute: World Greenhouse Gas Emissions in 2005 (2006). Available at: http://www.wri.org/resources/charts-graphs/us-greenhouse-gas-emissions-flow-chart (accessed April 9, 2019). Google Scholar.

Section 1

Global and Regional Views of Carbon Management

CHAPTER 1

Removing Carbon Dioxide from the Air to Stabilise the Climate

Richard C Darton* and *Aidong Yang*

1. Introduction

The rate at which greenhouse gases accumulate in the atmosphere can be reduced by cutting emissions, but elevated levels of CO_2 will persist for centuries. Climate models show that peak CO_2-induced warming depends mainly on cumulative emissions and not the emission pathway (Matthews, 2018). It is, therefore, expected to become necessary to remove CO_2 from the atmosphere in order to counter unacceptable climate change later in this century. Negative Emission Technologies (NETs) are the methods proposed for this. Nitrous oxide (N_2O) and methane (CH_4) mainly emitted by agriculture and industry are also significant contributors to global warming, but this chapter focuses on carbon dioxide. NETs for CO_2 are often termed carbon dioxide removal (CDR). It is a more efficient use of energy to capture carbon from flue gas than from the air, but even if applied quickly on a vast scale, it is expected that Flue Gas Capture (FGC) on its own will not be sufficient to meet policy goals—for example, peak warming of 1.5 °C above pre-industrial average temperature, the aspiration agreed at Paris in 2015 (Minx et al., 2018). The Intergovernmental Panel on Climate Change, which has reported the causes and effects of global warming in great detail, is clear on the need to apply CDR:

> "All pathways that limit global warming to 1.5 °C with limited or no overshoot project the use of carbon dioxide removal (CDR) on the order of 100–1000 $GtCO_2$ over the 21st century" (IPCC, 2018).

It is difficult to predict the precise timing and scale of CDR that might be needed, and many emissions scenarios have been considered by the IPCC. Figure 1 shows a typical scenario to the year 2100, in which peak warming is limited to 2 °C, and CDR is introduced from 2030 to ramp up to the scale needed. The ambition here is to draw-down around 810 Gt of atmospheric CO_2 (range 440–1020 Gt) by the year 2100 and sequester it somewhere securely (UNEP, 2017). Greater emissions reductions could reduce this targeted draw-down (IPCC, 2018). Fuss et al. (2018) indicate a total NETs deployment across the 21st century equivalent to net draw-down of 150–1180 $GtCO_2$ as necessary to meet the 1.5 °C target.

Whilst it is relatively simple to incorporate NETs into climate models, whether they can be applied at the scale needed or within the time frame assumed remains an open question. The impacts on people and ecosystems need to be considered, for these will be very large industries. Some NETs rely on well-

Department of Engineering Science, University of Oxford, Parks Road, Oxford OX1 3PJ, UK.
Email: aidong.yang@eng.ox.ac.uk
* Corresponding author: richard.darton@eng.ox.ac.uk

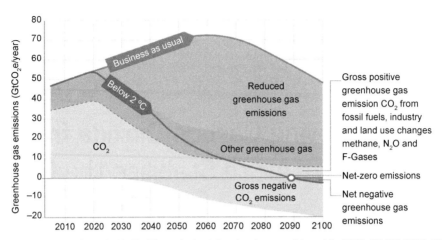

Figure 1. Emissions scenarios to 2100 (Gt CO_2 equivalent) showing the potential need for NETs (UNEP, 2017 - Figure 7.2).

known and proven technologies, but others still require a significant amount of research and development (Nemet et al., 2018). Moreover, recognising the objective of removing greenhouse gases from the air raises questions, such as how it will be paid for and managed. At the very least, NETs will compete with other desirable activities for resources and investment. Conjuring these industries into existence and then operating them for, say, a century, with the objective of managing the global climate, is an undertaking unlike any previously attempted.

It is important to distinguish between methods of capturing carbon dioxide from flue gas or other waste gases (Carbon Capture and Storage – CCS), and carbon dioxide removal from the atmosphere (CDR) followed by storage. Both CCS and CDR are responses to our failure to curb emissions of greenhouse gases sufficiently, but they have different characteristics:

a) CCS is a means of reducing emissions by using a process to capture carbon dioxide before it is released into the atmosphere, and storing the captured CO_2 securely. CCS is intended to reduce emissions particularly from large stationary sources of CO_2—power plants burning fossil fuel, cement factories, chemical plants and the like. Capturing CO_2 from distributed sources like motor cars, aeroplanes and agriculture would be more difficult to do, and is not really envisaged. Deploying CCS will slow the global rate of emissions but will not affect the CO_2 already present in the atmosphere.

b) CDR takes carbon dioxide out of the atmosphere into secure storage. It does not distinguish between CO_2 arising from natural processes and from man-made emissions. Chemical processes for CDR need more energy than chemical processes for CCS, because the concentration of CO_2 in the air is very low. Deploying CDR to an increasing extent will first decrease the rate at which CO_2 is accumulating in the atmosphere; then, when the rate of CDR exceeds the rate of CO_2 emission (i.e., net negative CO_2 emissions), the atmospheric CO_2 concentration will start to fall. When the rate of CDR equals the rate of emission of CO_2 and all other greenhouse gases (in terms of CO_2-equivalence), we will have net-zero emissions. In the scenario of Figure 1, this is projected to occur around 2090, after which we will have net negative greenhouse gas emissions.

Secure storage of CO_2, meaning that it is removed from the atmosphere for a long time, is sometimes termed "sequestration" and CCS can also be taken to stand for Carbon Capture and Sequestration. Requirements for storage longevity, monitoring and verification are discussed in section 3.8. Another phrase sometimes encountered is Carbon Capture and Utilisation (CCU), describing the manufacture of products using biomass or captured CO_2. The extent to which CCU can be regarded as either CCS or CDR depends on whether the product offers net long-term storage of CO_2—this is discussed in sections 3.7 and 3.8.

In this chapter, we first describe, in section 2, the technologies that might be used to capture CO_2 from the atmosphere, and indicate the chemistry involved. In section 3, we review options for storing the captured CO_2. Using abiotic sorbents to capture CO_2 is analysed in section 4, focussing in particular on the energy required and including an estimate of costs. Finally, we comment in section 5 on implications for setting policy.

2. Technologies for Capturing Carbon Dioxide from Air

Proposals to remove carbon dioxide from air have been reviewed by McLaren (2012), McGlashan et al. (2012), Fuss et al. (2018), Royal Society and Royal Academy of Engineering (2018) and the National Academies of Sciences, Engineering, and Medicine (2019). Broadly speaking, there are two main types of capture—by photosynthesis, which causes plant growth, or by means of an abiotic sorbent.

2.1 Capture by photosynthesis

In photosynthesis, growing biomass absorbs carbon dioxide from the atmosphere. It reacts with water in the plant to form carbohydrate and oxygen in a reaction which can be represented as

$$nCO_2 + nH_2O \xrightarrow{\text{sunlight}} (-CH_2O-)_n + nO_2 \tag{1}$$

Capture by photosynthesis is the basis of the coupled CDR system, BECCS (Bioenergy with Carbon Capture and Storage), which takes CO_2 from the air by growing crops that are used as fuel to provide useful heat or power. Carbon dioxide is then captured from the flue gas (a second capture of the CO_2) and sequestered. Photosynthesis has the key advantage that it is powered by sunlight. However, growing plants specifically to capture carbon dioxide, whether in the sea or on land, would compete for space with fishing, farming and other human activity and also with the need for preserving undisturbed ecosystems.

2.2 Capture by abiotic sorbent

Abiotic sorbents can be solid (adsorbents), or liquid (absorbents). In both cases, the sorbent has an affinity for carbon dioxide in the air. In Direct Air Capture (2.2.1), the sorbent is exposed to the air, then regenerated and re-used in a chemical process similar to industrial gas treating. The energy required to regenerate the sorbent is a major expense incurred in this route. Alternatively, in Enhanced Weathering and similar mineralisation schemes, air is contacted with rocks and minerals which react with CO_2 to form a stable product (2.2.2). The cost of regenerating the sorbent is avoided, though there will be other processing expenses.

2.2.1 Direct Air Capture (DAC)

The process parameters of DAC are rather different to most current industrial gas separations. Very large volumes of gas must be treated, potentially involving a lot of very large equipment; the pressure drop and efficiency of this equipment then become critical features of the design. The partial pressure of CO_2, the contaminant to be captured, is relatively low (~ 0.0004 bar), making it unsuitable for purely physical absorption processes—chemical reaction or chemisorption are needed. Fortunately, carbon dioxide is a fairly reactive chemical, and there are many proven industrial processes that can be used to remove it from various gases. Amongst these applications is flue gas CCS, usefully reviewed by the IPCC (Metz et al., 2005).

A well-known commercial solvent for carbon dioxide capture is aqueous potassium carbonate. The reaction is

$$CO_2 + K_2CO_3 + H_2O \rightleftharpoons 2KHCO_3 \tag{2}$$

This reaction however is rather slow, and to reduce the size of the equipment needed, the absorption is often speeded up by adding a homogeneous catalyst, such as arsenite or hypochlorite (Danckwerts and

Sharma, 1966; Astarita et al., 1981). Fast-reacting promoters, such as piperazine, can also be added to potassium carbonate to enhance the absorption rate and reduce equipment size (Cullinane and Rochelle, 2004).

Another important class of chemical absorbents much used in oil, gas and chemicals processing are alkanolamines in aqueous solution. For example, with monoethanolamine (MEA), the main reaction chemistry is a carbamate formation

$$CO_2 + 2RNH_2 \rightleftharpoons RNHCOO^- + RNH_3^+ \tag{3}$$

where R stands for the ethanolic group HOC_2H_4. The rate of reaction (3) is much greater than that of the uncatalysed reaction (2), which accelerates the absorption of CO_2 into the amine solvent.

In industrial processes for carbon dioxide removal, the sorbent is regenerated for use again by raising its temperature, which drives the equilibrium (e.g., reactions (2) and (3)) back towards the left-hand side. Regeneration requires energy which is commonly provided by low pressure steam, at around 120–130 °C, and it produces carbon dioxide as an off-gas at a pressure and purity depending on the process design. If a liquid solvent is used, it is pumped around in a cycle between absorber and regenerator vessels. In this cycle, the hot regenerated solvent is cooled against the loaded solvent, and then cooled further against ambient air or water in a "trim cooler" prior to entering the absorber again. Solid adsorbents are seldom moved around, so it is common to have several beds of adsorbent operating in parallel. When one is sufficiently loaded with CO_2, the inlet flow is switched to a fresh bed, whilst the loaded bed is regenerated and the CO_2 is driven off by raising the temperature and/or lowering the pressure. In this way, the adsorbent particles experience an adsorption/desorption cycle with pressure/temperature swing, albeit without moving.

Research has been directed to developing tailor-made Direct Air Capture technology. Conventional sorbents and existing proven process technology might be used, but novel sorbents, process line-ups and equipment can offer advantages for the specific requirements of air capture (e.g., Shi et al., 2016; Keith et al., 2018). Following the capture process, provisions must be made for utilising or storing the product carbon dioxide. If it is intended to compress the CO_2 for underground storage, a rather high purity is required, usually greater than 96%.

2.2.2 Enhanced Weathering (EW) and other mineralisation schemes

In these schemes, the sorbents are rocks or minerals generally in particulate form. They are not regenerated, and the reaction product comprises the long-term storage of the drawn-down CO_2.

For example, the "weathering" reaction of olivine, a silicate rock, with rainwater and CO_2, can be represented as

$$4CO_2 + Mg_2SiO_4 + 4H_2O \rightarrow 2Mg(HCO_3)_2 + H_4SiO_4 \tag{4}$$

and the weathering of carbonates, such as calcite, as

$$CO_2 + CaCO_3 + H_2O \rightarrow Ca(HCO_3)_2 \tag{5}$$

Once present in groundwater, the soluble products of these reactions can be stable for centuries, representing a net removal of carbon dioxide from the atmosphere. However, in a different environment, if the water flows into the sea, for example, further reactions will occur. These might typically be

$$2Mg(HCO_3)_2 + H_4SiO_4 \rightarrow 2MgCO_3 \downarrow + SiO_2 \downarrow + 2CO_2 + 4H_2O \tag{6}$$

$$Ca(HCO_3)_2 \rightarrow CaCO_3 \downarrow + CO_2 + H_2O \tag{7}$$

These mineralisation reactions release carbon dioxide back to the atmosphere, but in the case of silicates, reactions (4) and (6) still cause a net draw-down of CO_2. However, for carbonate rock, reaction (7) is the reverse of (5), and the net effect of both together is the transport of carbonate from one location to another (which may provide a temporary period of storage), with ultimately no net change in atmospheric carbon dioxide concentration. Engineered schemes to promote capture and storage through these reactions are known as Enhanced Weathering (EW).

Closely related to EW is the use of alkaline wastes from manufacturing or mining operations which similarly react with atmospheric carbon dioxide in the presence of water (Renforth, 2019). It is also possible to make a highly active particulate adsorbent by calcining carbonate minerals. For example, for calcium carbonate as present in limestone the chemistry is

$$CaCO_3 \xrightarrow{heat} CaO + CO_2 \tag{8}$$

$$CaO + H_2O \rightarrow Ca(OH)_2 \tag{9}$$

$$Ca(OH)_2 + 2CO_2 \rightarrow Ca(HCO_3)_2 \tag{10}$$

Reaction (8) is the calcining reaction which, in order to produce sufficient partial pressure of carbon dioxide in the off-gas, needs to take place at temperatures above 800 °C. In (9), the calcium oxide is "slaked" with water, a highly exothermic reaction. The resulting alkali can be exposed to atmosphere, and it will draw down CO_2 as in reaction (10). To maximise net positive removal, the carbon dioxide generated in (8) is captured and sequestered. The calcining operation can be engineered to produce rather pure CO_2, which facilitates CCS in this case. The alkali can be used in Ocean Liming (section 3.6), or perhaps in flue gas treating at coastal locations (Rau, 2011).

3. Storing the Captured Carbon Dioxide

Whether the carbon dioxide is captured by photosynthesis or abiotic sorption, it is necessary that it stays out of the atmosphere, preferably for at least several centuries. Clearly, there is no benefit to the climate in capturing carbon dioxide which is quickly released back into the atmosphere. Moreover, both the capture and storage should be feasible on the scale needed (Figure 1), and should not cause significant damage to the environment or society. The two questions, "how do we capture the CO_2?" and "what do we do with the CO_2?" are very closely linked. Table 1 lists a variety of possible locations for storing captured

Table 1. Schemes for removing CO_2 from the air and storing it.

		CO_2 captured from air by photosynthesis	CO_2 captured from air by abiotic sorbent
C-storage location	Plants on land	(1a) Afforestation and forest management (1b) Wetlands, peatlands and coastal ecosystem restoration and management	(2) DAC, using regenerable solid or liquid sorbent to produce CO_2-enriched air for improving agricultural yield – e.g. tomatoes
	Soil	(3) Increasing soil carbon by land management: Soil carbon sequestration (4) Biochar (from biomass pyrolysis)	(5) Enhanced Weathering (EW) of rock particles scattered on soil
	Above ground as mineral	(6) Biomass combustion to release energy with CO_2 captured from flue gas and reacted with mined mineral or suitable waste	(7) Mineral Carbonation reacting either air, or CO_2-enriched air (flue gas or DAC) with mined (processed) mineral or suitable waste
	Below ground as mineral	(8) Following biomass combustion to release energy, with flue gas treating and injecting CO_2 to react in rock such as silicate or basalt	(9) Following DAC, injecting CO_2 to react in rock such as silicate or basalt
	Below ground, as compressed CO_2	(10) Biomass combustion to release energy, with flue gas treating to produce high purity CO_2 for underground storage (BECCS)	(11) DAC to produce high purity CO_2 for underground storage (DACCS)
	Ocean	(12) As (10), with CO_2 dispersed into the ocean, or injected at a depth where it is not buoyant (13) Ocean Fertilization (OF) to stimulate growth of marine biota	(14) As (11), with ocean storage of CO_2 (15) Ocean Liming. Calcination of carbonate rock with CCS. Lime distributed at sea
	Human environment	(16) Buildings made with wood and similar biomaterials (17) Chemicals and products from biomass	(18) DAC, then utilizing produced CO_2 as feedstock for chemical products (CCU)

CO_2, together with the capture methods commonly suggested for those locations. They comprise some 18 different schemes for removing carbon dioxide from the air and storing it to mitigate climate change, though other suggestions and variations will also be found in the literature.

3.1 Plants on land (schemes 1a, 1b, 2)

Forests, wetlands, peatlands and coastal habitats provide significant storage of terrestrial biological carbon, but they are under threat from exploitation, degradation and land-use change. It is estimated that restoration and improved management of these ecologies might yield mitigation of up to 8 Gt/y CO_2 by 2030 (Griscom et al., 2017), as well as other benefits to biodiversity, water resources and so on. Planting new forests makes a significant contribution, but as they approach maturity, after some decades, net CO_2 removal declines. Thus, a long-term strategy for forest management is needed in order to maintain ecosystem health and ensure that captured CO_2 is not released prematurely. The use of carbon dioxide from DAC to promote growth in greenhouses of cash crops like tomatoes and aubergines has received much publicity. However, the storage lifetime of the carbon in a tomato is very short indeed, and the yearly production is only about 100 million tons. Some 93–95% of a tomato is water, so this is not a method of storage or carbon utilisation that will make any significant impact on climate change!

3.2 Soil (schemes 3, 4, 5)

There are many ways in which agricultural land, much of which has lost significant soil organic carbon (SOC), can be managed in order to both increase fertility and store more carbon. For example, the variety and rotation of crops can be improved, as can the use of manures, composts and fertilisers. Carbon (and water) loss can be reduced through a "no-till" policy that avoids ploughing. The potential benefits of restorative land use and the adoption of recommended management practice have been reviewed by Lal (2011). The storage promoted by these policies is known as soil carbon sequestration.

Biochar is a carbon-rich (65–90%) porous solid product of the pyrolysis of biomass (and also of animal waste). It can be added to soil which provides a stable storage environment. Biochar acts as a conditioner which increases SOC, water retention and fertility, and it can also reduce the emission from soil of the greenhouse gases methane and nitrous oxide. The production, properties and utilisation of biochar have been reviewed by Qambrani et al. (2017).

The weathering and mineralisation reactions (equations (4)–(7)) are part of the natural carbon cycle. They yield a stable (i.e., long-lasting) form of carbon dioxide storage as bicarbonate and carbonate ions, which are very common in soils and oceans. The reactions are spontaneous and remove about 1.1 Gt CO_2 from the atmosphere annually (Ciais et al., 2014). To capture extra carbon dioxide by these means at a rate sufficiently fast to help meet climate policy targets, the exposed rock needs a high surface area—that is, a small particle size. Mining and crushing rock require significant energy input, leading to higher costs of course. Various terrestrial EW schemes have been suggested, including spreading suitable mineral particles on agricultural land (which can also improve soil quality). For example, Strefler et al. (2018) estimate that spreading fine basalt particles on croplands might potentially capture 4.9 Gt/y CO_2, a rate facilitated by plant and root activity that accelerates the chemical reactions. Beerling et al. (2018) discuss the potential for biogeochemical improvement of croplands by amending soils with crushed fast-reacting silicate rocks as a strategy to address the threats of climate, food and soil security.

3.3 Above ground as mineral (schemes 6, 7)

Ex situ methods of reacting air or captured CO_2 with mineral might involve mining, transport and crushing of virgin rock, but attention has also been given to the estimated 7 billion tons of alkaline materials currently produced annually by industry as product or by-product. Such material—slag, mud and waste from steel, aluminium or cement manufacture, combustion ash, ultrabasic mine tailings—is often available as small particles, a suitable form for capturing carbon dioxide from the atmosphere. As materials for EW schemes, Renforth (2019) estimates their potential storage capacity at 2.9–8.5 Gt/y CO_2 by the year 2100.

3.4 Below ground as mineral (schemes 8, 9)

In situ methods envisage injection of carbon dioxide into permeable rock strata, where higher temperature and pressure will accelerate reaction with water and minerals. Once mineralised, there is a much smaller chance of the carbon dioxide escaping from the reservoir. Matter et al. (2016) report that in an experiment in which CO_2 was injected into a reservoir of basaltic rock, over 95% was mineralised to carbonate minerals within two years. This indicated a much higher rate of mineralisation than expected. Successful results were also obtained with a less pure stream of CO_2, an interesting result since secure storage of less pure carbon dioxide might enable the cost of separating CO_2 to be reduced.

3.5 Below ground as compressed carbon dioxide (schemes 10, 11)

BECCS (section 2.1) with CO_2 compressed for storage underground is the mitigation scheme most commonly assumed in climate modelling work. For example, Rogelj et al. (2018) analysed scenarios which would meet the Paris target limit of 1.5 °C global mean temperature rise this century using Integrated Assessment Models. In this study, capture of carbon from the atmosphere was mainly achieved through BECCS and afforestation, omitting other technologies. However, producing biofuel at the scale required would create serious competition for land that currently supports food production or diverse natural ecologies (Boysen et al., 2017).

Alternatively, CO_2 could be removed from the air by DAC (section 2.2.1), using sorption taking place in some more or less conventional processing equipment. The logistics are quite different to BECCS, as plantations of crops are replaced by regenerable sorbents which capture CO_2 from the air. The combined system (i.e., DAC with geological storage) is sometimes called DACCS.

Both BECCS and DACCS envisage carbon dioxide stored in a compressed state (super-critical if the pressure exceeds 74 bar) underground in geological formations where it would remain stable for a long period of time. This kind of storage is also expected to be used for storing the carbon dioxide captured from flue gas.

3.6 Ocean (schemes 12, 13, 14, 15)

An IPCC review of the potential of ocean storage of CO_2 pointed out that of the 1300 Gt anthropogenic emissions of CO_2 in the last 200 years, some 500 Gt have been taken up by the oceans as they approach chemical equilibrium with raised CO_2 concentrations in the atmosphere (Metz et al., 2005, Chapter 6, Ocean Storage). Compressed CO_2, captured on land by BECCS or DAC plants, could be deliberately stored in the ocean —pumped by pipeline or dispersed by ship, both of which have been considered. It is known however that acidification of the oceans by anthropogenic CO_2 poses a serious problem to marine ecosystems, as discussed in the IPCC review, so ocean storage has received little support. It remains a controversial topic (Goldthorpe, 2017).

Ocean Liming is a scheme that addresses both climate change and ocean acidification directly. Renforth et al. (2013) describe a typical case. The production of slaked lime follows the chemistry outlined in equations (8) and (9). The lime is then distributed at sea, facilitating further draw-down of carbon dioxide from the atmosphere, as in equation (10), though there will also be an equilibration with existing levels of carbonate and bicarbonate present in the oceans, according to equation (7).

Ocean Fertilisation (OF) was the first negative emission technology to receive much serious attention, from the early 1990s (Minx et al., 2018). The idea is to promote the growth of marine biota to draw-down carbon dioxide by the photosynthesis reaction, equation (1). A fraction of these marine plants would sink deeper, providing ocean-based sequestration. This is a natural process which could be enhanced by adding the nutrients nitrogen and phosphorus to surface waters in areas where their low availability currently limits plant growth. Alternatively, in about one third of ocean surface waters, where nitrogen and phosphorus are sufficiently present, the micronutrient iron is in limited supply, so fertilisation with this component would enhance plant growth. Williamson et al. (2012) reviewed the benefits and risks of ocean fertilisation and concluded that its contribution would be relatively modest, and extremely challenging to quantify on a long-term basis. *In situ* and far-field monitoring would be needed to verify the change in carbon flux caused by fertilisation and to check for possible rebound effects that might offset the initial change. There are also concerns about unintended side-effects, and the governance of a process involving the addition of such chemicals to open ocean waters.

3.7 Human environment (schemes 16, 17, 18)

Using plant products to replace materials made from fossil fuels or with the expenditure of much fossil energy will lead to a reduction in atmospheric carbon dioxide. Wood, for example, can be used in construction, and serves to store carbon for some decades at least. There is a similar benefit in replacing petrochemical products with ones made from biomass, when this leads to a net draw-down of CO_2. There is scope for the development of CO_2-based chemistry to replace petrochemistry, but again, careful accounting is needed in order to ensure that the net effect is beneficial when all processing and energy consumption, direct and indirect, is included. The direct use of biomass, or carbon captured from air or flue gas falls under the general description of Carbon Capture and Utilisation (CCU). Turning the captured carbon dioxide back into fuel by a chemical conversion consumes a lot of energy of course, and only results in minimal storage before the fuel is burnt again and CO_2 is returned to the atmosphere: This is not CDR. The desire to find a profitable use for the "unwanted" carbon dioxide in the atmosphere is understandable. However, except for the energy market itself, it is difficult to see a sufficiently large market for any carbon-based products which could take up even a small fraction of the 100–1000 Gt of CO_2 that we are looking to capture and store in this century. To illustrate this point, the current world plastics production is about 0.35 Gt per year (www.plasticseurope.org), which suggests that a very limited capacity of CO_2 utilisation would be offered by replacing materials currently derived from petrochemicals, a point made by Metz et al. (2005). Mac Dowell et al. (2017) concluded that "it is highly improbable the chemical conversion of CO_2 will account for more than 1% of the mitigation challenge".

3.8 Storage longevity, monitoring and verification

For the purpose of long-term climate stabilisation, CO_2 removed from the atmosphere needs to be stored securely, preferably for a duration of at least several centuries (Royal Society and Royal Academy of Engineering, 2018), The storage options listed in Table 1 exhibit varying degrees of permanence and certainty. Those options with biomass or soil organic matter as the stored form of carbon are considered to have the lowest (possibly sub-centennial) permanence; their storage longevity is subject to human interventions, like changes in land use and utilisation of biomass, and natural processes such as the occurrence of fire. As the RS&RAEng report states "…pathways that store carbon as impermanent organic materials with sub- centennial and uncertain lifetimes may prove ineffective without appropriate management."

Biochar and terrestrial weathering probably offer more stable storage, yet the generation of firm evidence and understanding remains a subject of research. At the other end of the longevity spectrum, mineral carbonation and ocean liming have the potential to store carbon for millennia or even longer. Storing CO_2 in deep sedimentary geological formations, which can potentially handle CO_2 captured in high purity by a range of schemes, is also considered to be among the most stable methods, though there may be leakage, depending on the geological characteristics of the storage site. This method is discussed in Chapter 7 of the report by the National Academies of Sciences, Engineering, and Medicine (2019).

Given the high reversibility of some of the stored forms of carbon and the uncertainty in permanence which is associated with virtually all options to some extent, the implementation of CDR will require careful monitoring and verification. The effectiveness and costs of monitoring and verification and their trade-offs also vary significantly between different CDR schemes. Bellasen et al. (2015) wrote in a review of existing regulatory systems that "The monitoring, reporting and verification (MRV) of greenhouse-gas emissions is the cornerstone of carbon pricing and management mechanisms". Establishing good MRV practice for CDR could utilise experience in the field of emissions control, in order to improve reliability and cost effectiveness. For example, as pointed out by the National Academies of Sciences, Engineering, and Medicine (2019), in the US, each year more than 2.5 Gt of brines are injected into deep underground formations for disposal. This experience helped underpin technical, administrative and regulatory approaches developed in the U.S. for sequestration of captured CO_2 in sedimentary basins.

4. Energy Considerations in Direct Air Capture

Of the many schemes for capturing and storing carbon dioxide from the air, DACCS (scheme 11 in Table 1) is the closest in terms of technology to a well-known and researched method of carbon management, namely CCS from flue gas. However, the deployment of FGC has been slow, for various reasons, including concern about its energy requirement, which must of course be paid for. Similarly, DACCS has been criticised as energy intensive, and therefore expensive, particularly compared to schemes utilising photosynthesis, where the energy for capture appears to be free. However, DACCS remains a potentially important weapon to tackle atmospheric carbon, so we devote this section to discussing its energy requirement in detail. Comparison is made with flue gas capture, and the approach is extended to estimating likely costs. Supporting calculations can be found in the Appendix to the chapter.

The calculation of energy required for FGC and DAC involves a number of steps, so it is useful to follow a clear calculation strategy. This is shown in Figure 2. We start with (i) the calculation of the (minimum) shaft work requirement w_{rev} for an idealised reversible separation process that involves no energy losses. Then calculation (ii) using the Carnot efficiency yields the (minimum) reversible heat requirement. The heat required by an actual process plant Q_{sep} is greater than the reversible amount due to losses, and this is calculated last (iii). We then calculate and compare values of Q_{sep}/w_{rev} for actual process plant data.

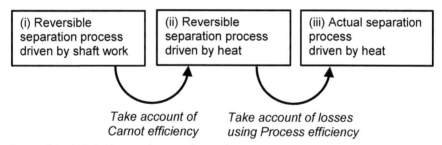

Figure 2. Strategy for calculating heat requirement of separation process. For many separations with regenerable sorbents, process efficiency is about ⅓ to ½.

Note: In this section quantities of thermal energy are designated MJ(th), and quantities of electrical energy or work are denoted MJ(e) when this distinction might otherwise be unclear.

4.1 Reversible work of separation

In Direct Air Capture, air is separated into a gas enriched in carbon dioxide and a second "reject" stream of inerts which may also contain some CO_2, as shown in Figure 3. The inerts are components which are not differentiated from each other in this particular separation. We analyse an arrangement in which all streams are at the same pressure, and at the same temperature T_0. The mole fraction of CO_2 in the feed is y_c and in the CO_2-enriched stream is y_{co}. The separation can be characterised by two further parameters, and we choose these to be α, the fraction of inlet CO_2 that is recovered in the CO_2-enriched stream, and

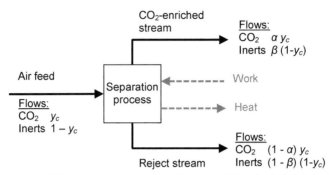

Figure 3. Separation of air into CO_2-enriched stream and reject stream. Basis of calculation: 1 mole of feed with mole fraction y_c of carbon dioxide.

β, the fraction of the inerts in the feed that slip through into the CO_2-enriched stream. The parameter β is found by mass balance to be given by $\alpha.y_c(1 - y_{co})/\{y_{co}(1 - y_c)\}$.

Shaft work is supplied in order to bring about the separation, and, by the first law, an equal quantity of waste heat is produced because the system is isothermal. The minimum quantity of work needed is that required by a perfectly reversible process, and this can be calculated from the change in Gibbs free energy between inlet and outlet streams. For this calculation, we need no information about the separation process used in Figure 3, beyond the constraint that it is thermodynamically reversible. This means that all gradients of temperature, pressure, velocity and concentration in the processing equipment are infinitesimally small, so that all internal transfer processes proceed infinitely slowly and there are no internal losses of energy. The advantage of considering such an idealised process is that its performance can be calculated unambiguously. It represents a clearly defined standard with which real processes can be compared. No separation process working with real equipment will be able to affect the same separation, at the same temperature and pressure, with less expenditure of work.

The reversible work,[1] which is, thus, the minimum required for the separation, w_{rev} per mole of CO_2 recovered in the CO_2-enriched stream, is given by

$$\frac{w_{rev}}{RT_0} = \ln \frac{\alpha}{\beta + (\alpha - \beta)y_c} + \frac{\beta(1 - y_c)}{\alpha y_c} \ln \frac{\beta}{\beta + (\alpha - \beta)y_c} + \frac{(1 - \alpha)}{\alpha} \ln \frac{1 - \alpha}{1 - \beta - (\alpha - \beta)y_c} +$$

$$\frac{(1 - \beta)}{\alpha} \cdot \frac{1 - y_c}{y_c} \ln \frac{1 - \beta}{(1 - \beta) - (\alpha - \beta)y_c}$$

(11)

The four terms on the right-hand side of equation (11) are the Gibbs free energy change of, respectively, (1) CO_2 in the CO_2-enriched stream, (2) inerts in the CO_2-enriched stream, (3) CO_2 in the reject stream, and (4) inerts in the reject stream. When concentrating dilute streams of carbon dioxide, the reversible work requirement is dominated by the first term. Allowing some inert to slip into the enriched stream reduces the work requirement, but the effect is initially slight, as shown in Figure 4. For air at 293 K and a CO_2 concentration of 400 ppm, the reversible work required for a complete separation ($\alpha = 1$ and $\beta = 0$) is 21.49 MJ/kmol CO_2, and only falls to half this value when the enriched stream purity drops to around 3% CO_2. Somewhat less work is required if the recovery is reduced, but again the effect is slight. Even restricting recovery (α) to 10%, also shown in Figure 4, only reduces the reversible work requirement by ~ 10% in the range of interest (purities > 96%). A low recovery fraction means that a lot more air must be treated per kmol CO_2 captured, incurring greater capital and operating expense.

---

[1] This derivation assumes the mixtures are ideal. For the treatment of non-ideal mixtures see, e.g., King (1980) Chapter 13.

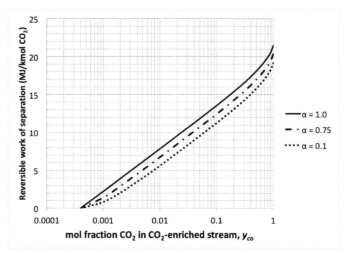

Figure 4. Reversible work of separating CO_2 from air. Mole fraction in air 0.0004. Calculated for $T_0 = 293$ K. α is the fraction of inlet CO_2 recovered in the CO_2-enriched stream.

4.2 Energy requirements in DAC process plant

The standard enthalpy of formation of CO_2 is $(-)393.5$ MJ/kmol, some 18 times the reversible work required for a complete separation of 400 ppm CO_2 from air, from equation (11). So it might seem that the energy cost of capturing an amount of CO_2 from the air would be modest compared to the heat that was obtained when it was formed during combustion. However w_{rev} is a work, not a heat, requirement. The sorbents used for DAC require regeneration, a step involving heat input, so the overall sorption process functions as a "separation engine". Heat is supplied to the engine and it produces work of separation as well as waste heat that must be removed. The Carnot efficiency of this separation engine, η_{Car}, can be estimated. For example for an amine treater with regeneration using low pressure steam at $T_H \sim 403$ K, and removing heat in the overhead condenser and trim cooler at a mean temperature of $T_C \sim 343$ K, η_{Car} (= $1 - T_C/T_H$) is 0.17 or $\sim 1/6$. Furthermore, the Carnot efficiency refers to a reversible process. In practice, when materials are processed at a commercial rate and scale, energy is dissipated by friction and in the diffusion of heat and mass across finite gradients of temperature and concentration, respectively. The efficiency of a real process might be one third to one half of the (reversible) Carnot value, though this will depend on the process and the design choices made. The heat requirement of an actual DAC plant producing pure CO_2 would then be about (2 to 3) × $(1/\eta_{Car})$ or 12–18 times w_{rev}, that is some 250–400 MJ/kmol CO_2. Thus, the energy needed for DAC might approach, and even exceed, the energy obtained by burning the carbon in the first place.

From published process data, it is possible to estimate the factor by which the energy required for carbon capture, Q_{sep}, exceeds the minimum work of separation, w_{rev}. The calculations are described in the Appendix, and summarised in Table 2. The data refer to four FGC plants (Metz et al., 2005) at the limit of their effectiveness, and the DAC process developed by Carbon Engineering (CE) (Keith et al., 2018).

Our initial supposition that the heat actually used to drive a capture process for carbon dioxide might be some 12–18 times the minimum reversible work is seen from Table 2 to be approximately true, though the spread of values is greater, about 5–24. Part of this spread is due to assumptions and simplification in the calculations and the degree of optimality of the process design. In particular, it is likely that some of the variation in Q_{sep}/w_{rev} represents different approaches to the trade-off between capital expenditure and energy efficiency, a well-known optimisation problem in process design (Smith, 2016, p460).

Energy efficiency can be improved in general by a greater degree of process integration, and by reducing temperature differences in exchange of heat between hot and cold streams. This, in turn, requires more heat transfer area, and thus, an increase in capital expenditure. Losses in mass transfer equipment can also be reduced by operating closer to equilibrium, but analogously with heat transfer,

Table 2. Reversible work and actual heat requirement for CO_2 capture processes, MJ/kmol CO_2 captured. Data and calculations for pulverised coal (PC) and natural gas combined cycle (NGCC) flue gas capture; for DAC by CE process. $T_0 = 21$ °C.

Case	y_c mol frac	y_{co} mol frac	α	β	w_{rev} MJ/kmolCO$_2$	Q_{sep} MJ/kmolCO$_2$	Q_{sep}/w_{rev}
PC 24%	0.12	0.96	0.9	0.005114	6.439	73	11.4
PC 40%	0.12	0.96	0.9	0.005114	6.439	156	24.2
NGCC 11%	0.04	0.96	0.9	0.001563	9.216	45	4.9
NGCC 22%	0.04	0.96	0.9	0.001563	9.216	141	15.3
DAC CE process	0.0004	0.958	0.745	$1.31 \ 10^{-5}$	19.98	337	16.9

this also requires more transfer area and, thus, a larger plant. Also, a more highly integrated design may have less flexibility with regard to accommodating summer/winter conditions, changes in flow-rate or composition and so on. Selection of the optimum design will then depend on a wide range of factors, such as the cost of capital, the cost of energy and consumed chemicals and the possibilities of integration with neighbouring process plant.

It seems reasonable to assume that currently optimised processes for both FGC and DAC should be able to operate at a ratio of approximately $Q_{sep}/w_{rev} \sim 14$. This ratio is suggested as appropriate for preliminary design and estimating purposes. It is not, however, a ratio that should be considered as permanently fixed. It obviously varies somewhat from design to design, and further optimisation and improvements in process and equipment will reduce energy requirements further. And importantly, new process concepts could lead to step changes, just as the introduction of combined cycle power generation schemes led to significant improvements in power plant efficiency that were once thought impossible. Such innovation might result, for example, from electrochemical schemes, where chemical energy is converted to electrical energy (work) without the need for heat generation which involves a Carnot efficiency. The requirement for shaft work or electricity within the process might be met by renewables (solar, wind, hydro), reducing the need to burn fossil fuel. Innovations in material science might lead to feasible membrane separation processes powered by renewable energy. Similarly, use of new adsorbents tailored to the specific FGC and DAC requirements might prove more efficient than existing schemes. It seems likely that such innovations could be applied both to FGC and DAC.

4.3 Chimney stack exit penalty

When flue gas flows out of a chimney stack into the atmosphere it mixes with the surrounding air, disperses and blows away from the stack. This process—mixing is indeed a process—is hard to observe, except when motion of the flue gas plume is marked by white condensation of water vapour as it cools down. The mixing process, though almost invisible, nevertheless causes a large increase in entropy, and thus, a loss in Gibbs free energy, which corresponds to a loss of potential work. In theory, a thermodynamic process which makes use of the difference in CO_2 concentration between flue gas and atmosphere to produce useful work could be created. Allowing the flue gas to mix irreversibly with the atmosphere destroys this possibility. As a result, once the flue gas has left the chimney stack and mixed with the atmosphere, it takes much more work to capture the carbon dioxide than it would have done if it had been captured from the flue gas where it is more concentrated.

This chimney stack exit penalty is quantified in Figure 5, for two flue gas concentrations representing power plants firing natural gas ($y_c = 0.04$) and coal ($y_c = 0.12$). Capturing carbon dioxide from air takes about twice as much reversible work as capturing it from a natural gas power plant flue gas. Compared with FGC at a coal-fired power plant, DAC requires three times as much reversible work. As we show in the Appendix, this energy penalty translates into a significant cost penalty–DAC costs around 72–88 US$/t$CO_2$ more than FGC.

4.4 Choice of solvent and cost of DAC

The IPCC's 2005 report on CCS excluded Direct Air Capture from consideration because the CO_2 concentration in ambient air was "…a factor of 100 or more lower than in flue gas. Capturing CO_2 from air by the growth of biomass and its use in industrial plants with CO_2 capture is more cost-effective based on foreseeable technologies…" (Metz et al., 2005). However, as we have seen, the energy required for separation varies inversely with the logarithm of the concentration, so that the reversible work cost of air capture (Figure 5) is only 2–3 times that of FGC. This still gives a great advantage to FGC, which also needs only to treat a much smaller volume of gas to catch a tonne of carbon dioxide. Air is a somewhat easier material to handle than flue gas, and the ability to site DAC plants almost anywhere, such as at locations where energy and/or storage are cheap, might outweigh some of the advantage that FGC would otherwise enjoy.

This chapter has considered the energy requirements of DAC in detail, as this process might become a huge industry later this century—it is based on known technology that could be applied if other more novel techniques cannot be successfully developed. However, there has been some confusion and disagreement about DAC in the literature, of which the very wide spread of cited costs is symptomatic. These vary from around 60 to 1000 US$ per tonne of CO_2, though it does rather depend on how the necessary energy is supplied (Fuss et al., 2018). In their review of a range of NETs, Fuss et al. (2018) state "…a significant amount of thermal energy is often required for DAC due to the requirement of strong binding of the captured material because of the extreme dilution of atmospheric CO_2." This is a misunderstanding. Strong binding of the sorbent with CO_2 would be required if it were necessary to reduce the concentration of CO_2 in the treated air to a low level—this would need a low equilibrium back pressure from the sorbent (low pK_b). This is not required in DAC. Atmospheric carbon dioxide currently has a concentration of about 400 ppm, and this is not "extreme dilution" as far as gas treating is concerned. In processing natural gas with counter-current contact, amines of intermediate strength are able to remove H_2S down to pipeline specifications (< 4 ppm). H_2S is a weaker acid than CO_2, but it reacts very rapidly in alkaline solutions.

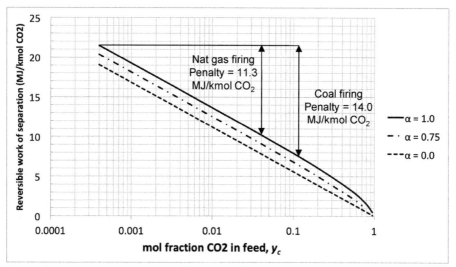

Figure 5. Chimney stack exit penalty. Penalties shown are the additional reversible work needed for DAC rather than FGC, for natural gas firing ($y_c = 0.04$) and coal firing ($y_c = 0.12$), calculated for $\alpha = 1.0$, $\beta = 0$ (100% recovery of pure CO_2). $T_0 = 293$ K and mole fraction in air is 0.0004.

It is true that the rate of absorption into solvents will be accelerated by reaction in the liquid phase, and CO_2 undergoes a second order reaction with the hydroxyl ion (Astarita et al., 1981), which is, thus, faster at high pH. This and other reactions in the liquid will enhance the absorption rate, enabling reduction in the size of the absorber, which is important in DAC. So, what is required is a fast reaction between CO_2 and sorbent, with a base of intermediate strength that can be regenerated with less energy than one with strong binding of CO_2. "Fast reaction" means that the characteristic time of the reaction is small compared with the characteristic time of mass transfer in the liquid phase. Speed of reaction should not be confused with base strength. Piperazine, for example, reacts quickly with CO_2, and is an effective promoter of absorption in potassium carbonate (section 2.2.1) and slower amines, such as methyldiethanolamine. It is a weak base. The development of DAC seems to have been unduly influenced by the consideration of strong (i.e., low pK_b) bases, which are costly to regenerate.

Although aqueous alkanolamines as a class are receiving a great deal of attention as possible solvents for carbon dioxide capture, the sourcing of these chemicals, potentially at a very large scale, requires consideration. They are commonly made by reacting ethylene oxide with ammonia or amines. Current global production of ethylene oxide (EO) is around 29 Mt/y, and scaling this up quickly by the order of magnitude that might be needed would be a challenge. EO is a flammable, explosive and toxic gas (normal boiling point 10.7 °C) and its manufacture and handling require great care. It is important that sorbents considered for large-scale application in carbon capture have a feasible supply chain. In some cases, this might require novel chemistry, and new manufacturing routes.

In the Appendix, we estimate the cost of capturing carbon dioxide from the air with a solvent, extrapolating cost and technical data for current flue gas capture processes. The estimated cost in 2013 for this illustrative case is US$ 155 per tonne of net carbon dioxide captured, including compression but excluding transport and storage. For every tonne of CO_2 captured from the atmosphere, 1.46 tonnes must be stored, assuming the energy required is supplied by burning natural gas; this ratio is similar to that for FGC at coal-fired power plants. These estimates are approximate and not based on any particular solvent, though they assume a process with characteristics like those using alkanolamine solvents, but with a very low pressure-drop contactor. For comparison, the levelised cost of carbon dioxide captured by the CE DAC process, when fired solely by natural gas as we have assumed, is reported to be in the range 168–232 US$/tCO$_2$, projected to fall to 126–170 US$/tCO$_2$ as plant and process improvements result from operational experience (Keith et al., 2018, Scenarios A and B).

Comparing FGC and DAC, the cost of net carbon dioxide avoided (~ 67 US$/tCO$_2$) in a coal-fired power plant is less than half the cost of DAC, and the cost in a gas-fired power plant (~ 83 US$/tCO$_2$) is only slightly more than half the cost of DAC. Including 1–19 US$/tCO$_2$ for transport and storage (Rubin et al., 2015), a carbon price can be calculated that would motivate the deployment of DAC, amounting to 156–174 US$/tCO$_2$, roughly double that required for Flue Gas Capture.

Costs discussed in this review should, as always, be treated with caution. Like all predictions, they incorporate many assumptions and uncertainties.

4.5 DAC: Meeting the energy demand

To capture the target amount of 810 Gt CO_2 by 2100 (Figure 1) would require some 387×10^{12} MJ of reversible work (at a rate 21 MJ/kmol CO_2, see Figure 4) yielding the purity required for pumping to underground storage. As we have seen, capturing this CO_2 by solvent-based DAC alone would require an amount of heat approximately 14 times as great: 5418×10^{12} MJ. Spread over a timescale of say 60 years, this is a heat requirement of ~ 90×10^{12} MJ/year. In comparison, the global consumption of primary energy of all types in 2017 was 566×10^{12} MJ (BP, 2018). Clearly, the use of DAC on a large scale will consume a significant fraction of the world's energy.

There is one potential mitigating factor, in the possibility that the direct air capture process could be driven by what is now classed as "waste heat" that is available from various industrial operations. For example, in this chapter, we have taken the average power plant thermal efficiency to be 40%. The remaining 60% of the energy input to electricity production is "waste heat" and is dispersed to the environment in either water- or air-cooling. In 2017, fossil fuel was used to produce 58.5×10^{12} MJ(e) globally, suggesting that waste heat totalling around 88×10^{12} MJ(th) was also produced. Some of this heat will be hot enough to drive a sorbent regeneration at perhaps 90–100 °C, and new sorbents might be developed which are more easily regenerated at a lower temperature; this would benefit both DAC and FGC, and the cost reductions could be substantial.

The possibility of utilising waste heat is supported by calculations of Rattner and Garimella (2011) who analysed data for 2007 published by the U.S. Energy Information Agency. They concluded that waste heat from U.S. power plants amounted to 60% of input energy. Total waste heat available above 30 °C from this source in 2007 amounted to 24.2×10^{12} MJ, with a waste heat weighted average temperature of 88 °C. This would provide usable heat for a significant quantity of DAC sorbent regeneration, but not for processes involving calcination, which needs much higher temperatures. In Rattner and Garimella's analysis only ~ 8% of the waste heat was furnished by exhausts at temperatures greater than 450 °C.

Hanak et al. (2017) describe the Origen Power process, in which a solid oxide fuel cell converts fuel to electricity, and the high temperature waste heat is used to calcine a carbonate solid. The calcination produces high purity carbon dioxide which can be pumped to storage, and the calcined solid can be used to draw down more carbon dioxide. Such processes for power generation become viable when carbon capture is mandated or incentivised.

5. Pointers to Policy

The scale of carbon dioxide capture and storage required to meet climate targets is a daunting challenge both for policy formulation and technology development; however, it makes no thermodynamic sense to contemplate removing carbon dioxide from the atmosphere later this century whilst continuing massive uninhibited emissions now. Capture from flue gas can be done using about half of the energy that will be required for air capture later, and at much less cost. At present, there is little evidence that other methods of CDR will prove to be significantly cheaper and more convenient than CCS—after all, BECCS actually relies on CCS technology. Capture from flue gas should, therefore, be incentivised and applied as soon as possible, and not delayed so that future generations face the worse problem of capturing even more CO_2 from the air, yet the IEA reports that industry is "woefully off-track" in installing CCS plant to meet even the very modest target of 500 Mt CO_2/y by 2030 (IEA, 2019).

The problem is that virtually cost-free discharge of flue gas to the atmosphere, as available at present, makes it almost impossible for commercial operations to justify the investment and operating expense of CCS. Future generations will regret this lack of action. One remedy for this would be for governments to incentivise the storage of carbon dioxide and penalise the failure to store it. It would be possible perhaps to adapt the Extended Producer Responsibility (EPR) approach which has been used in Europe since the 1990s for the reduction and management of various waste streams and to promote more sustainable use of resources (EU, 2014). Under EPR, regulations might require organisations which extract and/or import fossil fuels to pay the costs of capture and permanent storage of a fraction of the emissions for which they are responsible. Initially, the Removal Fraction (that proportion of emissions that is stored) should be set at a small fraction of the total production, as there is currently insufficient capability to store all the carbon dioxide emitted. Over time, as storage capacity increases, so too should the Removal Fraction. Ultimately, the goal is to match the whole national, and thus global, production of carbon dioxide with an equal amount of storage. The price of fossil fuels would rise under this regime, at first modestly as the removal fraction is small, but ultimately to meet the full cost of capture and storage. However, the scheme avoids the arbitrariness of permits and carbon tax, coupling the consumers' extra price burden to the actual cost of pollution control (Allen et al., 2009).

Such an EPR policy would achieve several ends—(i) the costs of clean-up are paid by those parties that are responsible for producing the emissions, an application of the ethical principle of 'the polluter pays'; (ii) it signals to emitting industries that in the long-term they will be responsible for the costs of clean-up. This will incentivise them to reduce emissions and also to find ways in which they can reduce the costs of capture and storage; (iii) it signals to those using developing techniques for capture and storage, working with flue gas or air, that doing so is a worthwhile endeavour; (iv) it acts as a price-discovery mechanism for the avoidance of harms caused by carbon dioxide emissions. Eventually, this could create a rational pricing of carbon dioxide emissions—the cost of 'cleaning up the mess'. Incentivising CCS in the UK context was the subject of a report by Oxburgh (2016).

The need for rapid movement on emissions reduction is underlined by Fuss et al. (2018) in their review of NETs. They looked at the characteristics, potential and limitations of BECCS, DACCS, Afforestation and forest management, EW, OF, Biochar, and Soil carbon sequestration (schemes described in Table 1 above). They concluded that "from a risk management perspective, the uncertainties and risks around large-scale NETs deployment suggest a need for swiftly ratcheting up emissions reductions over the next decade in order to limit our dependence on NETs for keeping temperature rise below 2 °C". A portfolio of several NETs would be necessary in order to meet the stated climate goal sustainably. They point out that there are many gaps in our knowledge concerning the claimed benefits, and that research on the side effects is basically non-existent. The technique DACCS emerged in this study as … "a relatively promising long-term option beyond 2050, being limited in potential only by the economic (and energetic) feasibility of scale-up" (Fuss et al., 2018). Doubts about the economic and energetic feasibility of the scale-up of DACCS really need to be addressed in order to develop new technology and schemes, ready for potential deployment. There is huge scope for novel chemistry and process engineering, but all NETs schemes also raise questions relating to environment and society that need considering.

The expected scale and cost of CDR suggest that, later this century, it could become a growing and innovative trillion US$/y industry, managing carbon. The magnitude of the investment required to develop this capability justifies much more effort on research, development and design than is currently evident. We have to improve and demonstrate CDR schemes and understand the full range of impacts. The shortening timescale of climate change lends urgency to this programme.

Appendix

A1. Process calculations

The IPCC Special Report on Carbon Capture and Storage (SRCCS) collected data for processes removing carbon dioxide from flue gas in power plants, mostly using amine solvents like MEA (Metz et al., 2005). In a power plant, the energy used by the carbon capture unit is a fraction taken from the energy that would otherwise be used to generate electricity. The SRCCS reported this fraction to be in the range 24 to 40% of Lower Heating Value (LHV) in new pulverised coal (PC) plants and for new natural gas combined-cycle (NGCC) power plants it was between 11 and 22% LHV (Metz et al., 2005). Table 2 shows data for PC and NGCC plants at the limits of the effectiveness ranges, and also, for comparison purposes, for the Carbon Engineering (CE) DAC process (Keith et al., 2018). The reversible work is calculated from equation (11) for the parameters shown.

For both FGC and DAC plants, the heat used for the separation is calculated from IPCC's Default Emission Factors using a net calorific basis, 0.0946 kg CO_2/MJ for bituminous coal, and 0.0561 kg CO_2/MJ for natural gas (Eggleston et al., 2006—Table 2.2). However the SRCCS reported energy demand includes the energy for post-capture compression, with the delivery pressure, suitable for piping the product to underground storage, being in the range 8–14 MPa. Since we wish to compare the thermodynamic performance of the separation processes with each other we have therefore subtracted the compression energy. We can then compare the heat requirement of the separation process, Q_{sep}, with the reversible work w_{rev} for the separation duty at the same temperature and pressure. The CE DAC process employs a 4-stage compressor with intercooling for the compression of almost pure carbon dioxide from atmospheric pressure (0.1 MPa) to 15 MPa. This includes a glycol dehydration system prior to the last stage, important for drying the gas before pipeline transport. The compression power consumption is 20.4 MJ(e)/kmol, which at a conversion efficiency of 40% requires 51 MJ(th)/kmol. We have subtracted this amount from the reported heat usage of the FGC units when calculating the net heat demand of the separation process, Q_{sep} shown in Table 2.

The CE DAC process absorbs carbon dioxide from air into aqueous potassium hydroxide. This is followed by a cation switch in a separate reactor, where calcium carbonate is precipitated. The carbonate is then calcined to produce the CO_2-enriched gas stream, and calcium oxide which is slaked to hydroxide and recycled to facilitate the cation switch. Data for this process included in Table 2 are those reported by Keith et al. (2018) based on process modelling, equipment testing and experience with a pilot plant capturing 1 t/day CO_2. The calculated value of the ratio Q_{sep}/w_{rev} for this process is 16.9.

Socolow et al. (2011) reported an analysis of DAC using aqueous sodium hydroxide followed by a cation switch and calcining of calcium carbonate. The process was similar in general outline to that described by Keith et al. (2018). The feed air composition was taken as 0.05% CO_2 and only 50% was captured ($\alpha = 0.5$). The carbon dioxide produced, almost pure, was compressed to 10 MPa, requiring 0.42 MJ(e)/kg, which can be subtracted from the overall power usage 1.78 MJ(e)/kg. The total energy consumed by the separation process is then 1.36 MJ(e)/kg of electrical power and an additional 8.1 MJ(th)/kg. Converting power to thermal energy at the usual efficiency (40%) gives a total requirement, Q_{sep}, of 506 MJ(th)/kmol. The minimum reversible work for this separation is 19.45 MJ/kmol, so the estimated value of Q_{sep}/w_{rev} is 26. This value is significantly greater than that found for the CE DAC process (16.9) and is at the top end of the range of Q_{sep}/w_{rev} for FGC processes in Table 2. For comparison purposes, Socolow et al. (2011) present study data from a United States' National Energy Technology Laboratory (NETL) PC plant with flue gas carbon capture using aqueous MEA (data: $y_c = 12.8\%$, $y_{co} = 99\%$, $\alpha = 0.9$, $T_0 = 40$ °C, delivery pressure 10 MPa (Ramezan et al., 2007). Again, using a heat-to-power conversion efficiency of 40%, the total thermal energy requirement was 3.33 MJ/kg CO_2; subtracting the compression energy yields a value $Q_{sep} = 100$ MJ/kmol and the ratio Q_{sep}/w_{rev} is 14.4 for the separation process. This is nicely within the range expected from Table 2.

At 40 °C the minimum work of compression of CO_2 from 0.1 to 15 MPa is 10.87 MJ/kmol, for a reversible isothermal process. From the simulation data for compressor power of Keith et al. (2018), we calculated that the heat required to generate this power was 51 MJ/kmol, giving a ratio for the compressor $(Q/w_{rev})_{comp} = 4.7$. This shows that compression is much more efficient than a separation process, which is why we prefer to analyse them separately as far as possible. In practice, some energy saving in the compression may be possible in an optimised compressor system (Jackson and Brodal, 2018).

A2. Cost estimates

These calculations enable us to make cost estimates for Direct Air Capture, assuming that the technology applied is similar in nature to the MEA-based post-combustion capture plants reported by Metz et al. (2005). The SRCCS costings, referring to a European or North American location, were adjusted and brought to a consistent 2013-basis by Rubin et al. (2015); some values for Pulverised Coal and Natural Gas combustion power plants are shown in Table A1. Note that these reported costs include compression of the CO_2 product, but exclude transport and storage.

The engineering duty of these capture plants is represented by the reversible work of separation W_{rev} (MJ/tCO_2), calculated from equation (11), shown in Table A1. Note that the spread in costs around the representative value is about +/– 25%. For process plant of a similar nature, where similar assumptions have been made in calculating fixed and variable costs, we might expect the performance cost to scale with the duty, and indeed the cost/duty ratio is found to be close to 0.326 US$/MJ(rev work) for both coal and natural gas plants. The reversible work for DAC (Figure 4 in main text) is some 21 MJ/kmolCO_2, or 477 MJ/tCO_2. Using the scaling described, we might therefore expect the cost of carbon captured by DAC from the air to be about 155 US$/t$CO_2$, exclusive of transport and storage.

Table A1. Scaling cost with reversible work. Range of cost of CO_2 is given as Low (Representative) High, indicating observed spread. Costs are adjusted SRCCS values, corrected to 2013 (Rubin et al., 2015), and exclude transport and storage.

Case	PC	NGCC
W_{rev}, MJ/tCO_2	146.4	209.5
Cost of CO_2 captured, US$/t$CO_2$	33 (48) 58	53 (68) 87
Representative cost of CO_2 captured, US$/MJ (rev work)	0.328	0.324

The estimate of 155 US\$/tCO$_2$ for DAC can be compared with the representative cost of CO$_2$ avoided by applying FGC at power plants. At pulverised coal plants the cost of avoided CO$_2$ is 67 US\$/tCO$_2$ so DAC is 88 US\$/tCO$_2$ (131%) more expensive; at NGCC (gas) plants the cost of avoided CO$_2$ is 83 US\$/tCO$_2$ so DAC is 72 US\$/tCO$_2$ (87%) more expensive.

The costs of transport and storage of carbon dioxide will vary with the location, and many other factors relating to specific projects (Metz et al., 2005). A range 1–19 US\$/tCO$_2$ can be taken as consistent with other costs given here (Rubin et al., 2015). An indicative range of cost of DACCS is then 156–174 US\$/tCO$_2$.

The amount of carbon dioxide that the DAC process sends to storage can be estimated. We estimate the DAC plant to require heat energy of $14 \times W_{rev}$, which is 6678 MJ/tCO$_2$. In addition there is 51 MJ/kmol or 1159 MJ/tCO$_2$ required to drive the compression, thus, 7837 MJ(th)/tCO$_2$ in total. If this energy is derived from natural gas with an emission factor of 0.0561 kg CO$_2$/MJ, then another 440 kg of CO$_2$ are generated for every tonne captured from the air. Assuming 90% of this extra CO$_2$ is captured by the DAC plant, the ratio of carbon stored to net carbon captured is (1 + 0.9*0.440)/(1–0.044), which is 1.46. This is similar to the ratio of carbon stored to carbon avoided for coal-fired power plants with post-combustion capture.

A large volume of air must be blown through the DAC contactor where CO$_2$ is absorbed, so it is very important that the pressure drop in this device be kept low, to minimise power consumption by the air blowers. The CE DAC process uses a novel structured packing arrangement with intermittent liquid flow (Keith et al., 2018), and its reported pressure drop is about 100 Pa, a very low value. Assuming this pressure drop and a fan efficiency of 70%, the work required is some 11.72 MJ(e)/kmolCO$_2$, or 266 MJ(e)/tCO$_2$ when the recovery α is 74.5%. Deriving this work from heat at an efficiency of 40%, gives a requirement for the air blowers of 665 MW(th)/tCO$_2$. This is a fraction 665/7837 of the total heat demand for the DAC process, or 8.5%. We note that in a study of post-combustion capture using MEA at a natural gas fired power plant, the flue gas booster fan required 16% of the energy required for capture and compression (Smith et al., 2013). The DAC plant estimate is significantly less than this, but does rely on novel low-pressure drop absorption technology. Using a conventional packed contactor would increase the pressure drop, power requirement and cost of DAC.

References

Allen, M.R., Frame, D.J. and Mason, C.F. 2009. The case for mandatory sequestration. Nature Geoscience, 2(12): 813–814.

Astarita, G., Savage, D.W. and Longo, J.M. 1981. Promotion of CO$_2$ mass transfer in carbonate solutions. Chemical Engineering Science 36(3): 581–588.

Beerling, D.J., Leake, J.R., Long, S.P., Scholes, J.D., Ton, J., Nelson, P.N., Bird, M., Kantzas, E., Taylor, L.L., Sarkar, B. and Kelland, M. 2018. Farming with crops and rocks to address global climate, food and soil security. Nature Plants 4: 138–147.

Bellassen, V., Stephan, N., Afriat, M., Alberola, E., Barker, A., Chang, J.P., Chiquet, C., Cochran, I., Deheza, M., Dimopoulos, C. and Foucherot, C. 2015. Monitoring, reporting and verifying emissions in the climate economy. Nature Climate Change 5: 319–328.

Boysen, L.R., Lucht, W., Gerten, D., Heck, V., Lenton, T.M. and Schellnhuber, H.J. 2017. The limits to global-warming mitigation by terrestrial carbon removal. Earth's Future 5: 463–474.

BP. 2018. BP statistical review of world energy, June 2018.

Ciais, P., Sabine, C., Bala, G., Bopp, L., Brovkin, V., Canadell, J., Chhabra, A., DeFries, R., Galloway, J., Heimann, M. and Jones, C. 2014. Carbon and other biogeochemical cycles. pp. 465–570. In Climate change 2013: The Physical Science Basis. Contribution of Working Group I to the Fifth Assessment Report of the Inter-governmental Panel on Climate Change. Cambridge University Press.

Cullinane, J.T. and Rochelle, G.T. 2004. Carbon dioxide absorption with aqueous potassium carbonate promoted by piperazine. Chemical Engineering Science 59(17): 3619–3630.

Danckwerts, P.V. and Sharma, M.M. 1966. The absorption of carbon dioxide into solutions of alkalis and amines (with some notes on hydrogen sulfide and carbonyl sulfide). The Chemical Engineer, October 1966, 244–280.

Eggleston, H.S., Buendia, L., Miwa, K., Ngara, T. and Tanabe, K. 2006. 2006 IPCC Guidelines for National Greenhouse Gas Inventories. Chapter 2, Stationary combustion. Japan.

EU. 2014. ec.europa.eu/environment/archives/waste/eu_guidance/introduction.html viewed on 23 May 2019.

Fuss, S., Lamb, W.F., Callaghan, M.W., Hilaire, J., Creutzig, F., Amann, T., Beringer, T., de Oliveira Garcia, W., Hartmann, J., Khanna, T. and Luderer, G. 2018. Negative emissions—Part 2: Costs, potentials and side effects. Environmental Research Letters 13(6): 063002.

Goldthorpe, S. 2017. Potential for very deep ocean storage of CO_2 without ocean acidification: a discussion paper. Energy Procedia 114: 5417–5429.

Griscom, B.W., Adams, J., Ellis, P.W., Houghton, R.A. et al. 2017. Natural climate solutions. PNAS 114(44): 11645–11650.

Hanak, D.P., Jenkins, B.G., Kruger, T. and Manovic, V. 2017. High-efficiency negative-carbon emission power generation from integrated solid-oxide fuel cell and calciner. Applied Energy 205: 1189–1201.

IEA. 2019. www.iea.org/tcep/industry/ccs/viewed on 23 May 2019.

IPCC. 2018. Summary for policymakers. *In*: Global Warming of 1.5 °C. An IPCC Special Report on the impacts of global warming of 1.5 °C above pre-industrial levels and related global greenhouse gas emission pathways, in the context of strengthening the global response to the threat of climate change, sustainable development, and efforts to eradicate poverty. Masson-Delmotte, V., Zhai, P., Pörtner, H.-O., Roberts, D., Skea, J., Shukla, P.R., Pirani, A., Moufouma-Okia, W., Péan, C., Pidcock, R., Connors, S., Matthews, J.B.R., Chen, Y., Zhou, X., Gomis, M.I., Lonnoy, E., Maycock, T., Tignor, M. and Waterfield, T. (eds.). World Meteorological Organization, Geneva, Switzerland, 32 pp.

Jackson, S. and Brodal, E. 2018, June. A comparison of the energy consumption for CO_2 compression process alternatives. In IOP Conference Series: Earth and Environmental Science 167(1): 012031. IOP Publishing.

Keith, D.W., Holmes, G., St Angelo, D. and Heided, K. 2018. A process for capturing CO_2 from the atmosphere. Joule 2(8): 1573–1594.

King, C.J. 1980. Separation Processes, 2nd edition. McGraw-Hill, New York.

Lal, R. 2011. Sequestering carbon in soils of agro-ecosystems. Food Policy 36: S33–S39.

Mac Dowell, N., Fennell, P.S., Shah, N. and Maitland, G.C. 2017. The role of CO_2 capture and utilization in mitigating climate change. Nature Climate Change 7(4): 243.

Matthews, H.D., Zickfeld, K., Knutti, R. and Allen, M.R. 2018. Focus on cumulative emissions, global carbon budgets and the implications for climate mitigation targets. Environmental Research Letters 13(1): 010201.

Matter, J.M., Stute, M., Snæbjörnsdottir, S.Ó., Oelkers, E.H., Gislason, S.R., Aradottir, E.S., Sigfusson, B., Gunnarsson, I., Sigurdardottir, H., Gunnlaugsson, E. and Axelsson, G. 2016. Rapid carbon mineralization for permanent disposal of anthropogenic carbon dioxide emissions. Science 352(6291): 1312–1314.

McGlashan, N., Shah, N., Caldecott, B. and Workman, M. 2012. High-level techno-economic assessment of negative emissions technologies. Process Safety and Environmental Protection 90(6): 501–510.

McLaren, D. 2012. A comparative global assessment of potential negative emissions technologies. Process Safety and Environmental Protection 90(6): 489–500.

Metz, B., Davidson, O., de Coninck, H., Loos, M. and Meyer, L. 2005. IPCC Special Report on Carbon Dioxide Capture and Storage. Cambridge University Press.

Minx, J.C., Lamb, W.F., Callaghan, M.W., Fuss, S., Hilaire, J., Creutzig, F., Amann, T., Beringer, T., de Oliveira Garcia, W., Hartmann, J. and Khanna, T. 2018. Negative emissions—Part 1: Research landscape and synthesis. Environmental Research Letters 13(6): 063001.

National Academies of Sciences, Engineering, and Medicine. 2019. Negative Emissions Technologies and Reliable Sequestration: A Research Agenda. Washington, DC: The National Academies Press.

Nemet, G.F., Callaghan, M.W., Creutzig, F., Fuss, S., Hartmann, J., Hilaire, J., Lamb, W.F., Minx, J.C., Rogers, S. and Smith, P. 2018. Negative emissions—Part 3: Innovation and upscaling. Environmental Research Letters 13(6): 063003.

Oxburgh, R. 2016. Lowest cost decarbonisation for the UK: The critical role of CCS. Report to the Secretary of State for Business, Energy and Industrial Strategy from the Parliamentary Advisory Group on Carbon Capture and Storage (CCS).

Qambrani, N.A., Rahman, M.M., Won, S., Shim, S. and Ra, C. 2017. Biochar properties and eco-friendly applications for climate change mitigation, waste management, and wastewater treatment: A review. Renewable and Sustainable Energy Reviews 79: 255–273.

Ramezan, M., Skone, T.J., Nsakala, N.Y., Liljedahl, G.N., Gearhart, L.E., Hestermann, R. and Rederstorff, B. 2007. Carbon dioxide capture from existing coal-fired power plants. National Energy Technology Laboratory, DOE/NETL Report, (401/110907).

Rattner, A.S. and Garimella, S. 2011. Energy harvesting, reuse and upgrade to reduce primary energy usage in the USA. Energy 36(10): 6172–6183.

Rau, G.H. 2011. CO_2 mitigation via capture and chemical conversion in seawater. Environmental Science & Technology 45(3): 1088–1092.

Renforth, P., Jenkins, B.G. and Kruger, T. 2013. Engineering challenges of ocean liming. Energy 60: 442–452.

Renforth, P. 2019. The negative emission potential of alkaline materials. Nature Communications 10, Article nr 1401.

Rogelj, J., Popp, A., Calvin, K.V., Luderer, G., Emmerling, J., Gernaat, D., Fujimori, S., Strefler, J., Hasegawa, T., Marangoni, G. and Krey, V. 2018. Scenarios towards limiting global mean temperature increase below 1.5 °C. Nature Climate Change 8(4): 325–332.

Royal Society and Royal Academy of Engineering. 2018. Greenhouse Gas Removal ISBN: 978-1-78252-349-9.

Rubin, E.S., Davison, J.E. and Herzog, H.J. 2015. The cost of CO_2 capture and storage. International Journal of Greenhouse Gas Control 40: 378–400.

Shi, X., Xiao, H., Lackner, K.S. and Chen, X. 2016. Capture CO_2 from ambient air using nanoconfined ion hydration. Angewandte Chemie–International Edition 55(12): 4026–4029.

Smith, N., Miller, G., Aandi, I., Gadsden, R. and Davison, J. 2013. Performance and costs of CO_2 capture at gas fired power plants. Energy Procedia 37: 2443–2452.

Smith, R. 2016. Chemical Process: Design and Integration (2nd edition). John Wiley & Sons.

Socolow, R., Desmond, M., Aines, R., Blackstock, J., Bolland, O., Kaarsberg, T., Lewis, N., Mazzotti, M., Pfeffer, A., Sawyer, K. and Siirola, J. 2011. Direct air capture of CO_2 with chemicals: A technology assessment for the APS Panel on Public Affairs. American Physical Society.

Strefler, J., Amann, T., Bauer, N., Kriegler, E. and Hartmann, J. 2018. Potential and costs of carbon dioxide removal by enhanced weathering of rocks. Environmental Research Letters 13(3): 034010.

UNEP. 2017. The Emissions Gap Report 2017. United Nations Environment Programme (UNEP), Nairobi.

Williamson, P., Wallace, D.W., Law, C.S., Boyd, P.W., Collos, Y., Croot, P., Denman, K., Riebesell, U., Takeda, S. and Vivian, C. 2012. Ocean fertilization for geoengineering: A review of effectiveness, environmental impacts and emerging governance. Process Safety and Environmental Protection 90(6): 475–488.

CHAPTER 2
Low-Carbon Technologies in Global Energy Markets

Yoram Krozer

1. Introduction

Do economies shift to low-carbon technologies? This question is discussed in terms of modern renewable energy based on geothermal, solar and wind resources in energy production which competes with fossil fuels (coal, oil, gas and nuclear) and traditional renewable energy (biomass and hydro). Justification of this focus is that modern renewable energy is infinite, nearly carbon-free throughout the life cycle, and it can be applied virtually all over Earth and in space on a mega-scale and individually. Applications of low-carbon technologies are comprehended as innovations and adaptations that enable nearly pollution-free energy consumption without large infrastructure but the production is variable in time, needs much space, and it can degrade nature and landscape.

These innovations were introduced in the 1970s, when activists were campaigning for self-reliance and pollution prevention, while businesses pursued stand-alone products in remote areas on demands of authorities; the societal perspectives on low-carbon technologies still differ. These applications grew fast when the international prices of fossil fuels increased during the periods from 1979 to 1986 and from 2005 to 2015. These prices in real US dollars of 2005 (US\$$_{2005}$) increased from US\$$_{2005}$ 20–30 to US\$$_{2005}$ 60–100 per barrel oil equivalent (b.o.e.). During high prices, policies supported low-carbon technologies with subsidies and taxes in a few countries, particularly much in the United States of America (USA), Europe Union (EU) and Japan (Haas et al., 2011). However, the policy support for low-carbon technologies was several times smaller than the support for fossil fuels (EEA, 2004; IEA/OEGCD, 2019). As the applications of low carbon technologies expanded, many adaptations entailing lower costs per energy unit in time, referred to as the cost-reducing technological change or technological learning, were introduced. The unit costs decreased faster than the unit costs of other energy technologies, whilst the unit cost of nuclear power increased (Rubin et al., 2015). The applications of low-carbon technologies, however, remained costly in many countries, when assessed based on standardized variables for investments and operations, called levelized costs of technologies (Levelized costs, 2019). Opinions about the future growth of low carbon technologies differ because fast dissemination due to the decreasing unit costs (Deng et al., 2012) and social initiatives in renewable energy (Sovacool, 2016) is obstructed by persistent energy systems (Grübler et al., 2016) and vested interests in energy markets (Fouquet, 2016).

Energy resources evolve toward lower carbon density quasi-autonomously; it is hypothesized, because more hydrogen and electrons per resource mass deliver higher energy performance (Herman

Sustainable Innovations Academy and University of Twente - CSTM, p/a Iepenplein 44, 1091JR Amsterdam, The Netherlands.
Email: krozer@xs4all.nl

et al., 1989). It is argued that high-carbon energy resources be substituted for lower carbon ones: Biomass and peat for fossil fuels based on coal, then oil and natural gas, as well as from low electron-density hydropower to high electron-density nuclear power, which is followed by geothermal, wind and solar energy (Grübler and Nakicenoviĉ, 1996). Note that biomass absorbs carbon dioxide (CO_2) from air when plants grow but as an energy resource it is more carbon-dense than coal. This hypothesis is referred to as the decarbonization of energy resources.

Another explanation of the shifts in energy resources refers to the "value added" of energy services; value added is income from sales of products minus costs of material purchases. In general, value added is generated when producers deliver novel qualities that are perceived as beneficial by consumers in businesses and households despite being costlier than the vested alternatives (Lancaster, 1966). Advances in energy services can deliver valuable qualities to consumers, such as refined fuels for high temperature combustion, electricity on grid for air conditioning, power off-grid for mobile applications and others. Applications of low-carbon technologies are costly but can be used in remote areas and stand-alone systems. They are also valued because they prevent pollution and are enable to generate income in communities. When the value added of energy services increase over several years, it is called the valorization of energy services. This viewpoint challenges the ideas that the countries' incomes grow mainly due to more cost-effective energy use (Ayres and Voudouris, 2014) or cheaper energy resources (Beaudreau and Lightfoot, 2015); cheap energy is important for basic industries and other energy-intensive activities but hardly relevant in services and energy-extensive businesses when compared to labor, capital and knowledge.

In this chapter, the question of whether the decarbonization and valorization can be observed as the global trends in energy markets is discussed. A trend is defined as annual average growth during 10 years or longer and the energy markets as transformations of energy resources starting from winning of the resources downstream, through processing into energy services for consumption in households and businesses. Based on a literature review (Krozer, 2017) and assessment of the value added in energy markets (Krozer, 2019), it is estimated whether those trends converge or diverge across countries, because converging trends are easier to manage in international policies, e.g., the policies on climate change. Changes from 1990 to 2015 are divided into the periods of low oil prices during 1990–2004 and high oil prices during 2005–2015; oil price is the conventional benchmark of all fuel prices (Energy expenditures, 2017). The period of high prices coincides with the financial crisis in 2008 and economic depression is several countries, as well as larger policy support for renewable energy and policies aiming to reduce carbon dioxide emissions (CO_2 emission), which is the largest mass of greenhouse gas that causes climate change. These periods indicate when conditions for low-carbon technologies were unfavorable and favorable, respectively.

Statistical data on income, energy consumption, energy resources, and CO_2 emissions are used. All countries above 100 million people in 2017 are considered, the European Union of 28 member countries (EU) as if it is one country. In order of purchase power (GDP–PPP), i.e., income for typical consumer purchases after inflation correction, those countries are: USA, Japan, and the EU, considered high income; Russian Federation (Russia), Mexico, Brazil, China, Indonesia, Philippines and India, considered mid income; Nigeria, Pakistan, Bangladesh and Ethiopia, considered low income. These countries together covered in 2015 about 70% of the global population, 78% of the global GDP, 65% of all energy production, 72% of the energy consumption, 76% of the electricity consumption, as well as 88% of the coal supplies, 63% of oil, 64% of gas, 83% of nuclear, 71% of biofuels, 66% of hydropower, 84% of applications of low-carbon technologies, and 73% of all CO_2 emissions. Trends in those countries indicate the global trends.

Solely authoritative, open sources databases are used in order to avoid opinions and private databases that can be biased by particular interests. Data on GDP–PPP, energy consumption and CO_2 are derived from the World Bank database and data on the energy resources are based on the International Energy Agency (IEA) database because they are absent in the World Bank data. It is given that the CO_2 emissions are not measured in the air but estimated with emission factors per resource after correction for absorption. The following factors are used based on IEA data: 3.80 ton CO_2 per ton coal, 2.53 per ton oil and 2.16 per ton oil equivalent of gas (IEA, 2017). Comparison of the World Bank and IEA data is also made. While the growth rates based on World Bank data are annual averages, the IEA data are annual averages per 5 years. The diverging or converging trends across countries are assessed based on the global growth rates and standard deviation of the growth rates across countries per period. A diverging trend is assumed when the standard deviation across countries is larger than the global growth rate and the converging trend when the standard deviation is smaller than that global growth rate. All regressions are Pearson correlation across fourteen countries and the world (n = 15).

After this introduction, the economic context of energy consumption is discussed. Thereafter, the decarbonization trend and the role of low-carbon technologies in it are presented. This is followed by discussion of the valorization trend. Finally, conclusions are drafted.

2. Economic Context

Are applications of low-carbon technologies affordable? To answer this question, one must consider the total incomes because a larger income enables more expenditure on those applications. A fairer income distribution is also relevant because more people can spend on energy, which is a basic good, and more people are able to afford low-carbon technologies if they perceived them beneficial. The countries' income growths and declining income disparities across countries and within countries enable more applications of low-carbon technologies. Herewith, GDP, GDP–PPP and Gini coefficients are used because they indicate the countries' economic output in nominal US$, the purchasing power of citizens and income distribution, respectively. The Gini coefficients show index per 10% income class in a country from 0 to 1; 0 meaning no income difference (all incomes equal) and 1 meaning one class has all income. This indicator is often used though far from perfect because do not cover wealth, consumption patterns, division between wages and capital, the richest few percent and the World Bank data do not cover all countries and only for a few years; however, more complete global statistics are unavailable.

2.1 Income for energy

The countries' economic output, citizens' purchasing power and income distribution in 2015, and growth during 1990–2015, as well as during periods 1990–2004 and 2005–2015 are assessed with the World Bank statistics. The income distribution is shown for the earliest and latest year between 1990 and 2015 because this data is scarce. All results are per capita, based on there being 7.4 billion inhabitants in the world in 2015 and the global population growth of 1.3% annual average throughout 1990–2015 though the growth rates vary from –0.1% in Russia to 3.0% in Ethiopia (note high correlation of income to population across countries: $R^2 = -0.61$). Table 1 shows the results.

The nominal average GDP per capita in 2015 was about US$ 11,000 globally, however, the US$ 640 average in Ethiopia was 88 times lower than the USA average. All countries' GDP grew but the growth rates were usually lower in high-income countries than in mid-income and low-income countries. A global citizen could purchase based on US$$_{2005}$ 17,000, but a citizen of the USA had a 35-times higher purchasing powers than an Ethiopian with an average of US$$_{2005}$ 1,600. The purchasing power of a global citizen grew five-fold from 1990 to 2015 but only twice in high income countries compared to nearly 14 times higher in China, 5 times in India and 4 times in Bangladesh and Ethiopia. The growth rates of GDP and purchase power converged across countries from 1990 to 2015; they converged more during high fuel prices from 2005 to 2015 than during low prices because they decreased in high income countries and increased in all mid and low-income countries, except Mexico. The income disparities within most mid-income countries and the USA (45 to 61 points) were twice as large as in Japan, the EU and low income countries (25 to 35 points), and the disparities decreased in most countries during 2005–2015 except the USA, Pakistan and Bangladesh. Regarding the converging purchase power across countries and larger disparities within the mid-income countries and USA than high- and low-income countries, the conventional division between the developing countries and developed ones hardly holds.

Regarding the growth of GDP and purchasing power in all countries, as well as the decreasing income disparities across countries and, within most countries, more people can afford low-carbon technologies.

Table 1. Economic indicators, global perspective.

Bold: higher growth during 2005–2015 than 1990–2004	GDP				GDP–PPP				Gini coefficient	
	USD nominal/ capita	Average growth bold: Lower growth rates during high fuel prices			US$2005/ capita	Average growth bold: Lower growth rates during high fuel prices			Earliest found 1990s to 2000s	2015
	2015	1990–2015	1990–2004	2005–2015	2015	1990–2015	1990–2004	2005–2015		
World	11,042	4.1%	3.5%	**4.9%**	15,694	4.4%	4.0%	**4.9%**	N.A.	N.A.
USA	56,611	3.5%	4.1%	**2.7%**	56,469	3.5%	4.1%	**2.7%**	38	41
Japan	35,764	2.1%	3.3%	**0.4%**	40,607	3.0%	3.2%	**2.6%**	32	-
EU	37,752	3.9%	4.5%	**3.0%**	38,447	3.9%	4.2%	**3.5%**	N.A.	N.A.
Russia	12,245	9.4%	4.7%	15.1%	24,692	5.5%	2.1%	10.1%	48	38
Mexico	10,761	6.2%	7.6%	**4.2%**	16,983	4.6%	4.9%	**4.4%**	50	46
Brazil	11,836	7.6%	3.3%	13.7%	15,617	3.8%	3.2%	4.5%	61	51
China	8,289	14.4%	12.0%	17.9%	14,450	11.5%	11.4%	11.7%	43	42
Indonesia	3,363	9.8%	7.9%	12.4%	11,040	5.5%	4.8%	6.4%	40	-
Philippines	2,926	6.3%	3.3%	10.4%	7,320	4.2%	3.2%	5.6%	N.A.	N.A.
India	1,700	6.7%	4.3%	10.2%	6,127	7.0%	6.0%	8.3%	35	-
Nigeria	3,446	13.4%	8.1%	20.9%	6,038	4.9%	4.6%	5.3%	45	43
Pakistan	1,362	5.7%	4.3%	7.6%	5,000	3.8%	3.7%	3.9%	33	34
Bangladesh	1,217	5.7%	3.3%	9.0%	3,336	5.7%	4.7%	7.0%	28	32
Ethiopia	640.2	4.7%	–3.6%	16.2%	1,633	5.6%	2.6%	9.9%	45	39

This is encouraging for the carbon management, though the affordability does not automatically enhance expenditures on energy consumption because other basic goods are also purchased.

2.2 Energy consumption

Table 2 shows that the global energy consumption per capita was about 22.6 MWh in 2015, but the consumption of 2.7 MWh in Bangladesh is 31 times lower than that in the USA. From 1990 to 2015, the energy consumption decreased in all high-income countries and Russia, and increased in all other countries. Although the growth of energy consumption is correlated to the purchase power across those countries ($R^2 = 0.78$ during 1990–2015, $R^2 = 0.76$ during 1990–2005 and $R^2 = 0.70$ during 2005–2015) the global consumption grew 7 times slower than that real income; it was 2 to 30 times slower in mid- and low-income countries. High prices of fossil fuels had little effect on the energy consumption that grew even faster in several mid- and low-income countries. The growing purchase power is apparently the main factor for the growth of energy consumption, not another way around. Moreover, the growth of energy consumption diverged across countries. This growth indicates that changes in the composition of businesses in economies, so called economic structure, and the use of energy-efficient technologies in businesses and households are also important factors in energy consumption.

Renewable energy in energy consumption was about 18% in 2015; this share varied from 3% in Russia and 6% in Japan up to 41% in Brazil, 46% in Pakistan, even 88% in Nigeria and 91% in Ethiopia. Although renewable energy is considered costly, its share in energy consumption is usually higher in mid- and low-income countries than in high-income countries. However, that share grew in high-income countries and declined in all mid- and low-income countries. High prices of fossil fuels invoked faster growth of the renewable energy share in high-income countries and Nigeria, but had hardly any impact in five mid- and low-income countries whilst the share declined in five others. Growth of the renewable energy share in energy consumption diverged across countries.

The energy consumption decoupled from income in nearly all countries during low and high prices of fossil fuels. More renewable energy was purchased in high-income countries where the share in energy consumption was low and less was purchased in mid-income and low-income countries where the share was high. The growth rates of energy consumption and the share of renewable energy in it diverged across countries.

3. Decarbonization of Economies

Is the decarbonization of energy resources a global trend? Along with the global growth of fossil fuels from 970 million tons in 1975 to nearly 12,000 million ton in 2005, the share of carbon-intensive coal declined from 98% to nearly 49% because it was replaced by the less carbon-intensive oil and gas (Krausmann et al., 2009). However, when the growth of energy consumption and the shares of individual fuels are combined, the global decarbonization is insignificant. Beneath, consumption of fossil fuels, renewable energy and low carbon technologies are assessed. The IEA data are used and compared to the World Bank data.

3.1 Resource composition

Table 3 shows the consumption of fossil fuels, all renewable energy and applications of low-carbon technologies in 2015, as well as growth during 1990–2015, during low prices of fossil fuels (1990–2004) and high prices (2005–2015).

Fossil fuels covered about 86% of all resources in 2015 which varied from 94% in Japan to 6% in Ethiopia; herewith, consumption per capita in Ethiopia was 104 times lower than in Japan and 213 times lower than in the USA. While the global consumption of fossil fuels grew by 1.8% annual average during 1990–2015, it declined by –0.8% in Russia and grew by 6.6% in Bangladesh. High prices of fossil fuels hardly influenced the global consumption of fossil fuels because lower consumption is observed in

Table 2. Energy consumption per capita and share of renewable energy in it.

Total and annual average growth	Energy consumption				Share renewable energy in total				
	kWh/capita	Average growth, bold: Lower growth or larger decline during high fuel prices			% all	Average growth, bold: Higher growth or smaller decline during high fuel prices			
	2015	1990–2015	1990–2004	2005–2015	2015	1990–2015	1990–2004	2005–2015
World	22,579	0.6%	0.3%	**1.0%**	18%	0.2%	0.0%	**0.5%**
USA	81,330	-0.4%	0.2%	**-1.2%**	9%	3.4%	2.4%	**4.9%**
Japan	39,137	-0.1%	1.0%	**-1.5%**	6%	1.2%	-0.6%	**3.7%**
EU	34,209	-0.4%	0.4%	**-1.6%**	17%	4.2%	2.3%	**6.8%**
Russia	55,763	-0.7%	-1.9%	1.0%	3%	-0.5%	-0.8%	**-0.1%**
Mexico	16,922	0.2%	0.6%	**-0.3%**	10%	-1.5%	-2.3%	**-0.4%**
Brazil	17,617	2.0%	1.4%	2.8%	41%	-0.7%	-0.6%	-0.8%
China	26,213	4.6%	3.8%	5.9%	13%	-4.1%	-3.5%	-4.8%
Indonesia	10,495	2.1%	2.8%	**1.2%**	37%	-1.8%	-2.4%	**-1.0%**
Philippines	5,773	0.2%	0.0%	0.4%	28%	-2.3%	-3.3%	**-0.8%**
India	7,749	2.5%	1.6%	3.8%	35%	-1.9%	-1.2%	-3.0%
Nigeria	8,664	0.4%	0.5%	0.2%	88%	-0.01%	-0.3%	**0.4%**
Pakistan	5,573	0.8%	1.5%	**-0.1%**	46%	-0.8%	-1.3%	**-0.2%**
Bangladesh	2,658	2.6%	1.9%	3.7%	37%	-2.6%	-2.2%	-3.2%
Ethiopia	5,787	0.2%	0.1%	0.3%	91%	-0.2%	-0.1%	-0.3%

Table 3. Consumption of fossil fuels all renewable energy and applications of low-carbon technologies.

Total and annual average growth	Fossil fuels				Total renewable energy including modern one				Applications of low-carbon technologies			
	TWh	Annual average growth **Bold: Lower growth during high fuel prices**			TWh	Annual average growth **Bold: Higher growth during high fuel prices**			TWh	Annual average growth **Bold: Higher growth during high fuel prices**		
	2015	1990–2015	1990–2004	2005–2015	2015	1990–2015	1990–2004	2005–2015	2015	1990–2015	1990–2004	2005–2015
World	137,092	1.8%	1.9%	**1.6%**	21,614	2.0%	2%	**2%**	2,333	7.1%	4%	**11%**
USA	24,280	0.6%	1.2%	**-0.4%**	1,763	1.7%	1%	**3%**	364	3.4%	-1%	**10%**
Japan	4,701	-0.2%	1.2%	**-2.2%**	291	2.4%	1%	**5%**	73	3.1%	2%	**5%**
EU	15,814	-0.6%	0.4%	**-2.0%**	2,610	4.5%	4%	**6%**	530	11.5%	10%	**14%**
Russia	8,008	-0.8%	-1.9%	**0.9%**	258	-0.6%	-1%	**0%**	2	11.2%	21%	**-4%**
Mexico	2,010	2.0%	2.8%	**0.7%**	179	0.4%	1%	**-1%**	49	0.3%	3%	**-3%**
Brazil	2,034	3.6%	3.4%	3.8%	1,394	2.4%	2%	**3%**	30	20.9%	9%	**39%**
China	31,610	5.6%	5.5%	5.6%	2,971	0.8%	1%	1%	537	37.4%	46%	25%
Indonesia	1,746	4.3%	5.4%	**2.6%**	876	2.0%	2%	2%	201	9.1%	12%	5%
Philippines	385	4.4%	4.6%	**4.0%**	220	0.6%	0%	**1%**	112	3.0%	4%	1%
India	7,421	5.5%	4.9%	6.4%	2,477	1.7%	1%	**2%**	56	29.1%	37%	17%
Nigeria	317	2.7%	3.1%	**2.2%**	1,304	3.1%	3%	3%	0	0.0%	0%	0%
Pakistan	677	4.0%	5.3%	**1.9%**	416	2.4%	3%	2%	1	0.0%	0%	0%
Bangladesh	330	6.6%	6.3%	7.0%	108	1.2%	1%	1%	0	0.0%	0%	0%
Ethiopia	35	5.9%	5.1%	7.1%	542	3.0%	3%	3%	0	0.0%	0%	0%

high-income countries but higher in several mid- and low-income countries. The growth of fossil fuels diverged across countries.

Renewable energy, including low-carbon technologies, covered about 14% of the global energy consumption. Biomass was the main resource in many mid- and low-income countries. It covered, for instance, 94% of all energy consumption in Ethiopia that hardly produced fossil fuels and 80% in Nigeria that was a large producer and exporter of fossil fuels. The renewable energy grew at a similar rate to fossil fuels during 1990–2015. The renewable energy growth was high in high-income countries whose total energy consumption declined but slow or declining in mid- and low-income countries where its share in energy consumption was high and the total energy consumption grew. In China, Bangladesh, India and Russia, the growth rates of renewable energy were even below their energy consumption growth. High prices of fossil fuels enhanced the renewable energy growth in high-income countries but rarely in other countries. The growth of renewable energy converged across countries.

Applications of low-carbon technologies covered about 1.5% of the global energy consumption in 2015. In a few countries, a lot of geothermal energy is used; for instance, 18.4% of energy consumption on Philippines, 7.7% in Indonesia and 2.2% in Mexico, according to energy balances in the IEA (2019). Solar and wind energy were applied in 2015 mainly in high-income countries (2.9% of all consumed energy in the EU, 1.5% in Japan, 1.4% in the USA) and a few mid-income countries (1.6% in China, 0.9% in Brazil and 0.6% in India). Throughout 1990–2015, those applications grew fast globally because 7.1% annual average and the growth rates were much higher in several countries: 37.4% in China, 29.1% in India, 20.9% in Brazil, 11.5% in the EU and 11.2% in Russia. During those twenty-five years, the global applications of low-carbon technologies increased 6 times, even 2816 times in China, but the applications were negligible and hardly grew at all in the low-income countries. High prices of fossil fuels invoked faster growth only in high-income countries. The growth of low-carbon technologies diverged across countries.

Trends differ across countries. Fossil fuel consumption was high in high-income countries and declined; it grew in most mid- and low-income countries whose energy consumption was low. The renewable energy consumption was a small energy resource in high-income countries, but grew; it was a large resource in mid- and low-income countries, but declined. Applications of low-carbon technologies grew fast in high-income countries and a few mid-income ones, but very little in low-income countries. The decarbonization is not the global trend in energy markets.

3.2 *Carbon dioxide emissions*

Assessments of CO_2 emission confirm that observation. Table 4 shows tons CO_2 emission per capita and kg CO_2 emission per kWh energy consumption, which indicate changes in economies and in energy resources, respectively. Also, the World Bank data and IEA data are compared.

A global citizen emitted on average about 5 ton CO_2 in 2015, but a USA citizen emitted 132 times more than an Ethiopian. The global CO_2 emission per capita grew during 1990–2015 at a rate similar to the energy consumption. It decreased in the USA and Russia where the emissions were high, as well as in the EU where they were lower. These emissions increased in Japan and in all mid- and low-income countries which reflect the growth of fossil fuels in their energy consumption. Higher prices of fossil fuels invoked lower CO_2 emissions per capita in all high-income countries but these emissions grew even faster in most mid- and low-income countries because oil and gas are replaced by cheaper, more carbon-intensive coal whilst the biomass consumption decreased. The growth of CO_2 emissions per capita diverged across countries.

The global CO_2 emission per kWh was 0.22 kg, but it was 13 times higher in China, where mainly coal is used, compared to Ethiopia, where a lot of biomass is used. Per kWh, these emissions hardly changed globally over the last 25 years, however, there are changes across countries. The carbon-intensity of energy resources decreased in the EU, Russia and the USA but increased in all other countries due to more coal and less biomass usage in energy consumption. Higher international prices of fossil fuels reinforced these trends toward less carbon-intensive resources in high-income countries but more carbon-

Table 4. Decarbonization of economies (per capita) and energy consumption (per kWh) measured by CO$_2$ emissions.

Total and annual average growth	ton/capita	Annual average growth Bold: Lower CO$_2$ during high fuel prices			kg/kWh	Annual average growth Bold: lower CO$_2$ during high fuel prices			WB/IEA data	
	2015	90–15	90–05	05–15	2015	90–15	90–05	05–15	kWh/cap	CO$_2$ t/cap
World	5.0	0.7%	0.4%	1.2%	0.22	0.1%	0.0%	0.2%	4%	12%
USA	16.7	-0.6%	0.1%	**-1.7%**	0.20	-0.2%	-0.1%	**-0.5%**	3%	7%
Japan	9.3	0.4%	0.8%	**-0.3%**	0.24	0.4%	-0.2%	1.3%	-1%	3%
EU	6.0	-1.2%	-0.3%	**-2.4%**	0.18	-0.8%	-0.7%	**-0.9%**	-6%	-4%
Russia	11.3	-0.6%	-1.8%	0.7%	0.20	-0.4%	-0.4%	-0.3%	-3%	11%
Mexico	3.7	0.2%	0.8%	**-0.6%**	0.22	0.0%	0.2%	**-0.3%**	-6%	2%
Brazil	2.7	2.7%	2.0%	3.7%	0.15	0.7%	0.6%	0.9%	6%	25%
China	7.5	5.5%	4.8%	6.5%	0.29	0.8%	0.9%	**0.6%**	4%	14%
Indonesia	1.7	4.1%	4.9%	**2.9%**	0.16	2.1%	2.1%	2.1%	3%	-1%
Philippines	1.1	2.0%	2.0%	2.1%	0.19	1.8%	2.0%	**1.6%**	-5%	7%
India	1.9	3.8%	2.7%	5.4%	0.24	1.3%	1.0%	1.6%	3%	19%
Nigeria	0.5	2.6%	6.6%	**-3.0%**	0.06	2.0%	5.8%	**-3.3%**	-3%	47%
Pakistan	0.9	1.5%	2.3%	**0.3%**	0.16	0.6%	0.8%	**0.4%**	-4%	15%
Bangladesh	0.5	5.0%	4.6%	5.6%	0.18	2.3%	2.6%	1.9%	-3%	9%
Ethiopia	0.1	3.2%	1.5%	5.6%	0.02	3.0%	1.4%	5.3%	-1%	23%

intensive ones in mid- and low-income countries; Nigeria, where coal and oil are substituted for gas, is an exception. The growth of CO_2 emissions per kWh also diverged across countries.

Those trends in energy consumption and CO_2 emissions are confirmed with the World Bank and IEA data. However, difference between the World Bank and IEA data on energy consumption and CO_2 emissions are noted. A few examples are mentioned, but more differences can be found. Global energy consumption per capita for the year 2015 is about 4% higher in the World Bank accounts than in the IEA but varies across countries from +6% for Brazil to –6% for the EU; this spread is as large as the annual Nigerian energy consumption. The CO_2 data for 2015 differ even more. The World Bank shows 12% larger global CO_2 emission per capita than the IEA and even 47% larger for Nigeria. Such differences in the authoritative databases pose risks for agreements on climate change because countries can dispute reports and manipulate data. The World Bank and IEA, herewith, have an important task to provide transparent, reliable, open access data on all issues related to energy.

Global CO_2 emissions are not reduced but growth rates diverged across countries. CO_2 emissions are stabilized and reduced in a few countries due to fast growth of low-carbon technologies.

4. Valorization

Does the valorization of energy services evolve across countries and contribute to lower carbon economies? As mentioned, the valorization of energy services means higher value added of energy services over a period of ten years or more, which means either higher sales of services or cheaper purchases of resources. The latter is observed: An estimate is that the international prices of energy resources decreased as a trend throughout the last century, except for coal, thanks to technological progress in exploration and exploitation (Shafiee and Topal, 2010). High prices during 1979–1987 and 2005–2015 are considered as deviations from that trend, caused by international cartels (Morriss and Meiners, 2016). In addition, consumer prices of energy services increased whenever innovations created breakthrough changes in energy market, such as supply of high calorific fuels, replacement of these fuels by electricity and, recently, introductions of low-carbon technologies and off-grid energy systems. For instance, the purchase price of natural gas by energy services in the USA in 2015 was on average US\$$_{2005}$ 0.009 per kWh compared to US\$$_{2005}$ 0.031 per kWh average sales price of natural gas to households and US\$$_{2005}$ 0.105 per kWh sales price of electricity, all excluding subsidies and taxes (EIA, 2019); the prices in the EU were about twice as high. Higher sales prices of energy services along with cheaper purchase of energy resources generate higher value added.

4.1 Energy services

Innovations are usually costly and perform imperfectly when they are launched but provide benefits to users and entail numerous adaptations which reduce costs and improve performance. This process, in time, also holds true for innovations in the energy business. In addition, these innovations in the energy services enable other businesses to increase their value due to costs reductions or performance enhancements. For instance, innovations in the fuel refining and electricity industries during the 1990s enabled larger pressures, finer mechanics, better ventilation, faster motion, more light, stronger sound and others performances in the energy consumption in nearly all businesses and households. More recent innovations in wind and solar power, as well as power storage in batteries, generated off-grid, non-polluting and flexible applications for energy consumption which are highly appreciated despite their high costs. For instance, the off-grid applications are presently many times costlier than the on-grid ones, but these innovations enable communication and sensing in space, mobility and housing in remote areas, distributed energy systems in communities and other valuable manners of energy consumption. Hence, larger expenditures on costlier energy services based on low-carbon technologies can be explained by higher performance of energy consumers.

Higher performance in energy consumption generates benefits for individuals, businesses and communities, which is observed throughout the past two centuries. Substitution of biomass for coal in

England and the USA enabled energy-efficient heating in households and powerful machines (Allen, 2012). They were major drivers of the income growth; herewith, scarce wood and pollution prevention were important incentives of that substitution (Rosenberg 1973 (1977)). Electricity networks delivered power on-grid which enabled income growth in manufacturing and commerce due to machines and tools for fine mechanics, light and sound during the last century (Fouquet, 2014). Electricity production emerged in the USA in the late 1800s despite the fierce opposition of the gas companies. The authorities invoked the introduction of the electric city lights in a bid to avoid city fires caused by the gas lights (DiLorenzo, 1996); note that the introduction of wind and solar power has some similarities. Electricity is meanwhile considered as a basic good though it is several times costlier per energy unit than other fuels. Such innovations are pursued by the civil society, scientists and entrepreneurs who sense pressing issues and envisage opportunities despite higher costs than the available alternatives, opposition of the vested interests and high risk of failures. Similar impediments are encountered in the low-carbon technologies.

More recently, various benefits or low carbon technologies and off-grid systems are mentioned in literature but systematic cost-benefit assessments of these innovative energy services in comparison to the vested ones are rare. Herewith, a few examples are mentioned. A broader review can be found in the Additional file in Krozer (2019). The benefits of the innovative energy services can be divided into ones for the individual interests of producers and consumers, and for their collective interests. For individual energy producers, the illustrative benefits are a spread of investment risks and deferring of costly infrastructure when off-grid applications are introduced. The collective producers' benefits are lower price volatility and fewer losses when power sources are diversified, as well as lower costs of pollution controls and others. Consumers can benefit individually due to combustion-free energy consumption, and collectively when new businesses are vested and jobs created in their community. In effect, many off-grid applications of low carbon technologies emerged and multiplied despite ten to hundred times higher costs per energy unit compared to electricity on grid or small-scale generators based on fossil fuels. These applications grow quickly, measured by the capacity of batteries, because they enable flexile, autonomous, clean activities (REN21, 2017).

An innovator generates high income due to its temporary monopoly in sales of applications, given quality demands on energy market. Unless that innovator is entitled by authorities to keep its monopoly position due to patents and other regulations of the sort, its high income is challenged by firms that developed alternatives and imitations at lower unit costs entailing a specialized business, such as the solar energy business. As the unit costs of applications decrease, entailing lower prices, the total business incomes fluctuate per year because the sales prices decrease whilst the sales volumes grow. The innovator often holds a high market share in the business due to its high income in the past because it can put efforts in the specialized manufacturing of adaptations and deliver low unit costs whilst the firms that maximize profit or that are unsuccessful in their specializations experience loss of the market share; the few firms that adapt their alternatives remain.

Herewith, the cost-reducing adaptations are illustrated for the low-carbon technologies in the USA during 2005–2015, based on the private data (Statista, 2019). Table 5 shows the annual consumption of wind and solar power in GWh and revenues due to that consumption in US\$$_{2005}$ million during 2005–2015 in the USA. These data are used for estimation of the costs per unit energy in US\$$_{2005}$ per kWh per year and annual income of the energy services in US\$$_{2005}$ per year. The fluctuations in business income are show by comparison of two subsequent years. The cost-reducing adaptation in low-carbon technologies evolved fast. As the sales volume of solar power increased 45 times, the unit costs decreased seven times and the volume of wind power increased nearly 11 times whilst the unit costs decreased by about 30% according to the data from Statista. Although the business income oscillated from highly positive to negative, the business income throughout those ten years was positive and grew. The energy services based on solar power increased their income by US\$$_{2005}$ 49 million annual average while the energy services based on wind power by US\$$_{2005}$ 164 million. Apparently, firms put efforts into the cost-reducing adaptations despite those fluctuations of income. A similar situation can be observed with respect to batteries, but good data over several years is not available. At present, batteries are costly but increasingly produced and used. As this production grows, the unit cost decrease and can meet the price on grid within a few decades (UCSUSA, 2019).

Table 5. Growth of solar and wind power along with decreasing unit costs (US$_{2005}$) in the USA.

Italic: own estimate	Solar power (PV)				Wind power				Share in the total renewable energy
	GWh	US$ mln	US$/kWh	Additional income of subsequent years*	GWh	US$ mln	US$/kWh	Additional income of subsequent years*	
2005	551	36	0.07		17,811	767	0.04		5%
2006	508	39	0.08	-5	26,589	1,169	0.04	-25	7%
2007	612	43	0.07	4	34,450	1,783	0.05	-268	10%
2008	864	47	0.05	14	55,363	2,718	0.05	146	15%
2009	892	54	0.06	-6	73,886	2,965	0.04	663	18%
2010	1,212	69	0.06	4	94,652	3,465	0.04	334	22%
2011	1,818	125	0.07	-22	120,177	3,997	0.03	403	24%
2012	4,327	150	0.03	148	140,822	4,301	0.03	383	29%
2013	9,036	165	0.02	148	167,840	4,691	0.03	434	34%
2014	17,691	172	0.01	151	181,655	5,213	0.03	-135	37%
2015	24,893	189	0.01	54	190,719	5,767	0.03	-295	40%
Sum	61,853	1,053		490	1,086,153	36,069		1,640	

*$s_t = (c_t - c_{t+1}) \cdot V_t$; s is additional value a year, c_t unit cost for $ct = R_t/V_t$, and V_t is volume of electricity a year.

Prices of energy services increase whenever innovations that are beneficial for consumers are introduced. Such innovations are followed by the cost-reducing adaptations when production grows. The growing production enables innovators to generate value added and maintain a competitive position during a long period of time if they are able to allocate efforts towards those adaptations.

4.2 Indicators of valorizations

The total benefit due to the innovative energy services in economy is often indicated by the energy-intensity, i.e., GDP per unit of energy consumption. That benefit increased regarding the global growth of energy-intensity by 0.6% to 2.2% per year throughout the last half century (WEC, 2010). However, it should be noted that the GDP growth is not primarily driven by better energy consumption, as shown by the decoupling from the income growth and energy consumption but better allocation of labor, capital and knowledge, which are prime resource in all modern economies. Another indicator of those benefits is the electrification, measured by the electricity consumption and access to it. Table 6 shows those indicators for 2015 and growth during 1990–2015, divided into periods 1990–2004 and 2005–2015; all data are per capita.

The energy-intensity was globally US$_{2005} 0.7 per kWh in 2015, which was 4 to 5 times higher in Philippines and Bangladesh than in Ethiopia, Russia and China and it grew in all countries, except Brazil. However, the energy-intensity grew slower than energy consumption in China, Indonesia, India and Bangladesh, which implies that energy is increasingly wasted in these large, growing economies; it is presumably related to large price subsidies for the energy consumption. Much carbon emission is caused by wasteful energy consumption. High fuel prices increased the energy-intensity in all countries, except Brazil and China, where costlier resources were even more wasted. Applications of low-carbon technologies contributes to these benefits as their growth correlates with the growth of energy intensity (the cross countries correlations were $R^2 = 0.50, 0.41$, and 0.81 during 1990–2015, 1990–2004 and 2005–2015, respectively). The energy-intensity converged across countries, albeit slowly.

The average electricity consumption was 3 mWh globally, which was 173 times higher in the USA than the 75 kWh in Ethiopia. Electricity consumption is in line with the income growth (the cross countries correlations were $R^2 = 0.74, 0.65$ and 0.82 during 1990–2015, 1990–2004 and 2005–2015, respectively). Faster growth of the electricity consumption than energy consumption during those periods indicates high appreciation. The appreciation varied, however, because the growth of electricity consumption in China was about 300 times higher than in Russia. During high prices of fossil fuels, the electricity consumption in high income countries declined and the growth slowed down in all other countries except in Brazil, China and India, which suggests that the electricity consumption has a high sensitivity to prices. The growth of electricity consumption diverged across countries.

Access to electricity was globally about 86%, but varied from 29% in Ethiopia to 100% in many countries. It grew in all countries except high-income ones and Russia. Higher income enlarged that access (the cross-countries correlations are high) but higher prices of fossil fuels impeded it. The growth of access to electricity also diverged.

Energy services provide benefits to economies in nearly all countries. In particular, the electricity consumption is highly appreciated in all countries. Higher benefits of energy and electricity consumption can be attained in the fast-growing economies when the energy consumption becomes less wasteful and higher value products are generated. The growth rates of energy consumption and electrification diverge.

4.3 Emerging markets

What markets for low carbon technologies can be expected? This question is discussed based on the assumption that the trends during 1990–2015 continue during 2015–2040, along with substitution of fossil fuels for renewable energy. The IEA data are used for the extrapolation of those trends.

The real income per capita would grow globally about three times and the energy consumption one and a half times. Most mid-income countries would catch up and surpass the high-income ones; China and India would become the largest economies and energy consumers. The energy production would

Table 6. Energy intensity and electrification of consumption in US$₂₀₀₅.

Total and annual average growth	Energy-intensity				Electricity consumption				Electrification	
	US$/kWh	Annual average growth Bold: Higher growth during high fuel prices			kWh/capita	Annual average growth Bold: Higher growth during high fuel prices			Access percent people	Electricity to energy growth
	2015	1990–2015	1990–2004	2005–2015	2015	1990–2015	1990–2004	2005–2015	1990–2015	2005–2015
World	0.69	1.5%	1.3%	**1.6%**	3,147	1.6%	1.4%	**2.0%**	86%	2.7%
USA	0.65	1.8%	1.8%	**1.9%**	12,975	0.5%	1.0%	**-0.3%**	100%	1.20%
Japan	0.91	0.9%	-0.1%	**2.3%**	7,651	0.6%	1.5%	**-0.6%**	100%	11.7%
EU	1.02	1.8%	1.3%	**2.4%**	5,752	0.6%	1.4%	**-0.5%**	100%	1.3%
Russia	0.45	1.7%	1.1%	**2.5%**	6,666	0.0%	-1.1%	**1.6%**	100%	0.0%
Mexico	0.98	1.0%	0.8%	**1.3%**	2,101	2.5%	3.5%	1.1%	98%	11.7%
Brazil	0.86	-0.3%	-0.2%	-0.3%	2,651	2.5%	2.1%	**3.0%**	100%	1.3%
China	0.51	4.5%	5.3%	3.4%	4,081	8.9%	8.5%	**9.6%**	100%	1.9%
Indonesia	0.99	1.4%	0.1%	**3.2%**	851	7.0%	8.0%	5.5%	98%	3.3%
Philippines	1.19	2.0%	1.2%	**3.2%**	713	2.9%	3.5%	2.0%	91%	17.9%
India	0.73	2.2%	2.1%	**2.3%**	846	4.6%	3.7%	**5.9%**	81%	1.8%
Nigeria	0.67	2.4%	2.0%	**3.1%**	146	2.8%	3.2%	2.2%	60%	6.8%
Pakistan	0.84	0.9%	0.1%	**2.0%**	460	2.3%	3.2%	1.0%	100%	2.8%
Bangladesh	1.17	0.9%	0.7%	**1.2%**	328	8.2%	9.1%	6.9%	63%	3.1%
Ethiopia	0.26	3.3%	0.4%	**7.5%**	75	5.0%	2.4%	**8.6%**	29%	29.1%

better balance consumption in all countries, except Japan, which would need more import. Electricity would become the largest global energy service measured by energy units, particularly large in China, India and the EU, though smaller in low-income countries. The composition of energy resources would change dramatically. Fossil fuels would decline globally by half. They would decline close to nil in China, India and Ethiopia but triple in Bangladesh, double in Pakistan and grow in Indonesia, Philippines, Nigeria and Mexico. All renewable energy would grow globally about 100 times but hardly at all in Russia and Mexico and its share in all energy consumption would decline in low-income countries, though grow in total. Within renewable energy, the applications of low carbon technologies would grow globally by a factor of 792. These would grow by a factor of 2794 in China, 589 in India, 15 in the EU and 14 in Russia, contrary to low-income countries, where the applications would remain scarce. CO_2 emissions would decline globally to 46% of the 2015 level. Whilst China, India and Ethiopia would become nearly emission-free due to low-carbon technologies and the EU reduce to 23% of its 2015 level, the CO_2 emissions would increase in other mid-income and low-income countries. The divergence in the growth of CO_2 emissions poses a challenge to the Paris Agreement on climate change.

Another issue is whether those large applications of low-carbon technologies are feasible. From the economic perspective there is reason for optimism because the unit costs decrease roughly 1.5 times per doubling of markets. If this continues, all available low-carbon technologies become cheaper than coal in energy production, whose costs would also decrease. Cheaper technologies attract investments. Scarce space can be an impediment in the populated countries but designs can resolve this issue. For instance, all Chinese consumption of 10,386 million ton oil equivalent in 2040, i.e., $1.2 * 10^{14}$ kWh, can be covered with the available solar technologies of 18% capture of irradiation and low average irradiation of 1,200 kW/m^2/year on about 100,000 km^2, which means on 0.1% of the Chinese land surface or on 4% of its urban surface (Demographia, 2019). The spatial allocation needs innovative solutions but if cities are adequately designed and developed most solar technologies can be integrated in buildings and roads, which are already proven technologies though still costly.

It should also be expected that the global energy market would be highly diverged by 2040. While the real incomes converge across countries, the energy consumption and resource composition diverge because fossil fuels are replaced by renewable energy in a few countries, but not in others. In addition, it should be expected that the power services on-grid and off-grid become by far the largest energy services measured by volume and value because the electricity consumption expands in nearly all countries and low-income countries catch up. Another expectation is that CO_2 emissions are substantially reduced due to fast growth of low carbon technologies in a few countries, but they increase in most mid- and low-income countries. This can cause a conflict of interests in the climate change policies. Higher global real income and energy intensity imply the valorization of energy services across countries. Lower fossil fuel consumption in a few countries due to fast growth of low-carbon technologies enhances low-carbon economies but the composition of energy resources diverges.

5. Conclusions

The question of whether economies shift to low-carbon technologies is answered based on the volume and growth rates of geothermal, solar and wind energy resources. Therefore, statistical data about the world and the 14 largest countries measured by population, the EU considered as one country, throughout 1990–2015 are used, covering the periods of low fuel prices, 1990–2004, and high fuel prices, 2005–2015.

Applications of low-carbon technologies become more affordable due to the growing real income in all countries, as well as smaller income disparities across countries and within most countries. The affordability also increased because energy consumption declined in high-income countries and grew at a rate slower than the income growth in mid- and low-income countries, and above all, because the costs of low-carbon technologies decreased.

However, opposing trends are observed in the resource composition. Fossil fuels declined in high-income countries but grew in other countries and globally. High-income countries with a low share of renewable energy in the energy consumption grew that share, contrary to mid- and low-income countries, where the share was high but declined. Within renewable energy, the applications of low-carbon technologies grew globally, even very fast in a few high- and mid-income countries, but much less in low-income ones. As a result of these trends, the decarbonization of energy consumption was negligible measured per capita and per energy unit, and CO_2 emission per capita increased. The diverging trends across countries pose a major challenge for international carbon management.

An encouraging trend is the growing value of energy services due to benefits for consumers in business and households. The applications of low-carbon technologies contribute to these benefits. The valorization of energy services is a global trend when measured by the increasing income per energy unit. This valorization is observed in nearly all countries and the growth rates converge across countries. Also, the electrification of energy consumption grows fast, but large differences between countries are observed. Hence, the substitution of fossil fuels for low-carbon technologies and CO_2 emission reduction are mainly driven by more valuable applications with the contribution of low-carbon technologies rather than low-carbon performance.

Low-carbon technologies become a dominant factor in energy markets because innovations and adaptation in their applications add value to energy services and provide benefits to consumers. This generates a reduction in CO_2 emissions. Policies enhance the valorization of energy services if they reduce subsidies for fossil fuels and support innovators in low-carbon technologies.

Acknowledgment

I am grateful to the editors of the journal Energy Society and Sustainability for the use of data. In memory of Cees van Leeuwen, environmental manager of Unilever, who directed this company to grow for sustainability.

References

Allen, R. 2012. Backward into the future, The shift to coal and implications for the next energy transition. Energy Policy 50: 17–23.

Ayres, R. and Voudouris, V. 2014. The economic growth enigma: Capital, labour and useful energy? Energy Policy 64: 16–28.

Beaudreau, B.C. and Lightfoot, H.D. 2015. The physical limits to economic growth by R&D funded innovation. Energy 84: 42–52.

Demographia World Urban Aras, http://demographia.com/db-worldua.pdf (accessed 19-4-2019).

Deng, Y.Y., Blok, K. and van der Leun, K. 2012. Transition to a fully sustainable global energy system. Energy Strategy Reviews 1: 109–121.

DiLorenzo, T.J. 1996. The Myth of Natural Monopoly. The Review of Austrian Economics 9(2): 43–58.

EEA. 2004. Energy Subsidies in the European Union, Technical Report 1/2004, Copenhagen, Denmark.

EIA. 2019. https://www.eia.gov/electricity/annual/html/epa_07_01.html (accessed 11-2-2019).

Energy expenditures. 2017. https://www.enerdata.net/publications/executive-briefing/world-energy-expenditures.html accessed 15-3-2017.

Fouquet, R. 2014. Long run demand for energy services: Income and price elasticities over two hundred years. Review of Environmental Economics and Policy 8: 186–207.

Fouquet, R. 2016. Historical energy transition: Speed, prices and system transformation. Energy Policy 22: 7–12.

Grübler, A. and Nakicenovič, N. 1996. Decarbonizing the global energy system. Technological Forecasting and Social Change 53(1): 97–110.

Grübler, A., Wilson, C. and Nemet, G. 2016. Appels, oranges and consistent comparison of temporary dynamics of energy transition. Energy Policy 22: 18–25.

Haas, R., Panzer, C., Resch, G., Ragwitz, M., Reece, G. and Held, A. 2011. A historical review of promotion strategies for electricity from renewable energy sources in EU countries. Renewable and Sustainable Energy Reviews 15: 1003–1034.

Herman, R., Ardekanin, S.A. and Ausubel, J.H. 1989. Dematerialization. pp. 50–69. *In*: Ausubel, J.H. and Sladovich, H.E. (eds.). Technology and Environment, National Academy Press, Washington DC.

IEA. 2017. http://www.iea.org/bookshop/729-CO2_Emission_from_Fuel_Combustion (accessed 26-7-2017).

IEA. 2019. https://www.iea.org/classicstats/statisticssearch/ (accessed 20-6-2019).

IEA/OECD. 2019. https://www.oecd.org/site/tadffss/data/ (accessed 19-6-2019).

Krausmann, F., Gingrich, S., Eisenmenger, N., Erb, K.-H., Haberl, H. and Fischer-Kowalski, M. 2009. Growth in global materials use, GDP and population during the 20th century. Ecological Economics 68: 2696–2705.

Krozer, Y. 2017. Energy markets: Changes toward decarbonization and valorization. Current Opinion in Chemical Engineering 17: 61–67.

Krozer, Y. 2019. Valorization of energy services: Essay on the value addition due to renewable energy. Energy, Sustainability and Society 9(9): 1–16.

Lancaster, K.J. 1966. A new approach to consumer theory. Journal of Political Economy 74(2): 132–157.

Levelized costs. 2019.

International: https://www.lazard.com/perspective/levelized-cost-of-energy-2017/ (accessed 5-2-2019) and https://www.irena.org/-/media/Files/IRENA/Agency/Publication/2018/Jan/IRENA_2017_Power_Costs_2018.pdf (accessed 5-2-2019).

US: https://www.eia.gov/outlooks/aeo/pdf/electricity_generation.pdf (accessed 5-2-2019).

EU: https://www.ise.fraunhofer.de/content/dam/ise/en/documents/publications/studies/EN2018_Fraunhofer-ISE_LCOE_Renewable_Energy_Technologies.pdf (accessed 5-2-2019).

India: https://www.energy.gov/sites/prod/files/2015/08/f25/LCOE.pdf (accessed 5-2-2019).

Morriss, A.P. and Meiners, R.E. 2016. Competition in Global Oil Markets: A Meta-Analysis and Review. http://secureenergy.org/wp-content/uploads/2016/02/SAFE_Competition-in-Global-Oil-Markets-A-Meta-Analysis-and-Review.pdf, (accessed 25-7-2018).

REN21. 2017. Renewables 2017 Global Status Report. Ren Secretariat. Paris, France.

Rosenberg, N. 1973 (1977). Innovative responses to material shortages. American Economic Review 63(2): 11–18, reprint in Dofman, R. and Dorfman, N.S. (eds.). Economics of the Environment, 1st ed. W.W. Norton & Company Inc., New York, pp. 390–399.

Rubin, E.S., Azevedo, I.M.L., Jaramillo, P. and Yeh, S. 2015. A Review of learning rates for electricity supply technologies. Energy Policy 86: 198–218.

Shafiee, S. and Topal, E. 2010. A long term view on fossil fuel prices. Applied Energy 87: 988–1000.

Sovacool, B.K. 2016. How long will it take, Conceptualizing the temporary dynamics of energy transition. Energy Research & Social Science 13: 202–215.

Statista. 2019. https://www.statista.com/statistics/289145/revenue-hydroelectric-power-industry-united-states/(accessed 11-2-2019).

UCSUSA. 2019. https://www.ucsusa.org/clean-vehicles/electric-vehicles/accelerating-us-leadership-electric-vehicles-2017 (accessed 11-2-2019).

WEC. 2010. Energy Efficiency: A Recipe for Success. World Energy Council, London, United Kingdom.

CHAPTER 3

Carbon Management

Forest Conservation and Management

Grace Ding[1,]* *and Nguyen Thuy*[2]

1. Introduction

The construction industry is dramatically growing and developing worldwide (Joseph and Tretsiakova-McNally, 2010). Construction activities are supporting the large upstream and downstream supply chains, including the building material products, such as tiles, wood, steel, cement, bricks and glass (Australian Industry Group, 2015). Simultaneously, the construction industry also contributes to the environmental impacts comprising of high energy usage, over-consumption of water and natural resources, and generation of waste to the surrounding environment. Each year, the construction industry in the world depletes approximately 25% of the global wood harvest, and is responsible for the use of 40% of stone, sand and gravel, and 16% of water usage (Joseph and Tretsiakova-McNally, 2010).

United Nations Environment Programme-Sustainable Building and Climate Initiative (UNEP-SBCI)'s report estimated that the building sector contributes up to 30% of annual global greenhouse gas (GHG) emissions and consumes up to 40% of all energy, both in developed and developing countries (UNEP-SBCI, 2009). Natural resources, such as iron ore, limestone, bauxite, coal, natural gas and petroleum, which are used to produce heavy building materials (such as concrete, steel and aluminium), are non-renewable resources. Australia and Brazil are the two leading countries of iron ore production for steelmaking and the production has decreased from 900 million metric tonnes in 2014 to 400 million metric tonnes in 2018 (US Geological Survey, 2018). The world's accessible iron ore reserves have now considerably declined approximately 44% from 150 billion metric tonnes in 2009 to 84 billion metric tonnes in 2018 (Crawford, 2011; US Geological Survey, 2018). Subsequently, with the average annual extraction rate of 4.9%, global iron ore reserves will be enough for only around 20 years.

The use of traditional heavy materials will not just depleting the non-renewable natural resources but also high in carbon dioxide (CO_2) intensity, which is considered to be one of the most significant contributing factors to climate change. Therefore, the selection of renewable building materials with low CO_2 emissions plays a crucial role in mitigating the environmental burdens and resource scarcity. The use of timber to substitute heavy materials in construction has now attracted escalating attention around the world.

[1] Faculty of Design, Architecture and Building, University of Technology Sydney, Building 5C, 1 Quay Street, Haymarket, NSW 2000.
[2] Faculty of Engineering and Information Technology, University of Technology Sydney, Building 11, 81 Broadway, Ultimo, NSW 2007.
   Email: lehongthuy.nguyen@student.uts.edu.au
* Corresponding author: grace.ding@uts.edu.au

Carbon emission is considered the main contributor to global climate change and it is closely related to the energy consumption of various activities in the economy. Anthropogenic emissions come from two principal sources, namely, the combustion of fossil fuels, and conversion of land use and cultivation of soil (Lal, 2008). Over the years, governments and organisations have developed policies, regulations and standards to reduce carbon emissions. The global forest is considered as an important way to mitigate climate change by absorbing and storing carbon from the atmosphere (Carroll, 2012; Mitchard, 2018). It is estimated that forests absorb approximately 10–15% of the global CO_2 emissions (Sample et al., 2015). However, recent research studies indicate that the carbon sink of forests is declining as a result of deforestation and forest degradation (Sample et al., 2015). Therefore, forest conservation and management potentially play an important role in carbon management in reducing atmospheric CO_2 concentration.

Timber is one of the most environmentally friendly and renewable building materials because the production of timber requires less energy and emits less CO_2 than other materials (Skullestad et al., 2016). Timber has been used for residential construction for centuries. More than 90% of residential housings in Australia use timber for the construction of wall and roof frames (FWPA, 2005). Chen (2012) states that the embodied energy used to construct a five-storey reinforced concrete (RC) building was higher than that of an equivalent engineered timber building by about 22%. Furthermore, building with timber saves almost 45 tonnes (out of 80 tonnes) of CO_2 per dwelling, which can be a significant difference compared with using traditional heavy materials at a global scale (Waugh Thistleton Architects, 2018). Wood waste, such as sawdust and offcuts produced during timber processing and timber components at the end of the building's lifespan, are potential bioenergy resources that could replace fossil fuels in the production of electricity. According to NAFI (2007), renewable energy produced using wood waste instead of coal-fired electricity generation reduces CO_2 emissions by 95–99% for each megawatt-hour (MWh).

This chapter begins with a review of the importance of forests, followed by the policies, initiatives and regulations concerning forest conservation and management at both national and international levels. This chapter examines the technological advancement of timber and engineered timber products as they replace traditional heavy material for constructions. Finally, the chapter discusses the challenges that timber and timber products may encounter in order to be in competition with the traditional materials.

2. The Importance of Forests in the Mitigation of Climate Change

Forests play an important role in the mitigation of climate change by absorbing CO_2 in the atmosphere. However, the land use conversion and fossil fuel combustion to support human activities have severely altered the global carbon cycle, leading to the continued increase of the CO_2 concentration in the atmosphere (Lal, 2008). Mitchard (2018) states that between 1960 and 2015, approximately 20% of anthropogenic carbon emissions were due to the change of land use of the tropical forests. Mitchard (2018) continues to state that the tropical forests are likely to become a source of carbon emissions owing to the continued vanishing of forests and the diminishing capacity of the remaining forests to assimilate the increasing CO_2 concentration in the atmosphere.

Forests cover about one-third of the earth's land and the combined temperate and boreal forests make up of approximately 49% of the total (Carroll, 2012; Knauf et al., 2015). Forests help in lessening climate change through their inherent ability to sequestrate and store CO_2 from the atmosphere to both above- and below-ground biomass through the processes of photosynthesis and tree growth. The problem of climate change is closely related to the increasing atmospheric concentrations of CO_2. Approximately a quarter of the anthropogenic CO_2 emissions are from the destruction of the forest ecosystem, farming activities and soil degradation, and the concentration will continue to rise if no immediate action is taken (FAO, 2018).

Deforestation and forest degradation caused by tree harvesting and disintegrating contribute to approximately 17% of global carbon emissions (UNREDD, 2015). Levashova (2011) states that illegal logging is one of the primary causes for the degradation of forest resources and more than 20% of timber entering the EU market comes from illegal sources. Therefore, the prevention of further deforestation and forest degradation plays a significant role in addressing the issue of global climate change.

The endurance of the forest ecosystem is vital in lowering CO_2 concentrations as it has the ability to store and sequestrate carbon from the atmosphere (Mackey et al., 2008; Lal, 2010; UN Climate Summit, 2017). It can improve the global carbon cycle by storing carbon above ground through trees and plants and below ground in soils (Stupak et al., 2011; Carroll, 2012). Therefore, managing forests sustainably is all about the maximising carbon sequestration, minimising carbon release, maintaining biodiversity, protecting water systems and providing environmental goods and services.

Carbon sequestration of forests is one of the essential ecosystem services and is defined as the net rate of carbon uptake by an ecosystem per annum (Yan, 2018). Carbon sequestration is the natural process of CO_2 being extracted from the atmosphere and stored in plants and soil for an extended period of time. Carbon sequestration in renewable products, including wood, bamboo or agricultural products, can be either at the product level or global level. The product level (biogenic CO_2) is related to the carbon stored in wood during the growth of a tree (van der Lugt et al., 2012). At the global system scale, CO_2 is captured in the forests, ocean and the soil. It is estimated that the amount of CO_2 emissions each year that are a result of burning fossil fuels and deforestation in tropical and sub-tropical forests is about 6.4 Gt and 1.93 Gt, respectively, whereas carbon sequestration each year due to reforestation on the Northern Hemisphere is only about 0.85 Gt (Vogtländer et al., 2014). Therefore, if there is no change in the area of forests and no change in using wood products, there will be no change for carbon sequestration.

The forest carbon cycle is where live forests sequestrate carbon from the atmosphere through the natural process of photosynthesis, the harvested forest, however, will need reforestation in order to continue with the process of carbon sequestration. Above ground, carbon is absorbed for growth and stored, while below ground, carbon is absorbed and stored in tree roots and soil. The harvested forest will turn into harvested wood and bioenergy. Managing young forests can maximise carbon uptake while conserving old forests and prolonging rotations lead to greater carbon storage in the below-ground soil pool (Carroll, 2012). Yan (2018) conducts research to assess carbon sequestration between living and harvested forest. Research results reveal that the impact of harvested forest on carbon sequestration can be improved by increasing growth rate, extending harvest period, and reducing harvest intensity.

Carbon can also be sequestered in wood products as long as they continue to be in use and will only be released when they are burned or decomposed at the end of the useful life. Approximately two million tonnes of timber and timber products are disposed of in landfills each year at the end of the useful life of wood products (Ximenes et al., 2013). Several studies have investigated the fate of carbon stored in wood products in landfills. In an early study by Micales and Skog (1997), the amount of methane and CO_2 generated from timber products in landfills was approximately 3% into the atmosphere. Ximenes et al. (2008) examine the decomposition of wood products in landfills in Australia. Research results reveal that after 46 years in the landfills, the loss of carbon was 8.7% for hardwood and 9.1% for softwood. Ximenes et al. (2013) calculate the amount of captured carbon in wood products in landfills based on the results of the bioreactor experiments. They concluded that particleboard and medium-density fibreboard (MDF) reactors stopped producing gas after two months, but no gas was produced in high-pressure laminate (HPL). In anaerobic reactors, in the laboratory under optimal decay conditions, the proportions of carbon loss were 1.65%, 0.65% and 0%, respectively, for particleboard, MDF and HPL, and carbon can be retained in storage indefinitely.

The use of traditional heavy materials in buildings is well recognised as environmentally and timber is proposed to be a suitable substituting material for the construction of buildings. While the interest in using timber in construction leads to more plantations and more carbon sequestered in harvested wood products, the negative or positive impact of the using wood in construction on the impact on the carbon cycle, emissions and sequestration will depend on the type of wood. For instance, the demand for tropical hardwood is more than the supply from plantations (35%–40% of FSC wood is from plantations). Consequently, the increasing demand for tropical hardwood may escalate deforestation if not managed sustainably and, thus, may disrupt the carbon sequestration cycle. On the contrary, the use of plantation wood may encourage afforestation and reforestation, resulting in an expansion of the carbon sequestration pool (Vogtländer et al., 2014). As a result, the use of timber products from well-managed forests will enhance the world's carbon sequestration and storage capacity.

3. Policies, Initiatives and Regulations of Sustainable Forest Management

Over the years, international collaborations and agreements to tackle deforestation and forest degradation in order to let forests recover and regrow have been made. Since the 1960s, the global forest development has attracted attention in its contribution to the well-being of humanity and the mitigation of climate change. In the 1970s to 1980s, the concerns over deforestation in the tropics triggered international action plans for the management and conservation of forests (FAO, 2018). Table 1 presents some of the key national and international policies, regulations and conventions for forest-related issues in chronological order.

3.1 International policies, regulations and conventions

As early as the 1970s, the importance of wildlife has been discussed at an international level. The Convention on International Trade in Endangered Species of Wild Fauna and Flora (CITES) launched in 1975 recognized the conservation and protection of wild fauna and flora as of global importance and their survival would not be threatened in international trade. However, it was not particularly focused on forestry issues.

The United Nations Framework Convention on Climate Change (UNFCCC) in 1992 marked the international discussion on addressing the problems of deforestation and forest degradation as an emissions reduction measure to mitigate climate change (Barbier et al., 2019). The Agenda 21 Chapter 11 initiated a global action to combat deforestation by strengthening international collaboration and recognising the roles and functioning of all types of forests, and effectively manage and conserve forests in both the developed and the developing countries (Bucknum, 1998; Stupak et al., 2011). The UNFCCC has adopted the "Non-Legally Binding Authoritative Statement of Principles for a Global Consensus on the Management, Conservation and Sustainable Development of all Types of Forests" (also known as the Forest Principles or Rio Forest Principles). The Principles are not legally binding demonstrating the high divergence of views during the negotiation (Ruis, 2001; Sample et al., 2015).

Following the UNFCCC, several panels were formed to focus on the discussion of forestry issues. The Intergovernmental Panel on Forests (IPF) (1995–1997) was established to elaborate the Forest Principles further. The Intergovernmental Forum on Forests (IFF) (1997–2000) was formed as a successor to the IPF to develop a framework for the implementation of the IPF proposals for action. The United Nations Forum on Forests (UNFF) (2000 until present) was established in order to develop international legally binding agreements for the long-term commitments in sustainable forest management (Humphrey, 1998, 2001; Ruis, 2001; Tegegne et al., 2018). The Collaborative Partnership on Forests (CPF) was established in 2001 among 14 international organisations to support the work of the UNFF.

Following the UNFCCC and the development of the CBD, there is a series of annual Conference of Parties (COPs) and other meetings organised worldwide to discuss and develop action plans for forest carbon management and the estimation and verification of carbon gains of forests (Knauf et al., 2015; Sample et al., 2015). International collaboration on forest-related matters was the REDD (reducing emissions from deforestation and degradation) and the REDD+ (reducing emissions from deforestation and forest degradation in developing countries, and the role of conservation, sustainable management of forests, and enhancement of forest carbon stocks in developing countries). As early as at the Kyoto Protocol in 1997, the vital role that forests play in reducing carbon emissions was first recognised and the idea of REDD was first discussed (Pistorius, 2012). The REDD was formally established in 2007 at the COP13 in Bali. The REDD is a UN programme in collaboration with the UN Food and Agriculture Organisation (FAO), UN Environment Programme (UNEP) and UN Development Programme (UNDP) to support developing countries in establishing the technical capacities needed to implement the goals (Pistorius, 2012; UNREDD, 2015).

The REDD was further discussed and expanded into REDD+ at the COP14 in Poznan to include the importance of conservation and improvement of forest and forest carbon stocks in developing countries (Pistorius, 2012; Kronenberg et al., 2015; Tegegne et al., 2018). REDD+ is a voluntary approach that

Table 1. Key national and international forest-related policies, regulations and conventions.

Year	Policies, regulations and conventions	Countries	Details
1900	Lacey Act	USA	• Landmark legislation as being the world's first ban on trade in illegally sourced wood products • Originally for prohibiting the trading of wildlife by making it a federal crime • Amendment in 2008 expands to include plants and plant products and to ban on illegal trading and logging into the market • Set requirements to provide plant and plant product declaration for entering the market
1975	Convention on International Trade in Endangered Species of Wild Fauna and Flora (CITES)	International	• Also known as the Washington Convention • Open for signature in 1973 and entered into force in 1975 • Developed as a result of a meeting at the International Union for Conservation of Nature in 1963 • A multilateral international agreement of 183 parties to protect endangered plants and animals • To ensure the survival of animals and plants in the wild will not be threatened by international trades
1983	International Tropical Timber Agreement (ITTA)	International	• First negotiated under the UN Conference on Trade and Development in 1983 • Open for signature in 1983 and entered into force in 1985 • Aims at promoting sustainable utilisation and conservation of tropical forests by providing an effective framework for the cooperation between producers and consumers • International Tropical Timber Organisation was established in 1986 to implement international sustainable forest management • ITTA 1994 replaced ITTA 1983 with a draft agreement on exports of tropical timber from sustainably managed sources and to establish funding to assist tropical timber producers in obtaining the necessary resources necessary to achieve goals • ITTA 2006 superseded ITTA 1994 in promoting the expansion and diversification of international trade of tropical timber from sustainably managed and legally harvested forests
1992	United Nations Conference on Environment and Development (UNCED), Rio de Janeiro	International	• Also known as the Earth Summit • Agenda 21, Chapter 11 on 'Combating deforestation'
1992	Statement of Principles for the Sustainable Management of Forests	International	• Also known as the Forest Principles or Rio Forest Principles • Adopted at the Earth Summit at Rio de Janeiro in response to global concerns about forestry practices • Non-legally binding document to make recommendations for the conservation and sustainable management of all types of forests to meet the social, economic, cultural and spiritual needs of present and future generations
1992	Convention on Biological Diversity (CBD)	International	• A multilateral treaty conceived in 1988 at the UNEP • Aim at developing national strategies for the conservation of biodiversity, the sustainable use, and the sharing of benefits arising from resources • Opened for signature at the Rio Summit and entered into force in 1993 • 168 Signatures and 198 parties in 2016 • The first time in international law that the conservation of biodiversity is a common concern of humankind • Two supplement agreements - Cartagena Protocol to regulate the transport of genetically modified organisms and Nagoya Protocol • On biosafety regulates the transport of genetically modified organisms

Year	Name	Scope	Description
1993	Forest Stewardship Council (FSC)	International	• A not-for-profit organisation to set international standards and to promote the world's forests are managed environmentally appropriate, socially beneficial and economically viable • An international certification and labelling system to ensure that paper and wood products are from responsible sources • Certified approximately 198 million hectares in 84 countries and 1,615 certificates
1995	Intergovernmental Panel on Forests (IPF)	International	• Developed by the United Nations Conference on Environment and Development (UNCED) by the 3rd session of the Commission on Sustainable Development (CSD) • Mandated to pursue international consensus and formulate actions on forest-related issues • To combat deforestation, forest degradation and to promote the management, conservation and sustainable development of all types of forests • Report finalised in 1997 that contained more than 150 proposals for action
1997	Kyoto Protocol	International	• First recognised the important role of forests in reducing carbon emissions from deforestation
1997	Intergovernmental Forum on Forests (IFF)	International	• Developed at the United National General Assembly 19th Special Session (UNGSS) • Mandated to promote and facilitate the implementation of the IPF proposals for action • Report finalised in 2000 at the 8th Session of the CSD and contained more than 120 additional proposals for action
1999	Programme for the Endorsement of Forest Certification (PEFC)	International	• The world's largest forest certification and labelling system of forest-based products based in Geneva, Switzerland • Includes 35 worldwide independent national forest certification schemes • Aims at promoting sustainable forest management through independent 3rd party certification • Certified approximately 2/3 of the world's certified forest
2000	United Nations Forum on Forests (UNFF)	International	• Certified approximately 300 million hectares as at September 2018 • Developed by the United Nations Economic and Social Council as a high-level intergovernmental policy forum on forest-related matters based on the Rio Forest Principles and the outcomes of the IPF and IFF
2001	Collaborative Partnership on Forests (CPF)	International	• Established in 2001 by the Economic and Social Council of the United Nations (ECOSOC) • Supports the work of UNFF and in partnership with 14 international organisations aims at improving forest management and conservation and issues in relation to the production and trade of forest products
2003	Forest Law Enforcement, Governance and Trade (FLEGT) Action Plan	European Union	• The earliest policy direction to combat illegal logging • Mainly voluntary • Developed the FLEGT Regulation for implementation in 2005
2005	FLEGT Regulation	European Union	• Establishment of bilateral FLEGT Voluntary Partnership Agreement (VPA) between the EU & timber-exporting countries • Entail the setting up of a licensing scheme to identify legal products exported from producer countries and approve them for importation into the EU • Leading to the introduction of green procurement policies and implementation of sustainable procurement policies • Motivated to the development of the Timber Regulation
2007	Reducing emissions from deforestation and forest degradation (REDD)	International	• The idea of REDD was first discussed at the Kyoto Protocol in 1997, first recognised the importance of forest in reducing carbon emissions • UNFCCC-COP13 in Bali formally established the REDD • Commitments from international communities to reducing forest degradation through REDD activities

Table 1 contd. ...

... Table 1 contd.

Year	Policies, regulations and conventions	Countries	Details
2008	Conservation, sustainable management of forests, and enhancement of forest carbon stocks (REDD+)	International	• UNFCCC-COP 14 in Poznan REDD was expanded to REDD+ • Expanded to include the conservation, sustainable management of forests and enhancement of forest carbon stocks in developing countries and develop funding system to help developing countries to conserve biodiversity and protect vital ecosystems • Key guidance and framework largely completed in 2013 (COP 19 in Warsaw)
2010	EU Timber Regulations (EUTR) EU Regulation No. 995/2010	European Union	• Entered into force in 2013 and binding to all member states • Ban on illegal logging and importation of illegal timber products into EU countries • Undertake 'due diligence' in the tracking of imported timber products
2014	New York Declaration on Forest	International	• UN Secretary-General's Climate Summit held in New York (also known as New York Climate Summit) • A voluntary and non-legally binding declaration supported by 40 governments, 192 organisations and 16 indigenous peoples • Aims to reduce deforestation in half by 2020 and end by 2030, and to restore 150 million hectares by 2020 and 350 million hectares by 2030
2015	Paris Agreement	International	• COP 21 of UNFCCC reached a landmark agreement to combat climate change • Opened for signature in 2016 and entered into force in the same year, ratified by 176 out of the 197 countries • Final adoption of REDD+ and encourage countries to engage • Aims to conserve half of the terrestrial space for biodiversity by 2050 • Propose to use bioenergy with carbon capture and storage
2015	2030 Agenda for Sustainable Development	International	• UN Sustainable Development Summit in New York, formulating plans for the year 2030 • Adopted by the General Assembly of the UN and agreed by 193 members • Aims at ending poverty and four objectives—eradicate policy, heal the planet, secure prosperity, and foster peace and justice • Developed 17 sustainable development goals and 169 targets (28 targets are forest-related)
2017	UN Strategic Plan for Forests 2017–2030	International	• A special session of the UNFF in 2017 to provide vision for global forests in 2030 and agreed by 197 members • Adopted by the UN Economic and Social Council and the UN General Assembly in 2017 • The first UN strategic plan for forests • Set 6 global forest goals and 26 targets to increase forest area by 3% by 2030 (approx. 120 million hectares)

Sources: Bucknum, 1998; Humphrey, 1998, 2001; Curran et al., 2005; Flejzor, 2005; Levashova, 2011; Stupak et al., 2011; Pistorius, 2012; Kronenberg et al., 2015; Prestemon, 2015; Sample et al., 2015; Gratzer and Keeton, 2017; Holl, 2017; UNSPF, 2017; Guan et al., 2018; Mitchard, 2018; PEFC, 2018; Tegegne et al., 2018; Barbier et al., 2019; FSC, 2019.

provides results-based payments for verified emission reductions from participating countries. It aims at reducing carbon emissions from deforestation and forest degradation, conserving forest carbon stocks and managing the forest sustainably (Newton et al., 2015; REDD+, 2016).

A more recent international agreement was the New York Declaration on Forest (NYDF) in 2014, which aims to restore 150 million hectares of forest by 2020 and 350 million hectares by 2030 (Holl, 2017; NYDF, 2017). The NYDF advocates incentive schemes of forestry restoration and protection to participating countries. The Paris Agreement in 2015 was established for the final adoption of the REDD+ and aims to conserve half of the terrestrial space for biodiversity by 2050 (Mitchard, 2018; Barbier et al., 2019). The 2030 Agenda for Sustainable Development (2030 Agenda) was adopted at the General Assembly of the UN in order to address global environmental, social and economic challenges and to provide a healthy and peaceful living environment (Gratzer and Keeton, 2017). The 2030 Agenda contains 17 sustainable development goals and 169 targets, of which 28 targets are forest-related (FAO, 2018). The UN Strategic Plan for Forests 2017–2030 was initiated and met in 2017. The UN strategic plan was adopted by the UN Economic and Social Council, and the UN General Assembly in the same year (UNSPF, 2017). The UN strategic plan, although only voluntary, has a solid forest-focused direction for member countries to set up national implementation goals and achieve the target. All these policy regulations are shown in Table 1.

3.2 National policies and regulations

In addition to the international conferences, conventions and meetings that represent a global collaboration of preventing deforestation and forest degradation, policies and regulations have also been established at the national level. The most important one is the Lacey Act of 1900 in the US. The Act has been marked as landmark legislation to be the world's first ban on trade in illegally sourced wood product. The Lacey Act was originally developed in order to prohibit the trading of wildlife, but it was expanded to cover illegal timber and made it unlawful to place any illegal timber or related products on the US market without proper declaration (Prestemon, 2015; Guan et al., 2018).

Similarly, in the European Union (EU), the Forest Law Enforcement, Governance and Trade (FLEGT) Action Plan and the FLEGT Regulation were established in 2003 and 2005, respectively, aiming to combat illegal logging and ban the trading of illegally-sourced timber (Levashova, 2011; Tegegne et al., 2018). The FLEGT, under the Voluntary Partnership Agreement between the EU member countries and timber exporters, has set up a licensing scheme in order to identify legal timber and timber products in trading (Levashova, 2011; Guan et al., 2018). In 2010, under the motivation of the FLEGT, the EU Timber Regulation (EUTR 995/2010) was developed. The FLEGT licensed timber meets the due diligence requirements of the EUTR (Levashova, 2011; Prestemon, 2015; Tegegne et al., 2018). Over the years, the importance of forest management has encouraged the development of country-specific policies and regulations to suit their forest management.

3.3 Forestry labelling and certification programmes

Third-party certification and labelling systems have also been developed internationally in order to set standards and to promote responsible forest management for local implementation in individual countries. The Forest Stewardship Council (FSC) was first established in 1993 as a not-for-profit organisation which certifies timber and timber products from responsible sources (Guan et al., 2018; FSC, 2019). The FSC certified approximately 1/3 of the world's forests. The Programme for the Endorsement of Forest Certification (PEFC) was developed in 1999 and was also known as the Pan-European Forest Certification in 2004 (PEFC, 2018). The PEFC is the world's most extensive forest certification and labelling system of forest and forest-related products that certified approximately 2/3 of the world's forests (Stupak et al., 2011; Guan et al., 2018; PEFC, 2018). FSC and PEFC together account for almost 100% of certified forests in the world. Timber sourced from certificated forests is well recognised and accepted in the environmental assessment of buildings in BREEAM, LEED and GreenStar.

The national and international collaborations play a significant role in the conservation and management of forests. The collaborative efforts help in broadening and engaging different stakeholders to support sustainable land-use and forest management. In addition, the collaborative actions help raise awareness for the importance of forest protection and promote the use of timber and timber products in the design and construction of low impact building. As part of the climatic change mitigation approach, research and development of prefabricated engineered wood products has gained momentum as they are increasingly used as alternative structural materials to replace traditional steel and concrete and reduce the energy consumption and greenhouse gas emissions related to the manufacturing process.

4. Emergent Timber Technologies

Timber and timber products have attracted much attention in the design and construction of sustainable buildings as replacements for traditional heavy materials, as they are a renewable resource and have a low-impact nature. Conventional timber and timber products are used for architectural and non-load bearing structures. However, with the advancement of technologies, timber can now also be used structurally in buildings and for high-rise construction. Timber from sustainably managed forests and plantations can be utilised as lumber or manufactured into engineered products. The international collaborative efforts recognise the value of forests for the continued survival and well-being of humankind and have been important drivers for the research and development of timber and timber products that can replace heavy materials in buildings. The benefits of using timber in buildings include low weight, high strength to weight ratio, easy to adjust on site, simple connections and high efficiency in erection, in addition to architectural features and natural characteristics inherent in the product.

The development of engineered wood products (EWPs) is emergent timber technologies that enable more widespread use of timber and timber products in buildings. Popular EWPs include laminated veneer lumber (LVL), glue-laminated timber (glulam), cross-laminated timber (CLT) and oriented strand board (OSB). They have been developed over the years in order to enable the construction of prefabricated timber structures that compete with steel and concrete in mid- and high-rise buildings. Mass timber construction (MTC) is an innovative construction method, which mainly utilises EWPs as key structural materials. The adoption of digital design and prefabrication in MTC allow building elements such as beams, columns, floors and walls to be pre-cut, prefabricated and transported to the construction site for immediate installation.

The use of EWPs in construction has been developed since World War II, mainly for non-structural elements, but it has now also been developed for structural applications (Manninen, 2014). In North America, the number of mass timber building per year has increased from under 20 projects in 2014 to more than 200 projects in 2018 (The Beck Group, 2018). Most EWPs are typically made of softwood, such as spruce, pine or fir. The density of softwood is approximately 500 kg/m^3 whereas the densities for steel and concrete are 7800 kg/m^3 and 2400 kg/m^3, respectively. Therefore, the use of EWPs in construction can significantly reduce the weight of the structure, simplify foundation design, and reduce embodied energy and associated CO_2 emissions.

The light weight and flexibility characteristics of EWPs also mean that building components are simple and safe to construct and the prefabrication method can be applied relatively quickly. Incorporating prefabrication in construction, in turn, can considerably reduce building time as well as labour cost, delays due to adverse weather conditions and environmental impacts. CLT, LVL, glulam and OSB are the most common EWPs used for prefabricated structural applications. In addition, timber can also be combined with RC to form composite structures; timber-concrete composite structure is a typical example.

4.1 Cross-laminated timber (CLT)

CLT is the leading innovation among EWPs. It is made of at least three cross-bonded layers (usually three, five or seven) of solid sawn timber. CLT was first developed in Switzerland in the early 1970s and widely used in the 2000s when some European countries, such as Austria, Germany, Switzerland,

Sweden, Norway and the United Kingdom, changed building code to permit multi-storey timber buildings. Approximately 90% of worldwide CLT production volume (around 800,000 m³) is in Central Europe, particularly in the alpine area (Fink et al., 2018). Outside Europe and North America, the CLT market is relatively young and it is estimated that North America consumes 45,000 m³ CLT for buildings each year (Schwarzmann et al., 2018). In New Zealand, the first commercial manufacturer of CLT, XLam, started production in 2012. Since the market for CLT has increased in terms of both the local and the global demand, the Australian CLT production has reached the capacity of 60,000 m³ per year.

CLT is mainly made from softwood, such as Radiata pine (*Pinus radiata*) or Spruce (genus *Picea*). However, hardwood is also a potential for CLT production. It is also possible to replace single layers of CLT with other engineered timber products, such as LVL, oriented strand board (OSB), plywood or multi-layer solid wood panels (Brandner et al., 2016). CLT manufacturing offers the possibility of utilising lower-grade dimensional lumber. Hence, low-grade lumber and forest mortality caused by insect, disease and fire could be effectively used (Karacabeyli and Douglas, 2013). According to the United States net annual growth between the harvest removals and mortality from 1952 to 2012, even though the rate of forest mortality slightly goes up, the rate of growing stock is still more than the total amount of timber harvested and forest mortality. This means that the mass timber development demand does not overwhelm the raw material supply (The North American Mass Timber State of the Industry, 2019).

Phenol resorcinol-formaldehyde (PRF), emulsion polymer isocyanate (EPI) and one-component polyurethane (PUR) are the three types of adhesives mainly used for CLT production. PUR, which is a formaldehyde free and light-coloured adhesive, is mostly used in CLT manufacturing. A low required amount of PUR with no additional hardener is used for CLT production (Mohammed and Munoz, 2011). It is important to ensure that adhesives meet specific requirements, such as strength, durability, moisture resistance, heat performance. The overall yield rate of CLT production is around 43%, which means every 1 m³ of the log can produce 0.43 m³ of CLT. However, the production of CLT is a circle process where waste, such as wood chips, sawdust and offcuts, is efficiently reused, mainly to generate the energy for factory equipment, kiln drying and local communities (Waugh Thistleton Architects, 2018).

CLT has been used in the construction of housing, multi-storey residential and non-residential buildings. Using CLT in construction offers several benefits, including short construction time, lightweight (20% the weight of concrete), minimal waste and noise during competitive construction cost, good lateral and seismic load resistance, adequate fire performance, stiffness, and high aesthetic value. CLT can be used for large panel prefabrication to provide floor slabs, roofs, beams, columns, load-bearing walls or shear walls. Typical dimensions of CLT is 20 m length, 50–300 mm thick and up to 4800 mm width. Non-residential and commercial/office buildings need long spans of up to 7 m, mainly for parking spaces at the basement and desired open office layout. CLT has a high strength-to-weight ratio, which allows expanding floor span without increasing weight, and CLT is, therefore, a competing alternative to concrete and steel.

4.2 Laminated veneer lumber (LVL)

The manufacturing of LVL is done by bonding multiple rotary-peeled veneers with their grain parallel to the longitudinal axis of the section under heat and pressure. LVL provides a wide range of structural applications, such as beam, column, truss, portal frame post and beam structure, structural decking, I-joist flanges and stressed skin panels. The length can be up to 20 m, the width is from 19 to 200 mm, and the depth is from around 90 to over 2500 mm.

The log will be cut to length, debarked and soaked or sprayed with hot water before peeling in a rotary lathe in order to ensure the quality of veneers. Since LVL is a veneer-based wood product, the quality of veneer is, therefore, one of the driven factors. In the last few decades, technological improvements in wood processing minimise waste and allow smaller diameter logs from young and fast-grown plantation forests (Leggate et al., 2017). Good quality veneers are used for the production of LVL, and offcuts with defects, such as knots, wane, voids or end of log's veneer sheet, can be used as strands for laminated strand lumber (LSL), oriented strand lumber (OSL) or parallel strand lumber (PSL) manufacturing. This helps to optimise material usage.

However, in order to optimise material usage and increase the mechanical properties of LVL, more research on the potential of using secondary quality wood is necessarily undertaken. Purba et al. (2019) have recently shown that the knot proportions on the veneer surface do have a negative influence on mechanical properties of LVL produced from secondary quality hardwood. Although thick veneers consume less adhesive and production time, it may reduce the modulus of elasticity and modulus of rupture of LVL. On the other hand, thinner veneer increases LVL strength by better distributing the defects. Using thin veneer leads consumption that is much more adhesive, effort, and time in LVL production. The 3 mm thick veneer optimises mechanical properties of LVL (Purba et al., 2019).

4.3 Glued laminated timber (Glulam)

Glulam is an EWP manufactured by gluing several graded timber laminations with their grain parallel to the longitudinal axis of the section. Members can be straight or curved, horizontally or vertically laminated and can be used to create different structural forms. Solid 20 to 50 mm thick laminates are typically finger-jointed into lengths and clamped together by adhesive under pressure. The use of glulam can be in large structural elements, beams, columns, trusses, bridges, portal frames, post and beam structures. Common size of glulam ranges from 60 to 250 mm wide and 180 to 1000 mm deep. There are no limits for length or shape. The dimension is, therefore, determined by transportation capacity (Glued Laminated Timber Association of UK).

4.4 Oriented strand board (OSB)

OSB is another engineered structural product, manufactured from thin strands of wood glued together with adhesive under heat and pressure in specific orientations. Wood strands are 100 mm long and 1.2 mm thick and dried to a moisture content of 6–8% before they are bonded together with resin base adhesive (Alldritt et al., 2014). OSB is largely used for wall panels, beams, I-joist webs, floor sheeting and roof panels (Mekonnen et al., 2014). OSB was originated and first produced in the early 1980s in North America. The production of OSB grows originally from 719,000 m^3 to significantly 26,632,305 m^3 in 2014 in North America (Jin et al., 2016). OSB is typically made from low-density trees, such as aspen or southern yellow pine, that are relatively fast-growing species. Other types of strand boards such as OSL, LSL and PSL are also manufactured from wood strands in a similar method by bonding with resins under high pressure (Cai and Ross, 2010).

4.5 Prefabricated composite structures

Timber can also be combined together with RC to form a composite structure. Timber-concrete composite structure (TCC) is a composite design that combines the best properties of timber and traditional RC in the design of slab structure. It is a composite structure that connects a concrete topping with timber joists or beams (FWPA, 2016). Timber is connected to overcome the tension stress while the concrete topping is installed in the compression zone using various connection methods such as dowel type fasteners, notches, friction-based connection and adhesives (Yeoh et al., 2011; Dias et al., 2018). The TCC slab was first developed in Germany in 1922 in response to the shortages of reinforcement in concrete after the first and second world wars (Yeoh et al., 2011) and it is now widely used in the US for short and medium span bridges, structural application of new buildings and refurbishment of old historical buildings (Yeo et al., 2011; Rodrigues et al., 2013).

The composite structure overcomes the shortcomings of using only timber or RC in slab design. The TCC design increases the stiffness and load bearing capacity. In addition, the design reduces the volume of concrete and encourages the use of timber to reduce CO_2 emissions and increase CO_2 sequestration. The fire performance of TCC structure is competitive with the conventional RC structure, and research results indicate that TCC structure has the performance increased from 60 to 90 minutes (Yeoh et al., 2011). The TCC slab design can also take advantage of reducing the time for site operation of concrete.

The concrete slab can be prefabricated off-site, integrating connectors for the fixing of timber beams on site (Lukaszewska et al., 2008).

The technique of using pre-tension or post-tension to improve structural performance using unbonded tendon was originally found in a concrete structure in connecting columns and beams in the structure, but now the technology has been applied to EWPs (Iqbal et al., 2010; Negrao, 2012). The technique called Press-Lam has been developed to allow EWPs to use high strength unbonded steel cables to create connections between timber beams and columns or columns and walls to the foundations in order to improve the net strength of a timber element (Wanninger, 2015). Research studies indicate that a reduction of the deflections can be achieved with an increase of the bending strength (Negrao, 2012). The prestressed timber structure can be used in multi-storey buildings with large structural timber members made from LVL, CLT or glulam.

I-Joist is another composite wood structure that is designed to replace structural lumber in construction. Composite I-joist is an EWP manufactured from OSB to form the web, and the flanges are made from two LVLs. The I-joist can also be produced by using plywood as web and OSL or sawn structural lumber as flanges. Due to the declining availability of high quality and large dimension lumber, composite I-joist has been developed as a structural element to carry the load for floor and roof systems (Wilson and Dancer, 2005; Islam et al., 2015). The web and flanges of an I-joist are routing and shaping from OSB and LVL. All components are assembled together with resin, pressed mechanically and heated to accelerate resin cure. The I-joist is less expensive, lighter in weight, stronger and more efficient than a solid sawn lumber beam and has become an important substitution of structural steel and RC beams. I-joist is a common construction element in North America and Europe and can be used for commercial and residential buildings (Islam et al., 2015).

5. Engineered Timber Case Studies

The development of EWPs has enabled structural building elements to be constructed and achieve a similar performance to traditional RC. EWPs have also been applied in medium and high-rise construction. The Forté building in Docklands, Melbourne, Australia was completed in 2012. It was a 10-storey CLT apartment and was the world tallest timber residential building until 2016. The timber structure weighted 485 tonnes, connected with 5,500 angle brackets using 34,550 screws. Ground and first floor slab were constructed from geopolymer concrete due to the larger spans required in the retail space and as well as wet and termite resistance purposes. Prefabricated CLT panels were shipped from Europe and transported to the construction site, craned into positions and then screwed together. By using the platform-frame system, each floor was set on the walls below and then another storey of walls was raised, and so on up the building (Durlinger et al., 2013).

The International House Sydney at Barangaroo showcased the possibility of long span mass timber floor and exposed mass timber system. The project was completed in 2017 and was the first 6-storey exposed engineered timber commercial/office building in Australia. Notably, all six levels above the concrete retail ground floor are engineered timber. The office building, completed in late 2018 in Brisbane, was the world tallest and largest mass timber building, with a six by eight-metre grid of exposed glulam columns with CLT cladding and CLT flooring system.

In Canada, the UBC Brock Common building, completed in 2016, is currently the tallest mass timber building for student accommodation at the University of British Columbia in Vancouver. It is an 18-story building, in which 17 storeys are mass timber structures. It took only ten weeks to complete the mass-timber levels. The building has a flooring system with no less than 2 hours of fire-resistance rating and a sprinkler system throughout. The structural system of the building is a hybrid configuration of concrete podium and cores, CLT/LVL columns and floors, and a steel roof system. Results of life cycle analysis from cradle to gate showed that the use of mass timber instead of concrete has a positive impact on the environment. For instance, compared to the original concrete building, mass timber design has negative global warming potential due to the possibility of carbon sequestration in mass timber, even beyond the building lifetime (Connolly et al., 2018).

The use of EWPs has also been considered for buildings in seismic regions. The NMIT Arts and Media Building in Nelson, New Zealand, completed in 2011, demonstrated that timber could be successfully used in multi-storey commercial buildings. The building utilised LVL for many structural components, including columns, beams, floor systems and shear walls. Its shear walls were designed to resist lateral load, especially seismic load. The system relies on coupled pairs of LVL shear walls, incorporating high strength post-tensioned steel tendons. The shear walls are centrally fixed so that they can rock during a seismic event; a series of U-shaped steel plates placed between the walls form a coupling mechanism and act as dissipaters to absorb seismic energy. This allows the primary structure to remain essentially undamaged while these replaceable connections act as plastic fuses. LVL has strength properties that allow fabrication of beams, columns and walls at dimensions similar to concrete and steel design. Spanning 9.6 m, the primary LVL floor beams provide a large open floor plate, comparable to traditional commercial structures (John et al., 2011).

6. Timber and Timber Products for Carbon Management

With the increased focus on the issues of climate change, timber has become an important contemporary building material in construction. The renewable nature and low impact characteristics of timber have attracted much attention in the building sector. Timber and timber products are lower in embodied energy and CO_2 emissions than the equivalent design using traditional building materials (Fraisse et al., 2006; John et al., 2011; Knauf et al., 2015). In addition, they provide a carbon sink as a construction material, resulting in a potential reduction in the CO_2 concentration in the atmosphere (Schmidt and Griffin, 2012).

The low embodied energy characteristic of timber is important and case studies have recently been undertaken in order to compare the embodied energy and carbon impact of timber against heavier materials, such as concrete, bricks and steel, for structural elements in buildings. Some of the case studies are summarised in Table 2. From the case studies, research results demonstrate that timber construction consumes less embodied energy and has a carbon reduction benefit over those buildings using heavier materials.

The problems of climate change can be addressed by managing forests sustainably and using timber and timber products in buildings. Sustainable forest management (SFM) is a management process to ensure that trees continue to grow, absorb and store for an extended period of time (Lippke et al., 2011; Yan, 2018). The forest ecosystem is well recognised as critical in altering the atmospheric carbon concentrations of the global environment. Carbon emissions are highly receptive to land use change and proper management of forests can help to change the forest from a source to a sink (Carroll, 2012). The SFM is about managing and sustainably using the forestry in order to maintain the social, environmental

Table 2. Selected case studies of comparing embodied energy between timber and traditional materials.

Main research themes	Results	References
Life cycle energy use of timber, concrete and steel office structures	Embodied energy of steel building is 1.61 times higher than the concrete structure; embodied energy of timber structure is 1.27 times lower than the concrete structure	Cole and Kernan, 1996
CO_2 balance wood v concrete in multi-storey building	Embodied energy for concrete 60–80% > than timber	Borjesson and Gustavsson, 2000
LCA case study of home materials—Compare concrete, timber, aluminium, glass, etc.	Concrete has highest embodied energy %	Asif et al., 2007
Compare embodied energy in homes (Mixed weight materials v heavyweight)	50% < Embodied enegy light construction	Mendoca and Braganca, 2007
Primary energy—8 storey timber building case study	Negative CO_2 balance for timber building due to sequestration	Gustavsson and Joelsson, 2010
LCA Australian case study. Timber v brick veneer/concrete floor	Timber outperforms brick veneer	Carre, 2011
LCA brick v timber optimised design-Sydney	GHG savings with timber design	Ximenes and Grant, 2012

and economic functions for all types of forests for present and future generations (UNFCCC, 2002; UNFF, 2007). SFM is crucial in ensuring a steady supply of raw materials for the production of timber products, however, the large scale of deforestation has hampered the timber reserves in the forest. According to FAO (2018), the world's forest areas have decreased from 31.6% in 1990 to 30.6% in 2015. Therefore, as part of the SFM, afforestation and reforestation are two crucial strategies to restore the degraded forest and the related environment.

The use of timber and timber products can also be used to address climate change. Timber products convert forest into harvested wood so that they can continue storing carbon in large quantities for the duration of their useful life (Lehmann, 2013). According to Lippke et al. (2011), it is preferable that the timber be regularly harvested rather than left as a growing forest, as the harvested timber stores carbon for its entire useful life. Growing forests will remain carbon neutral as they absorb CO_2 during their growth, but after they die and decay the carbon returns to the atmosphere. In addition, old forests slow the absorption of CO_2.

The use of timber products, such as EWPs, can replace high emission intensity materials and alleviate environmental impacts (Lippke et al., 2011). The development of EWPs has shifted the use of timber from architectural to structural uses that are potential alternatives to traditional methods in the design and construction of buildings, in particular low impact buildings. EWPs such as LVL and CLT are now considered as sustainable building materials and systems that have the potential to turn buildings into carbon sinks if they are used on a large scale, replacing concrete and steel in construction (Lehmann, 2013).

The traditional RC structure is a high embodied energy design. The production of cement is a major energy consumption process in producing the clinker (Holtzhausen, 2007; Hossain et al., 2017). The production of cement contributes to approximately 5–10% of the total CO_2 emissions and consumes approximately 12–15% of total industrial energy use (Hossain et al., 2017). For the production of 1 tonne of cement clinker, approximately 0.87 tonne of CO_2 is released into the atmosphere. Steel production is also high in both embodied energy and emissions. The embodied energy required in the manufacturing of steel includes the extraction of iron ore then melting it in a furnace with oxygen to remove impurities and reduce carbon content. According to Quader et al. (2016), the production of 1 tonne of steel generates approximately 1.8 tonnes of CO_2.

In addition, there is significant scope for higher carbon storage in houses by increasing the use of timber and timber products for the construction of sub-floor and wall cladding systems. As an example, doubling the wood used in houses to 0.14 m³ per m² of floor area would result in additional annual carbon storage in houses in Australia from 1.6 Mt CO_2-e in 2008 to 4 Mt CO_2-e in 2050 (Kapambwe et al., 2008).

Holtzhausen (2007) estimated that only around 0.098 tonnes of CO_2 are emitted from every gigajoule of embodied energy consumed in timber. The displacement of CO_2 has been estimated for the use of timber instead of material alternatives to the order of 3.9 tonnes per tonne of timber used (Sathre and O'Connor, 2010). Estimates of carbon sequestration are around 1 tonne per metre cube of timber (Lehmann, 2013). Therefore, in comparison to the traditional heavy materials used in the construction of buildings, timber and timber products are more environmentally friendly materials. Lippek et al. (2011) state that replacing steel joists with approximately 1 tonne of engineered wood joists can cut down CO_2 emissions by about 10 tonnes. In addition, for every tonne of wood flooring used to replace concrete floor. CO_2 emissions can be reduced by approximately 3.5 tonnes.

Availability and advancement of EWPs, powerful lifting, transport equipment and use of high precision production facilities enable the use of prefabricated timber elements, which can span longer than traditional systems (Kolb, 2008). However, compared to concrete and steel building construction, there are also several challenges. Some of the significant challenges are fire performance of timber building, lack of information and evidence relating to constructability and technical guidelines, i.e., (pre)fabrication and standardised connection details, particularly for mid- to high-rise non-residential buildings. In addition, the supply chain for timber is not as well-organised as for other heavy building materials.

The permitted number of storeys or storey height of mass timber building is different in different countries, depending on the local regulations. Before 1994, some countries in Europe restricted the use of timber for load-bearing structures of more than two storeys due to fire regulations (Falk, 2005). However, several countries in Europe now have no specific regulations or do not limit the number of storeys in timber buildings (Östman and Källsner, 2011). In the United States, the development of new technologies and innovation in wood framing designs have permitted the construction of timber buildings up to four storeys, depending on the occupancy classification and the presence of automatic sprinkler systems (American Wood Council & International Code Council, 2015). Before 1992, timber structures were limited to three storeys in New Zealand, but changes in the Building Code since then have increased both the number and size of multi-storey timber buildings (Banks, 1999). Since May 2016, the Australian National Construction Code has permitted the use of timber framing for buildings in Class 2 (apartments) and 3 (hotels) up to 25 metres height (approximately 8 storeys) under deem-to-satisfy provisions. This code change creates a significant opportunity for the development of the commercial timber building industry in Australia.

There is no doubt that timber is environmentally, aesthetically and health-wise an ideal choice of material. However, its usefulness will rely on the sustainable supply of raw materials. Therefore, forestry and acquisition of raw materials are important elements in the long term to ensure logs are sourced from sustainable plantation forests. Davidson and Hanna (2004) state that softwood timbers, such as Radiator Pine, are cheap and easy to process and are the primary product used in the production of EWPs. As such, their supply will, in turn, influence the potential output for the EWP industry.

The case studies in Section 6 have demonstrated the possibility of using EWPs in mid- and potentially large-scale buildings. However, research studies also indicate that there are limitations of EWPs and these may have hindered the uptake of the EWPs. The main issues are related to the lack of confidence and knowledge among professionals in the design and construction of large buildings using timber (Holmes et al., 2011; Thomas et al., 2014). In addition, there is a negative perception of timber in relation to durability, fire, acoustic, thermal performance and durability (Gold and Rubik, 2009; Nolan, 2010; Roos et al., 2010). These obstacles are viewed as difficult to overcome when timber construction is specified. Furthermore, the lack of confidence in using timber stems from a multitude of issues, such as lead times, cost implications, connection details, availability, commercial risk, lack of assistance and poor marketing (Holmes et al., 2011). However, some of these negative perceptions conflict with research into timber performance in areas of durability, fire, acoustic and thermal performance (Papadopoulos et al., 2008; Lennon et al., 2010; Ruben et al., 2011). Structural performance and termite resistance are governed by building codes and standards that dictate minimum performance for all buildings regardless of building material selection.

7. Conclusion

The importance of forests is well recognised as being connected to the well-being of the present and future generations. The conservation and protection of forests have been significant agenda items at both the national and international levels. Deforestation and forest degradation have posed serious problems for the well-being of humanity and the natural environment. The loss of forestland has caused the diminishing of the biological diversity and is jeopardising economic benefits to people relying on the forest for survival. Over the years, international collaborative efforts towards managing forest resources have been established, achieving the goal of maintaining an effective carbon sink for climate change mitigation, addressing the problem of deforestation and forest degradation. It is no doubt that, for the forest protection and conservation to be successful, it will require international agreements to develop plans of action and implement them at the national level.

Timber is a sustainable building material and is well accepted for its renewable nature and low embodied energy characteristic, compared with the traditional heavy materials. The development of engineered wood products has shifted the use of timber from architectural to structural uses in order to compete with steel and concrete in the design and construction of buildings. The low impact nature and structural capacity of EWPs can turn buildings into carbon sinks and store carbon for an extended period of time. The advancement of technologies and prefabrication techniques have enabled EWPs to span longer than traditional systems for mid- and high-rise buildings. EWPs are relatively new and not as mature as steel and concrete in the construction industry. The uptake of EWPs faces significant challenges of misperception with regards to performance, i.e., fire safety and acoustics, and lack of confidence among design professionals in relation to constructability and design details. There is no doubt that EWPs have the potential and possibility to replace traditional heavy materials and promote low impact design and construction of buildings. However, for the uptake of EWPs to be successful, more research and development may be required in order to generate more data and improve the perception and reputation of timber among design professionals.

References

Alldritt, K., Sinha, A. and Miller, T.H. 2014. Designing a strand orientation pattern for improved shear properties of oriented strand board. Journal of Materials Civil Engineering 26(7): 04014022.

American Wood Council & International Code Council 2015, 2015 Code conforming wood design.

Asif, M., Muneer, T. and Kelley, R. 2007. Life cycle assessment: A case study of a dwelling home in Scotland. Building and Environment 42: 1391–1394.

Australian Industry Group. 2015. Australia's Construction Industry: Profile and Outlook. July. <http://www.aigroup.com.au/portal/binary/com.epicentric.contentmanagement.servlet.ContentDeliveryServlet/LIVE_CONTENT/Economic%2520Indicators/Construction%2520Survey/2015/Construction%2520industry%2520profile%2520and%2520Outlook.pdf>.

Banks, W. 1999. Multi-Storey Timber Construction-A Feasibility Study, New Zealand Timber Design J 8: 14–23.

Barbier, E.B., Burgess, J.C. and Dean, T.J. 2019. How to pay for saving biodiversity. Science 360(6388): 486–488.

Brandner, R., Flatscher, G., Ringhofer, A., Schickhofer, G. and Thiel, A. 2016. Cross laminated timber (CLT): overview and development. European Journal of Wood and Wood Products 74(3): 331–51.

Bucknum, S. 1998. The U.S. commitment to Agenda 21: Chapter 11 combating deforestation—The ecosystem management approach. Duke Environmental Law & Policy 8(2): 305–343.

Cai, Z. and Ross, R.J. 2010. Mechanical properties of wood-based composite materials. Wood handbook: Wood as an engineered material: Chapter 12. Contennial ed. General technical report FPL; GTR-190 Madison, WI: U.S. Department of Agriculture, Forest Service, Forest Products Laboratory, 12.1–12.12.

Carre, A. 2011. A comparative Life cycle assessment of alternate constructions of a typical Australian house design. RMIT University, Melbourne.

Carroll, M., Milakovsky, B., Finkral, A., Evans, A. and Ashton, M.S. 2012. Managing carbon sequestration and storage in temperate and boreal forests. Managing forest carbon in a changing climate. Ashton, M.S., Tyrrell, M.L., Spalding, D. and Gentry, B. (eds.). Springer 10: 205–226.

Chen, Y. 2012. Comparison of environmental performance of a five-storey building built with cross-laminated timber and concrete. Sustainable Building Science Program, University of British Columbia-Department of Wood Science, Vancouver, BC, Canada.

Cole, R.J. and Kernan, P.C. 1996. Life-cycle energy use in office buildings. Building and Environment 31(4): 307–17.

Connolly, T., Loss, C., Iqbal, A. and Tannert, T. 2018. Feasibility study of mass-timber cores for the UBC tall wood building. Buildings 8(8): 98.

Crawford, R. 2011. Life Cycle Assessment in the Built Environment. Routledge, London.

Curran, M.P., Miller, R.E., Howes, S.W., Maynard, D.G., Terry, T.A., Heninger, R.L., Niemann, T., van Rees, K., Powers, R.F. and Schoenholtz, S.H. 2005. Progress towards more uniform assessment and reporting of soil disturbance for operations, research, and sustainability protocols. Forest Ecology and Management 220(1): 17–30.

Davidson, A. and Hanna, D. 2004. Engineered wood products: Prospects for Australia. ABARE Report 04.14. Forest and Wood Products Research and Development Corporation, Canberra.

Dias, A., Schänzlin, J and Dietsch, P. 2018. Design of timber-concrete composite structures: A state-of-the-art. European Cooperation in Science & Technology, Cost Action FP/1402/WG4, Shaker Verlag Aachen.

Durlinger, B., Crossin, E. and Wong, J. 2013. Life cycle assessment of a cross laminated timber building.

Falk, A. 2005. Architectural Aspects of Massive Timber: Structural Form and Systems. Doctoral dissertation, Luleå tekniska universitet.

FAO. 2018. The state of the world's forests 2018—Forest pathways to sustainable development. Food and Agriculture Organisation, United Nations.

Fink, G., Kohler, J. and Brandner, R. 2018. Application of European design principles to cross laminated. Engineering Structures 171: 934–43.

Flejzor, L. 2005. Reforming the international tropical timber agreement. Reciel. 14(1): 19–27.

Fraisse, G., Johannes, K., Trillat-Berdal, V. and Achard, G. 2006. The use of a heavy internal wall with a ventilated air gap to store solar energy and improve summer comfort in timber frame houses. Energy and Buildings 38(4): 293–302.

FSC. 2019. FSC facts & figures. Forest Stewardship Council, FSC Global Development, Bonn, Germany.

FWPA. 2005. Build in timber. Forest and Wood Products Research and Development Corporation, Canberra, Australia.

FWPA. 2016. Timber concrete composite floor. Forest and Wood Products Australia, Australia.

Gold, S. and Rubik, F. 2009. Consumer attitudes towards timber as a construction material and towards timber frame houses—selected findings of a representative survey among the German population. Journal of Cleaner Production 17(2): 303–309.

Gratzer, G. and Keeton, W.S. 2017. Mountain forests and sustainable development: The potential for achieving the United Nations' 2030 Agenda. Mountain Research and Development 37(3): 246–253.

Guan, Z., Xu, Y., Gong, P. and Cao, J. 2018. The impact of international efforts to reduce illegal logging on the global trade in wood products. International Wood Products Journal 9(1): 28–38.

Gustavsson, L. and Joelsson, A. 2010. Life cycle primary energy analysis of residential buildings. Energy and Buildings 42(2): 210–20.

Holl, K.D. 2017. Research directions in tropical forest restoration. Annals of the Missouri Botanical Garden 102(2): 237–250.

Holmes, M., Crews, K. and Ding, G. 2011. The influence building codes and fire regulations have on multi-storey timber construction in Australia. World Sustainable Building Conference, 18–21 October: 224–235 Helsinki, Finland.

Holtzhausen, H. 2007. Embodied energy and its impact on architectural decisions, WIT Transactions on Ecology and the Environment. 102.

Hossain, M.U., Poon, C.S., Lo, I.M.C. and Cheng, J.C.P. 2017. Comparative LCA on using waste materials in the cement industry: A Hong Kong case study. Resources, Conservation and Recycling 120: 199–208.

Humphreys, D. 1998. The report of the intergovernmental panel on forests. Environmental Politics 7(1): 214–221.

Humphreys, D. 2001. The creation of the United Nations Forum on Forests. Environmental Politics 10(1): 160–166.

Iqbal, A., Pampanin, S. and Buchanan, A.H. 2010. Seismic performance of prestressed timber beam-column sub-assemblies. New Zealand Society for Earthquake Engineering Conference, Wellington, New Zealand (http://db.nzsee.org.nz/2010/Paper27.pdf).

Islam, M.S., Shahnewaz, Md. and Alam, M.S. 2015. Structural capacity of timber I-joist with flange notch: Experimental evaluation. Construction and Building Materials 79: 290–300.

Jin, J., Chen, S. and Wellwood, R. 2016. Oriented strand board: Opportunities and potential products in China. BioResources 11(4): 10585–10603.

John, S., Mulligan, K., Perez, N., Love, S. and Page, I. 2011. Cost, time and environmental impacts of the construction of the new NMIT Arts and Media building.

Joseph, P. and Tretsiakova-McNally, S. 2010. Sustainable non-metallic building materials. Sustainability 2(2): 400–427.

Kapambwe, M., Ximenes, F., Vinden, P. and Keenan, R. 2008. Dynamics of carbon stocks in timber in Australian residential housing. Forest and Wood Products Australia Project No. PN07. 1058.

Karacabeyli, E. and Douglas, B. 2013. CLT: Handbook Cross-laminated Timber, FPInnovations.

Knauf, M., Köhl, M., Mues, V., Olschofsky, K. and Frühwald, A. 2015. Modeling the CO_2 effects of forest management and wood usage on a regional basis. Carbon Balance and Management 10: 13.

Kolb, J. 2008. Systems in timber engineering: Loadbearing structures and component layers, Walter de Gruyter.

Kronenberg, J., Orligóra-Sankowska, E. and Czembrowski, P. 2015. REDD+ and institutions. Sustainability 7: 10250–10263.

Lal, R. 2008. Sequestration of atmosphere CO_2 in global carbon pools. Energy & Environmental Science 1: 86–100.

Lal, R. 2010. Managing soils and ecosystems for mitigating anthropogenic carbon emissions and advancing global food security. BioScience 60(9): 708–721.

Leggate, W., McGavin, R.L. and Bailleres, H. 2017. A guide to manufacturing rotary veneer and products from small logs.

Lehmann, S. 2013. Low carbon construction systems using prefabricated engineered solid wood panels for urban infill to significantly reduce greenhouse gas emissions. Sustainable Cities and Society 6: 57–67.

Lennon, T., Hopkin, D., El-Rimawi, J. and Silberschmidt, V. 2010. Large scale natural fire tests on protected engineered timber floor systems. Fire Safety Journal 45(3): 168–82.

Levashova, Y. 2011. Review of European Community & International Environmental Law 20(3): 290–299.

Lippke, B., Oneil, E., Harrison, R., Skog, K., Gustavsson, L. and Sathre, R. 2011. Wood products reduce carbon dioxide emission levels. Carbon Management 3: 303–333.

Lukaszewska, E., Johnsson, H. and Fragiacomo, M. 2008. Performance of connections for prefabricated timber–concrete composite floors. Materials and Structures 41: 1533–1550.

Mackey, B., Keith, H., Berry, S. and Lindenmayer, D. 2008. Green Carbon Part 1: The role of natural forests in carbon storage. Canberra: ANU Press (Retrieved from www.jstor.org/stable/j.ctt24hcnf).

Manninen, H. 2014. Long-term outlook for engineered wood products in Europe. Technical Report 91. European Forest Institute.

Mekonnen, T.H., Mussone, P.G., Choi, P. and Bressler, D.C. 2014. Adhesive from waste protein biomass for oriented strand board composites: Development and performance. Macromolecular Materials and Engineering 299: 1003–1012.

Mendonca, P. and Braganca, L. 2007. Sustainable housing with mixed weight strategy—a case study. Building and Environment 42(9): 3432–43.

Micales, J.A. and Skog, K.E. 1997. The decomposition of forest products in landfills. International Biodeterioration & Biodegradation 39(2-3): 145–58.

Mitchard, E.T.A. 2018. The tropical forest carbon forest and climate change. Nature 559: 527–559.

Mohammed, M. and Munoz, W. 2011. CLT Handbook: Connections in cross-laminated timber buildings. FPInnovations, Quebec, Canada.

NAFI. 2007. Forest Industries and Climate Change. National Association of Forest Industries.

Negrao, J.H.J.O. 2012. Prestressing systems for timber beams. World Conference on Timber Engineering, 16–19 July, Auckland, New Zealand.

Newton, P., Schaap, B., Fournier, M., Cornwall, M., Rosenbach, D.W., DeBoer, J., Whittemore, J., Stock, R., Yoders, M., Brodnigc, G. and Agrawal, A. 2015. Community forest management and REDD+. Forest Policy and Economics 56: 27–37.

Nolan, G. 2010. Opportunities and constraints for timber in non-residential construction in Australia. World Conference Timber Engineering (WCTE), Italy.

Östman, B. and Källsner, B. 2011. National building regulations in relation to multi-storey wooden buildings in Europe, SP Trätek and Växjö University, Sweden.

Papadopoulos, A.N., Avtzis, D. and Avtzis, N. 2008. The biological effectiveness of wood modified with linear chain carboxylic acid anhydrides against the subterranean termites. Holz Roh Werkst. 66: 249–252.

PEFC. 2018. Sustainable forest management—Requirements. PEFC Council, Geneva, Switzerland.

Pistorius, T. 2012. From RED to REDD+: The evolution of a forest-based mitigation approach for developing countries. Current Opinion in Environmental Sustainability 4: 638–645.

Prestemon, J.P. 2015. The impacts of the Lacey Act Amendment of 2008 on U.S. hardwood lumber and hardwood plywood imports. Forest Policy and Economics 50: 31–44.

Purba, C.Y.C., Pot, G., Viguier, J., Ruelle, J. and Denaud, L. 2019. The influence of veneer thickness and knot proportion on the mechanical properties of laminated veneer lumber (LVL) made from secondary quality hardwood. European Journal of Wood and Wood Products 1–12.

Quader, M.A., Ahmed, S., Ghazilla, R.A.R., Ahmed, S. and Dahari, N. 2016. Evaluation of criteria for CO_2 capture and storage in the iron and steel industry using the 2-tuple DEMANTEL technique. Journal of Cleaner Production 120: 207–220.

REDD+. 2016. About REDD+. UN-REDD Programme, Geneva, Switzerland.

Rodrigues, J.N., Dias, A.M.P.G. and Providência, P. 2013. Timber-concrete composite bridges: State-of-the-art review. BioResources 8(4): 6630–6649.

Roos, A., Woxblom, L. and McCluskey, D. 2010. The influence of architects and structural engineers on timber in construction-perception and roles. Silva Fennica. 44(5): 871–84.

Ruben, B., Bjorn, P.J. and Gustavson, A. 2011. Aerogel insulation for building applications: A state-of-the-art review. Energy and Buildings 43: 761–769.

Ruis, B.M.G.S. 2001. No forest convention but ten tree treaties. An International Journal of Forestry and Forest Industries 52(3) Unasylva - No. 206 (http://www.fao.org/3/y1237e/y1237e00.htm).

Sample, V.A., Birdsey, R.A., Houghton, R.A., Swanston, C., Hollinger, D., Dockry, M. and Bettinger, P. 2015. Forest carbon conservation and management: Integration with sustainable forest management for multiple resource values and ecosystem services. Discussion Paper. Pinchot Institute for Conservation, USA.

Sathre, R. and Connor, J. 2010. Meta-analysis of greenhouse gas displacement factors of wood product substitution. Environmental Science & Policy 13: 104–114.

Schwarzmann, G., Hansen, E. and Berger, G. 2018. Cross-laminated timber in North America: What can we learn? BioProducts Business 81–91.

Skullestad, J.L., Bohne, R.A. and Lohne, J. 2016. High-rise timber buildings as a climate change mitigation measure—a comparative LCA of structural system alternatives. Energy Procedia 96: 112–123.

Stupak, I., Lattimore, B., Titus, B.D. and Smith, C.T. 2011. Criteria and indicators for sustainable forest fuel production and harvesting: A review of current standards for sustainable forest management. Biomass and Bioenergy 35: 3287–3308.

Tegegne, Y.T., Cramm, M. and van Brusselen, J. 2018. Sustainable forest management, FLEGT, and REDD+: Exploring interlinkages to strengthen forest policy coherence. Sustainability 10: 4841–4853.

The Beck Group. 2018. Mass Timber Market Analysis, Council of Western State Foresters, Portland Oregon.

The North American Mass Timber State of the Industry. 2019. Mass Timber: Industry Report.

Thomas, D., Ding, G. and Crews, K. 2014. Sustainable timber use in residential construction: Perception versus reality. WIT Transactions on Ecology and the Environment 186: 399–410.

NYDF. 2017. New York Declaration on Forests: Declaration and action agenda, United Nations (www.undp.org/content/dam/undp/library/Environment%20and%20Energy/Forests/New%20York%20Declaration%20on%20Forests_DAA.pdf), accessed 22 April 2019.

UN Climate Summit. 2017. New York declaration on forests—Declaration and action agenda. United Nations.

United Nations Forum on Forests (UNFF). 2007. Report of the Seventh Session (24 February 2006 and 16 to 27 April 2007). Economic and Social Council, United Nations: New York, USA.

United Nations Environment Programme-Sustainable Building and Climate Initiative (UNEP-SBCI). 2009. Buildings and climate change: Summary for decision-makers. United Nations Environmental Programme, Sustainable Buildings and Climate Initiative, Paris. 1–62.

UNFCCC. 2002. The Marrakesh accords and the Marrakesh declaration. In report of the conference of the parties on its seventh session, Marrakesh, 29 Oct–10 Nov 2001.

UNREDD. 2015. UN-REDD programme strategic framework 2016–2020. United States, Washington, US.

US Geological Survey. 2018. World reserves of iron ore as of 2018, by country (in million metric tons). Statista.

UNSPF. 2017. United Nations Strategic Plan for Forests 2017-2030: Brief Notes. United Nations Forum on Forest, UN, New York (https://www.un.org/esa/forests/wp-content/uploads/2017/09/UNSPF-Briefing_Note.pdf).

van der Lugt, P., Vogtländer, Y., van Der Vegte, J. and Brezet, J. 2012. Life cycle assessment and carbon sequestration: The environmental impact of industrial bamboo products. Proceedings of the 9th World Bamboo Congress, Belgium, 10–15 April. Gielis J. and Potters G. (ed.). 73–85.

Vogtländer, J.G., van der Velden, N.M. and van der Lugt, P. 2014. Carbon sequestration in LCA—a proposal for a new approach based on the global carbon cycle: Cases on wood and on bamboo. International Journal of Life Cycle Assessment 19(1): 13–23.

Wanninger, F. 2015. Post-tensioned timber frame structures. IBK Bericht Nr. 364. ETH Zurich Research Collection, Switzerland.

Waugh Thistleton Architects. 2018. 100 Projects UK CLT. The Softwood Lumber Board & Forestry Innovation Investment.

Wilson, J.B. and Dancer, E.R. 2005. Gate-to-gate life-cycle inventory of I-joist production. Wood and Fibre Science 37 Corrim Special Issue. 85–98.

Ximenes, F., Gardner, W. and Cowie, A. 2008. The decomposition of wood products in landfills in Sydney, Australia. Waste Management 28(11): 2344–54.

Ximenes, F.A. and Grant, T. 2012. Quantifying the greenhouse benefits of the use of wood products in two popular house designs in Sydney, Australia. International Journal of Life Cycle Assessment 18: 891–908.

Ximenes, F., Brooks, P., Wilson, C. and Giles, D. 2013. Carbon Storage in Engineered Wood Products in Landfills (PRB 180-0910). Forest & Wood Products Australia.

Yan, Y. 2018. Integrate carbon dynamic models in analysing carbon sequestration impact of forest biomass harvest. Science of Total Environment 615: 581–587.

Yeoh, D., Fragiacomo, M.M., de Franceschi, M. and Boon, K.J. 2011. State of the art on timber-concrete composite structures: Literature review. Journal of Structural Engineering 137(10): 1085–1095.

CHAPTER 4

Reducing Carbon Footprint of Products (CFP) in the Value Chain

Annik Magerholm Fet* and Arron Wilde Tippett

1. A Systemic Description of a Product and its Value Chain

"System" is derived from the Greek word "*systema*" meaning an organised whole. A system constitutes a complex combination of subsystems and interacting system elements. A product, therefore, can be described as a system. A system must have a purpose: It must be functional and able to respond to some identified needs and requirements over its entire life cycle. The system boundaries describe the interface between the system under study and the environment, the system under study and other interrelated systems. Material and energy crossing the system boundaries are defined as inputs to, or outputs from, the system. Man-made systems consume energy and operate within the natural system, referred to as the environment. The effect of man-made systems on natural systems are a subject for study as these effects are often undesirable. To understand these effects, and their impacts, the system's interactions and interchanges with the environment must be analysed at each system level for all the life cycle phases of the system. This approach has not been taken into consideration sufficiently in earlier analyses of systems (Kellogg, 1981).

This chapter presents a model of how to analyse and communicate the Carbon Footprint of a Product (CFP). Systems Engineering (SE) provides a complete overview that helps to understand the interactions between the systems, subsystems and system elements and the environment throughout the entire system life cycle, which gives a holistic perspective (Hitchings, 2008). The process of bringing a product into being starts with an understanding of the stakeholder needs and description of the requirements the product should fulfil. The preparatory stages of bringing this product into being consist of a conceptual design followed by a detailed design specifying the performance of the product that ensures the product meets the original needs and requirements. Most often, the life cycle of the product is defined as the stages beginning with raw material acquisition, continuing through processing, material manufacturing, product fabrication, use, operation, maintenance and concluding with a variety of product retirement (End-of-Life) options.

Decisions made during the early design phase can significantly impact the product's overall life cycle performance. Both the life cycle costs and the life cycle environmental impacts should be measured against the system's total performance or related to the overall function of the system. When comparing

Department for International Business, NTNU, Norway, Larsgårdsveien 2, 6009 Ålesund.
Email: arron.w.tippett@ntnu.no
* Corresponding author: annik.fet@ntnu.no

the environmental impact of systems, the results are related to a defined functional unit or declared unit of the system.

2. The LCA-Methodology and Classification of Emissions

There are several methods for determining the specific environmental characteristics of a given product. The most extensive method for studying environmental impacts throughout a product's life cycle is the Life Cycle Assessment (LCA) methodology. LCA was first developed in Switzerland in the sixties (Fink, 1997) and then further developed by the Society of Environmental Toxicology and Chemistry (SETAC; Consoli et al., 1993). These practices are standardised in the ISO 14040-documents (ISO, 2006b; ISO, 2006c; ISO, 2012a; ISO, 2012b). The methodology includes the following steps: Goal and scope definition, inventory analysis, impact assessment, and interpretation, as illustrated in Figure 1 (ISO, 2006b). The LCA methodology is an iterative process, whereby the LCA practitioner is able to move repetitively between the four steps to improve the study, if required.

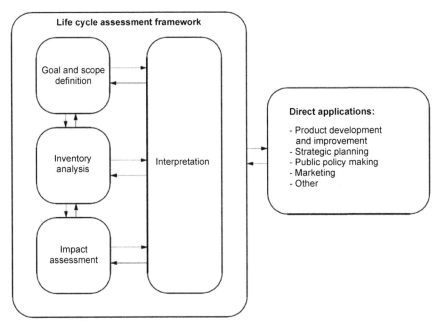

Figure 1. Phases of an LCA (ISO 14040:2006, ISO, 2006b).

2.1 Goal and scope definition

In this first stage of an LCA, the application, depth and subject of the study are defined. This includes a determination and description of the functional unit and the specification of the system boundaries. Key activities for the LCA practitioner are familiarizing themselves with the system under study, and working directly with the client to agree on the goal and scope of the study.

2.2 Inventory analysis

This stage requires that all emissions and raw material consumption during each process, throughout the entire life cycle, are identified and recorded. The result is a long list of emissions and raw materials, known as the inventory table. The key activities at this stage include preparing process flow-charts, collecting data and processing the data.

2.3 Impact assessment

This is a technical, quantitative and/or qualitative process to analyse and assess the effects of the environmental burdens identified in the inventory analysis.

Figure 2 illustrates the process of impact assessment through a model starting with a list of environmental aspects linked to air, water, soil, waste and noise. Among the aspects linked to air, different emissions are listed, among these CO_2 and VOC, which are referred to as greenhouse gases (GHG) and have an impact on climate change. The impact can further be calculated using the ISO 14064 standards and communicated by the ISO 14067 as CFP (ISO, 2018c). Key tasks at this stage, as described above, are classification and characterisation of all the aspects described in the inventory.

There are many, but no generally accepted, methodologies for consistently and accurately associating inventory data with specific potential environmental impacts (Finnveden et al., 2009). It is difficult to find weighting factors that can be commonly adopted all over the world. The methodological and scientific framework for impact assessment is still being developed, and although not fully developed and validated, impact assessment generally includes classification, characterisation and valuation. The main purpose of the classification is to briefly describe which potential environmental effects the inputs and outputs may cause. During classification, the different aspects from the inventory table are noted under the relevant impact categories. For example, all emissions contributing to global warming are noted under the heading Global Warming Potential (GWP). The characterisation is a quantitative step in which the relative contributions of each input and output to its assigned impact categories are assessed, and the contributions are aggregated within the impact categories.

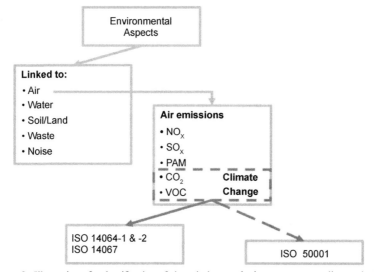

Figure 2. Illustration of a classification of air emissions to the impact category climate change.

2.4 Interpretation

In line with the defined goal and scope, interpretation is the phase of an LCA in which a synthesis is drawn from the findings of either the inventory analysis or the impact assessment, or both. The findings of this interpretation may form conclusions and recommendations to decision-makers. The interpretation phase may involve the iterative process of reviewing and revising the scope of the LCA. In this chapter's context it is about finding potential for the reduction of GHG-emissions along the value chain of systems.

3. The Standards for PCR, EPD and CFP

The emission of greenhouse gasses contributes to GWP (Sproul et al., 2019). The purpose in this chapter is to demonstrate a methodology to determine the carbon footprint in particular subsystems along the entire life cycle value chain of a product, with the intention to reduce the Carbon Footprint of the Product (CFP). However, communication of CFP should be done according to a set of rules, such as those described in international standards. There are different types of environmental information on products. According to the ISO 14020-series, these are Type I-programmes (ISO 14024:2018, ISO, 2018a) (multiple criteria-based, third party programmes awarding labels claiming high overall environmental performance), Type II-programmes (ISO 14021:2016, ISO, 2016) (self-declared environmental claims, requires that life cycle considerations be taken into account) and Type III-programmes (ISO 14025:2006, ISO, 2006a). Type III requirements are used both to conduct an LCA of the product in accordance with the ISO 14040-standards (ISO, 2006b; ISO, 2006c; ISO, 2012a; ISO, 2012b) and to get an approval of the LCA, and a third-party verification of the declaration. The information should enable comparisons between products fulfilling the same function.

To this end, an environmental product declaration (EPD) is developed, based on product category rules (PCRs), in accordance with the requirements in the ISO/TS 14027:2017 (ISO, 2017b). PCRs are generally developed by industry in collaboration with a national or international EPD program operators, such as the Norwegian EPD Foundation (EPD Norge, 2019). The PCR sets out the information that must be included in EPDs for a specific product category, in addition to the general rules for EPDs (Fet and Skaar, 2006). According to ISO/TS 14027:2017 the PCR-document must describe: (1) the product category, (2) materials and substances to be declared (e.g., specification of materials and chemical substances that can affect human health and the environment during production, use and disposal of the product), in addition to (3) specification of goal and scope, inventory analysis, impact assessment category selection (for example, GWP as CO_2-equivalent) and interpretation rules. Instructions for the content and format of the Type III environmental declaration, or EPD, should also be stated in the PCR.

The LCA results may include information about raw material acquisition, energy use and efficiency, content of materials and chemical substances, pollutant emissions to air, soil and water, waste generation and the environmental impact associated with the product. For a CFP the impact from GHG-emissions should be stated. All EPDs in a product category shall follow the same format and include the same data as identified in the PCR provided by the program operator.

3.1 The CFP-standard

The CFP reflects the potential effect from the sum of GHG-emissions expressed as CO_2-equivalents, which are associated with the environmental aspect of the life cycle of a product affecting climate change. There are different ways of communicating CFP. Common to all of them is that GHG emissions should be quantified and converted into CF based on standardised methods, such as the ISO 14060 family of GHG standards and the ISO 14026-standard for Environmental labels and declarations—Principles, requirements and guidelines for communication of footprint information.

According to the ISO 14060-standards, the family of these standards provides clarity and consistency for quantifying, monitoring, reporting and validating or verifying GHG-emissions to support reduction of GHG-emission along the value chain of a product. In addition, the ISO 14040-standards for LCA, the ISO 14020-standards for product declaration and the ISO/TS 14071:2014 (ISO/TS, 2014) standard for conducting critical review of the results are a family of standards that should be considered in the documentation of CFP. An overview of the principles of each of the relevant standards, and diagrams showing how they are connected, is given in Annex 1 to this chapter.

4. Quantification of the CFP

As illustrated by the Table A1 and Figures A1 and A2 in the Annex, the GHG-statement should be aligned with the needs and requirements of the intended user of the statement. As for an EPD, the quantification and communication should be conducted according to the requirements in the PCR. The methodology for a CFP follows the same 4 steps as for an LCA.

4.1 CFP goals and scope

For a CFP, the overall goal is to calculate the potential contribution of a product to global warming expressed as CO_2-equivalents. In defining the scope of the CFP study, the system under study, the subsystems and the system elements, the system boundaries, its functional unit, data and data quality requirements should be described. The system boundary shall be the basis used to determine which unit processes contribute to the CFP study. As for EPDs, a product flow diagram, including system elements and an overview of the life cycle stages of the product as shown under the LC-Inventory (LCI), is helpful. For large products, such diagrams can be quite complex and they will vary according to the type of product.

In addition to a diagram, the full documentation of each part of the life cycle that is likely to contribute to GHG-emissions should be undertaken. This can include GHG-emissions during raw material extraction, which often spans several locations worldwide; the manufacture of the product, including transport between production sites and to the retailer and user; and the transportation and energy used during the end-of-life treatment. Different scenarios might be a part of the analyses.

Depending on the type of product, the use-phase can be quite unpredictable. For products that consume energy, it is important that this is described in a consistent way. Similarly, an overview of end-of-life processes should be included, e.g., collection, packaging and transport, preparation for recycling and reuse, dismantling of components, shredding and sorting material for recycling, energy recovery or incineration.

4.2 CFP inventory analysis

During the LCI, the collected data (measured, calculated or estimated), shall be used to quantify the inputs and outputs of unit processes for each system element. The CFP study shall include an identification of actual processes shared with other product systems and deal with them in accordance with relevant allocation procedures.

4.3 CFP impact assessment

Figure 2 illustrates a model for impact assessment. According to the ISO 14067-standard (ISO, 2018c), the impact of each GHG-emissions shall be calculated by multiplying the mass of GHG released or removed by the 100-year Global Warming Potential (GWP 100) given by the International Panel for Climate Change (IPCC) in units of kg CO_2-equivalents per kg emission (Mhyer et al., 2013). The CFP is the sum of the calculated impacts.

According to the IPCC, the 100-year global warming potential (GWP) is used to represent shorter-term impacts of climate change, reflecting the rate of warming. 100-year global temperature potential (GTP 100) is used as an indicator for the longer-term impacts of climate change, reflecting the long-term temperature rise. There is no scientific basis for choosing a 100-year time horizon compared to other time horizons, but the time horizon is a value judgement of an international convention that weighs the effects that are likely to occur over different time horizons (Mhyer et al., 2013). Furthermore, ISO 14067:2018 (ISO, 2018c) recommends how to calculate and assess, e.g., the impact of removals of CO_2 into biomass or emissions of biogenic CO_2. Similarly, the standard recommends that the most recent IPCC report be used to set the characterisation factors for fossil and biogenic methane.

4.4 CFP interpretation

The final stage, the interpretation of a CFP study, should include an identification of the significant issues regarding the life cycle stages, unit processes, subsystems and flows based on the results of the quantification of the CFP. According to ISO 14026 (ISO, 2017a), the results should be presented in a CFP study report and be further used in footprint communications. The CFP study report shall include information about the system boundary, including the type of inputs and outputs of the system as elementary flows, considering their importance for the conclusions of the CFP study. The results shall be documented in mass of CO_2-equivalents per functional unit as specified in the PCR. The results of the quantification of the CFP shall be further interpreted according to the defined goal and scope of the CFP study.

5. Case Study—Galvanised Steel Staircase

This case study considers the carbon foot printing of a Steel staircase, see Figure 3, manufactured in Norway, using LCA methodology. The material flow diagram is illustrated by Figure 4, and the results can be found in the EPD for Lonbakken Mek. Verksted (Fet, 2014).

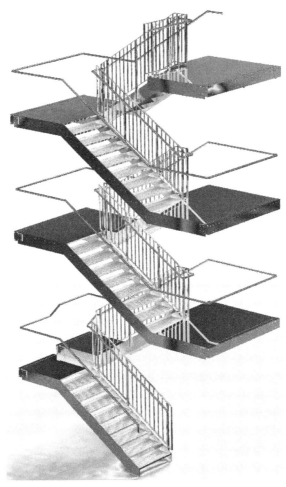

Figure 3. Steel staircase (Fet, 2014).

5.1 Technical information

The total weight, including fastening brackets and bolts, is approximately 4375 kg. The galvanised steel staircase requires indoor assemblage. The product should last for 100 years, and the results from the

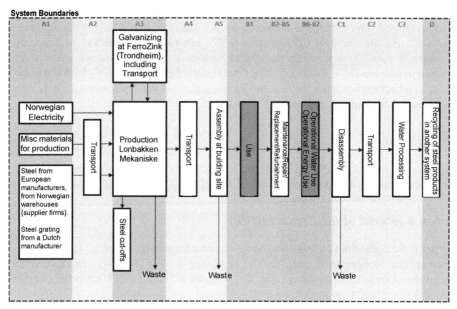

Figure 4. Flow diagram for a steel staircase (Fet, 2014).

analysis are presented per 1 kg steel (declared unit), while the functional unit is defined as per kg building steel structure with an expected service life of 100 years. All calculated environmental impacts (GHG-emissions) from the products life cycle are adjusted to the declared or functional unit.

5.2 System boundaries

The life cycle phases in Figure 4 are described by A1–A3 for the production phase, A4–A5 for the transport and on-site assembly, B1–B7 for the use-phase, C1–C3 the end-of life treatment, and finally, D represents a possible system in which the steel output from the end-of-life is recycled, and from which an environmental credit is attained and given to this system. System boundaries are found in the flow diagram. Waste flows, especially cut-offs from the steel manufacturing in A3, are treated within the module they occur.

5.3 Cut-off rules and allocation

All major raw materials and all the essential energy is included. Raw materials and energy flows that are included with very small amounts (< 1%) are to be omitted, however, this cut-off rule does not apply for hazardous materials and substances. The allocation is made in accordance with the provisions of EN 15804 (CEN, 2012). Incoming energy, water and waste production in-house are allocated equally among all products through mass allocation. Since the recycled materials used in the product system have no inherent difference in chemical properties from the virgin material they are replacing, they can be treated as per the ISO 14044:2006 guidelines for the closed-loop allocation in an open system. Therefore, all environmental impacts, energy and resource use associated with the production of the recycled materials are allocated to the original product that generated the recycled material.

5.4 Data assumptions

The scenarios in the different life cycle stages are described by the following information. Truck transport is modelled using a generic 32 t truck dataset for European conditions, as truck sizes will vary depending on the delivery. A conservative approach has therefore been taken, in order to ensure that emissions are not underreported. The distance is measured by the manufacturer, which is the distance from the production site to the site where the steel staircase will be assembled. Transport in A4 is calculated as the actual estimated distance. To account for the impacts generated in the construction phase, electricity has been allocated to the phase by a fraction of 1/3 of the manufacturing phase (A3). With a standard coating thickness in an inland environment, the scenario for the use stages B2–B5 does not include maintenance, repair, replacement or refurbishment over a 100-year horizon.

5.5 Electricity mix

Electricity used in the manufacturing processes was accounted for using an electricity mix process specific to Norway, giving a GHG-emission factor 0,0172 kg CO_2-equivalent per MJ used energy.

5.6 Carbon Footprint of the Product (CFP)

As seen from Table 2, the main contributions to GWP come from the production stage A1–A3. A smaller amount comes from transportation, given for this particular scenario. There is a negative value at the D (recovery) phase. This is an environmental credit, whereby we observe CO_2-equivalent uptake into the system through the recycling of steel.

6. CFP as an Instrument for Reducing Carbon Footprint in the Value Chain of a System

This chapter started with the concept of "bringing a system into being", and using system as a synonym for a product. Products in this case are interpreted as physical man-made systems; however, it can also be interpreted as a service. Through the principles of SE, a specification of the product performance along the entire value chain of the product should be prepared based upon a set of defined needs and requirements.

The life cycle value chain of a system might be quite complex, consisting of a set of subsystems, which again represent a combination of many processes and material flows as demonstrated by the classes A1-5, B1-7, C1-4 and D, as shown in Table 1 of the case study. The CFP-study should report the results of the analyses of each subsystem and, by summing up the report, give information about where in the life cycle of the product the GHG-emissions occur, which module or subsystem gives the highest impact to the CFP, and where there might be potential for GHG-reductions.

Low hanging fruit can be realised by identifying alternative sub-systems, or modules, which have reduced GWP whilst still meeting the needs and requirements in terms of the product's design. Incentives to investigate where in the system GHG-reduction can be realised become stronger in the face of carbon taxing and stricter requirements in environmental documentation. On the other hand, the recycling phase of the system can help to reduce GHG-emissions across the whole life cycle of the product.

If the requirements described in the early design phase also include requirements to a CFP below a certain level, then a CFP-report can be used as a decision supporting tool for setting the priorities for which module of the product and where in the product life cycle the potential of reducing the CFP is highest. Results from a CFP-report can be fed back to the design phase where the choices of materials in the actual product take place. Similar for the design of the entire value chain, a CFP-report will contain an overview of the impact caused during raw-material extraction, during pre-production, transportation, etc., and, thereby, give input to the process of designing the supply chain of the product. The ring or feedback structure is the most typical of the patterns in the analytical methods in SE. In the design of a product, the availability of suppliers, the costs of materials, etc., will determine the design of

Table 1.   Illustration of the activities in each subsystem during the life cycle stages.

System boundaries (X = included, MND = module not declared, MNR = module not relevant)

Product stage				Construction installation stage	Use stage							End of life stage				Beyond the system boundaries
Raw materials	Transport	Manufacturing	Transport	Construction installation stage	Use	Maintenance	Repair	Replacement	Refurbishment	Operational energy use	Operational water use	De-construction demolition	Transport	waste processing	Disposal	Reuse-Recovery-Recycling-potential
A1	A2	A3	A4	A5	B1	B2	B3	B4	B5	B6	B7	C1	C2	C3	C4	D
X	X	X	X	X	MNR	X	X	X	X	MNR	MNR	X	X	X	MNR	X

Table 2.   Calculated contribution to GWP from each of the subsystems per 1 kg of steel staircase produced.

Parameter	Unit	A1-A3	A4	A5	B1	B2	B3	B4	B5	B6	B7	C1	C2	C3	C4	D
GWP	kg CO_2-eqv	2,68E+00	2,71E-02	5,07E-03	-	0	0	0	0	-	-	2,54E-03	5,34E-03	3,87E-02	-	-1,30E+00

the value chain of the products. However, the market situation can often change this pattern, since the choice of suppliers may vary according to new market availability. Similarly, the energy mix, the access to infrastructure and to transport solutions will change over time. To get an optimal solution regarding CFP, all changes in the system must be included in the analyses for GHG-emissions or reductions, and information should be fed back to the decision maker. Such structures are typical in trade-off analyses where iteration is done until an optimal solution or design according to specified needs and requirements is accepted.

The CFP is only one of a variety of environmental impacts that can arise from a product's life cycle. CFP should not be the sole component of a decision-making process for product improvement. However, in this chapter the focus is on impacts from GHGs, and not the full set of impact categories as seen in an EPD.

7. Concluding Remarks

The life cycle assessment and carbon foot-printing methodologies can be applied to undertake comparative studies. These are important for decision making as they can relate different product options directly to their environmental impact. Comparative studies can be used in order to assess the viability of production in countries with alternative energy mixes; a country with a high proportion of fossil-fuel power plants will contribute to a higher CFP. They can also suggest new options for product design by comparing the choice of materials in sub-systems or modules in terms of their GHG-emissions. It is essential to include a full product life cycle in comparative CFP studies unless the function of the product is included in a partial CFP and all omitted processes from the system are identical for the compared products. Furthermore, when CFP-PCRs are adopted in comparative studies, the identical CFP-PCR shall be utilised for each of the products assessed. Finally, comparative CFP studies should use the identical functional units, system boundaries, descriptions of the data, criteria for the inclusion or exclusion of any inputs or outputs from the system and assumptions regarding the use or end-of-life stages.

This chapter has described the LCA methodology and the standards required to communicate the LCA result as an Environmental Product Declaration (EPD). It has further described the standards required to convert this information into the CFP, providing an overview of options for GHG-emissions reductions. The challenge is to feed this overview back to the product designer for use in the planning and design of low carbon product systems, subsystems, system elements and value chains. The decision-making process can become complex for larger systems. According to Systems Engineering, the process of bringing a system into being is by a stepwise approach starting with a specification of the needs and associated requirements to the life cycle performance of the systems, followed by trade-off analyses of alternative solutions for reduction of carbon emissions. Finally, the SE process considers ways to evaluate that in meeting the initial requirements, the product also incorporates the options for reducing carbon footprints in the value chain of the product.

Annex: Overview of ISO-standards for conducting LCA and transforming the results into Environmental Product Declarations (EPDs) and Carbon Footprint of Products (CFP)

Table A1. Principles covered by the ISO 14060-standards .

ISO 14064-1:2019, *Greenhouse gases—Part 1: Specification with guidance at the organisation level for quantification and reporting of GHG emissions and removals*	Principles and requirements for designing, developing, managing and reporting organisation-level GHG inventories.
ISO 14064-2:2019, *Greenhouse gases—Part 2: Specification with guidance at the project level for quantification, monitoring and reporting of GHG emission reductions or removal enhancements*	Principles and requirements for determining baselines, and for the monitoring, quantifying and reporting of project emissions.
ISO 14064-3:2019, *Greenhouse gases—Part 3: Specification with guidance for the validation and verification of GHG assertions*	Requirements for verifying GHG statements related to GHG inventories, GHG projects, and carbon footprints of products.
ISO 14067:2018, *Greenhouse Gases—Carbon footprint of products (CFP)—Requirements and guidelines for quantification*	Principles, requirements and guidelines for the quantification of the carbon footprint of products. The aim of this document is to quantify GHG emissions associated with the life cycle stages of a product.
ISO/TR 14069:2013, *Greenhouse gases—Quantification and reporting of greenhouse gas emissions for organisations—Guidance for the application of ISO 14064-1*	Assists users in the application of ISO 14064-1, providing guidelines and examples for improving transparency in the quantification of emissions and their reporting.
ISO 14026:2018, *Environmental labels and declarations—Principles, requirements and guidelines for communication of footprint information*	Principles, requirements and guidelines for: footprint communications for products addressing areas of concern relating to the environment; footprint communication programmes, as well as requirements for verification procedures.
ISO/TS 14027:2017, *Environmental labels and declarations—Development of product category rules*	Principles, requirements and guidelines for developing, reviewing, registering and updating PCR within a Type III environmental declaration or footprint communication programme based on life cycle assessment (LCA) according to ISO/TS 14067.
ISO 14044:2006, *Environmental management—Life cycle assessment—Requirements and guidelines*	Covers two types of studies: life cycle assessment studies (LCA studies) and life cycle inventory studies (LCI studies). LCI studies are similar to LCA studies but exclude the LCIA phase.
ISO/TS 14071:2014, *Environmental management—Life cycle assessment—Critical review processes and reviewer competencies: Additional requirements and guidelines to ISO 14044:2006*	Provides additional specifications to ISO 14040:2006 and ISO 14044:2006. It provides requirements and guidelines for conducting a critical review of any type of LCA study and the competencies required for the review.

LCA requirements from ISO 14044
Requirements for establishing PCR from ISO/TS 14027
ISO 14067
Requirements for footprint communication from ISO 14026
Requirements for communication and critical review from ISO 14044
Additional requirements for critical review from ISO/TS 14071
Footprint communication
Communication
CFP study report
No communication
Internal use of the report, e.g., decision support

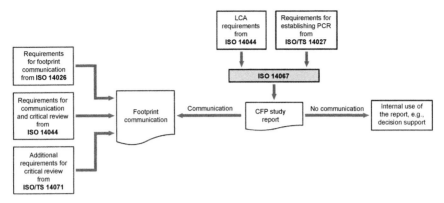

Figure A1. Relationship between ISO 14067 and standards beyond the GHG management family of standards (ISO 14067:2018, ISO, 2018c).

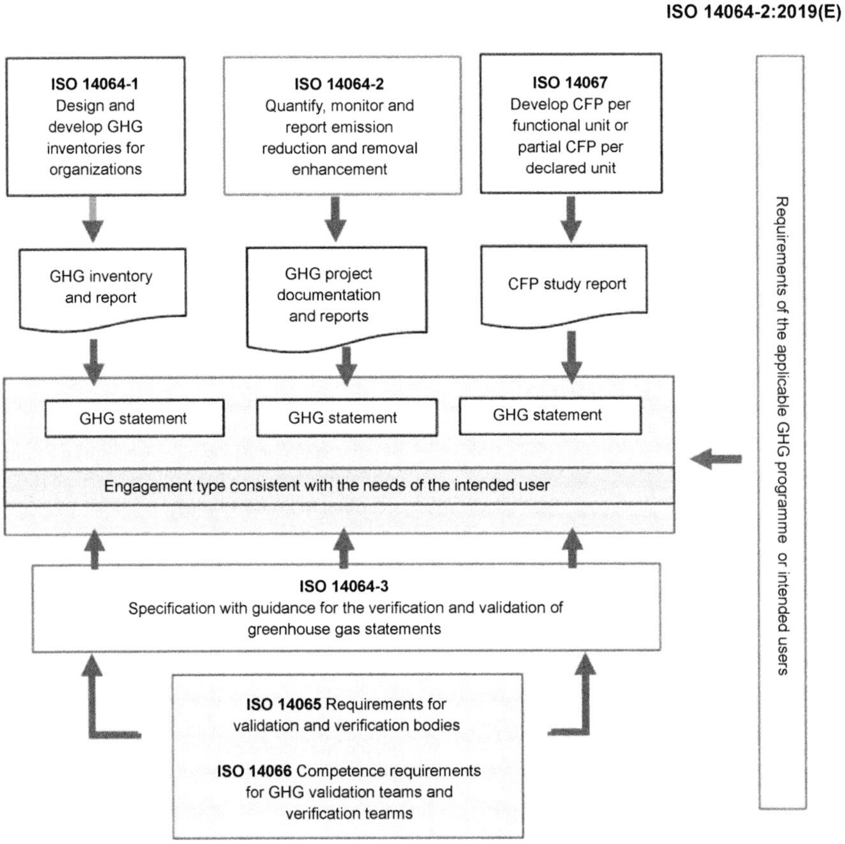

ISO 14064-2:2019(E)
ISO 14064-1 Design and develop GHG inventories for organizations
ISO 14064-2 Quantify, monitor and report emission reduction and removal enhancement
ISO 14067 Develop CFP per functional unit or partial CFP per declared unit
Requirements of the applicable GHG programme or intended users
GHG inventory and report
GHG project documentation and reports
CFP study report
GHG statement
GHG statement
GHG statement
Engagement type consistent with the needs of the intended user
ISO 14064-3 Specification with guidance for the verification and validation of greenhouse gas statements
ISO 14065 Requirements for validation and verification bodies
ISO 14066 Competence requirements for GHG validation teams and verification tearms

Figure A2. Relationship among the ISO 14060 family of GHG standards (ISO 14064-2:2019, ISO, 2019a).

References

CEN. 2012. EN 15804:2012, Sustainability of construction works—Environmental product declaration—Core rules for the product category of construction products. CEN-CENELEC Management Centre, Brussels, Belgium.

Consoli, F., Allen, D., Boustead, I., Fava, J., Franklin, W., Jensen, A.A., de Oude, N., Parrish, R., Perriman, R., Postlethwaite, D., Quay, B., Siéguin, J. and Vigon, B. 1993. Guidelines for Life Cycle Assessment. A Code of Practice, SETAC Press, Pensacola, FL.

EPD Norge. 2019. The Norwegian EPD Foundation Official Website. Available at: https://www.epd-norge.no/?lang=no_NO [accessed: 9th August 2019].

Fet, A.M. and Skaar, C. 2006. Eco-labeling, product category rules and certification procedures based on ISO 14025 requirements (6 pp). The International Journal of Life Cycle Assessment 11(1): 49–54.

Fet, A.M. 2014. Environmental Product Declaration: Lonbakken Mek. Verksted AS Steel Staircases. The Norwegian EPD Foundation. Oslo, Norway.

Fink, P. 1997. The roots of LCA in Switzerland—Continuous learning by doing. International Journal of Life Cycle Assessment 2(3): 131–134.

Finnveden, G., Hauschild, M.Z., Ekvall, T., Guinee, J., Heijungs, R., Hellweg, S., Koehler, A., Pennington, D. and Suh, S. 2009. Recent developments in life cycle assessment. Journal of Environmental Management 91(1): 1–21.

Hitchins, D.K. 2008. Systems Engineering: A 21st Century Systems Methodology. Wiley Blackwell, New York, NY, USA.

ISO. 2000. ISO 14020:2000 Environmental labels and declarations—General principles.

ISO. 2002. ISO/TR 14062:2002, Environmental management—Integrating environmental aspects into product design and development. ISO International Organization for Standardization, Geneva, Switzerland.

ISO. 2006a. ISO 14025:2006, Environmental labels and declarations—Type III environmental declarations — Principles and procedures. ISO International Organization for Standardization, Geneva, Switzerland.

ISO. 2006b. ISO 14040:2006, Environmental management—Life cycle assessment—Principles and framework. ISO International Organization for Standardization, Geneva, Switzerland.

ISO. 2006c. ISO 14044:2006, Environmental Management—Life cycle assessment—Requirements and guidelines. ISO International Organization for Standardization, Geneva, Switzerland.

ISO. 2010. ISO 26000:2010, Guidance on social responsibility. ISO International Organization for Standardization, Geneva, Switzerland.

ISO. 2011. ISO 14066:2011, Greenhouse gases—Competence requirements for greenhouse gas validation teams and verification teams. ISO International Organization for Standardization, Geneva, Switzerland.

ISO. 2012a. ISO/TR 14047:2012, Environmental management—Life cycle assessment—Illustrative examples on how to apply ISO 14044 to impact assessment situations. ISO International Organization for Standardization, Geneva, Switzerland.

ISO. 2012b. ISO/TR 14049:2012, Environmental management—Life cycle assessment—Illustrative examples on how to apply ISO 14044 to goal and scope definition and inventory analysis. ISO International Organization for Standardization, Geneva, Switzerland.

ISO. 2013a. ISO 14065:2013, Greenhouse gases—Requirements for greenhouse gas validation and verification bodies for use in accreditation or other forms of recognition. ISO International Organization for Standardization, Geneva, Switzerland.

ISO. 2013b. ISO/TR 14069:2013, Greenhouse gases—Quantification and reporting of greenhouse gas emissions for organizations—Guidance for the application of ISO 14064-1. ISO International Organization for Standardization, Geneva, Switzerland.

ISO. 2014. ISO/TR 14071:2014, Environmental management—Life cycle assessment—Critical review processes and reviewer competencies. ISO International Organization for Standardization, Geneva, Switzerland.

ISO. 2015. ISO 14001:2015, Environmental management systems—Requirements with guidance for use. ISO International Organization for Standardization, Geneva, Switzerland.

ISO. 2016. ISO 14021:2016, Environmental labels and declarations—Self-declared environmental claims (Type II environmental labelling). ISO International Organization for Standardization, Geneva, Switzerland.

ISO 2017a. ISO 14026:2017, Environmental labels and declarations—Principles, requirements and guidelines for communication of footprint information. ISO International Organization for Standardization, Geneva, Switzerland.

ISO. 2017b. ISO/TS 14027:2017, Environmental labels and declarations—Development of product category rules. ISO International Organization for Standardization, Geneva, Switzerland.

ISO. 2018a. ISO 14024:2018, Environmental labels and declarations—Type I environmental labelling—Principles and procedures. ISO International Organization for Standardization, Geneva, Switzerland.

ISO. 2018b. ISO 14064-1:2018, Greenhouse gases—Part 1: Specification with guidance at the organization level for quantification and reporting of greenhouse gas emissions and removals. ISO International Organization for Standardization, Geneva, Switzerland.

ISO. 2018c. ISO 14067:2018, Greenhouse gases—Carbon footprint of products—Requirements and guidelines for quantification. ISO International Organization for Standardization, Geneva, Switzerland.

ISO. 2019a. ISO 14064-2:2019, Greenhouse gases—Part 2: Specification with guidance at the project level for quantification, monitoring and reporting of greenhouse gas emission reductions or removal enhancements. ISO International Organization for Standardization, Geneva, Switzerland.

ISO. 2019b. 14064-3:2019, Greenhouse gases—Part 3: Specification with guidance for the verification and validation of greenhouse gas statements. ISO International Organization for Standardization, Geneva, Switzerland.

Kellogg, W.W. 1981. Climate Change and Society: Consequences of Increasing Atmospheric Carbon Dioxide. Routledge, New York, NY, USA.

Mhyer, G., Shindell, D., Bréon, F.-M., Collins, W., Fuglestvedt, J., Huang, J., Koch, D., Lamarque, J.-F., Lee, D., Mendoza, B., Nakajima, T., Robock, A., Stephens, G., Takemura, T. and Zhang, H. 2013. Anthropogenic and natural radiative forcing. *In*: Stocker, T.F., Qin, D., Plattner, G.-K., Tignor, M., Allen, S.K., Boschung, J., Nauels, A., Xia, Y., Bex, V. and Midgley, P.M. (eds.). Climate Change 2013: The Physical Science Basis. Contribution of Working Group I to the Fifth Assessment Report of the Intergovernmental Panel on Climate Change, Cambridge University Press, Cambridge, United Kingdom and New York, NY, USA.

Sproul, E., Barlow, J. and Quinn, J. 2019. Time value of greenhouse gas emissions in life cycle assessment and techno-economic analysis. Environmental Science & Technology 53(19): 6073–6080.

CHAPTER 5

Significance of Greenhouse Gas Measurement for Carbon Management Technologies

James R Whetstone

1. Introduction

Carbon management systems aim to reduce atmospheric warming and climate effects by controlling warming agent flows to the atmosphere. Atmospheric warming is primarily driven by its greenhouse gas concentrations. Therefore, greenhouse gases are a primary focus of carbon management efforts. The main greenhouse gases (GHGs) of interest are carbon dioxide (CO_2), methane (CH_4), and nitrous oxide (N_2O), the halogenated hydrocarbon gases not covered by the Montreal Protocols (often termed the F-gases—fluorinated hydrocarbons), nitrogen tri-fluoride and sulfur hexa-fluoride. Reduction strategies are informed by estimation and measurement of greenhouse gas quantities emitted from and taken up by a wide range of processes and economic activities occurring at Earth's surface. Reliable quantitative information is critical to assessing the performance of reduction/mitigation efforts. Greenhouse gas inventory reports are widely-accepted sources used as mitigation policy performance metrics to assess efforts often focused on energy production and usage or efficiency and on land use. The UN Framework Convention on Climate Change (UNFCCC) and the Task Force on National Inventories of the Intergovernmental Panel on Climate Change (IPCC, TFI) have established requirements and guidelines to provide a uniform and robust inventory reporting framework for quantifying greenhouse gas emissions (IPCC-TFI, 2006).

Carbon management strategies aim to reduce atmospheric greenhouse gas emissions both locally and globally. Mitigation actions are primarily aimed at reductions in energy usage and are most often implemented at regional and local scales (Gurney, 2011). Mitigation effort implementation will benefit from information at the scales over which carbon managers have purview and exercise control. Evaluation of the effectiveness of these efforts will benefit from more reliable and accurate information on both emissions and uptake across global to local geospatial scales. Independent emissions measurements are useful for demonstrating progress toward mitigation goals; they support fairness in trading, permitting, and regulation, and utilize and guide financially efficient decisions about potential reduction pathways. Additionally, improving the reliability of emissions and uptake information advances the reporting accuracy both nationally and sub-nationally. This chapter will survey measurement systems and

Special Programs Office, National Institute of Standards and Technology, 100 Bureau Drive, Gaithersburg, Maryland, USA.
Email: james.whetstone@nist.gov

approaches underpinning inventory data compilation and reporting, physical and chemical processes upon which measurement technologies rely, and their application from global to local scales. Measurement and estimation performance needs at sub-national, national, and international levels will be discussed in order to give context in carbon management quantification requirements.

2. The Atmosphere, Quantities, Measurements and Standards

Measurement systems use approaches and configurations reflecting their application, available scientific knowledge, and technological capabilities, and to some degree, their cost of realization and implementation. Emission and uptake processes result in greenhouse gas exchange flows, or fluxes, between Earth's surface and its atmosphere, from sources like power plants and vehicles and from uptake sinks like vegetation on the land and marine life and seawater in the ocean. Measurements to quantify exchange fluxes are heavily dependent upon their nature, i.e., the physical characteristics and conditions under which the processes function and within which measurement must be accomplished. For some systems, direct measurements of emission mass flows can be realized. In others, estimation models are needed in order to infer emissions based upon process characteristics and reference measurements or data. Here, quantification will be discussed in two methodological classes, termed "top-down" and "bottom-up" methods.

Bottom-up approaches quantify exchange flows with the atmosphere based upon emission process characteristics and activity parameters that are often tabulated by economic sector and found in publicly available datasets. This type of information is the basis for national emissions reports to the UNFCCC and is discussed later on in this chapter. Examples include refrigerant (fluorinated gas) leakage from air conditioning systems and from their production processes, carbon dioxide (CO_2) from fossil fuel combustion systems used for applications ranging from electricity generation to heating buildings, and methane emissions from leakage in natural gas production, transportation, and distribution to users or enteric fermentation from cattle, to name a few. A variety of methods are used, ranging from emission-activity factor models and data to direct measurement.

Top-down approaches use quantification means based on observation of atmospheric properties and dynamics coupled with greenhouse gas concentration observations expressed as dry atmosphere concentration values. The remainder of this chapter will summarize some of the more widely-used methods, including brief process descriptions illustrating quantification techniques and discussion of quantification performance needs to inform carbon management efforts.

2.1 *Quantities and units*

In this chapter, the term carbon measurement relates to the determination of greenhouse gas amounts moving between Earth's surface and atmosphere. Quantification approaches are governed by the physical nature of emission sources and sinks and determine which of two quantification methods may be applicable. Here, the term "flow" will be applied to the relatively restricted case of gases moving within process structures and the rate of gas movement in them, e.g., flow of exhaust gases from fossil fuel combustion in power plants or vehicles. Combustion products, mainly CO_2 and water vapor, are confined within the process, such that measurement of exhaust gas constituency and bulk gas flowrate quantifies the amount of CO_2, or other minor constituent, emitted. Flowrate units, generally mass per unit time, e.g., kilograms/second or kilograms/minute, are often converted to emission factors in order to be consistent with IPCC emission reporting methodologies, as discussed below. Although these measurement strategies are employed in power plants, they are not currently in practice for vehicle emissions. Characterization of vehicle and engine types or classes benefit from direct constituent and flow measurement over the range of operating conditions to quantify emissions and emission factors (NVFEL, 2019) used for emissions estimation.

Some emission quantification methodologies, discussed later in this chapter, are based on quantification of flows of greenhouse gases to the atmosphere, based upon observation of greenhouse gas dry air concentration, measured as mole fraction, and combined with methods to estimate their atmospheric dynamics. A flux parameter involves the movement of items per unit area per unit time. For example, the mass of a greenhouse gas per meter squared-second is a commonly used unit, i.e., $kg/(m^2\text{-}sec)$ used to describe flows of atmospheric trace gases, e.g., greenhouse gases. Concentration is a measure of the amount of one substance mixed with others. Common usage in the atmospheric and greenhouse gas observing communities is the statement of greenhouse gas mole fraction values in units of $\mu mol/mol$ (10^{-6} moles/mole). Mole fraction is inextricably connected to the mole, the unit of the amount of substance of the International System of Units, the SI, and denoted as mol (BIPM, 2019). Just as one dozen consists of 12 things, one mole contains $6.022\ 140\ 76 \times 10^{23}$ elementary entities, i.e., Avogadro's Constant of elementary entities, such as molecules of CO_2.

Dry Air Constituent Mole Fractions	
Nitrogen	78.08%
Oxygen	20.95%
Argon	0.9%
Carbon Dioxide (CO_2)	405–411 μmol/mol
Methane (CH_4)	1.858–1.867 μmol/mol
Nitrous Oxide (N_2O)	≈ 330 nmol/mol
F Gases	~ 1 to 600 pmol/mol

Greenhouse gases often have atmospheric mole fraction values in this and smaller ranges relative to dry air. The convention of stating mole fraction values on a dry air basis is used to properly account for variations in atmospheric conditions. Perhaps as importantly, or more so, this is a means to exclude effects of the highly variable dilution of water vapor in the atmosphere, which varies from near zero to ≈ 4%. As shown in the box above (NASA, 2019; NOAA, 2019), greenhouse gas mole fractions are considerably smaller than daily water vapor variations and have been shown to remain unchanged as atmospheric water vapor content varies.

2.2 Radiative physics, atmospheric warming, and greenhouse gas measurements

Earth's atmosphere transmits and absorbs both solar radiation and the thermal radiation emitted by Earth's surface. The atmosphere is composed primarily of the two-atom molecules, nitrogen (N_2) and oxygen (O_2). The remaining ≈ 1% of the dry atmosphere is largely composed of chemically inert atomic argon, the long-lived, 3- and 4-atom greenhouse gas molecules, the large molecule halocarbons, and the chemically reactive species impacting human health. Incoming solar radiation, occurring across the ultraviolet, visible, and infrared (IR) spectral regions, is partially reflected back to space, absorbed by atmospheric components, or is absorbed by Earth's surface, warming the planet to levels that support life. Radiative models of the Earth predict that absorption of incoming solar radiation by Earth's surface alone warms it on average to ≈ –20 ºCelsius (≈ 0 ºF). Earth's warmed surface emits thermal radiation across the infrared region at wavelengths ranging from ≈ 1 to 20 micrometers (μm). This additional warming, supporting life on Earth, is due to retention of part of the thermal energy radiated by Earth's surface by some atmospheric gases, i.e., the greenhouse gases. The selective absorption of GHGs retains thermal energy in the atmosphere. The outsized capability of low concentration greenhouse gases to warm the atmosphere results from their capacity to absorb thermal radiation much more strongly than oxygen and nitrogen, as discussed later in this chapter, a capacity several orders of magnitude greater. The thermal energy absorbed by GHGs is transferred as kinetic energy, temperature, through molecular collision processes to the atmosphere's most abundant constituents, nitrogen and oxygen, thereby increasing atmospheric temperatures.

Measurement of greenhouse gas dry air mole fraction (moles of a greenhouse gas per mole of dry air), in or sampled from the atmosphere or from processes on Earth's surface, is a central observable parameter for quantifying exchange fluxes in top-down or bottom-up methods. A variety of measurement methods are used. Their selection for a particular application is dependent upon considerations of sensitivity, accuracy, application conditions, and cost, among others. Mole fraction measurement methods utilize several of the chemical or radiative properties of greenhouse gases. Application areas range from the standards laboratory environments to those used in field observation. The highest accuracy capabilities of laboratory environments are needed in order to assign mole fraction values to standard gas mixtures.

Field applications may involve direct measurements of samples drawn from the atmosphere or an emitting process stream. Atmospheric sampling occurs, continuously, daily, and weekly while sampling of power generation stacks occurs on a near real-time basis. These observations are often referenced to standard gases of known and certified composition as a means of ensuring consistency and accuracy in measurement results.

2.3 Global atmospheric greenhouse gas observations

The trace concentration of greenhouse gases in the atmosphere has been, and continues to be, a measurements and standards challenge when monitoring their behavior both globally and locally. Scientific interest, and concern for atmospheric GHG impact on Earth's atmosphere, began in the early and mid-19th century. Initial measurements (Foote, 1856) of the heating of CO_2 by solar radiation were followed by robust, higher-accuracy measurements and investigations of atmospheric CO_2 by Tyndall (Jackson, 2018) in the mid-19th century. These results and experiments regarding the effect of solar radiation on the Earth led to quantitative investigations on solar heating of the atmosphere by CO_2 and water vapor, the most abundant greenhouse gases in the atmosphere (Arrhenius, 1896). Continuous measurements of atmospheric CO_2 mole fractions were initiated in 1957 by C.D. Keeling of Scripps Institution of Oceanography (SIO) (Scripps-CO_2, 2019) at the Mauna Loa, Hawaii observatory of the National Weather Service and subsequently the National Oceanic and Atmospheric Administration (NOAA). Figure 1 shows this high-accuracy, atmospheric CO_2 data record based on instrumental and flask sampling measurements. These results illustrate that stable, highly accurate measurement standards, initially developed by Keeling, are the basis for underpinning this and other long-term data records of atmospheric greenhouse gas behavior.

The widely recognized Keeling Curve serves as one of the most highly regarded assessment tools for long-term monitoring of atmospheric CO_2 mole fraction. It is a proxy for assessing the atmosphere's warming capability globally. In addition to following the yearly increase in atmospheric CO_2 mole fraction, the sensitivity of the measurements, and the stability of the standards upon which these are based, also clearly show the annual variation (4 to 6 µmol/mol) in the northern hemisphere's atmosphere. This 1% to 1.5% annual variation in the global atmospheric signal, as measured at Mauna Loa, is a demanding measurement supported by standard gas mixtures with mole fraction uncertainties of 0.1 to 0.2 µmol/mol. Early in his research efforts, Keeling recognized the need for stable and accurate standard gas mixtures to support the Mauna Loa CO_2 observations, and subsequently those made at several locations roughly along a longitude reaching from Pt. Barrow in Alaska to near the South Pole to monitor

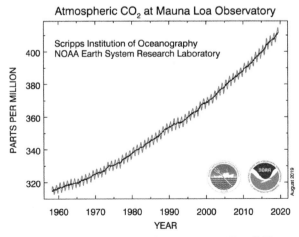

Figure 1. Global greenhouse gas mole fraction measurement data: Mauna Loa, Hawaii Observatory. https://www.esrl.noaa. gov/gmd/ccgg/trends/full.html.

global atmospheric behaviors. This type of observing capability continues and has been expanded under the international efforts of the World Meteorological Organization's Global Atmospheric Watch (GAW) program (GAW, 2019) aimed at monitoring atmospheric constituent behaviors.

As discussed below, the standards and measurements underpinning long-term, high-accuracy data records were pioneered by Keeling and continue to rely on efforts of the Scripps CO_2 Measurements Program (Scripps-CO_2, 2019) and that of the Global Monitoring Division (GMD) of NOAA's Earth Systems Research Laboratory (GMD, NOAA, 2019) (Zhao, Estimating uncertainty of the WMO Mole Fraction Scale for carbon dioxide in air, 2006). Sustaining a consistently accurate measurement standards capability has been, and continues to be, a central feature supporting reliability and confidence in monitoring of atmospheric greenhouse gas concentrations. GMD, designated as WMO's Central Calibration Laboratory (CCL) for these gases, provides real air mole fraction standards to the GHG monitoring and research community worldwide (see discussion of the term 'real air'). As WMO's CCL, GMD participates with others in the atmospheric monitoring community and in the international metrology community (BIPM/CCQM, 2017) to ensure consistency in the gas mixture standards upon which atmospheric monitoring is based. These standards comparison activities are largely conducted under the aegis of the WMO/GAW organization. Such joint community efforts ensure international recognition of these standard gas mixtures through linkage to the International System of Units, the SI (BIPM, 2019) for consistency and accuracy among affected international communities.

Annual variations in data records such as these reflect the role of CO_2 exchange processes between Earth's atmosphere, land and oceans. As the primary carbon source for photosynthetic chemical pathways in vegetation, CO_2 is both removed from the atmosphere through uptake by plant material (photosynthesis) and respired by plants, particularly when photosynthesis ceases during the night. Animals, soils and water-borne microbes also respire CO_2 as part of their normal biologic function. The globally aggregated CO_2 uptake processes on land and in the oceans account for approximately half the CO_2 currently emitted yearly from anthropogenic sources (Allen, 2014). Similar annual cycles in mole fraction are also observed for other atmospheric greenhouse gases, e.g., methane (Basso, 2016; Christensen, 2003), although the process mechanisms giving rise to these differ. The differing magnitude of variations reflects GHG emission source and sink complexity globally, regionally, and locally, as do estimation and measurement system challenges for quantifying GHG atmospheric exchange fluxes over this broad and complex range of spatiotemporal scales and processes.

2.4 Greenhouse gas mole fraction standards—History and methods

The need for a stable and accurate frame of reference upon which to base atmospheric observations became clear early in Keeling's atmospheric CO_2 concentration investigations. He established high-accuracy standard gas mixtures as a reference upon which to base observations taken at widely differing locations and over extended time periods (Scripps-CO_2, 2019). Keeling's development of these standards underpins the near continuous observations of atmospheric CO_2 content taken at Mauna Loa and subsequently expanded to other locations in the 1960s and 70s by SIO, NOAA's Earth Systems Research Laboratory and others to better understand atmospheric behavior globally. The Mauna Loa data record is the longest term, high-accuracy CO_2 record in existence. Recognized by the WMO early on, the Scripps CO_2 project became the forerunner of global atmospheric greenhouse gas observations. As part of this effort, Scripps served as the WMO's Central Calibration Laboratory (CCL) for CO_2 and some related trace atmospheric constituents until the late 1990s when ESRL/GMD assumed those responsibilities (Zhao, 1997).

2.4.1 Standards realization methods

The making of greenhouse gas mole fraction standard gas mixtures is generally based on the two methods described below. These use bulk gas that, in some procedures, are combined with high purity greenhouse gases as additives in order to reach target mole fraction values. The bulk gas is most often drawn directly from the atmosphere and is termed "real air" as opposed to bulk gases derived by mixing pure components, nitrogen, oxygen, and argon, to nominal atmospheric constituency. As dry air mole fraction measurement standards, water vapor is removed at the inlet to the compression system using the atmosphere as a bulk gas source in standard gas mixture manufacturing. Currently, aluminum cylinders are widely used as these have been shown after decades-long studies to contain high purity gas mixtures for extended time periods with minimal trace constituent mole fraction change. Before addition to a mixture, high-purity GHGs are generally subjected to extensive chemical analyses in order to identify and quantify any contaminants. Either the gravimetric or manometric method is used to make (gravimetric) or measure (manometric) and assign concentration values to individual or groups of standards.

The gravimetric technique is an additive method involving several high-accuracy mass measurements of the cylinder ultimately containing the gas mixture standard (Rhoderick, 2016; Hall, 2019). The procedure begins with several evacuations of the cylinder to remove any gas that would be unquantified components of the final mixture. The cylinder valve is then closed, and the cylinder weighed in the evacuated state. With mixture GHG target values determined, high-purity CO_2, CH_4, N_2O and/ or other trace gases of interest are added to the evacuated cylinder in targeted amounts. The cylinder is weighed, then the bulk gas is added to the cylinder in an amount necessary to attain the final target mass. Finally, the cylinder is weighed in its filled state. These cylinder mass values along with their uncertainty and other factors are combined in order to calculate the final mole fraction value(s) and uncertainty estimates. The accuracy of gravimetrically prepared standards depends on GHG purity analysis and mass measurement accuracy. Accuracy of environmental parameters, such as temperature, pressure, and humidity of the weighing laboratory environment, must also be considered in order to achieve accurate mass measurements. These standards are then the basis for assigning mole fraction values to secondary and tertiary standards or reference materials used to disseminate field standards. The great advantage of the gravimetric technique is its universal applicability. However, it has the disadvantage of lacking the internal procedural capability to check the primary standards for drift over time. As with any contained mixture, a variety of processes can change the mole fraction of gas delivered from the container. For example, absorption/desorption processes at cylinder walls or slow chemical reactions occurring in the contained gas have been shown to change mole fraction values over time (Miller, 2015; Schibig, 2018). Intensive study of drift in mole fraction values has resulted in the widespread use of aluminum alloy cylinders, as these have been demonstrated to minimize change in these mixtures. However, surveillance measurements, particularly for primary standards, as well as their descendants, is required to ensure the integrity of assigned mole fraction values over extended time periods.

The manometric method is a separation method. It quantifies a gas mixture's CO_2 mole fraction based on accurate measurements of gas volume, pressure and temperature during separation process steps (Zhao, 2006). Carbon dioxide is separated from the bulk gas by passing a known volume of the gas mixture through cold traps in order to condense the minor constituents that have boiling point (B.P.) temperatures higher than those of the nitrogen (B.P. = 77.3 K) and oxygen (B.P. = 90.2 K) that are the bulk of the mixture. However, for CO_2 quantification, the boiling temperatures of CO_2 (B.P. = 194.65 K), and N_2O (B.P. = 184.67 K) are sufficiently similar to warrant a separate measurement of N_2O content in the final sample, generally with gas chromatography, to account for its presence, if any. Residual water vapor (B.P. = 373.15 K) may also be condensed in the separated sample even though measures are taken to dehumidify the starting gas sample. Measures may be taken to either eliminate or quantify its contribution also (Meyer, 2018).

To quantify the mole fraction of CO_2 in the gas mixture, an initial sample is transferred to the largest (approximately 6 to 10 liters) of three volume-calibrated glass containers. After it is allowed to temperature equilibrate, the number of moles of the starting mixture is determined based on accurate pressure and temperature measurements of this known volume of gas. The largest container is then connected to a smaller container via a manifold. The second container's temperature is held at temperatures sufficient to condense the CO_2, and perhaps some N_2O and residual water vapor, into this second container. This container is slowly warmed so that the remaining CO_2 evaporates and is re-condensed as it transfers into a third smaller volume, which is then closed to the manifold connecting the containers and associated instruments. This final gas sample is then allowed to thermally equilibrate to near room temperature. As with the initial gas sample, accurate pressure and temperature measurements of the gas in this final volume-calibrated container determine the number of moles of CO_2 of the sample. Nitrous oxide concentration measurement for this final sample is a final correction in the mole fraction value assignment procedure.

Comparison of gas mixtures prepared by gravimetric or manometric methods with one another is a means of assessing consistency between the methods and the accuracy of the mixtures as standard artifacts. Recent comparisons using reference mixtures compared with primary standards by NOAA/GMD (manometric method) and NIST (gravimetric method) demonstrate the level of agreement currently obtainable between the two methods and laboratories. Agreement within 0.05%, 0.13%, and 0.06% respectively for CO_2, CH_4, and N_2O (Rhoderick, 2016) has been demonstrated for GHG reference materials of northern hemisphere air. Periodic comparison efforts involving members of the international metrology community organized under the Mètre Convention (BIPM, 2019) and members of the atmospheric monitoring community organized under the WMO's GAW contribute to the stability and international recognition of these standards.

3. Assessing Measurement System Performance Needs

Carbon management system performance metrics will likely include compatible quantitative and societal measures. Quantitative information on total emissions, for example, from national and sub-national geographic or political regions, cities, states, or provinces, provide a means to gauge the overall effectiveness of reduction efforts. Information at multiple temporal and spatial scales assist in identifying locations and times within a system where actions may be taken or modified to optimize management efforts and gauge progress of individual activities, as well as to assess the level of target attainment at particular times. The Paris Accords (COP21, 2015) of the UNFCCC established the Intended Nationally Determined Contribution mechanism where nations across the globe have pledged greenhouse gas reductions. At this writing, 164 nations have documented their intent (INDC, 2019). Similar statements and legislation developed by sub-national governments illuminate local and regional needs. Non-governmental and private sector communities have focused on these issues and are becoming closely involved in carbon management-related activities, usually contained within sustainability and resiliency efforts.

Assessing quantification capability can benefit from some of these governmental statements. Reduction targets taken from selected INDCs and greenhouse gas reduction legislation of some U.S. states are shown in Table 1. Where carbon neutrality is a goal, the reduction target is taken as 100% relative to the baseline year. Measures of the degree of attainment are dependent upon baseline year quantification uncertainty and on understanding of measurement system performance. One- and five-year timeframes reflect measurement system performance needs consistent with yearly national reporting cycles, with the 5-year cycle of the stocktake concept defined within the Paris Agreement. Stocktaking is intended to be a means for analysis of national progress toward an INDC goal at 5-year intervals, beginning in 2023. It will give national governments the opportunity to assess implementation effort effectiveness and allow collective efforts to assess progress towards achieving the purposes of the Paris Agreement. At this writing, the mechanisms to be used for the global stocktake are under negotiation. Although procedures to be used are not agreed upon, carbon management system informational needs can be estimated from INDC information. Meeting such performance targets is challenging for some current measurement capabilities, but these targets provide useful goals for design and development of carbon measurement capabilities.

4. IPCC Inventory Reporting Methodologies

In cooperation with the UNFCCC, the Task Force on National Inventories of the International Panel on Climate Change (IPCC-TFI, 2019) has produced extensive guidance on national data acquisition,

Table 1. National or U.S. state greenhouse gas emission reduction targets.

Nation/State	Relative reduction	Target year	Base year	Yearly reduction	5 Year reduction
United States[1]	26%–28%	2025	2005	1.35%	6.75%
European Union	≥ 40%	2030	1990	1.0%	5%
China	60%–65%/unit GDP	2030	2005	2.4%–2.6%/unit GDP	12%–13%/unit GDP
California[2]	15%	2020	1990	0.5%	2.5%
	40%	2030		1.0%	5.0%
	100%	2050		1.7%	8.5%
Maryland[3]	25%	2020	2006	1.8%	9.0%
	40%	2030		1.7%	8.5%
New York[4]	40%	2030	1990	1.7%	8.5%
Washington State[5]	0%	2020	1990	*	*
	25%	2035		0.5%	2.5%
	50%	2050		0.8%	4.0%

compilation, and reporting methods. These are widely used by nations developing and making greenhouse gas emissions inventory reports and assist in producing reliable, accurate, consistent and comparable inventories of emissions and removals of greenhouse gases. The initial task force guideline documents were published in 1996 and subsequently revised, with the most recent revision having been published in 2006 (IPCC-TFI, 2006). A subsequent revision is underway, with release planned for 2019. TFI methods have relied upon a straightforward and widely-accepted methodological approach that combines information of the extent of a human activity, activity data, with a parameter that quantifies emissions or removals on a mass basis for that activity.

This is often expressed in simple mathematical form by the following:

$$\text{GHG Emissions} = \text{Activity Factor} \times \text{Emission Factor}, \tag{1}$$

where: GHG Emissions   =   mass of emissions,
         Emission Factor   =   GHG emissions mass (kilograms) per unit activity, and
         Activity Factor   =   the number of activity units involved.

This model is applied over many economic activities that emit or remove atmospheric greenhouse gases. Implementation can be straightforward using references, economic, and/or measurement data. Activity data can be taken from a range of sources derived from largely economic information concerning pertinent activities, e.g., fuel importation and consumption data or transportation data, generally from publicly available sources. Emission factor data can range from default values to very specific values based upon well-documented methods associated with specific process characteristics within a particular nation.

Emission factors are routinely expressed in terms of the mass of greenhouse gas emitted per unit activity, e.g., weight, volume, distance, or duration. In physical terms, these are a rate of flow to/from the

---

[1] https://www4.unfccc.int/sites/submissions/INDC/Published%20Documents/United%20States%20of%20America/1/ U.S.%20Cover%20Note%20INDC%20and%20Accompanying%20Information.pdf.
[2] https://www.arb.ca.gov/cc/ab32/ab32.htm; https://www.gov.ca.gov/2015/04/29/news18938/.
[3] http://www.mde.state.md.us/programs/Air/ClimateChange/Pages/index.aspx.
[4] https://www.dec.ny.gov/energy/99223.html.
[5] https://ecology.wa.gov/Research-Data/Scientific-reports/Washington-greenhouse-gas-limits.

Table 2. Examples of emission/activity factor types and units.

Activity type	Activity factor	Emission factor	Emission mechanism
Power Generation	Terajoules (TJ)	kg of CO_{2eq}/TJ	Fossil Fuel Combustion
Natural Gas Distribution	Miles of pipeline	Grams of CO_{2eq} per mile of distribution piping	Methane Leakage
Vehicle Transportation	Vehicle Miles Traveled	Grams of CO_{2eq} per mile traveled	On-Road Emissions[6]

atmosphere per unit activity, e.g., kilograms of CO_2 or CO_2 equivalent[7] ($CO_{2\,eq}$) per activity unit. Often the averages of available data of acceptable quality for an activity are used as emission factors, examples are given in Table 2. The IPCC-TFI methods only apply at national scales, although the underlying methodology is applicable sub-nationally, and the Paris Agreement recognized the need for efforts by sub-national governments and the private sector. To facilitate flexibility and applicability of this approach to emissions data compilation and reporting over a wide range of activities, a three-Tier approach is used, where Tier 1 uses default emission factors with Tiers 2 and 3 progressing to a greater specificity as determined by a nation's situation.

4.1 Quantification technologies and methods

Emissions and uptake quantification fall into two classes: The top-down and bottom-up classes mentioned above. Bottom-up approaches primarily use the emission/activity factor model. Top-down approaches are based on atmospheric observations of GHG mole fraction, using methods that range from physical sampling to remote sensing coupled in some way with determination of pertinent dynamics of the atmosphere generally or specifically within a specified region. There is a rich literature describing various aspects of both quantification methods. The references contained in this chapter are rich sources for further investigation and research, if so desired. The remainder of this chapter will summarize some of the more widely used, including brief descriptions of the processes where they are often applied.

4.1.1 The energy sector

Fossil fuel combustion for energy generation, whether for thermal or electrical power generation, and for transportation, comprises approximately 80% of the U.S. 2017 inventory (EPA, 2019). Equation (2), derived from equation (1), reflects the use of fossil fuel consumption as the main parameter describing energy production and utilization activity. Fuel consumption associated with several types of energy sector process activities are defined by the IPCC-TFI. The four main categories of energy sector emissions are: (1) primary energy source exploration and exploitation, (2) conversion into more useful energy sources, e.g., refining of crude petroleum to products such as fuels for vehicles or energy production, (3) fuel transmission and distribution, and (4) fossil fuel usage in stationary and mobile applications.

$$\text{GHG Emissions} = \text{Fuel Consumption} \times \text{Emission Factor}, \qquad (2)$$

where:
	GHG Emissions for a fuel type	=	emissions mass (kg)
	Fuel Consumption	=	the amount of fuel combusted (TJ)
	Emission Factor (EF)	=	emissions produced per of fuel combusted (kg/TJ)

Economic statistics most often provide the source for fuel consumption data. It should be noted that fuel consumption has units of Terajoules (TJ), the amount of energy produced by a fuel's combustion. Use of fuel consumption as the activity parameter assumes that fuel carbon content is directly equated to the

---

[6] The presumption is that the vehicles are internal combustion engine-driven vehicles. As electric vehicles continue to penetrate the marketplace, a means of distinguishing between the two will likely be needed in order to more accurately discriminate between emitting and non-emitting vehicles.

[7] https://stats.oecd.org/glossary/detail.asp?ID=285.

complete conversion of fuel carbon to CO_2, i.e., full oxidation. This necessitates the use of two additional fossil fuel parameters, heat energy or calorific content for a specific fuel and a unit's conversion factors. Although combustion is usually the end result, the fossil fuel supply chain has recently been investigated for leakage to the atmosphere. These studies have used a variety of Tier 3 methods to quantify emissions of methane to the atmosphere (Alvarez, 2018).

Commerce in fuel generally involves the measurement of either fuel volume or mass at the point where statistics are collected, for example, liters of fuel oil or tons of coal. Conversion to mass units allows calorific value conversion factors to determine the amount of energy produced by combusting a unit fuel mass or volume, i.e., allowing computation of energy content in units of Terajoules. The full oxidization assumption is generally sound, particularly for stationary sources (power plants for example), where process equipment design supports it. Extensive determinations support the fact that CO_2 emission factors are relatively insensitive to the combustion process technology itself. As a result, estimated values based on emission factors representing a global mean are used in situations where more detailed information on the combustion process emissions may not be well known. However, because national situations may differ from a globally averaged value, particularly in situations where full oxidation may not be the case, TFI guidelines provide flexibility to account for a range of situations. Such cases could include the transportation sector, where incomplete combustion may occur across a vehicle fleet due to tradeoffs in the operation of internal combustion engines, in those cases where incomplete combustion is a consequence of the reduction of unwanted pollutants. Given the broad range of fossil fuel combustion conditions and the complete oxidation condition assumption, the TFI has provided a means to compensate for incomplete combustion cases via a three-tiered structure that allows emission factor adjustment to more accurately reflect use conditions.

Tier 1 methods use nationally determined fuel consumption statistics and reference emissions factors published by the TFI Emission Factor Database (EFDB) (IPCC, 2018). These data reflect average properties of similar combustion systems. To facilitate use, the EFDB contains emission factor values based on global averages of energy produced by the combustion of fuel types, e.g., residual or other grades of fuel oils, shale oil, pulverized bituminous coal, or natural gas. Each of these fuel sub-types may have firing condition dependencies. The TFI Tier structure provides the means to adjust emission factor values according to more detailed fuel specifications that for a nation may differ from global averages, e.g., where combustion process conditions lie outside the range of conditions corresponding to EFDB values. In such cases, Tier 2 and Tier 3 methods are provided. Tier 2 methods use the same fuel consumption statistics as used in Tier 1 but may be combined with country-specific emission factors that more closely reflect that nation's practice. In the use of the higher tier methods, IPCC notes that it is good practice in inventory compilation to provide well-documented information describing the methodology used to develop the emission factors used in developing a nation's yearly emission report.

Tier 3 methods allow the most latitude in determining emission factor and activity data. A nation may choose to include additional data or methods that more closely reflect its usage and practices. For example, a nation may have more accurate fuel characterization data based on conditions of use, or direct emissions measurement data for individual plants that may use Tier 3 methods. Use of energy content values for a fuel whose reference value may differ significantly from reference values due to national circumstances is another example where a nation may choose not to use EFDB values. Tier 3 methods may also account for uncombusted fuel leaving the combustion process, e.g., coal found in ash based on ash carbon content analysis or emissions associated with the combustion process or plant-specific process, e.g., emissions of methane or nitrous oxide, or use of direct measurement with the Continuous Emissions Monitoring systems (CEMs), as is often the case in the U.S. (CEMs are described in a subsequent section of this chapter.) As mentioned above, the TFI guidelines stress the need to document procedures and data sources as good practice.

4.1.2 *Emissions determinations via fuel calculations*

Emission quantification via the fuel calculation method relies on fuel mass or volume and calorific value determination. Mass or volume quantification accuracy for fuel combusted in individual plants is largely dependent upon the capability, condition, and maintenance of the quantification equipment itself. When operating within manufacturer's specifications, emissions control technologies generally operate with stated emissions performance. This may vary among plants due to maintenance and operations practices.

Fuel calorific value measurement of solid fuels depends upon several issues associated with the sampling methods. Sources of variability in the sampling of coal streams are complicated and influenced by mining and cleaning processes meant to remove extraneous, non-combustible materials. Relatively small (1 cm^3 to 10 cm^3) coal samples are required for calorific value measurement via complete carbon oxidation in an O_2 atmosphere. The means of obtaining relatively small samples representative of a large mass of solid fuel raises difficult to resolve questions. Potentially, the fuel that is fired may differ significantly from a sample drawn from the bulk of the fuel itself. These sources of uncertainty in calorific content of the mass of fuel finally burned are difficult to quantify without extensive auxiliary observations of the firing process that generally lie far outside common practice.

Unlike solid fuels, quantification of natural gas volumes and calorific value determination can be accomplished with excellent accuracy for carbon management. Extensive thermochemical property data (REFPROP, 2019) developed in recent decades has resulted in uncertainty in carbon content below 1% based upon determination of natural gas mixture composition. Gas composition information is routinely obtained from constituent analysis in natural gas distribution systems. Combustion equipment is designed to achieve complete fuel carbon oxidation using either gas turbine or gas combustor systems found primarily in electrical generation. This information supports high accuracy CO_2 emissions estimates based on the fuel calculation method for natural gas.

4.2 Direct emissions measurement—continuous emissions monitoring technology

Production of steam for industrial use, e.g., manufacturing of paper, chemical processing, or electricity generation, relies strongly on fossil fuel combustion. Steam boiler combustion control methods (Babcock and Wilcox, 2018) are longstanding and have been adapted to emissions quantification in the form of Continuous Emissions Monitoring Systems (CEMs). Historically, this measurement technology was used for acid gas emission control as mandated in the U.S. by the Clean Air Act amendments implemented in the early 1990s. CEMs technology is a bottom-up, tier 2 or 3 method as it utilizes emission process characteristics for quantification, directly measuring bulk flue gas flow rate and mole fraction of pertinent flue gases in a plant's stack.

$$\dot{M}_{CO_2} = \rho_{CO_2} Q_{flue} \tag{3}$$

$$\dot{M}_{CO_2} = X_{CO_2} \left(\frac{M_{CO_2}}{M_{air}}\right) \rho_{air} Q_{flue} \tag{3a}$$

where:
$\dot{M}_{CO2}$	=	mass flow of CO_2 (kg/s)
X_{CO2}	=	mole fraction of CO_2,
M_{CO2}	=	relative molecular mass of CO_2,
M_{air}	=	relative molecular mass of CO_2,
ρ_{CO2}	=	density or mass fraction of CO_2 (kg/m^3),
ρ_{air}	=	air density (kg/m^3), and
Q_{flue}	=	volumetric flow of flue gas emitted from the stack (m^3/s).

Equation (3) relates the measured parameters to CO_2 mass flow and 3a uses the CO_2 mole fraction. As these measurements are always made in powerplant stacks, temperature and pressure measurements of the flue gas are required in order to properly determine the gas densities involved.

CEMs were originally equipped with SO_2 and NO_X mole fraction measurement channels. The advent of CO_2 reporting requirements resulted in the addition of another mole fraction measurement channel to existing CEMs installations as a cost-effective means of CO_2 emissions self-reporting by individual power plants. For CEMs methods flue gases are continuously sampled for real-time analysis, primarily using non-dispersive infrared (NDIR) analyzers periodically calibrated with gas mixtures of known mole fraction values. (A description of NDIR technology is given in a subsequent section of this chapter.) To ensure consistency in the accuracy of flue gas mole fraction analysis across the emissions reporting system, EPA requires that these calibration gases be traceable to U.S. national measurement standards and used at specific frequencies in field installations. NIST supports accurate component concentration measurements by providing certified gas mixtures via the NIST Traceable Reference Materials Program (NIST-NTRM, 2019). The NTRM program works with specialty gas manufacturers to supply standards of certified accuracy to plants and the stack testing community. NTRM reference gas mixtures are traceable to NIST primary mole fraction standards over specific ranges of field use at uncertainty levels well below the 2% EPA requirement. The specialty gas industry uses NTRMs as internal working standards in its field standard certification procedures. These practices indicate that flue component mole fraction quantification may contribute uncertainty in CEMs quantification of CO_2 in stack gases, but that contribution is likely significantly below $\approx 20\%$ seen in comparing fuel calculation/CEMs-based emissions.

4.3 Comparing fuel calculation and CEMs measurements

Emissions reporting data compilation for stationary sources may use either of two methods, direct measurement (Tier 2 or 3) or estimation of emissions based on the fuel calculation method (Tier 1). Most U.S. electrical generation plants are equipped with CEMs technology for direct CO_2 emissions measurement. These plants also report the amount of energy produced and fuel used to the U.S. Energy Information Administration (EIA). Several comparisons investigating differences arising from use of the two methods for individual plants within U.S. plants have been published (Ackerman, 2008; Borthwick, 2011; Quick, 2013; Gurney, 2016). Although the average difference for all is reported to be 1 to $\approx 3\%$, individual generation plant values reportedly differed by as much as 25%. Figure 2 illustrates data taken from publicly available EPA and EIA databases for approximately 800 plants using a range of solid fuels. Although the distribution is not Gaussian, for the purposes of estimation, a Gaussian distribution, shown as the red curve, was used as a simple way to approximate a mean difference of $\approx 1.5\%$ and a half-width of $\approx 15\%$ (Borthwick, 2011). These studies raised questions concerning the accuracy of both methods. Since mole fraction measurements in U.S. power plant stacks must demonstrate traceability to NIST mole fraction standards at the 2% level or below, bulk flue gas flow determination accuracy was investigated in order to determine whether its measurement may be a contributing factor to these differences at levels comparable to that of fuel calculation methods.

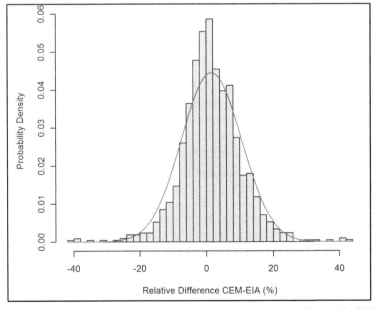

Figure 2. Individual plant emissions differences, CEMs/Fuel calculation (Borthwick, 2011).

4.4 Flue gas flow rate measurement

Uncertainty in stack flue gas flow measurement is the other CEMs parameter to consider as a possible source of the difference between fuel calculation and CEMs methods. Stack flow measurement technology has evolved over the last several decades with the advent and application of non-intrusive, ultrasonic flow metering technology. It has become the dominant stack flow measurement method in the U.S., with installations using either one or two ultrasonic flow meters (USMs), the latter usually in a crossed beam configuration. Of the two CEMs parameters, it may have the greatest impact as it affects the calculation of emitted trace constituent mass for all components, e.g., CO_2, SO_2 or NO_x.

A USM is composed of two ultrasonic transceivers mounted on either side of a stack. In operation, one USM transceiver (consider the red transceiver pair in Figure 3) transmits ultrasound to the meter's other transceiver, USM 1a to b, along a diametral chord across the stack as shown. This chord is arranged at a fixed inclination angle, φ, relative to the stack's vertical axis. USMs use a Doppler shift flow measurement strategy, in that the ultrasonic wave's velocity relative to the transceiver is either enhanced or retarded by stack gas flow velocity. The USM controller alternates send/receive functions between transceivers to measure transit time with and against the main flow direction so as to minimize or eliminate sensing artifacts. The transit time difference between ultrasound pulse bursts as they transit with and against the flow direction is a measure of the average flow velocity along the ultrasonic beam path. Using absolute transit times, t_{up} and t_{down}, measured pathlength between the transceivers, and inclination angle value, both the average flow velocity and the speed of sound in the gas can be calculated.

Recent research (Johnson, 2019; Bryant, 2018) and testing results have shown that this approach can achieve accuracy levels of ≈ 1% for the crossed-beam configuration shown in Figure 3. When only a single meter was used in these tests to calculate flow velocity, errors in the range of 10% to 17% occurred, depending upon flow velocity. These performance levels have been obtained in NIST flow testing systems that emulate stack flow characteristics, i.e., strongly asymmetric axial velocity profiles, significant cross-axis velocities (swirl or turbulence), over axial velocity ranges normally found in stacks, up to ≈ 25 m/s. Flow standards used in these investigations are traceable to U.S. national standards having uncertainties of ≈ 0.7%. Some industrial stacks in the U.S. are equipped with crossed-beam systems. Currently, collaborative efforts between NIST and industry partners are investigating crossed-beam method performance in stacks of coal-fired electrical generating plants in order to assess their

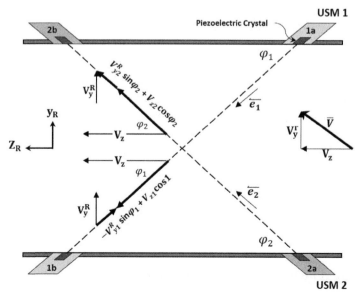

Figure 3. Transceiver arrangement and velocity vectors for two ultrasonic flow meters configured in a crossed beam metering configuration. Diagram courtesy of A. Johnson, NIST.

performance in industrial settings. This assessment is anticipated to result in similar results in coal-fired plant stacks. Use of the cross-beam method can significantly improve accuracy if applied through coal-fired plants.

5. Atmospheric Measurement of Greenhouse Gas Fluxes

Unlike observation strategies useful for direct GHG flowrate measurement, like CEMs, flux quantification for GHG fluxes originating from surface processes where direct measurements are not possible must be accomplished using atmospheric observations of mole fraction coupled with inferential analysis methods. As would be expected, these require detailed information describing atmospheric properties and motions in addition to greenhouse gas concentration observations. Atmospheric motions at fine spatiotemporal scales are realized with simulation models, whether of the global atmosphere or more locally focused numerical weather prediction (NWP) models. Atmospheric parameters, such as wind speed and direction, temperature and pressure, are needed throughout the estimation and modelling domain. These, coupled with mole fraction observations at various domain locations, provide a means to track atmospheric greenhouse and other observable trace gases through the atmosphere. From the carbon management point of view, these methods allow for the estimation of source/sink locations and the magnitude of the flux at those locations.

Equation (5) relates greenhouse gas, or any relatively long-lived trace gas, atmospheric mass flux to observable parameters, or, in the case of air density, to a parameter directly computed from observed parameter values. Mass flux is a vector quantity due to its dependence on velocity. Both must be represented by a magnitude and direction.

$$\overrightarrow{M_G} = x_G \overrightarrow{M_A} = x_G \rho_A \overrightarrow{V_A} \tag{4}$$

where: $\overrightarrow{M_G}$ = Mass flux of the greenhouse gas of interest,

x_{GHG} = Greenhouse gas mole fraction,

$\overrightarrow{M_A}$ = Mass flow or flux of the atmosphere,

ρ_A = Density of the atmosphere, and

$\overrightarrow{V_A}$ = Atmospheric velocity.

Atmospheric velocity vectors, composed of wind speed and direction values, across the geographic domain of interest impart these properties to mass flux with scalar quantities of greenhouse gas concentration and air density. Descriptions of atmospheric motions both at global scales, known as global circulation models, and at more local scales, NWP models, are used to simulate greenhouse gas flows in the atmosphere. Although these modelling approaches rely heavily on fundamental descriptions/models of the fluid dynamic system of the atmosphere, they are aided by the use of meteorological observations. In most cases, meteorological parameter measurement data is available at a relatively few points in the domain relative to the number of cells in a computational domain. Advances in and assimilation of radar-based observation, both from the surface and space, have substantially improved the skill of atmospheric property simulation from the surface to the top of the atmosphere. As discussed in more detail below, NWP and global atmospheric circulation models can provide simulated atmospheric parameter values throughout a domain of interest with varying degrees of temporal and spatial resolution. These, coupled with GHG mole fraction measurement data, are used in the inferential method described below for tracking atmospheric GHG flows from their sources on or near the surface, through the planetary boundary layer, and to the global atmosphere.

5.1 Greenhouse gas dynamics in the global atmosphere

The movement of CO_2 and methane at continental and global scales has been estimated with global atmospheric circulation models, coupled with mole fraction measurements, through the contributions of several nationally-sponsored, mole fraction observing systems (ECMWF, 2018; Rodenberg, 2017). The CarbonTracker (CT) model, developed by NOAA's GMD (NOAA-CT, 2017), estimates atmospheric carbon uptake and release globally. The observed mole fractions used are predominantly taken from northern hemisphere locations, which tends to make simulated values there somewhat more accurate. CT, and the associated long-term monitoring of atmospheric CO_2 upon which it depends, are tools to advance the understanding of carbon uptake and release from land ecosystems and the oceans. Providing a means of studying the movement of atmospheric CO_2 globally, CT and other global models can be tools for monitoring environmental changes, including human management of land and oceans and the potential impact of carbon management efforts globally. A companion model tracking global methane emissions (NOAA-GMD, 2018), provides methane emission flux simulations from North American and global natural and anthropogenic methane emissions. Other global-scale GHG inversion models also produce flux estimates annually (Wageningen Univ. & ICOS Netherlands, 2019; ECMWF–CAMS, 2019; Max Planck Institut for Biogeochemistry, 2019). Many assimilate satellite observations either in addition to *in situ* data or alone. Various Bayesian data assimilation methods are utilized, including Ensemble Kalman Filter and 4D-Var methods.

The CarbonTracker system is an example of atmospheric inversion analysis combining statistical optimization, global atmospheric circulation simulation, and CO_2 and CH_4 observational data. The underlying global transport models have similarity to finite element/finite difference analyses used widely in many science and engineering applications. These methods enable the investigation of the behavior of complex systems using spatial and temporal gridding of the domain containing the item of interest and embedding fundamental process descriptions or parameterizations of them within each grid cell. For example, a spatial grid encompassing a body where thermal and/or mechanical stresses are to be investigated or that of a fluid in a flow field confined to a specified geometry, pressure and temperature regime can be simulated and compared with measured parameters. In meteorological models, three-dimensional grids beginning at Earth's surface and extending to the top of the atmosphere are used to simulate atmospheric dynamics and properties over specific geographical regions and time periods. Physical and hydrodynamic principles and parameterizations of individual Earth system processes, e.g., cloud formation, planetary boundary layer dynamics, land-atmosphere gas exchange, and solar radiation intensity at the surface, are used at each grid cell along with pertinent input data. At each time step, the model provides an array of atmospheric parameter values, e.g., pressures, temperatures, wind direction and speed, and surface energy exchange. As described below, application of NWP models to simulate atmospheric dynamics at local scales is a tool to investigate greenhouse motions and source and sink characteristics in urban areas. These are important to carbon management as the smaller geographic regions of cities (small relative to the total land surfaces of the earth) have outsized contributions to national and global emissions, and, therefore, are likely to be a focus of carbon management efforts.

5.2 Urban and regional greenhouse gas emissions and observing networks

Cities and metropolitan areas aggregate populations and energy usage and are major contributors to a nation's total GHG emissions inventory. In 2018, it was estimated that approximately 55% of the world's populations live in urban areas and megacities (cities with populations > 10 million) with urban populations projected to reach 68% by 2050 (UN-DESA, 2018). Further concentration of human activities and concomitant emissions are anticipated. Measurement systems that better quantify urban emissions are likely to become critical as a means of quantifying the effectiveness of carbon management efforts.

Prior to ≈ 2010, most atmospheric greenhouse gas measurement capabilities were focused on continental- and global-scale observations and analyses. A notable U.S. exception is the research effort in Salt Lake City (Strong et al., 2011; Lin, 2018) that began in the mid-2000s. In recent years, urban greenhouse gas observing networks have been established in six U.S. cities: Salt Lake City, Utah; Boston, Massachusetts (Sargent, 2019); Indianapolis, Indiana (Elementa, 2018); Los Angeles, California (Yadav, 2019); Oakland and the Bay Area of California (Cohen, 2019); and the Baltimore, Maryland-Washington, D.C. region of the Northeast Corridor (Martin, 2019). These observing networks, and the research efforts they support, differ from one to the next in the number of observing nodes, GHGs observed (although

CO_2 and methane are the most prevalent) and in the range of emission estimation methods brought to bear, although most if not all utilize versions of Bayesian inference methods. These efforts are aimed at developing and demonstrating emissions quantification capabilities using combinations of bottom-up and top-down techniques. The effort in the Oakland/San Francisco Bay Area (Kim, 2018) is focused on both air quality and greenhouse gas measurements, using a network of low to moderate cost/low to moderate performance sensors in a relatively high-density observing strategy, with $\approx$ 50 observing nodes, at this writing. The Boston project is focused on whole-city emissions estimates. Three of these cities form NIST's Urban Greenhouse Gas Measurements Test Bed System, Indianapolis (the Indianapolis Flux Experiment, INFLUX), the LA Megacity Carbon Project, and the Northeast Corridor. The Baltimore/ Washington (the NEC/BW project) is the beginning of this effort that is planned to reach to the Boston area given sufficient resources. All use surface-based observing networks of designs specific to the urban character of the locale. Aircraft-based whole-city observations and methods of analysis are also used. Most employ high-sensitivity GHG instrumentation and standards referred to mole fraction standards disseminated by WMO's Central Calibration Laboratory (NOAA's GMD) in order to ensure consistency in measurement data globally. Emissions modelling provides a means of transforming bottom-up emissions data and estimates at spatial and temporal scales consistent with the variety of atmospheric observation and analysis methods applied (Gurney, 2011). These transformations of inventory reporting information combined with atmospheric observation and analysis, described subsequently, are a means of moving toward an as yet unrealized reference framework supporting quantitative comparison of emissions measurement and estimation results for a range of independent emissions quantification methods.

As with global emissions dynamics simulation, urban GHG mole fraction observational data is combined with atmospheric transport simulation and statistical optimization methods in order to infer source/sink properties of greenhouse and other longer-lived trace gases. Aircraft observations are often used in a mass-balance analysis to provide whole-city emissions estimates. Figure 4 is a cartoon showing the major elements of an urban region that includes residential, commercial, and industrial buildings, power plants, transportation and communications systems, and geo-political boundaries. Situated within the region is a communications tower-based, two-segment GHG concentration observing network. One segment is located within the more densely urbanized section of the city and the other located along its boundaries. The wind moves GHGs into and through the urban and regional domain. Incoming atmospheric mole fractions are modified by emission and removal processes in and around the city.

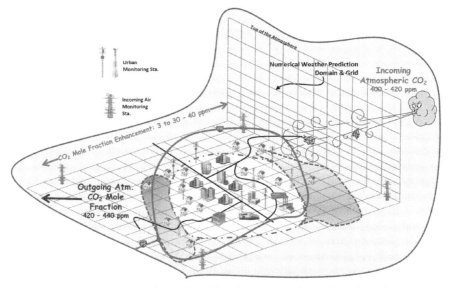

Figure 4. Elements of the atmospheric inversion method of locating and estimating the flow of greenhouse gases to and from the atmosphere in an urban setting. The NWP model establishes the three-dimensional gridded domain depicted by the red (surface/horizontal) and green (vertical) meshes.

Incoming and outgoing GHG concentrations are measured by the outer segment nodes of the observing network. Its inner segment is more sensitive to GHG fluxes within the city. NWP models simulate atmospheric motions within its spatial and temporal grid system, an essential element for inferring the location and magnitude of source emissions and uptake. Mole fraction values shown in Figure 4 are those typically found in moderately-sized U.S. cities. The atmosphere entering the domain is laden with CO_2 resulting from partial mixing of emissions upwind. Typically, incoming air concentrations vary due to differing upwind emissions, causing ever-changing incoming values dependent upon wind speed and direction and temporal change of upstream GHG source and sink fluxes. Emission enhancements are differences between observed incoming concentrations and those observed by network nodes. Estimates of incoming mole fraction loading may also use other sources, e.g., global or continental-scale models, for incoming mole fraction estimation, although direct measurement is likely to give more realistic estimates across the domain's entry plane. The outgoing flux from a city increases in mole fraction by a few to several tens of µmol/mol depending on atmospheric dynamics and varying emissions and uptake from the parts of the urban area with significant vegetation.

5.2.1 Atmospheric inversion modeling of urban and regional greenhouse gas fluxes

Bayesian inference methods are applied to estimation of GHG fluxes and their dynamics across global, continental, regional, or urban modeling domains. The observed quantity is the GHG concentration. Statistical optimization methods are based on Bayes' concept that the probability associated with an event, or initial data set, can be updated by additional information. Stated more formally, Bayesian inference is a method of statistical inference based upon Bayes' theorem,[8] where it is assumed that the probability for a hypothesis can be modified or updated as more evidence or information becomes available. In the case of evidence-based GHG flux estimation, an initial estimate, termed a prior, is updated by additional observations in order to inform or refine the prior estimate with more information and construct a posterior flux estimate. In estimating urban source and sink fluxes, the hypothesis is based upon emissions inventory data for the region of interest. In some cases, an analysis may begin with a so-called "flat prior" that might be derived from a whole-city emission estimate which is then sub-divided equally among the surface grid cells of the NWP applied domain, as illustrated in Figure 4.

Figure 5 illustrates the major components of the Bayesian inference approach as applied to atmospheric observation and analysis:

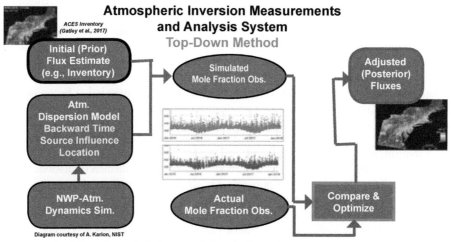

Figure 5. Atmospheric inversion analysis frame work diagram illustrating the major components of flux estimation.

---

[8] Although Bayes' theorem originated in the early 19th century, its implementation is, in most cases, computationally intensive. Only with the advent of high capacity computing capabilities has it become widely used in science, technology, and some social science fields.

1. The initial, or prior, flux estimate for the domain over the time of the analysis,
2. An NWP and a transport/dispersion model for simulating atmospheric transport dynamics,
3. Simulation of network node observations based on a flux prior coupled with atmospheric transport,
4. The observed mole fraction data, and
5. Statistical optimization.

NWP simulation of atmospheric dynamics, particularly those within the planetary boundary layer (PBL), where emissions first begin mixing with the atmosphere, coupled with Lagrangian backward-time dispersion modelling provides a means to simulate mole fraction observations. Another approach is the development of a sensitivity function, or adjoint, by the NWP model itself using its atmospheric turbulence and dispersion models. The latter approach is rarely implemented due to the complexities of forming the adjoint.

With sensitivity or influence functions, Greenhouse gas mole fraction signals can then be simulated by combining prior emissions information with the sensitivity, or influence function, to obtain a modelled observed signal (orange, upper, data trace). These can then be compared with the GHG mole fraction observations (purple, lower, data trace). Optimization of the simulated data by adjusting prior emissions values relative to the measured data yields an updated mole fraction data set for each network node along with estimates of fluxes at each grid cell in the domain. In essence, mole fraction measurements conflated with atmospheric dynamics simulation and initial emissions estimates inform an evidence-based measurements and analysis outside of traditional ones.

The next few sections discuss the major features of each of these atmospheric analysis method components used widely for greenhouse gas flux estimations from the global to the local scale and based on a variety of atmospheric observation methods and initial flux prior estimates.

5.2.1.1 The prior flux estimate—Emissions modelling

Emissions data sources contributing to inventory compilation cover broad ranges of temporal and spatial scales. The purpose of emissions models is to transform these data using emission-activity factor methods to scales consistent with the NWP scales driving atmospheric inversion analysis. Development of these transformations has been underway and in use since the mid-1980s (Marland, 1985; Anders, 1996; Gurney, 2003; Gregg, 2009) and used as prior estimates or to compare with various other types of atmospheric analyses. Initial efforts were focused on anthropogenic emissions from fossil fuel combustion, particularly by power plants in the U.S. that are Continuous Emissions Monitoring System-equipped and coal-fired as these are some of the largest emissions sources.

Emissions modelling methods have been extended to deal with other emission sectors relying on a range of data sources. Because CO_2 emissions are closely related to the fossil fuel combustion process itself in many of these processes, emissions flow directly to the atmosphere without a practical means of direct measurement. Rather, some kind of indirect analyses involving proxy parameters associated with the originating processes are needed for determining activity and emission factor values. Here, a general description of proxy parameter uses, and the underlying operational information needed in order to estimate vehicular emissions will be used as an example of emissions estimation from a complex emissions sector, transportation emissions. Similarly, complicated analyses are applicable to other sectors and their accompanying proxies and operational conditions.

Since vehicles change location continuously and are not equipped with direct emissions measurement, emissions models have been developed primarily for air quality emissions. These also provide CO_2 emissions values (EPA-NVFEL, 2019) that rely on proxies, such as vehicle type and operating parameters, to estimate activity/emission factor values coupled with spatiotemporal information. Emissions maps that include time varying behavior, such as that of Figure 6, can then be made (Gately, 2015). Vehicle miles travelled is a primary activity proxy. Emission factor proxies are somewhat more complex as they must account for engine emission characteristics that vary with engine, vehicle type, and load. Vehicle fleet models and engine types and their emissions properties are characterized as a function of load and speed by both the U.S. EPA's National Vehicle and Fuel Emissions Laboratory (NVFEL, 2019) and auto manufacturers using extensive engine and vehicle type testing to obtain both activity and emission factor

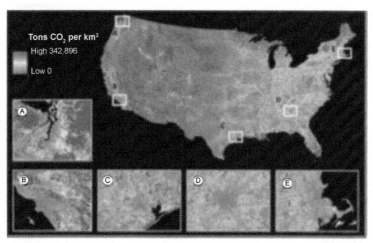

Figure 6. On-road CO_2 emissions: Coterminous U.S. with selected urban areas at 1 km resolution (Insets) maps showing details of metro areas surrounding Seattle (A), Los Angeles (B), Houston (C), Atlanta (D), and Boston (E). Image from (Gately, 2015).

data. Engine load, and therefore emitted amounts of GHGs and air quality gases, is vehicle speed, weight, and road characteristic dependent. Road properties, primarily vertical incline, influence vehicle engine load, as does vehicle speed. To facilitate the function of these and similar models, road information is given in the form of segments of known length with additional descriptive parameters related to vehicle loading. These properties (location and road segment length) and load requirements to maintain a given speed are among the parameters needed in order to provide spatiotemporal vehicle emissions data. Road type and location, miles travelled on each type, average speed, vehicle type and similar information can be obtained from U.S. statistical data collected and disseminated by local, state, or national transportation agencies. Similar rationale and methods are applied to train, marine, and plane emissions.

5.2.1.2 Simulating atmospheric transport dynamics

Inversion analyses require the use of dispersion models coupled with NWP models, e.g., the Weather Research and Forecasting model (Skamarock, 2008) or those of the European Centre for Medium-Range Weather Forecasts (ECMWF, 2019), to simulate the transport and dispersion of trace gases in the atmosphere. Typically, NWP simulations output atmospheric dynamics and related information with spatiotemporal resolutions of 1 km^2 and larger and at sub-hourly scales in urban domains. In addition to using fundamental physical principals, NWP models rely on meteorological observations taken both at individual observing stations on Earth's surface and from both surface and satellite-based radar observations, to update simulations and forecasts of atmospheric dynamics (Kalnay, 2003). For North America, an alternative to NWP simulations for a given domain are meteorological data products from the U.S. National Weather Service, such as the North American Mesoscale Forecast System[9] (NAM). NAM data is available at a range of spatiotemporal scales, the smallest currently being 12 km, beginning in 2003 and 2004. Similar meteorological data products are available from Europe and other parts of the world.

5.2.1.3 Influence functions, footprints, and simulated emissions—Linking source/sink location with observations

A means is needed to relate contributions to observed mole fraction signals from source or sink fluxes upwind of the observation location, whether the observation is from a surface-based network, a satellite

---

[9] https://www.ncdc.noaa.gov/data-access/model-data/model-datasets/north-american-mesoscale-forecast-system-nam.

track giving path-integrated mole fraction observations, or an aircraft flight path. Identifying these source/sink locations and estimating their flux magnitude at an observing node location is a two-step process utilizing wind fields throughout the computational domain. Approaches based on Eulerian or Lagrangian dynamical methods are used to determine influence functions, or footprints, linking network observing nodes to flux locations.

Lagrangian methods take the view of a reference frame attached to marker particles that move through the atmosphere under the influence of wind fields provided by Eulerian NWP simulation data. These are principally three-dimensional wind speed and direction, turbulence intensity, and thermodynamic atmospheric state parameters defining the varying meteorological conditions encountered during of a marker's trajectory. In simulations run backward in time, marker particles are released from a receptor, an observing location, to determine the particles' trajectories through the domain during the time of the simulation. For example, an air mass with straight-line winds of 5 m/s ($\approx$ 11 miles/hr) transits a domain of 200 km ($\approx$ 125 miles) extent in approximately 11 hours. It is not unusual for modelling computations to span several times this transit time as meteorological conditions, such as shifting wind direction and speed, may cause marker particles to remain within the computational domain significantly longer. Typically, several hundred to thousands of particles may be released, generally uniformly, over the period of one to several hours. The resulting collection of trajectories allows determination of marker residence time in grid cells of the atmospheric column above each surface grid cell in the domain (Lin, 2012; Lin, 2003). The assumption that each atmospheric column exists in a vertically well-mixed state below a specified altitude, i.e., the GHG mole fraction below that altitude is constant, allows association of residence time with surface or selected grid cell emission and uptake locations. Common practice is to use PBL height or one half that value as the height below which the well-mixed assumption holds. Residence times are normalized to the total time and number of markers to develop the fraction of GHG mole fraction enhancement that source/sink grid cells contribute to, or influence, mole fraction signals. Influence functions have units of mole fraction per unit flux, e.g., $\mu mol/mol/(mol/(m^2\text{-}s))$. Convolving influence functions with prior flux estimates provide simulated mole fraction signals at an observing location, as illustrated by the orange data trace shown in Figure 5.

Eulerian or forward-time methods, e.g., WRF-CHEM (Grell, 2005; Fast, 2006) of describing particle motions use a spatially fixed reference frame to move particles within the computational domain. Influence functions are determined in the same general way, i.e., through particle release and trajectory tracking. However, markers travel forward in time and must be released from all potential flux origination locations, tracing out trajectories that may or may not intercept receptors. These require multiple simulations to link flux origination locations with those of receptors, have much higher computational costs, and are, therefore, rarely used for influence function determination.

6. Measuring Mole Fraction of Atmospheric Greenhouse and Trace Gases

Mole fraction determination is pervasive in measurements used in bottom-up and top-down quantification both for greenhouse gases and other atmospheric trace gases. Two approaches are used, either direct *in situ* observation or determination via samples collected in the field and measured in a laboratory. The latter is accomplished with a variety of well-known laboratory analytical methods involving gas chromatography and mass spectrometry and the combination of the two. Some field applications use research-grade instruments that have been modified for field usage, particularly for quantification of the very low atmospheric concentrations of the fluorinated gases (MIT, 2019; Prinn, 2018). Various methods based on infrared spectroscopy have been developed and widely used in both laboratory and field applications.

6.1 Infrared spectroscopy

Molecular spectroscopy involves the study of the motions of the atoms comprising molecules. It provides a means of identifying and measuring mole fraction of molecular species in gas mixtures such as the atmosphere. Internal molecular motions, and their resulting spectra, are well-described by a branch of quantum mechanics developed and verified over much of the 20th century and with spectroscopic research continuing today. Spectroscopic properties of molecules are rooted in the number and kind of atoms comprising a molecule. Motions of these atoms, rotation and vibration (bending and stretching motions) about the molecule's center of mass, give rise to quantized energy levels comprising absorption bands that occur primarily across the infrared (IR) region of the optical spectrum. Absorption of a photon whose frequency is the same as that of an energy level raises the total energy of the molecule resulting in change of rotational or vibrational motion. The molecule is then said to be in an excited state above that it normally occupies. It will revert to its original state, its ground state, upon emission of a photon of similar energy or through collisions with nearby molecules.

Molecules having different constituent atoms have spectroscopic absorption bands that cover different wavelength regions. Internal chemical bonds between atoms have characteristic spectra depending mainly on the mass values of a molecule's atoms. For example, the three atom molecules of water vapor (H_2O) and CO_2 have different chemical bond strengths, resulting in absorption bands occurring differently in wavelength regions of the IR. Bands characteristic of different molecules may or may not overlap in wavelength. An example of a CO_2 band structure in the 1.56 to 1.59 micrometer (µm) wavelength region is shown in Figure 7. The data are from the widely-used, high-resolution transmission molecular absorption database, HITRAN (HITRAN, 2016; Gordon, 2017). The 1.57 µm CO_2 band is one of several CO_2 IR bands. Others occur near 2.1 µm, 4.23 µm and at longer wavelengths in the thermal IR, each with differing structure and absorption strength. At the longer wavelengths, line strength increases significantly. As discussed below, measuring instrument design strategies utilize infrared absorption bands, or some portion of them, as a means of determining mole fraction values. Depending upon factors such as instrument target performance accuracy and range, application, cost, and available technology, absorption strength, either of a band or single line within a band, may also be a design factor.

Energy absorbed by a GHG molecule in a collision with a photon whose energy corresponds to that of one of the molecule's absorption band transitions changes its motion state to a higher energy level. Depending upon photon energy and details of the absorption process, this energy is partitioned between the rotational and vibrational states of the GHG molecule. It sheds excess energy through collisions with nearby molecules in the atmosphere. This quickly transfers energy to the collision partner allowing the GHG molecule to return to its pre-absorption state. Since oxygen or nitrogen molecules make up 99% of the atmosphere and neither have energy level structures that allow them to accept the GHG molecule's excess energy as rotational or vibrational motions, the added energy becomes kinetic energy or heat. By rapidly absorbing thermal energy radiated from the earth and transferring that energy to the main

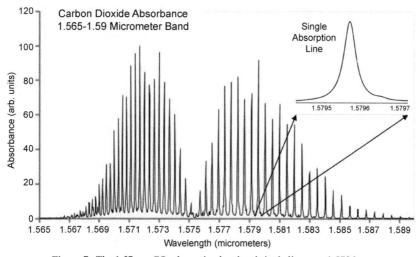

Figure 7. The 1.57 µm CO_2 absorption band and single line near 1.5796 µm.

atmospheric constituents via the molecular collisions continuously occurring in the atmosphere, GHG molecules are highly efficient in raising the atmosphere's temperature.

6.2 Spectroscopic mole fraction analyzer designs

Greenhouse gas mole fraction measurement techniques for gas mixtures based upon IR spectroscopy use either absorption bands or individual transitions within bands of the GHG of interest. Absorption bands of individual molecules are well separated in wavelength. This separation lends itself to instrument designs aimed at a particular molecular species. However, where bands of different molecules overlap in wavelength, either these regions are avoided, or a strategy must be used to separate contributions from different molecular species. For example, water vapor is ubiquitous in atmospheric measurement situations as it has bands at several IR wavelengths and can be a confounding element in instrumental designs. Some field installations avoid or minimize water vapor interferences by drying the sample air prior to introduction to an instrument's measuring volume, while others do not, often depending on instrument application. An instrument's molecular specificity is realized by confining its detector response to radiation of wavelength absorption by the species to be measured. For the generalized NDIR instrument design shown in Figure 8, this is achieved by the band pass filter. For example, CO_2 has a very strong absorption band in the 4.2 μm to 4.3 μm region with no other molecule having a significantly interfering band. This band, the so-called "fundamental band" with the strongest absorption, is the basis for many instrument designs. These use a filter that transmits this range of radiation wavelengths to the detector while rejecting all others. Typically, such a filter would have a passband with center wavelength of 4.26 μm and width of 100 nm to 200 nm.

In all instrumental designs however, source photon removal via molecular absorption along the instrument's sensitive path is a fundamental objective. Since absorption band strengths of molecules differ substantially, instrument designs are adjusted to meet application requirements. Although maximizing photon absorption by the pertinent GHG molecule is a main instrumental goal, several other processes may be simultaneously at work, resulting in additional intensity reductions at the detector. Such interference effects may be minimized with a blend of instrumental design and calibration procedures using the appropriate gas mixture standards discussed above.

Figure 8 shows a form of the Beer-Lambert law and a general instrument design incorporating an IR source, absorption path, bandpass filter, and detector. This general design uses tungsten filament electrical lamp sources that emit broadly across the infrared, but mostly outside the wavelength band of interest. (Recent advances in infrared light emitting diode technologies are an emerging alternative appearing in some research grade and commercial devices.) Band pass filter technologies for GHG molecules are well developed and widely available. The majority of instrumental designs utilize a single detector and bandpass filter, although in the last few years multi-detector assemblies with integrated electronic amplification and wavelength selectable filters have become available. These support strategies of integrated absorption signal/reference signal detection using a single gas sample path. The reference filter's pass band is selected to be in a wavelength region where molecular absorption does not occur. In this way, the reference channel signal compensates for parasitic effects that impact both detectors.

The form of the Beer-Lambert relation shown in Figure 8 assumes multiple absorbing species. Source radiation transmitted to the detector, T_d, is the ratio of the radiant flux transmitted through the gas, I_t, and measured at the detector, to the radiant flux incident on the measurement path, I_o. T_d depends

$$T_d = \frac{I_t}{I_o} = -le^{\Sigma_i \sigma_i n_i}$$

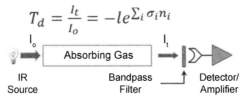

Figure 8. IR absorbance-based mole fraction measurement, Non-Dispersive IR (NDIR).

on the attenuation parameter, σ_i, for the ith gas species observed, the attenuation pathlength, l, and the number density, n_i, of the attenuating species. The number density is the value desired for mole fraction determination. In most instrument designs, absorbing pathlength is fixed and the attenuation parameter, σ_i, can be determined from spectral reference data, i.e., absorbance information, obtainable from references such as HITRAN.

The Beer-Lambert relation is a widely-used instrument design foundation for mole fraction measurement, however, complications may arise for some applications. Although T_a, I_o, and the pathlength may be measured directly, they are more often inferred from the ratio of the signal to reference channels, or in single channel instruments by turning the lamp off periodically. Other instrumental effects that increase the uncertainty in NDIR mole fraction measuring instrument designs must be compensated. These include drift characteristics in IR detector response and source emission. Compensation for various scattering processes due to elements comprising the optical path may also be needed. All these effects can limit NDIR measurement accuracy, although perhaps not its sensitivity, in most cases. Such deficiencies can be compensated considerably or eliminated by instrument calibration procedures based on mole fraction standards for the gas or gases of interest. This has been the practice in many applications where field measurement requirements approach those of standard lab capabilities.

6.2.1 Long absorption pathlength instrument designs

Typical NDIR instrument designs have pathlengths ranging from a few to ≈ 30 centimeters and usually employ an analyte molecule's strongest absorption bands for maximum signal amplitude. Spectroscopic investigation of weakly absorbing molecular bands in the early to mid-20th century resulted in the development of several optical cavity designs aimed at substantially increasing the absorption pathlength with multi-pass designs. These are often based on reflections between two or more mirrors to fold the absorption path within tractable cavity lengths. An early design still in current use is the White cell (White, 1942), a basic version is shown in Figure 9. By adjusting the orientation of the two smaller concave mirrors appropriately, multiple passes of the input beam transit the cavity, thereby increasing pathlength. The configuration shown supports eight transits with precise adjustment resulting in more passes. Succeeding cell designs feature more passes due to greater mechanical stability (Herriott, 1965).

With the advent of high-reflectance mirror technology, reflectance > 0.9995, and lasers, resonant cavities, such as those illustrated in Figure 10, came in to widespread use. Technological innovations driven significantly by the telecommunications revolution resulted in frequency-stable, tunable infrared lasers, optical fiber components, and IR detector technologies that have advanced molecular spectroscopy research and measurement instrument designs. Due to effective pathlength increase and other strategies, these typically have performance superior to NDIR instrument designs. The optical cavities used lengthen absorption paths from fractions of meters to multiple meters and kilometers at bench scale or smaller physical sizes through the application of a number of strategies (Romanini, 2014).

High reflectivity, resonant and non-resonant optical cavities have shifted measurement strategies from use of all or most of an absorption band to that of a single or a few adjacent individual transitions within a band. This wavelength, or frequency, selectivity also provides means to compensate for overlap of absorption features of multiple molecules in the same wavelength region by direct observation and compensating analysis.

Analytical instrument designs based on these technological innovations and applied to concentration measurement instrument designs for CO_2, CH_4, N_2O, CO, and other small molecular species at high

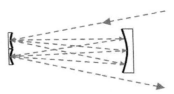

Figure 9. White cell optical paths

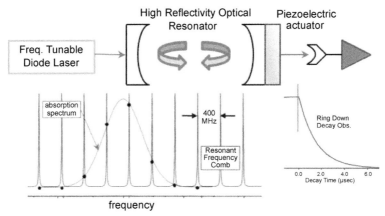

Figure 10. CRDS-based mole fraction measuring instrument design. Image courtesy of J. Hodges and D. Long, NIST.

sensitivities also featuring good temporal stability began appearing commercially in the late 1990s and early 2000s. Resonant optical cavities were combined with the exquisitely fine control of emission frequency of tunable IR lasers. This resulted in designs with long absorption pathlengths and much greater control of the frequency of the light used to probe molecular absorption features. These designs have effective path lengths of 100s to 1,000s of meters. One approach, and variations on it, based on the use of resonant cavity configurations, is used in several commercial instruments as well as in research applications and is known as cavity ringdown spectroscopy (CRDS). Figure 10 shows the main components of a CRDS design along with a conceptual diagram of the spectral properties of the resonant cavity using a Fabry-Perot interferometer cavity (Busch, 1999). The high reflectivity optical resonator (mirror reflectance of 0.999 and higher) not only provides the means to substantially lengthen the absorption path, but it also replaces the bandpass filter of NDIR designs. A cavity's filtering properties result in a comb of frequencies, illustrated as the series of transmission peaks shown in blue in Figure 10. These equally-spaced, comb frequencies allow the cavity to accept laser light only at those exact frequencies or wavelengths. Frequency spacing of cavity resonances, termed the free spectral range, Δf, of the cavity is dependent upon mirror spacing and the speed of light.

$$\Delta f = c/2l$$

where: c = speed of light and
l = mirror spacing.

The 400-megahertz (MHz) resonant frequency separation shown in Figure 10 corresponds to a mirror separation of $\approx$ 37 centimeters ($\approx$ 12 inches). Cavity length tuning via the piezoelectric actuator shown in the figure mainly changes the frequency position of comb teeth.

The combination of laser frequency tuning and cavity length control gives exceedingly fine tuning of light either entering the cavity at a comb frequency, or strongly rejecting if the laser is tuned off a comb frequency. Measurement of absorption feature properties, both amplitude and frequency, is accomplished by successively tuning the excitation laser to comb frequencies. As an example, an idealized molecular absorption feature (the red line labelled 'absorption feature' in Figure 7) is shown superimposed over the cavity's resonant frequency comb. The black dots indicate cavity resonance overlap with the absorption line. Adjustment of laser frequency and cavity resonance conditions are the means of probing absorption feature shapes and amplitudes, a widely-used method for accurate measurement of absorption line shapes/amplitudes. Cavities designed for high-sensitivity measurement of spectral reference data often have $\approx$ 100 MHz cavity resonance separations and widths in the tens of kilohertz range, giving them the ability to measure absorption features with significantly higher accuracy than previously (Yi, 2018).

Long-pathlength instrument designs transform signal acquisition from one that relies on long-term temporal stability of the detector, as in NDIR designs, to one that relies on short-term temporal stability where drift effects are negligible. In the case of CRDS-based designs, incident radiation from the laser is tuned to the center frequency of a comb tooth, allowing the cavity's optical field to increase. The detector monitors interior cavity field strength and when it reaches a preset level, the laser frequency or the comb tooth frequency is changed. Without further laser input, the cavity optical field intensity decays exponentially due predominantly to molecular absorption, although other cavity loss mechanisms are present but tend not to have the temporal nature of the molecular absorption process. The detector monitors the exponentially decreasing signal (the 'ring-down' time depicted at the right in Figure 10). The cavity's optical field is allowed to decay far below its original value, after that time, the procedure can be repeated. Detector voltage is digitized throughout, and this data is fit over the ringdown time to an exponential in order to obtain a ring-down time value. Since ring-down times typically have durations of a few to tens of microseconds, drift in detector response is effectively eliminated. Ring-down times with the cavity evacuated and filled with the gas mixture to be measured allow calculation of mole fraction values based upon line strength and frequency properties intrinsic to the analyte molecule. Reference information for the molecular absorption feature being used by the instrument design is the basis for this analysis. Auxiliary instrument parameters values, e.g., gas mixture pressure and temperature, must also be measured in order to compute mole fraction values.

Commercially available CRDS instruments used for atmospheric monitoring of greenhouse gases have demonstrated excellent stabilities over several weeks or longer, based upon repeated observations of GHG mole fraction standards (Kwok, 2015; Karion, 2013). Uncertainty in instrument response of $\approx$ 0.1 µmol/mol at nominal CO_2 values ranging from $\approx$ 400 µmol/mol to 800 µmol/mol while supported by periodic observations of standard gas mixtures are routine.

6.2.2 High-accuracy spectral reference data

Spectral reference data of line shape and amplitude are determined using research-grade CRDS instruments measuring mole fraction of primary gas concentration standards whose mole fraction value has been assigned using the gravimetric methods discussed previously. CRDS determinations of molecular absorption line and band spectral reference data for the greenhouse gas molecules and other species of interest, such as oxygen, carbon monoxide, and water vapor provide spectral reference data (Polyansky, 2015) are used in applications such as remote sensing. The combination of these measurement methods has been shown to provide line strength and shape data at previously unattained accuracy levels. These data have been incorporated in widely available spectral reference datasets, such as the HITRAN database, as a primary means of disseminating them to both the instrument design community and to the atmospheric remote sensing community. Such information is widely used in remote sensing applications.

6.3 Remote sensing of mole fraction

Remote sensing-based measurements of atmospheric greenhouse gas mole fraction rely on spectroscopic features of greenhouse and air quality gas molecules. Two general approaches are used: Methods that disperse the observed light spatially over a relatively broad wavelength range in order to observe single or multiple absorption bands or those that are based on relatively narrow laser emissions. Dispersive methods range from the grating spectrometers used in the satellite instruments of NASA's Orbiting Carbon Observatory (Crisp, 2017) to the Fourier Transform IR spectrometer used by the Japan Aerospace Exploration Agency's GHG observing Satellite 'IBUKI' (Nakajima, 2017). These specialized instruments feature limited spectral range resulting in good resolution in regions of individual absorption bands. Instruments with broader spectral range, e.g., the TROPOMI instrument on the European Space Agency's Sentinel-5P satellite (ESA, 2019), have a much wider spectral range to observe a larger number of analytes with less spectral resolution. These are focused on the air quality trace gases, ozone, SO_2, NO_2, CO, CH_4, and atmospheric particulates.

The frequency tunability and stability of single-frequency laser-based methods most often observe features of a single absorption line in a band, similar to the approach described above for CRDS instruments. Remote sensing instruments project the laser beam over paths of hundreds of meters to several kilometers in the atmosphere to make either path-integrated or path-segregated measurements using on-line center, off-line laser frequency tuning. Optical frequency combs are laser sources having a spectrum of emissions that is spread over an octave or a wavelength range of a factor of 2 (Cundiff, 2003; J. Hall, 2006; Hänsch, 2006). In the near infrared, this range can extend from ≈ 1 μm to ≈ 2 μm, covering absorption features of several small molecules of interest to both greenhouse and air quality gas measurements. Frequency comb observations involving the measurement of multiple absorption bands simultaneously for multiple molecular species (Coddington, 2016) over kilometer ranges have been reported. To date, laser-based methods have only been used in surface and airborne platform applications.

6.3.1 Low Earth Orbit (LEO) satellite instruments

Several satellite-borne CO_2 observing instruments are currently on-orbit: NASA's Orbiting Carbon Observatory (OCO) satellites (Eldering, 2017; Osterman May, 2015), the Japan Aerospace Exploration Agency's Greenhouse Gases Observing Satellite (GOSAT) series (JAXA—Japan Aerospace Exploration Agency JAXA, 2019), GHGSat–Claire (GHGSat, Inc., 2019) that monitors both CO_2 and CH_4, and China's TanSat or CarbonSat (China Meteorlogical Administration, 2018). Recently, there have been several announcements of intent to put additional greenhouse gas observing satellites or groups of them into orbit (EDF, 2018; Bloomberg, 2019). These observing instruments are carried on satellite platforms in LEO, sun-synchronous, polar orbits such that the instruments view Earth's surface in sunlight constantly with a near constant illumination angle. LEO satellites revisit the same surface location with fixed periodicity. For example, the OCO-2 satellite revisits the same spot on Earth every 16 days.

On-orbit observing instruments use sunlight reflected from Earth's surface as their light source, resulting in two transits through the atmosphere. This increases observational sensitivity of column-averaged mole fraction. Designed to be sensitive over restricted wavelength regions, such instruments resolve the individual lines across all or most of an absorption band in order to improve mole fraction measurement accuracy. Interference of on-orbit observations of atmospheric column-averaged mole fractions arise for the presence of clouds and/or significant atmospheric particulate concentration in the atmospheric observing path. To account for these variations in effective atmospheric optical depth, the OCO series of instruments contain three spectrometers tuned to two CO_2 absorption bands, the 1.61 μm and 2.05 μm bands, and one to oxygen's A absorption band near 765 nm. As an observational calibrant, molecular oxygen is well mixed and of nearly constant concentration and distribution in the atmosphere. This observing channel gives the means for simultaneous observation of atmospheric optical depth/pathlength and detection of atmospheric scattering effects that interfere significantly with observations in the two CO_2 observing bands. The OCO series of instruments have associated optical systems that momentarily lock their view onto specific surface locations while flying overhead. For the OCO-2 instrument, a target track pass can last for up to 9 minutes while acquiring as many as 12,960 samples at local zenith angles that vary between 0° and 85°. These samples are obtained from $\approx 2 \times 3$ km footprints on Earth's surface. Target mode allows these to be closely spaced in a selected region, such as a city, thereby providing spatially dense mole fraction data over selected regions of ≈ 10 kilometers.

6.3.2 Geosynchronous Orbit (GEO) satellite instruments

A primary characteristic of greenhouse gas observing satellites in LEO is the relatively sparse coverage of the surface and atmospheric column over which measurements are made with a $< 1.5°$ separation between ground tracks, or ≈ 160 km (100 miles). Geostationary satellites travel in an orbit that matches the speed of Earth's rotation so that they remain above the same place on Earth's surface constantly. Platforms in GEO provide observing instrument designs with the opportunity to cover all of the Earth's surface viewed from that orbital location. Geostationary satellites are mainly used for communications, TV broadcasting, and weather observation. NASA, NOAA, and ESA all have satellite missions in geostationary orbit, e.g., the Geostationary Operational Environmental Satellites GOES-16 and 17 observe weather patterns and environmental parameters in the Northern Hemisphere. Although there are no greenhouse gas observing satellites in GEO, NASA has begun planning an observing instrument design with the potential to advance understanding of the global carbon cycle by mapping key carbon gases in the atmosphere from the GEO vantage point, a satellite designated GeoCARB (Moore, 2018). Placed above the Americas, such an instrument in GEO could provide sustained mole fraction observations from between $\approx 50°$ North to $\approx 50°$ South latitude, stretching from the southern tip of Hudson Bay to a few degrees north of Tierra del Fuego, South America. Column averaged mole fraction measurements of CO_2, CH_4, and CO would be available across the hemisphere. This instrument design will be based on OCO spectrometer designs with the addition of a CH_4 channel and using the oxygen A-Band total column calibration strategy of the OCO satellite series. A-Band capability also supports observation of solar induced fluorescence (Sun, 2018; Koffi, 2015) that would provide additional information about photosynthetic activity of surface vegetation, potentially providing significantly improved quantification of CO_2 uptake across the Western Hemisphere. Instrument designs that could supply data at spatial resolutions of 5 to 10 kilometers at an estimated rate of ten million samples per day are being investigated. Such data is anticipated to allow determination of both the exchange of carbon between the atmosphere and the land masses of the Americas and the distribution and dynamics of CO_2, CH_4, and CO in the atmosphere not available otherwise. The GeoCarb launch date is currently estimated to be in the early 2020s.

6.4 Surface-based instruments

Surface-based instruments that measure atmospheric greenhouse and air quality gas use the similar classes of techniques as do LEO satellite instruments, dispersive spectrometers coupled with solar viewing collection optics and laser-based sources having both narrow and broadband emission capability with specialized detection. Fourier Transform IR (FTIR) spectrometers are available in a range of wavelength resolutions. Those used in atmospheric observing applications clearly resolve individual lines of absorption bands supporting mole fraction determination. Laser-based sources fall into the two categories discussed previously.

6.4.1 Dispersive, solar-viewing instruments

Solar viewing Fourier Transform IR (FTIR) spectrometers are used for total atmospheric column observation of mole fractions from Earth's surface over broad wavelength ranges using the sun as the illumination source (Chen, 2016). NASA's Total Column Carbon Observing Network (Wunch, 2011) has been established to support the calibration and validation of greenhouse gas observing satellites, such as the OCO and GOSAT series. The TCCON instruments are very high-resolution spectrometers located globally near or on satellite overpass tracks. Although this network has stations located globally, coverage of Earth's surface is sparse. Recently, the COllaborative Carbon Column Observing Network (COCCON) has been established for use in remote regions of the world and utilizes portable sun-observing FTIR instruments (Frey, 2019). If realized, a COCCON observing network could add substantially to existing mole fraction observing capabilities. The portability and lower cost of these devices hold potential for uses ranging from urban to remote settings.

6.4.2 Laser-based, path-integrated methods application

Laser-based mole fraction measurements use two long path methods, integrated path differential absorption (IPDA) and Light Detection and Ranging (LIDAR). Both rely on signals returned to a transceiver set from reflections. In the IPDA case, the reflection comes from a reflective surface located at some distance from the transceiver and the laser illumination is constant over the period of the measurement. IPDA methods employ both single-frequency lasers using the on-line/off-line diode laser tuning method (Johnson, 2013; Dobler, Greenhouse Gas Laser Imaging Tomography Experiment (GreenLITE), 2016) and frequency comb methods (Reiker, 2014; Waxman, 2019). Path-integrated measurements often use retroreflectors to return the incident laser beam to a co-located receiver, often a small telescope. Applications of these methods is spatially limited by the effective range of the laser system used. Some examples of these are given later in the chapter.

6.4.2.1 Single-frequency laser example

An example single-frequency laser application using a tunable, intensity-modulated laser source initially developed to measure CO_2 mole fraction over pathlengths of several hundred meters covering an area of ~ 1 km² to generate near-real-time two-dimensional estimates of mole fraction spatial distribution. Sparse tomographic reconstruction analysis was used to compute 2-D, CO_2 mole fraction maps using chords of pathlength of 50 to 200 meters to demonstrate feasibility of the technique, the GreenLITE[10] system (Dobler, Greenhouse Gas Laser Imaging Tomography Experiment (GreenLITE), 2016). Subsequent developments extended the range to more than 5-kilometer pathlengths suitable for application over urban settings involving complex emissions sources.

As a demonstration in urban applications, a prototype system was operated over the center of Paris for approximately a year, beginning in November 2015 (Dobler, 2017), to capture seasonal emissions variations. Using two transceivers separated by ≈ 2 km and 15 retroreflectors separated from each transceiver by 3 to 5 km, the spatial distribution of CO_2 emissions over the center of Paris was mapped on a continuous basis. This year-long deployment demonstrated the feasibility of using such a system in an urban environment. The top image in Figure 11 illustrates transceiver and retroreflector positions in Paris, showing the chords between each transceiver and the array of reflectors and a mapping of CO_2 mole fraction. Each transceiver was mounted on a remotely controlled pointing system that acquired a

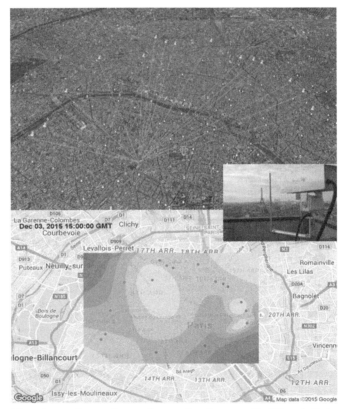

Figure 11. The disposition of transceivers and retroreflectors placed on top of buildings in Paris roughly arranged in a hemisphere. Orange and Green rays are chords, or paths, between a transceiver and one of the 15 retroreflectors. Images courtesy of J.T. Dobler, Spectral Sensor Solutions, LLC & T.S. Zaccheo, Atmospheric and Environmental Research.

---

[10] Certain commercial equipment, instruments, or materials are identified in this paper in order to specify the experimental procedure adequately. Such identification is not intended to imply recommendation or endorsement by the National Institute of Standards and Technology, nor is it intended to imply that the materials or equipment identified are necessarily the best available for the purpose.

return from individual reflectors. Each transceiver cycled through the reflector array with at a rate of $\approx$ 10 seconds per reflector, transiting the complete array in less than 4 minutes including a calibration observation.

An integrated column mole fraction measurement is measured along each chord. These IDPA data are input for a sparse 2-D tomographic reconstruction algorithm of the mole fraction field. Each chord crossing the reconstruction returns a mole fraction estimate at that location. These values then allow interpolation between chord crossings, thereby giving continuous estimation of mole fraction over the measurement domain. Because the cost of additional chords is that of a retroreflector and its installation, a relatively small additional cost, a system architecture containing many mole fraction values could be realized where doing so with discrete instruments could be prohibitive in both cost and access to chord crossing locations. In this 30-chord design, $\approx$ 98 chord crossings result. Modification of transceiver design for 360° rotation with similar pointing capability, rather than the $\approx$ 180° rotation used in this demonstration, could support a substantial increase in reflectors and chords depending upon reflector placement, which is strongly influenced by local building/structure access in a given urban area. Such a design could significantly increase the size of the sampled domain.

Initial comparisons with an *in situ* CRDS instrument calibrated relative to WMO standards indicates that the system precisely tracks the CO_2 concentrations within an urban environment. However, measurements made both before and after the Paris demonstration indicate that although the precision of the system is $\approx$ 0.5 µmol/mol or smaller, offsets of the order of 2 µmol/mol to 3 µmol/mol may be present. Systems such as these continue to evolve as research continues to improve their performance and methods developed to better establish standards needed to assess long-path, laser-based methods.

6.4.2.2 Dual-frequency comb example

Dual-frequency comb spectroscopy (DCS) methods have recently been demonstrated over the city of Boulder, Colorado (Waxman, 2019). The DCS makes simultaneous measurements of multiple species and path-integrated temperature with low systematic uncertainty and without the need for instrument calibration. Additionally, the eye-safe, high-brightness, single transverse-mode DCS output allows for beam paths exceeding 10 km, while the speed and parallelism of the measurement suppress any spectral distortion from the inevitable turbulence-induced power fluctuations over such a path. These measurements were made along two paths, as shown in Figure 12, one a reference path and the other over the city itself. The reference path traversed a very low population/industrial activity region to the west-southwest of the city while the over-city beam traversed a path covering a substantial fraction of the city. Data at 5-minute resolution for CO_2, CH_4, and H_2O and its isotopologues were acquired for 7½ weeks as a demonstration of the method for emissions quantification as a means to estimate traffic emissions of CO_2. The data were filtered for conditions where wind direction was from the west at wind speeds sufficiently low to allow the increase in mole fraction from traffic in the over-city path to have a measurable enhancement relative to the reference path. Using a Gaussian plume model, combined with anticipated traffic emissions and accounting for non-traffic sources, the measured emission values relative to a bottom-up city inventory estimate agreed within 25% to 32% relative to the measured value. As discussed above, IDPA methods of higher chord count provide spatial resolution.

6.4.2.3 Laser-based, spatially resolved methods

Spatially resolved methods are generally based on Lidar techniques that use short laser pulses reflected back to a receiver by atmospheric particulates. These pulses are much shorter that the round-trip time necessary for a reflected pulse to return to the receiver. For example, a laser pulse of $\approx$ 50 ns width travels approximately 300 meters in a microsecond. A detected roundtrip time, measured from the beginning of the pulse from the laser to the time of detection, determines the reflecting particle's distance from the laser source. Temporal binning of detected signals effectively segregates the range of the laser pulse into path segments between the source and reflecting particles located along the laser's path. This temporal gating, range gating, effectively segments the laser beam's path into segments over its practical range.

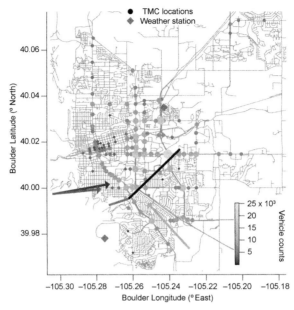

Figure 12. Reference and over-city DCS IDPA chords for Boulder, CO. Green diamonds indicate weather stations providing wind speed and direction. Colored circles are turning movement traffic counts used as a proxy for traffic source locations. Color and size represent traffic counts. Dominant wind directions are shown for prevailing and test days (22 Oct.–purple and 25 Oct.–blue). The dominate wind direction is indicated by the aqua arrow. Image courtesy of E. Waxman, NIST.

The combination of a short pulse-length laser tuned to frequencies either coincident or non-coincident with an absorption line (on-line/off-line) with range gating of signals reflected from atmospheric particulates is often termed Differential Absorption Lidar (DIAL). Recent developments of CO_2 DIAL have shown promising demonstrations of this method while complying with eye safety requirements for laser beams (Wagner, 2018). DIAL is free of constraints of IPDA methods that rely on retroreflectors located across a domain at fixed locations. Applying CO_2 DIAL for real-time vertical mole fraction profiling has the potential to add considerably to our knowledge of vertical mole fraction distribution in the PBL and perhaps above it. Applied in a horizontal or near horizontal mode, DIAL functions similarly to the IDPA methods described above. However, DIAL does not depend upon detection of signals reflected from fixed objects. Rather DIAL detects photons reflected back to its receiver/detector from atmospheric particles, much the same as particle and Doppler LIDARs. Although, current embodiments of eye-safe CO_2 DIAL show good agreement with CRDS measurements made near the beam path in research level experiments, DIAL detector sensitivity limitations must be overcome before broader applications can be supported.

6.5 *Trace gas and energy flux measurement at micro-meteorological scales*

Measurement of the exchange of trace gases, water vapor, and energy between the atmosphere and Earth's surface has seen extensive development in the research community for several decades, driven mainly by technological developments in wind velocity measurement due to advances in ultrasonic anemometry. The significantly improved frequency response in 2 and 3 dimensions up to ≈ 100 Hz has advanced the development of the eddy covariance (EC) measurement technique. EC quantifies exchange fluxes of trace gases, water vapor, particulates, and sensible heat energy between the surface and the PBL via the constant flux layer communicating with surface processes. EC measurement has been important to better understand ecosystems and the processes forming their function and to the field of micro-meteorology with industrial applications, e.g., in agriculture (Parker, 2015), being an important research element. The information and insights provided by the EC and related methods is an important means of testing ecosystem, crop, and land surface models useful in predicting response both to changes in average values of climate-related parameters, and to their anticipated extremes.

The eddy covariance technique is based on the measurement of atmospheric turbulence properties. A fundamental assumption is that of homogenous and fully turbulent atmospheric transport in the region of interest at and near the surface. Primarily driven by circular eddy structures in the lower atmosphere, these eddies transport atmospheric constituents primarily horizontally but with vertical components that carry gases or energy from the surface into the PBL via a constant flux layer. Mathematically the vertical flux, or net ecosystem exchange, F (in units of mol m^{-2} s^{-1}), can be represented as the covariance of the vertical velocity and dry air molar density of the species of interest, e.g., CO_2 (Burba, 2013; Gu, 2012).

$$F \approx \overline{c_d} \ \overline{w's'}$$

where:

F	=	vertical flux or net ecosystem exchange (mol m^{-2} s^{-1}),
$\overline{c_d}$	=	mean dry molar density (mol m^{-3}),
w'	=	instantaneous deviation from the mean of the vertical velocity (turbulent component),
s'	=	instantaneous deviation from the mean of the species of interest dry air mole fraction, and
$\overline{w's'}$	=	mean of the instantaneous deviations of the vertical velocity and dry air mole fraction covariance.

This simple mathematical relation rests on numerous assumptions. Some of the primary ones are the following.

- Homogenous and fully turbulent atmospheric flow with mean vertical flow assumed to be small to negligible (in some theoretical treatments).
- No sources or sinks sufficient to interrupt this homogeneity exist in the domain.
- Air density fluctuations are negligible, i.e., air density is constant or closely so over the period of the observations, generally on the order of a few minutes to hours.
- The species of interest are chemically stable during the time of the observations.

Experimental conditions must also account for these assumptions; some of these are the following:

- Terrain over which the flow is established is homogenous.
- Observations at the measurement site are representative of an area upwind, i.e., a fetch or footprint (this chapter in a larger geographical sense, sec 5.2.1.3, and in the next section).
- Wind velocity, atmospheric temperature, and pressure instrumentation must be sensitive to small changes at high frequencies.
- Mean air flow and turbulence at the measurement point are not appreciably distorted by the observing installation structure or the instruments themselves.

6.5.1 Observations and footprint modeling—the field of view

In the eddy covariance technique, micrometeorological instruments are usually placed on a tower at a height chosen for sensitivity to the region of interest. Instrumentation generally includes ultrasonic anemometers, positioned to measure the three atmospheric velocity components, air temperature and pressure, and specific auxiliary parameters pertinent to the investigation. Frequency response of the instruments observing EC-related parameters should be comparable to that of the anemometry, if possible, they should have similar frequency response. Because this may not be the case, correction methods have been developed to account for these effects in the analyzing observations (Horst, 2000). Observations are usually taken over hourly or sub-hourly periods when atmospheric dynamics are relatively constant, such that the EC assumptions of constant mean air density and atmospheric velocity are met. The term "footprint" summarizes the concept of the region, or effective fetch, that influences data at the observation station. Much like the influence or sensitivity functions discussed previously (sec 5.2.1.3) which determine the location of sources or sinks that influence or enhance the signals observed in the much taller towers associated with urban networks. Footprint models seek to describe the originating locations of EC signals. Several footprint modeling approaches have been developed in order to establish formal connections between micrometeorological flux measurements of trace gases above a vegetation region and their mass conservation in a surface-vegetation-atmospheric context. Footprint modelling is a means of investigating limitation of the concept, particularly in the design and analysis of experimental arrangements (Schmid, 2002; Vesala, 2008).

6.5.2 EC applications in carbon management

Although in many cases carbon management focuses on emissions, one should keep in mind that, on a global scale, approximately one half of yearly anthropogenic emissions are removed from carbon sinks, either by the oceans or the land. EC methods have been widely utilized as a critical resource to increase our knowledge of vegetative exchanges with the atmosphere, i.e., CO_2 uptake or removals. For example, agricultural applications use EC methods to better quantify greenhouse gas emissions from cropped and grazed soils under current management practices and to identify and further develop improved management practices that will enhance carbon sequestration in soils, decrease emissions and uptake of CO_2 due to crop and soil photosynthetic activity. Methane emissions from various livestock-related process, and N_2O emissions, e.g., from fertilization process, are also investigated using EC methods. This information can promote sustainability and provide a sound scientific basis for carbon credits and GHG trading programs (Follet, 2010).

Several research networks that use EC as a central method for the study of ecosystem processes have been established. In the U.S., NSF's National Ecological Observatory Network (NEON) (Batelle, 2011) has been designed to collect data that characterize and quantify complex, rapidly changing ecological processes across the continental U.S. The Ameriflux research network relies upon principal investigator-managed sites uses EC methods to measure ecosystem fluxes of CO_2, water, and energy in North, Central, and South America (Ameriflux/DOE, 2015). The FluxNet network is a global network of micrometeorological tower sites that use eddy covariance methods to measure the exchanges of carbon dioxide, water vapor, and energy between terrestrial ecosystems and the atmosphere (ORNL DAAC, 2002). The Integrated Carbon Observing System Research Infrastructure (ICOS, 2008) aims to provide long-term, continuous observations of greenhouse gas sources and sinks. ICOS is a pan-European research organization focused on carbon cycle science, greenhouse gas budgets, and the perturbation thereof. The EC technique is a principal method in the national ICOS observing stations that span applications to ecology studies and atmospheric and ocean stations. To ensure consistency in data and analysis, ICOS has established a system of standards and instrumental configurations primarily for atmospheric stations (Rebmann, 2018).

6.6 Mass balance and tracer gas methods for emissions determination

Mass balance flux estimation methods are usually conducted using measurements of trace gas mole fraction and horizontal wind velocity to estimate emissions flux using a conservation of mass between upwind and downwind observation points of the atmosphere. Wind vector information can be obtained using either direct measurement, often the case with airborne experiments, or simulated values obtained from Numerical Weather Prediction Models. The method has been applied to atmospheric trace gases of interest to both the air quality and greenhouse gas measurement communities. The mass balance method is based on measurement of the difference in flux entering and leaving a region and has been applied to cities (Salmon, 2018), to large sources within urban areas (Conley, 2017), and more generally to individual or small groups of sources (EPA GHGRP, 2015; White, 1976). The method may be applied to individual emission processes or facilities or to defined geographical domains using either fixed or mobile observing methods located on the surface or from airborne platforms. Flux quantification from a source, such as a city or an isolated emission source, e.g., power plant located far from other sources, using this approach is a difference measurement based on mole fraction observations taken both upwind and downwind of the source using various types of mobile platforms.

6.6.1 Airborne platforms

Aircraft-based greenhouse gas mass balance measurements are quite useful for obtaining a snapshot of emissions from domains over which an aircraft can fly multiple transects in and out of the source's concentration plume. Measurements of the trace gas concentration are made in real-time of air typically pulled into a sampling port as the aircraft transits the region. The spatial resolution and transect length of such measurements depends on the aircraft's speed and the instrument sampling time. Given the variability in wind direction and speed over the area flown, instantaneous aircraft velocity measurement and trajectory information are required in order to determine complementary wind vector values during the course of an observation. The use of aircraft GPS instrumentation or fast-response wind probes has improved the accuracy and reliability of these measurements in recent decades.

As shown in Figure 13, flight planning is used to orient flight paths to be as near perpendicular to the wind direction as practicable. Figure 10 shows a diagram of a domain around the Baltimore-Washington D.C. region with CO_2 mole fraction measurement over multiple horizontal and vertical transects. These snapshots of mole fraction spatial distribution are taken both horizontally and vertically in the planetary boundary layer where emission plumes, to some degree, remain intact as they move vertically through it. The variation in mole fraction both horizontally and vertically is sampled, clearly showing the downwind plume from the urban areas. In this case, both upwind and downwind transects are used to determine the flux difference due to city emissions. In some cases, horizontal downwind transects that sample the atmosphere well away from the urban plume may be used, as they are perhaps more representative of background conditions (Karion, 2015; Cambaliza, 2014).

The data shown in Figure 13 illustrates the requirements placed upon mole fraction measuring instrument response, sensitivity, and accuracy in this application. Here, the CO_2 mole fraction range is 10 µmol/mol to 15 µmol/mol. To support these accuracy requirements, airborne instrumentation measurement strategies often include the use of on-board reference gas standards to characterize instrument performance and accuracy before, after, and at times during a flight.

6.6.2 Surface mobile platforms

Observations of emissions using ground vehicles have been used to characterize emissions of methane from the oil and gas production, transportation, and delivery systems. Similar in concept to airborne methods, IR imagers and mole fraction measurement instruments mounted on cars and trucks are used to observe trace gas plumes downwind of point sources. These mobile sensing approaches have been used extensively to identify emission sources and fugitive emissions, and to characterize known emission sources (Albertson, 2016; Yacovitch, 2017; Weller, 2018). Unlike airborne platforms, where the vertical extent of a plume can be observed, surface vehicle-based surveys are limited in altitude to a few meters above ground level. Wind speed determinations are made with anemometry, either at a fixed location near the point of investigation or mounted on the vehicles involved. In the latter case, compensation for vehicle motion relative to wind direction and speed is required.

Another technique often combined with mobile observing platforms uses a surrogate gas that can be observed experimentally as a tracer to infer the emission rate of another gas not measured directly. Tracer experiments often involve releasing tracer gases at measured rates near a suspected emission location. A mobile measuring vehicle transits downwind of the locations of interest, observing the enhancements of the tracer gases. Given that the ratios of measured species to non-measured species mixing ratios are known, these combined with measured tracer gas release rate(s) provide the means to determine emission rates (Yacovitch, 2017).

7. High Spatial Density Greenhouse Gas Observing Networks—The INFLUX Example

The Indianapolis Flux Experiment (INFLUX) began in 2010 to develop and demonstrate urban-scale or whole-city GHG flux measurement capabilities, and particularly to compare results from top-down and bottom-up methods. Indianapolis was chosen in part because of its moderate size ($\approx$ 1.8 million inhabitants in the city and surrounding area), relative isolation from other strong emissions sources, and location in a landscape of flat terrain where NWP models are anticipated to have optimal performance. Critically, Indianapolis was the development site for an urban-scale, high spatiotemporal resolution, bottom-up emissions model or data product, known as Hestia (Gurney, 2017).

The initial INFLUX project concept to compare three independent and complementary whole-city emissions quantification methods: Two based on top-down atmospheric observing methods and the bottom-up Hestia emissions data product. Two top-down atmospheric flux measurement approaches were employed: The aircraft mass balance measurement and a two-station, surface-based, mole fraction observing network for the gases CO_2, CH_4, CO, and water vapor. The initial surface observing network had two tower-based nodes located to the southwest and the northeast of the city, respectively, along the prevailing wind direction.

The objectives of INFLUX were broadened shortly after the project began to investigate the capability of a spatially dense, surface observing network coupled with Bayesian inference analysis to identify CO_2 source locations and estimate flux magnitude at those locations (Davis, 2017). The second INFLUX phase aims to develop, assess, and minimize uncertainties of methods for quantifying greenhouse gas emissions at the urban scale, using the Indianapolis urban area (Figure 14) as a testbed. Target performance criteria for the expanded INFLUX goals were demonstration of emission source location identification within 1 to 2 km^2, relative flux magnitude uncertainty of $\approx$ 10%, all at a weekly or sub-weekly temporal scale. Similar expectations were put in place for whole-city flux determinations. The observing network density within the city was increased from 2 to 12 nodes. High-accuracy measurement of CO_2, CH_4, and CO mole fraction from continuous atmospheric sampling made at heights from 39 m to 136 m above ground level use existing communication towers fitted with a multi-line sampling manifold to sample the atmosphere at several elevations. Cavity ring-down spectrometers (CRDS) are used at all twelve sites. Each network node has at least one reference gas mixture contained in high pressure aluminum cylinders that is traceable to WMO standards (Zhao, 2006) and to the SI (Rhoderick, 2016). These measures are taken to ensure data quality and consistency within the observing network, based on a reference framework closely tied to global mole fraction references disseminated by the WMO Central Calibration Laboratory (Richardson, 2017). To assess CRDS instrument performance periodically and conserve valuable gas mixtures, gas mixtures in cylinders of known (calibration cylinders) or of unknown (target cylinders) mole fraction are used by a network nodes' control hardware to periodically present to the CRDS instrument a gas of stable mole fraction. These procedures are used to quantify instrument drift and detect failure conditions. Network-wide round robin tests performed every 1 to 2 years detected possible drift in mole fraction of both target and reference gases.

Flask sampling systems deployed at six network nodes drew samples from the same manifold as the node's CRDS analyzer. Flasks, containers made of specialty glass so as to not contaminate gas samples and having a volume of approximately one liter, were filled approximately monthly and sent to NOAA-GMD for analysis. Laboratory gas analysis for approximately 40 trace gas species was made by NOAA-GMD on flask samples (Sweeney, 2015; Turnbull, 2012). Analyses also included the greenhouse gases and particularly the carbon 14 radio-isotope in the form of $^{14}CO_2$. Radio-carbon data assist in estimating usage of fossil fuels in the city. Data from flask analyses provides a means to directly compare results obtained from the node's CRDS spectrometer. These comparisons are a means of periodic verification of analyzer performance for CO_2, CH_4, and CO mole fraction. Comparison levels were less than or equal to 0.18 μmol/mol CO_2, 1.0 nmol/mol and 6 nmol/mol for CH_4 and CO, respectively (Turnbull, 2015).

To better characterize atmospheric dynamics and energy transfer between the surface and the atmosphere, additional measurement capabilities were added. Doppler LIDAR measurements located to the Northwest of the city center have provided three-dimensional wind vector and turbulence data and PBL depth information continuously for the last several years. Four network towers were fitted with eddy covariance measurements in order to investigate surface-atmosphere energy exchange behavior across the typology of the city. These towers sampled energy flux properties and behaviors of land surface for urban typologies ranging from rural to the heavily urban center of Indianapolis (Sarmiento, 2017). This rather diverse set of measurements is indicative of the many processes operating in urban environments

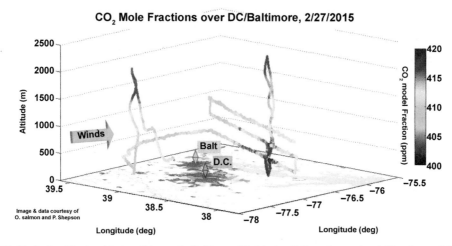

Figure 13. Vertical and horizontal aircraft transects; Baltimore–Washington D.C. regions and coded by observed CO_2 mole fraction. Image and data courtesy of O. Salmon and P. Shepson.

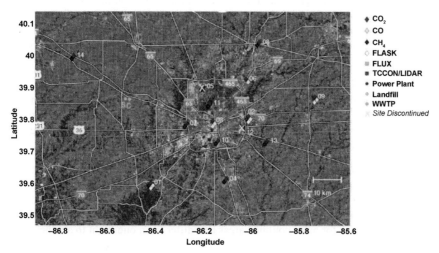

Figure 14. Locations and designations of the INFLUX observing network. Each communications tower-based, network observing node measures CO_2, CO, and/or CH_4 continuously, ≈ 5 second cycle time, depending on the type of CRDS analyzer installed. Six nodes have flask sampling capability and several have eddy flux measurement capability. LIDAR, both Doppler and particulate, systems monitor PBL dynamics to the Northwest of the city. http://sites.psu.edu/influx/site-information/.

found in U.S. cities of moderate size. This information allowed assessment of various processes impacting atmospheric transport within and around the city as a means of better understanding how those processes interact and impact GHG emissions estimation.

These various information sources provided an opportunity for extensive observation-to-model comparison, e.g., the WRF model, to assess simulation errors in meteorological variables such as latent heat and sensible heat[11] fluxes, air temperature near the surface and in the PBL, wind speed, direction, turbulence, and PBL height (Deng, 2017). Although results such as these have yet to be used to modify NWP models and additional information is needed, they represent progress toward a better understanding of PBL dynamics to improve NWP capabilities, which in turn is anticipated to have a positive effect

---

[11] Sensible heat in a thermodynamic system is an exchange that results in a temperature change in the system.

on reduction of uncertainties in top-down flux estimation. Extensive descriptions of the methods used and some results derived from INFLUX are published in a journal special issue (Elementa Special Feature, 2016).

Frequently updated information on emission rates and their change will considerably enhance evaluation of the effectiveness of carbon management approaches in cities. Recent results from four of the independent CO_2 emissions methods of the INFLUX project were analyzed in order to address two key questions: (1) what the magnitude of the whole-city emission is and (2) what the uncertainty of these quantification results is. As one of the few urban areas where multiple emission assessment methods have been implemented, Indianapolis provides a unique opportunity to compare different methods directly. To address these questions, a comparison of the whole-city CO_2 emission rate was made, based upon different approaches: A science-driven high-resolution urban inventory-based emission data product (Gurney, 2011), an atmospheric transport model inversion based on *in situ* tower observations, and two different mass balance flux estimates from aircraft observations. To evaluate differences between the methods, discrete flask-based measurements of $^{14}CO_2$ were used to determine fossil fuel CO_2 separately from biogenic CO_2 flux contributions. To minimize complications associated with biospheric CO_2 fluxes in this first attempt to compare differences between methods, only winter-time emissions were considered (Turnbull, 2019).

CO_2 emission rate values reported initially for the methods used in this comparison ranged from 14,600 to 22,400 mol/s with a 1 sigma variation of 21% and a maximum difference 42%. Although an improvement over previous uncertainty estimates for urban emissions of 50–100% from other studies (Gately, 2017; Pecala, 2010), these initial INFLUX flux estimate uncertainties are insufficient for detection of emissions trends on the order of 10% per decade. Re-analysis of the initial estimate involved the following adjustments:

1. Adjustment for fossil fuel CO_2 emissions alone relative to total CO_2 emissions.
   The Hestia data product compiles data for anthropogenic CO_2 emissions derived from combustion of fossil fuels that include bioethanol. The atmospheric inversion and aircraft mass balance methods both estimate the net total urban enhancement in CO_2, which includes the influence of both anthropogenic and biogenic CO_2 fluxes.

2. Adjustment of geographic region sampled by the various methods.
   The geographic area for which emissions are sampled differ among the different methods. For example, the Hestia data product incorporates information from the 8 surrounding counties and Marion County that contains the geopolitical boundaries of Indianapolis. The aircraft mass balance method samples fluxes for more confined areas that differ from one flight to another.

3. Adjustment for the time of day and time period of an estimated flux.
   Hestia is the mean flux over all hours of the day. The inversion analysis only uses observational data from the 11 am to 4 pm local time period. The nine aircraft mass balance flights used in the analysis were only conducted on weekdays in November and December 2014, in the daytime. Adjustments to Hestia and the inversion method were made to coincide as much as practicable with the mass balance observation times.

4. Accounting for CO_2 background or atmospheric concentration choice in the incoming atmosphere.
   In principle, the background CO_2 signal would be the mole fraction that would have been observed by at tower or aircraft location with no urban emissions present. Each method treats estimation of this background concentration differently. Consistent methods are needed that properly account for differences in choice of background for different methods.

Results were compared in winter, only to avoid the substantial biogenic CO_2 flux contributions both coming into the Indianapolis domain from the large agricultural regions outside it during the growing season and those occurring within the city from urban vegetation. Hestia did not contain a biogenic component, which is typical of emissions models in their current state, although efforts are underway to include urban domain vegetation estimates (Sargent, 2018). As shown in Figure 15, by using this

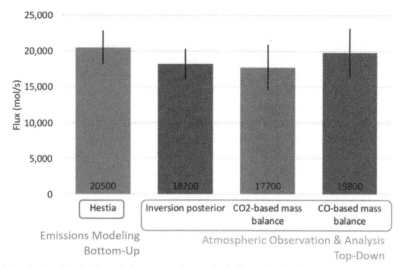

Figure 15. Whole-city fossil fuel CO_2 emissions rate estimates for Indianapolis in winter agree within 7%. This is achieved by accounting for differences in spatial and temporal coverage, as well as the trace gas species measured. Image courtesy of J. Turnbull.

reconciliation procedure to account for differences in spatial and temporal coverage, agreement among the wintertime whole-city fossil fuel CO_2 emission estimates falls within 7% (Turnbull, 2019). Error bars on each method indicate uncertainty estimates for each.

Relative to previous urban-scale comparisons, this result is a significant advance. This study represents the first comprehensive, multiple-method assessment of urban fossil fuel CO_2 emissions and demonstrates that agreement across these can be useful to those tasked with carbon management of whole city emissions.

The complementary application of multiple scientifically driven, evidence-based emissions quantification methods enables and improves confidence levels and demonstrates the strength of the joint implementation of rigorous inventory and atmospheric emissions monitoring approaches. However, this advance has been made with significant restrictions from the point of view of general application, e.g., exclusion of biogenic emissions, data from the midday only, and through the use of highly accurate measurements, costly equipment and research-grade analysis methods. INFLUX represents the application of considerable research resources, not likely to be available in most urban regions on a routine basis. Continued advances in measurement capabilities are needed in order to move research and implementation of results toward broad application. For example, replication of convergence among methods as demonstrated by the initial success of the INFLUX research should be demonstrated over all seasons, times of day, in different urban regions, and at considerably reduced resource expenditure to make them practically applicable. Doing so requires continued efforts by both the scientific and the stakeholder communities and requires considerably expanded engagement and effort.

8. International Engagement

The atmosphere is a global entity impacting all nations with its warmth. Quantification of carbon is a common thread within any carbon management approach, whether that be based on regulatory or market-based methods or a combination thereof. Consistently accurate quantification methods rely on application of measurement methods with demonstrated performance and correspondence in their results as a means of achieving uniformity. Engagement among parties ranges from UNFCCC/IPCC requirements and structures to those useful for local governments to manage their carbon emissions and private sector involvement in those endeavors. Recent international efforts have seen the development of a new entity that can increase the focus on advancing consistency and accuracy of greenhouse gas measurements.

The Integrated Global Greenhouse Gas Information System (IG^{3}IS) is a component of the WMO's Global Atmospheric Watch program. IG^{3}IS aims to promote development of evidence-based greenhouse gas emissions information and the methodologies that produce it at quality levels consistent with carbon management quantification needs (GAW–WMO Global Atmospheric Watch, 2019; IG3IS, 2018). With a 50-year history of providing research and information on the atmosphere, particularly atmospheric trace gases, GAW and now its IG^{3}IS component, is a driver for consistency in data and measurement and monitoring methods supporting international engagements under the Paris Agreement. GAW's archive of GHG measurement, standards, and monitoring method documentation is an important archive of information aimed at improving atmospheric trace gas quantification capabilities.

The UN Environment Program is closely involved with IG^{3}IS because its efforts will further UNEP's global interests in caring for the world's environment where quantification of process impacting Earth is providing universally recognized bases for a wide variety of activities that encompass carbon management. The General Conference on Weights and Measures (CGPM), established by the Convention of the Mètre (French: Convention du Mètre, also known as the Treaty of the Mètre) in 1875, coordinates measurement standards efforts of the international metrology[12] community and the development and use of the International System of Units, the SI. Its efforts establish the international measurements framework utilized by member and non-member governments to act in common accord on all matters relating to units of measurement, a key ingredient in facilitating equity in commerce and regulation (BIPM-WMO, 2002; BIPM/CCQM, 2017). Working across these organizations and the private sector, IG^{3}IS structures, objectives, and methodologies can provide carbon management systems with consistent, and internationally recognized information and measurement methodologies useful in supporting successful implementation of carbon reduction strategies. The meshing of systems to provide consistently accurate information based on internationally recognized methodologies providing quantitative data of undisputed quality is an important element of carbon management implementation.

9. Summary

The natural philosophers and scientists of the late 18th and 19th centuries laid the foundations of our knowledge of the Earth's atmosphere and the potential impacts that the relatively small concentrations of greenhouse gases have upon it. The analysis of Arrhenius, based substantially upon Tyndall's extensive observations during the mid to late 19th century, was the first quantitative analysis of the impact of water vapor and 'carbonic', CO_2, on atmospheric warming. In the mid-20th century, Keeling's measurements of CO_2 mole fractions propelled the evolution of scientific knowledge and investigation underpinning advances in understanding of the carbon cycle and its impact on atmospheric warming. Keeling's close attention to and the development of highly accurate measurements was an important contribution to the development of highly accurate gas mixture standards. These continue to support scientific innovations and carbon management policies and practices. Advances in understanding of the carbon cycle are based on quantifying and measuring greenhouse gas exchanges between the atmosphere and Earth's surface. As nations individually and collectively move toward managing the emission and uptake of atmospheric carbon, those efforts will substantially benefit from quantitative, evidence-based, and internationally recognized information of sufficient accuracy.

This chapter has discussed some of the measurements and standards systems needed to quantify reductions in greenhouse gas emissions based upon current estimates and statements of reduction requirements forecast for this century. Much of the measurements and standards technology reviewed here is in varying stages of development in anticipation of the need for its use in implementing carbon management practices. Significant measurement science and technology challenges remain alongside those of developing and implementing effective mechanisms to move the results of these investments into practice, thereby providing carbon management system practitioners with access to the quantitative information required for effectiveness and efficiency of management efforts.

---

[12] The scientific study of measurement.

Acknowledgements

The author thanks Dr. Anna Karion for her many suggestions concerning the creation of this chapter and her critical technical review of it.

References

Ackerman, K.V. 2008. Comparison of two U.S. power-plant carbon dioxide emissions databases. Environ. Sci. Technol. 42: 5688–5693.

Albertson, J.D. 2016. A mobile sensing approach for regional surveillance of fugitive methane emissions in oil and gas production. Environ. Sci. & Tech. 50: 2487–2497. doi:10.1021/acs.est.5b05059.

Allen, R. 2014. Climate.Gov. https://www.climate.gov/news-features/blogs/enso/role-ocean-tempering-global-warming.

Alvarez, R.A. 2018. Assessment of methane emissions from the U.S. oil and gas supply chain. Science 361(6398): 186–188. doi:10.1126/science.aar7204.

Ameriflux/DOE. 2015. Ameriflux. Accessed August 2, 2019. https://ameriflux.lbl.gov/.

Anders, R.J. 1996. A 1° × 1° distribution of carbon dioxide emissions from fossil fuel consumption and cement manufacture. Global Biogeochemical Cycles 48(12): 419–429. doi:10.1029/96GB01523.

Arrhenius, S. 1896. On the influence of carbonic acid in the air upon the temperature of the ground. The Philosopical Magazine and Journal of Science 41(Series 5): 237–276.

Babcock and Wilcox. 2018. Steam: Its Generation and Use. Babcock and Wilcox.

Basso, L.S. 2016. Seasonality and interannual variability of CH_4 fluxes form the eastern Amazon Basis inferred from atmospheric mole fracton profile. Jour. Geophys. Res. 121(1): 168–184. doi:10.1002/2015JD023874.

Batelle. 2011. NEON: The National Ecological Observatory Network. Accessed August 2, 2019. https://www.neonscience.org/.

BIPM. 2019. Bureau Internationale des Poids et Mesures. Accessed July 15, 2019. https://www.bipm.org/en/about-us/.

BIPM/CCQM. 2017. GAW Partnership with BIPM. Accessed July 10, 2019. http://www.wmo.int/pages/prog/arep/gaw/documents/GAWSymp2017P3Viallon.pdf.

BIPM-WMO. 2002. CIPM-WMO Agreement. June 28. Accessed June 20, 2019. https://www.bipm.org/en/worldwide-metrology/liaisons/wmo.html.

Bloomberg. 2019. New Wave of Satellites Could Pinpoint Greenhouse Gas Offenders. April 17. Accessed July 6, 2019. https://www.bloomberg.com/news/articles/2019-04-17/new-wave-of-satellites-could-pinpoint-greenhouse-gas-offenders.

Borthwick, P. 2011. Examination of United State Carbon Dioxide Emissions Databases. EPRI CEMs User Group. Chicago, IL.

Bryant, R. 2018. Improving Measurement for Smokestack Emissions—Workshop Summary. NIST Special Publication.

Burba, G. 2013. Eddy Covariance Method—for Scientific, Industrial, Agricultural, and Regulatory Applications. Lincoln, Nebraska: Li-COR Biosciences, Inc.

Busch, K.W. 1999. Cavity-Ringdown Spectroscopy. An Ultratrace-Absorption Measurement Technique. Oxford University Press.

Cambaliza, M.O.L. 2014. Assessment of uncertainties of an aircraft-based mass balance approach for quantifying urban greenhouse gas emissions. Atmos. Chem. & Phys. 14(17): 9029–9050.

Chen, J. 2016. Differential column measurements using compact solar-tracking spectrometers. Atmos. Chem. Phys. 16: 8479–8498. doi:10.5194/acp-16-8479-2016.

China Meteorlogical Administration. 2018. Overview of TanSat. April 4. Accessed July 6, 2019. http://www.cma.gov.cn/en2014/news/Features/201804/t20180424_466965.html.

Christensen, T.R. 2003. Factors controlling large scale variation in methane emissions from wetlands. Geophys. Res. Lett. 30(7): 67–1 to 4. doi:10.1029/2002GL016848.

Coddington, I. 2016. Dual comb spectroscopy. Optica 3: 414–426. doi:10.1364/OPTICA.3.000414,2016.

Cohen, R. 2019. The Berkeley Atmospheric CO_2 Observation Network. http://beacon.berkeley.edu/about/.

Conley, S. 2017. Applications of Gauss's theorem to quantify localized surface emissions from airborne measurements of wind and trace gases. Atmos. Meas. Tech. 10(9): 3345–3358.

Crisp, D. 2017. The on-orbit performance of the Orbiting Carbon Observatory-2 (OCO-2) instrument and its radiometracally calibrated products. Atmospheric Measurement Technology 10: 59–81. doi:10.5194/amt-10-59-2017.

Cundiff, S.T. 2003. Colloquium: Femtosecond optical frequency combs. Rev. Mod. Phys. 75(1). https://doi.org/10.1103/RevModPhys.75.325.

Davis, K.J. 2017. The Indianapolis Flux Experiment (INFLUX): A test-bed for developing urban greenhouse gas emission measurements. Elem. Sci. Anth. 5. doi:10.1525/elementa.188.

Deng, A. 2017. Toward reduced transport errors in a high resolution urban CO_2 inversion system. Elem. Sci. Anth. 5: 20. doi:http://doi.org/10.1525/elementa.133.

Dobler, J.T. 2016. Greenhouse Gas Laser Imaging Tomography Experiment (GreenLITE). https://www.osti.gov/servlets/purl/1301861.

Dobler, J.T. 2017. Demonstration of spatial greenhouse gas mapping using laser absorption spectrometers on local scales. Jour. Appl. Rem. Sensing 11. doi:10.1117/1.JRS.11.014002.

ECMWF. 2018. CAMS Greenhouse Gases Flux Inversions. https://apps.ecmwf.int/datasets/data/cams-ghg-inversions/.

ECMWF. 2019. User guide to ECMWF forecast products. https://www.ecmwf.int/en/about/media-centre/focus/user-guide-ecmwf-forecast-products.

ECMWF–CAMS. 2019. CAMS Greenhouse Gases Flux Inversions. Accessed July 6, 2019. https://apps.ecmwf.int/datasets/data/cams-ghg-inversions/.

EDF. 2018. EDF Announces Satellite Mission to Locate and Measure Methane Emissions. April 11. Accessed July 6, 2019. https://www.edf.org/media/edf-announces-satellite-mission-locate-and-measure-methane-emissions.

Eldering, A. 2017. The Orbiting Carbon Observatory-2 early science investigations of regional carbon dioxide fluxes. Science 358 (#6360). doi:10.1126/science.aam5745.

Elementa Special Feature. 2016. Quantification of Urban Greenhouse Gas Emissions: The Indianapolis Flux Experiment. Elem. Sci. Anth. https://collections.elementascience.org/quantification-of-urban-greenhouse-gas-emissions.

Elementa, Special Feature. 2018. Quantification of Urban Greenhouse Gas Emissions: The Indianapolis Flux Experiment. Accessed July 6, 2019. https://collections.elementascience.org/quantification-of-urban-greenhouse-gas-emissions.

EPA GHGRP. 2015. Emission Calculation Methodologies—GHG Reporting Program. Accessed July 6, 2019. https://www.epa.gov/sites/production/files/2015-07/documents/ghgrp_methodology_factsheet.pdf.

EPA. 2019. Greenhouse Gas Emissions. Accessed May 2, 2019. https://www.epa.gov/ghgemissions/inventory-us-greenhouse-gas-emissions-and-sinks.

EPA-NVFEL. 2019. MOVES and Other Mobile Source Emissions Models. Accessed April 20, 2019. https://www.epa.gov/moves/latest-version-motor-vehicle-emission-simulator-moves.

Fast, J. 2006. Evolution of ozone, particulates, and aerosol direct forcing in an urban area using a new fully-coupled meteorology, chemistry, and aerosol model. J. Geophys. Res. 111(D21305). doi:10.1029/2005JD006721.

Follet, R.J. 2010. GRACEnet. Nov. 4. Accessed July 29, 2019. https://www.ars.usda.gov/ARSUserFiles/np204/GRACEnetpage/Follett_GRACEnet.pdf.

Foote, E. 1856. Circumstances affecting the heat of the Sun's rays. Annual Meeting of the American Association for the Advancement of Science, August 23. https://thinkprogress.org/female-climate-scientist-eunice-foote-finally-honored-for-her-contributions-162-years-later-21b3cf08c70b/.

Frey, M. 2019. COllaborative Carbon Column Observing NEtwork (COCCON): Long-term stability and ensemble performance of the EM27/SUN Fourier transfor spectrometer. Atm. Meas. Tech. 12(3): 1513–1530. doi:10.5194/amt-12-1513-2019.

Gately, C.K. 2015. Cities, traffic, and CO_2: A multidecadal assessment of trends, drivers, and scaling relationships. Proc. Nat'l. Acad. Sc. 4999–5004. doi:www.pnas.org/cgi/doi/10.1073/pnas.1421723112.

Gately, C. 2017. Large uncertainties in urban-scale carbon emissions. Jour. Geophys. Res. Atm. 122(20): 11242–11260. doi:10.1002/2017JD027359.

GAW–WMO Global Atmospheric Watch. 2019. IG3IS—Integracted Global Greenhouse Gas information System. https://ig3is.wmo.int/en/who-we-are/our-organization.

GAW. 2019. Greenhouse Gases. Accessed August 4, 2019. http://www.wmo.int/pages/prog/arep/gaw/ghg/ghgbull06_en.html.

GHGSat, Inc. 2019. GHGSat, Global Emissions Monitoring. Accessed March 13, 2019. https://www.ghgsat.com/who-we-are/our-satellites/claire/.

GMD, NOAA. 2019. Global Greenhouse Gas Reference Network. Accessed December 20, 2018. https://www.esrl.noaa.gov/gmd/ccgg/.

Gordon, I. 2017. The HITRAN2016 molecular spectroscopic database. Spectrosc. Radiat Transfer 3–69.

Gregg, J.S. 2009. The temporal and spatial distribution of carbon dioxide emissions from fossil fuel use in North America. Jour. Appl. Meteor. and Clim. 48(12): 2528–2542. doi:10.1175/2009JAMC2115.1.

Grell, G.A. 2005. Fully coupled 'online' chemistry in the WRF model. Atmos. Envirn. 39: 6957–6976.

Gu, L. 2012. The fundamental equation of eddy covariance and its application in flux measurements. Agricultural and Forest Meteorology (Elsevier) 152: 135–148. doi:10.1016/j.agrformet.2011.09.014.

Gurney, K.R. 2003. TransCom 3, CO_2 inversion intercomparison: 1, Annual mean control results and sensitivity to transport and prior flux information. Tellus B 55(2): 555–579. doi:10.3402/tellusb.v55i2,16728.

Gurney, K.R. 2011. Quantification of fossil fuel CO_2 emissions on the building/street scale for a large U.S. city. Chan, F., Marinova, D. and Anderssen, R.S. (eds.). 19th International Congress on Modelling and Simulation (MODSIM2011). Perth, Australia: Modeling & Simulation Soc. Australia & New Zealand, Inc. 1781–1787.

Gurney, K.R. 2016. Bias present in U.S. federal agency power plant CO_2 emissions data and implications for the U.S. clean power plant. Env. Res. Lett. doi:10.1088/1748-9326/11/6/064005.

Gurney, K.R. 2017. Reconciling the differences between a bottom-up and inverse-estimated $FFCO_2$ emissions estimate in a large U.S. urban area. Elementa, Sci. Anth. 5(44). doi:10.1525/elementa.137.

Hall, B.D. 2019. Gravimetrically prepared carbon dioxide standards in support of atmospheric research. Atmos. Meas. Tech. 12(1): 517–524. doi:10.5194/amt-12-517-2019.

Hall, J.L. 2006. Nobel Lecture: Defining and measuring optical frequencies. Rev. Mod. Phys. 74(4): 1279–1295. doi:10.1103/revmodphys.78.1279.

Hänsch, T.W. 2006. Noble lecture: Passion for precision. Rev. Mod. Phys. 78(4): 1297–1309. doi:10.1103/revmodphys.78.1297.

Herriott, D.R. 1965. Folded optical delay lines. Applied Optics 4(8): 883–889. doi:10.1364/AO.4.000883.

HITRAN. 2016. High-Resolution Transmission Molecular Absorption Database. Accessed March 5, 2019. https://hitran.org/.

Horst, T.W. 2000. On frequency response corrections for eddy covariance flux measurements. Boundary Layer Meteorology (Kulwer Acad. Pub.) 94: 517–520.

ICOS. 2008. Integrated Carbon Observation System. Accessed Aug 5, 2019. https://www.icos-ri.eu/.

IG3IS. 2018. IG3IS Science Implementation Plan. June. http://www.wmo.int/pages/prog/arep/gaw/documents/IG3ISImplementationPlanEC70.pdf.

IPCC. 2018. EFDB—emission factor database. Accessed June 14, 2019. https://www.ipcc-nggip.iges.or.jp/EFDB/main.php.

IPCC-TFI. 2006. 2006 IPCC Guidelines for National Greenhouse Gas Inventories. Accessed June 13, 2019. https://www.ipcc-nggip.iges.or.jp/public/2006gl/index.html.

IPCC-TFI. 2019. Task Force on National Greenhouse Gas Inventories. Accessed June 12, 2019. https://www.ipcc-nggip.iges.or.jp/.

Jackson, R. 2018. The Ascent of John Tyndall. Oxford University Press.

JAXA—Japan Aerospace Exploration Agency JAXA. 2019. EORC-JAXA. March. Accessed July 6, 2019. https://www.eorc.jaxa.jp/GOSAT/index.html.

Johnson, A.N. 2019. Non-nulling measurements of flue gas flows in a coal fired power plant stack. FLOMEKO Conference Proceedings. Lisbon, Portugal: IMEKO—International Measurement Confederation.

Johnson, W. 2013. Micropulse differential absorption lidar for identification of carbon sequestration site leakage. Applied Optics 52: 2994–3003. doi:10.1364/AO.52.002994.

Kalnay, E. 2003. Atmospheric Modeling, Data Assimilation and Predictability. Cambridge University Press.

Karion, A. 2013. Long-term greenhouse gas measurements from aircraft. Atmospheric Measurement Techniques 6(3): 511–526.

Karion, A. 2015. Aircraft-based estimate of total methane emissions from the Barnett Shale region. Environmental Science & Technology 49(13): 8124–8131.

Kim, J. 2018. The Berkeley atmospheric CO_2 observation network: Field calibration and evaluation of low-cost air quality sensors. Atmos. Meas. Tech. 11: 1937–1946. doi:10.5194/amt-11-1937-2018.

Koffi, E.N. 2015. Investigating the usefulness of satellite-derived fluorescence data in inferring gross primary productivity within the carbon cycle data assimilation system. Biogeosciences 12: 4067–4084. doi:10.5194/bg-12-4067-2015.

Kwok, C.Y. 2015. Comprehensive laboratory and field testing of cavity ring-down spectroscopy analyzers measuring H_2O, CO_2, CH_4 and CO. Atmospheric Measurement Techniques 8(9): 3867–3892.

Lin, J.C. 2003. A near-field tool for simulating the upstream influence of atmospheric observations: The Stochastic Time-Inverted Langrangian Transport (STILT) model. Jour. Geophys. Res. (AGU) 108(D16): 4493–4411. doi:10.1029/2002JD003161.

Lin, J.C. 2012. Langragian modeling of the atmosphere: An introduction. Chap. Intro in Lagrangian Modeling of the Atmosphere, Volume 200, Brunner, J., Gerbig, D., Stohl, C., Luhar, A., Webley, A. and Lin, P. (eds.). Amer. Geophysical Un. doi:10.1029/GM200.

Marland, G. 1985. CO_2 from fossil fuel burning: Global distribution of emissions. Tellus 37B(4-5): 555–579. doi:10.1111/j.1600-0889.1985.tb00073.x.

Martin, C.R. 2019. Investigating sources of variability and error in simulations of carbon dioxide in an urban region. Atmos. Environ. 199: 55–69. doi:10.1016/j.atmosenv.2018.11.013.

Max Planck Institut for Biogeochemistry. 2019. JENA CarboScope. Accessed July 6, 2019. http://www.bgc-jena.mpg.de/CarboScope/?ID=s)).

Meyer, C., interview by J. Whetstone, 2018. NIST Physical Scientist

Miller, W.R. 2015. Investigating adsorption/desorption on carbon dioxide in aluminum compressed gas cylinder. Analytical Chemistry 87(3): 1957–1962. doi:10.1021/ac504351b.

MIT. 2019. AGAGE. Accessed June 14, 2019. https://agage.mit.edu/instruments/medusa-gas-chromatography-mass-spectrometry-medusa-gc-ms.

Moore, B. 2018. The potential of the geostationary Carbon Cycle Observatory (GeoCarb) to provide multi-scale constraints on the carbon cycle in the Americas. Frontiers in Environmental Science 6: 13.

Muntean, M. 2018. Fossil CO_2 emissions of all world countries—2018 Report. EU Joint Research Commission, Luxembourg: Publ. Off. European Union. doi:10.2760/20158,JRC113738.

Nakajima, M. 2017. Fourier transfer spectrometer on GOSAT and GOSAT-2. Proceedings Volume 10563, International Conference on Space Optics—ICSO 2014; Tenerife, Canary Islands, Spain: SPIE Digital Library. https://doi.org/10.1117/12.2304062.

NASA. 2019. Earth Fact Sheet. April 22. Accessed August 15, 2019. https://nssdc.gsfc.nasa.gov/planetary/factsheet/earthfact.html.

NIST-NTRM. 2019. The NIST Traceable Reference Materials (NTRM) Program for Gas Standards. Accessed December 19, 2018. https://www.nist.gov/mml/csd/sensing/ntrmprogram.

NOAA. 2019. Earth Systems Research Laboratory, Global Monitoring Division, The NOAA Annual Greenhouse Gas Index (AGGI), https://www.esrl.noaa.gov/gmd/aggi/aggi.html.

NOAA-CT. 2017. CarbonTracker 2017. Accessed June 20, 2019. https://www.esrl.noaa.gov/gmd/ccgg/carbontracker/index.php.

NOAA-GMD. 2018. CarbonTracker-CH_4. Accessed June 20, 2019. https://www.esrl.noaa.gov/gmd/ccgg/carbontracker-ch4/.

NVFEL. 2019. About the National Vehicle and Fuel Emissions Laboratory (NVFEL). https://www.epa.gov/aboutepa/about-national-vehicle-and-fuel-emissions-laboratory-nvfel.

Oda, T. 2018. The open-source data inventory for anthropogenic CO_2, version 2016 (ODIAC2016): A global monthly fossil fuel CO_2 gridded emissions data product for tracer transport simulations and surface flux inversions. Earth Syst. Sci. Data 10: 87–107. doi:10.5194/essd-10-87-2018.

ORNL DAAC. 2002. FLUXNET. Accessed August 2, 2019. https://daac.ornl.gov/cgi-bin/dataset_lister.pl?p=9.

Osterman, G.B. 2015. Orbiting Carbon Observatory-2 (OCO-2): Data Product User's Guide, Operational L1 and L2 Data Versions 6 and 6R (PDF). NASA. OCO D-55208. Retrieved 14 May 2015.

Palmer, P.I. 2018. A measurement-based verification framework for UK greenhouse gas emissions: an overview of the Greenhouse gas UK and Global Emissions (GAUGE) project. Atm. Chem. Phys. 11753–11777. doi:10.5194/acp-18-11753-2018.

Parker, D.J. 2015. Science Direct–Micrometeorology. In Encyclopedia of Atmospheric Sciences, by Pyle and Zhang North, 2998. Elsevier Ltd. Accessed August 2, 2019. https://www.sciencedirect.com/topics/agricultural-and-biological-sciences/micrometeorology.

Pecala, S.W. 2010. Verifying greenhouse gas emissions: Methods to support international climate agreements. Washington, DC: National Academies Press.

Polyansky, O.L. 2015. High-accuracy CO_2 line intensities determined from theory and experiments. Phys. Rev. Lett. 114(24).

Prinn, R.G. 2018. History of chemically and radiatively important atmospheric gases from the Advanced Global Atmospheric Gases Experiment (AGAGE). Earth System Science Data 10(2): 985-1018. doi:10.5194/essd-10-985-2018.

Quick, J.C. 2013. Carbon dioxide emission tallies for 210 U.S. coalfired power plants: A comparison of two accounting methods. J. Air & Waste Management Assoc. doi:10.1080/10962247.2013.833146.

Rebmann, C. 2018. ICOS eddy covariance flux-station site setup: A review. International AgroPhysics 471–494. doi:10.1515/intag-2017-0044.

REFPROP. 2019. NIST Reference Fluic Thermodynamic and Transport Properties Database. https://www.nist.gov/srd/refprop.

Reiker, G.B. 2014. Frequence-comb-based remote sensing of greenhouse gases over kilometer air paths. Optica 290–298. http://dx.doi.org/10.1364/OPTICA.1.000290.

Rhoderick, G.C. 2016. Development of a Northern continental air standard reference material. Anal. Chem. 88(6): 3376–3385. doi:10.1021/acs.analchem.6b00123.

Richardson, J.S. 2017. Tower measurement network of *in-situ* CO_2, CH_4, and CO in support of the Indianapolis FLUX (INFLUX) experiment. Elem Sci Anth 4. doi:10.1525/elementa.140.

Rodenberg, C., MPI Biogeochemistry, Jena. 2017. Atmospheric CO_2 Inversion. http://www.bgc-jena.mpg.de/CarboScope/?ID=s.

Romanini, D. 2014. Introduction to cavity enhanced absorption spectroscopy. Gagliardi, G. and Loock, H.P. (eds.). Cavity-Enhanced Spectroscopy and Sensing. Springer Series in Optical Sciences, vol. 179. Berlin: Springer. doi:10.1007/978-3-642-40003-2_1.

Salmon, O.E. 2018. Top-down estimates of NO_x and CO emissions from Washington, DC-Balitmore during the WINTER campaign. J. Geophys. Res-Atm. 123(14): 7705–7724.

Sargent, M. 2018. Anthropogenic and biogenic CO_2 fluxes in the Boston urban region. Proc. Nat. Ac. Sci. 7491–7496. doi:10.1073/pnas.1803715115.

Sarmiento, D.P. 2017. A comprehensive assessment of land surface-atmosphere interactions in a WRF/Urban modeling system for Indianapolis, IN. Elem. Sci. Anth. 5(23). doi:10.1525/elementa.132.

Schibig, M.F. 2018. Experiments with CO_2-in-air reference gases in high-pressure aluminum cylinders. Atmos. Meas. Tech. 11(10): 5565–5586. doi:10.5194/amt-11-5565-2018.

Schmid, H.P. 2002. Footprint modeling for vegetation atmosphere exchange studies: A review and perspective. Agricultural and Forest Meteorlogy (Elsevier) 159–183.

Scripps-CO_2. 2019. Scripps CO_2 Program. Accessed March 10, 2019. http://scrippsco2.ucsd.edu/publications/scientific_literature.

Skamarock, W.C. 2008. A Description of the Advanced Research WRF Version 3. NCAR Technical Note, NCAR/TN–475+STR, National Center for Atmospheric Research. https://opensky.ucar.edu/islandora/object/technotes%3A500/datastream/PDF/view.

Smith, I.A. 2019. Evidence for edge enhancements of soil respiration in temperate forests. Geo. Res. Lett. 46(8): 4278–4287. doi:10.1029/2019GL082459.

Sun, Y. 2018. Overview of Solar-Induced chlorophyll Fluorescence (SIF) from the Orbiting Carbon Observatory-2: Retrieval, cross-mission comparison, and global monitoring for GPP. Remote Sens. Env. 209: 808–823. doi:https://doi.org/10.1016/j.rse.2018.02.016.

Sweeney, C.A. 2015. Seasonal climatology of CO_2 across North America from aircraft measurements in the NOAA/ESRL Global Greenhouse Gas Reference Network. Jour. Geo Res: Atmospheres 120(10): 2321–2327.

Turnbull, J.C. 2012. An integrated flask sample collection system for greenhouse gas measurements. Atmos. Meas. Tech. 5(9): 2321–2327.

Turnbull, J.C. 2015. Toward quantification and source sector identification of fossil fuel CO_2 emissions from an urban area: Results from the INFLUX experiment. Jour. Geophys. Res-Atm. 120(1): 292–312.

Turnbull, J.C. 2019. Synthesis of urban CO_2 emission estimates from multiple methods from the Indianapolis Flux Project (INFLUX). Environ. Sci. & Tech. 53: 287–295. doi:10.1021/acs.est.8b05552.

UN-DESA. 2018. UN Dept. of Economic and Social Affairs. May 16. Accessed February 22, 2019. https://www.un.org/development/desa/en/news/population/2018-revision-of-world-urbanization-prospects.html.

Vesala, T. 2008. Flux and concentration footprint modelling: State of the art. Environmental Pollution (Elsevier) 152: 653–666.

Wageningen Univ. & ICOS Netherlands. 2019. CT—Europe. Accessed July 6, 2019. http://www.carbontracker.eu.

Wagner, G.A. 2018. Optics Express 26(15): 19420–19434. doi:10.1364/OE.26.019420.

Waxman, E.M. 2019. Estimating vehicle carbon dioxide emissions from Boulder, Colorado, using horizontal path-integrated column measurements. Atm. Chem. and Phys. 19(7): 4177–4192. doi:10.5194/acp-19-4177-2019.

Weller, Z.D. 2018. Vehicle-based methane surveys for finding natural gas leaks and estimating their size: validation and uncertainty. Environ. Sci. & Tech. 52(20): 11922–11930. doi:10.1021/acs.est.8b03135.

White, J.U. 1942. Long optical paths of large aperture. Journal for the Optical Society 32(5): 285–288. doi:10.1364/JOSA.32.000285.

White, W. 1976. Formation and transport of secondary air pollutants: Ozone and aerosols in the St. Louis urban plume. Science 194(4261): 187–189.

Wunch, D. 2011. The total carbon column observing network. Phil. Trans. Royal Soc. A. doi:10.1098/rsta.2010.0240.

Yacovitch, T.I. 2017. Natural gas facility methane emissions: Measurements by tracer flux ratio in two U.S. natural gas producing basins. Elem. Sci. Anth. 5: 69. doi:10.1525/elementa.251.

Yadav, V. 2019. Spatio-temporally resolved methane fluxes From the Los Angeles megacity. Jour. Geo. Res-Atm. 5131–5148. doi:10.1029/2018JD030062.

Yi, H. 2018. High-accuracy 12C16O2 line intensities in the 2 micron wavelength region measured by frequency-stabilized cavity ring-down spectroscopy. Jour. Quant. Spec. and Rad. Trans. 206: 367–377. doi:10.1016/j.jqsrt.2017.12.008.

Zhao, C. 1997. A high precision manometric system for absolute calibrations of CO_2 in dry air. J. Geophysical Res. 102(D5): 5885–5894.

Zhao, C. 2006. Estimating uncertainty of the WMO Mole Fraction Scale for carbon dioxide in air. J. Geophysical Res. 111(D08). doi:10.1029/2005JD006003.

Section 2

Fossil Sector: Coal/Petroleum/ Natural Gas

CHAPTER 6

Carbon Policies for Reducing Emissions in Power Plants through an Optimization Framework

Aurora del Carmen Munguía-López and *José María Ponce-Ortega**

1. Introduction

Power generation systems have improved in both efficiency and profitability throughout the years. Nevertheless, addressing the environmental impact has not been a priority. Over 66% of the global electricity production is from fossil fuel combustion (The World Bank, 2014). Consequently, the electricity and heat generation sector is the largest source of carbon dioxide emissions, accounting for about 42% of world emissions from fuel combustion (IEA, 2017). Since the beginning of the Industrial Revolution, the atmospheric CO_2 concentration has increased by more than a third and continues to do so. Specifically, in November 2017, its global concentration was 405.12 ppm and one year later it had already increased to 408.02 ppm (NOA, 2018). According to the Intergovernmental Panel on Climate Change, the human influence on the current warming trend is clear and the scientific evidence for warming of the climate system is unequivocal (IPCC, 2013). The rise in carbon dioxide concentration in the atmosphere is a worldwide matter of concern due to its impact on global warming. The heat-trapping nature of carbon dioxide and other gases was already demonstrated in the mid-19th century. As fossil fuels are expected to remain the primary resource for producing electricity in the immediate future, there is a need to develop strategies to tackle this problem. These strategies must include alternatives to reduce emissions and simultaneously satisfy the increasing electricity demand.

From an optimization perspective, several macroscopic systems involving the integration of processes and the evaluation of different configurations have been proposed. These approaches consist of developing mathematical model formulations to find the optimal design of the system. The optimization model can be formulated considering a single objective (typically economic) or multiple objectives (e.g., economic, environmental and social). Through multi-objective optimization, tradeoffs between the objective functions can be found. Thus, the profitability and the other objectives of the system can be analyzed simultaneously by a set of optimal solutions. Particularly for reducing emissions in power plants, integrated systems involving processes to capture and utilize carbon have been proposed. Furthermore, the combined use of fossil and non-fossil fuels along with novel configurations to satisfy power demands has been widely reported. This type of system is usually analyzed at a macroscopic level since the

Universidad Michoacana de San Nicolás de Hidalgo, Departamento de Ingeniería Química, Francisco J. Mujica S/N, Ciudad Universitaria, Morelia, Michoacán, México, 58060.
* Corresponding author: jmponce@umich.mx

formulation can account for various possibilities, such as distinct energy resources or technologies, as well as its corresponding inlet and outlet flow rates, revenues and costs. To model this, the formulation can involve mass balances, discrete decisions and both linear and non-linear logical relationships. All of which can result in formulations of thousands of equations as well as continuous and integer variables.

When innovative technologies or other alternatives to reduce emissions are required, there can be high investment costs as well as variable costs involved. Hence, the relevance of carbon policies that aim to move towards these systems. In this regard, certain economic instruments involving penalizations and compensations have been developed by several governments (Vatn, 2015). Those policies have been addressed worldwide, for instance in China (Liu and Lu, 2015), South Africa (Alton et al., 2014) and the United States (Kaufman et al., 2016). The approaches include similar prices for the economic penalties. Moreover, other findings involve negative impacts, such as an increase in the urban-rural gap (Liang and Wei, 2012) and wealth redistribution if the penalization scheme is not well-designed (Chen et al., 2015).

The carbon tax is defined as a penalization for generated emissions given as a monetary cost per ton of carbon dioxide produced (Avi-Yonah and Uhlmann, 2009). An opposite parameter is the carbon tax credit, which consists of considering the avoided emissions and, based on the amount of reduction, some compensation that can be institutional, public, or private is given. The reductions occur after a change in technologies or in the production process (Graefe et al., 2011). These instruments have been approached in economics as the "polluter pays principle" and the "provider gets principle", which means that those affecting the environmental qualities should pay while those improving them should be compensated (Vatn, 2015). Both carbon monetization strategies have been criticized for some challenges, including the correct setting of the tax rate, collecting the tax and using the resulting revenue (Marron and Toder, 2014). Additional issues causing uncertainty are that the tax credit has an arbitrary economic value (Hoel, 1996) and that the tax is standardized without accounting for economic sectors, industrial development and regional conditions (Newell et al., 2013). Whether the final decision is investing to reduce emissions or paying the economic penalty, there are associated costs (Clarkson et al., 2015). This is the importance of analyzing the effect of carbon penalizations and compensations.

2. Carbon Policies through Optimization Approaches

Recently, several optimization studies have addressed different energy systems through multi-objective models and monetization of carbon externalities. In this regard, Sánchez-Bautista et al. (2017) reported the tradeoffs among economic, environmental and social objectives for an integrated energy system including fuel and biofuel production along with carbon capture through forest plantations. Other mathematical models have addressed the impact of carbon policies on promoting the generation of clean energy (Wong et al., 2010) and on fostering investment for carbon capture in coal-fired power plants (Guillén-Gosálbez et al., 2012). Similarly, comparative scenarios with penalizations or compensations have been evaluated in the design of combined heat and power systems involving biogas use by means of a mathematical programming model, which gives compromise solutions (Fuentes-Cortés et al., 2017). Moreover, Pascual-González et al. (2016) proposed a decision-support tool including a multi-objective model, environmentally extended input-output tables and life cycle assessment, with the objective of minimizing emissions at a global macroeconomic scale through modifications in the economic sectors. Galán-Martín et al. (2018) proposed an optimization approach based on the study of the effect of interregional cooperation in meeting the emissions targets with cost-effective solutions. A mixed-integer linear programming (MILP) model has been developed by Ren and Gao (2010) for the integrated planning and evaluation of distributed energy resources, where different effects (economic, energetic and environmental) as well as economic policies (carbon tax and energy prices) are considered.

An analysis of carbon policies has been proposed for a macroscopic system that integrates power plants involving chemical looping combustion systems with algae cultivation to utilize CO_2. This optimization work includes the formulation of a general MILP model and the obtained results show important economic benefits as well as reductions in emissions, specifically with the economic compensations (Munguía-López et al., 2018). Furthermore, the impact of economic penalties and compensations on a system of water distribution networks involving power-desalination plants has been reported. The study includes the use of renewable fuels to reduce emissions, while through an optimization framework economic, environmental and social benefits are obtained. The results refer to a case study related to a water and energy management problem in Sonora, Mexico (Munguía-López et al., 2019). Both approaches will be further described in the next sections.

The carbon penalizations and compensations evaluated in the aforementioned studies have been previously reported and are the following. For the carbon taxes, predicted values for monetizing the externalities in the USA in the future are considered (Figure 1) as well as the prices of 10 and 15 $/ton CO_2 that are expected to be included in future Mexican regulations (SEMARNAT and INECC, 2012). The perspective of carbon prices in the USA is based on assuming a policy that applies a fee on carbon dioxide emissions starting at 25 $/ton in 2015 and rising by 5% per year (Kaufman et al., 2016). On the other hand, the values for the carbon tax credits vary from 0.3 to 130 $/ton CO_2 avoided. These extreme prices are evaluated along with intermediate compensations including 1, 4, 7, 10, 80 and, 120 $/ton CO_2, which represent the variation of the carbon price for world emissions. The average carbon price for tax credits has changed in the last years from 4 to 7 $/ton CO_2 and most of the world emissions are priced at 10 $/ton CO_2. Furthermore, economic models have estimated that the prices 80 and 120 $/ton CO_2 are needed to meet climate stabilization goals (Kossoy et al., 2015). Note that the considered taxes and tax credits have been already used or predicted to be used in the future. However, the impact of carbon policies that change over time has not been evaluated. As future work, an approach including the effect of variable penalizations and compensations over a time horizon would be useful to identify optimal policies and recommend carbon prices for the next years.

Besides carbon taxes and tax credits, alternative policies for reducing emissions in the electricity sector have been proposed and actively implemented, such as Cap and Trade and Renewables Portfolio Standards. Cap and Trade is a market-based policy tool that establishes an emission cap (the maximum quantity of authorized emissions) and allowances to emit a specific quantity (e.g., 1 ton) of a pollutant. The total number of allowances equals the level of the cap. Allowances can be bought or sold (traded) in an allowance market (Schreifels and Kruger, 2003). There are some differences between the Cap and Trade program and the taxation approach. For instance, the first one would establish a price on emissions indirectly by limiting total emissions and issuing tradable emissions allowances, while a carbon tax would directly establish a price on emissions (Kaufman et al., 2016). On the other hand, Renewables Portfolio Standards (RPS) are regulations that require electricity supply companies to obtain specific amounts of renewable energy generation (such as wind, solar, biomass and geothermal) over time. It has been suggested that these policies have played an important role in driving U.S. renewable electricity growth (Barbose et al., 2015).

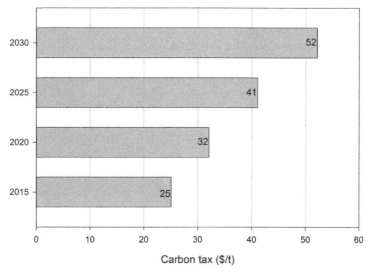

Figure 1. The perspective of carbon prices in the USA for different years.

3. Carbon Policies in the Optimal Design of Power Plants Involving Chemical Looping Combustion and Algae Systems

In search of power generation systems with minor environmental impact, the use of chemical looping combustion (CLC) in the generation of electricity has been proposed. This novel technology is classified as a variety of post-combustion, oxy-combustion and pre-combustion. It enables to capture the carbon dioxide with small penalties in thermal efficiency (Anheden and Svedberg, 1998). The basic principles of CLC systems and their potential application in power generation cycles have been widely reported (Ishida and Jin, 1996a; Ishida and Jin, 1997b; Ishida and Jin, 2001c). Previous simulation studies have addressed the technical and economic feasibility as well as the sensitivity analysis of several parameters involved in this type of systems (fuels, oxygen carriers, CLC configurations, and power cycles) (Petriz-Prieto et al., 2016; Zhu et al., 2015). After the CO_2 is captured, its utilization is desirable. An alternative that has recently gained attention is sending the gas to an algae cultivation system. A microalgae system consists of different stages (cultivation, harvesting, extraction and production) to process the algal biomass and produce biofuels along with other valuable byproducts such as glycerol, ethanol and proteins (Dickinson et al., 2017). Some optimization approaches have addressed the integration of power plants with algae systems. For instance, Gutiérrez-Arriaga et al. (2014) reported an analysis including energy integration, a conventional power plant (without modifying combustion systems) and a single-process evaluation for each stage of the algae cultivation system. The model involves economic and environmental objectives to obtain tradeoff solutions.

3.1 Model formulation

In this context, a mathematical model for the optimal integration of power plants involving chemical looping combustion with algae systems has been proposed (Munguía-López et al., 2018). The model accounts for the optimal selection of the processes for each stage of the algae cultivation system as well as the optimum operation conditions, combustion systems and power cycles for the power generation plant. Economic and environmental objectives are taken into account including the maximization of the profit and the minimization of the emissions. The sales of electricity, biodiesel, glycerol, ethanol and proteins as well as the costs associated with the global system are considered. The generated carbon dioxide in the power plant and the amount sent to the algae-to-biodiesel process are considered as well. Besides, the impact of carbon taxes and tax credits on the objective functions and on the optimal configuration is evaluated. The problem was formulated as a MILP model that represents the global system at a macroscopic level. The proposed superstructure to represent the potential configurations is shown in Figure 2.

The mathematical model formulation includes several relationships to compute the flow rates, requirements, costs, revenues and the selection of the optimum technologies. The profit is computed considering the total sales (IN), capital costs (CAP) and operating costs ($COPER$). When the carbon tax ($CCTAX$) is evaluated, the profit (P^{TAX}) is estimated by equation (1). Equation (2) is required in order to compute the profit (P^{COMP}) when the carbon tax credit ($CCOMP$) is analyzed. The corresponding penalization and compensation are given depending on the total emissions and on the avoided emissions, respectively. To obtain the total emissions, the CO_2 mitigated by the algae system is subtracted from the CO_2 generated in the power plant, as indicated in equation (3). The avoided emissions are defined by the difference between the emissions generated in a conventional system and the total emissions in the integrated system.

$$P^{TAX} = IN - CAP - COPER - CCTAX \tag{1}$$

$$P^{COMP} = IN - CAP - COPER + CCOMP \tag{2}$$

$$EM = prod^{CO2} - fixed^{CO2} \tag{3}$$

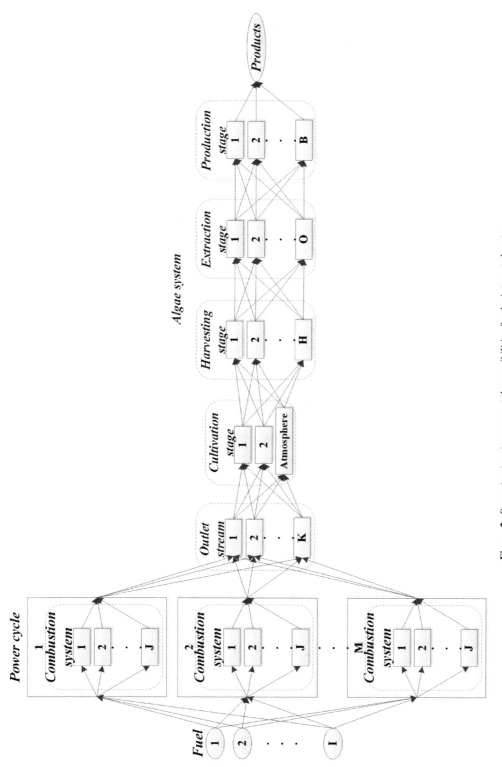

Figure 2. Superstructure to represent the possibilities for the integrated system.

3.2 Results

Although the objectives are opposite, the constraint method (Diwekar, 2008) is employed as solution procedure to generate the set of optimal solutions or Pareto front. The optimization results were obtained by solving the model using parameters reported in the literature. When the carbon taxes were evaluated, the Pareto curves presented in Figure 3 were obtained. The tradeoffs between the economic and environmental objectives can be identified through the Pareto front. As expected, with solutions involving higher profits the emissions increase as well. However, there are solutions where the generated emissions are low, and the system is still profitable. Each optimal solution of the Pareto set refers to

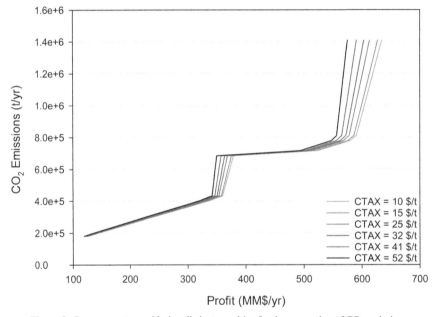

Figure 3. Pareto curves considering distinct penalties for the generation of CO_2 emissions.

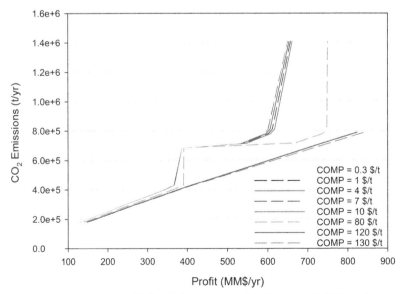

Figure 4. Pareto curves considering distinct compensations for the avoided CO_2 emissions.

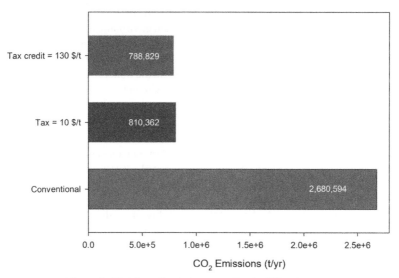

Figure 5. Variation of emissions according to different cases.

a different configuration in terms of the selection of energy sources and technologies, including their required specific conditions. Note that there is no further reduction in emissions despite increasing the tax. This occurs because the evaluated penalizations are not enough to promote the use of processes that contribute to further reducing emissions.

On the other hand, the results involving the carbon tax credits are shown in Figure 4. When the highest compensation values are evaluated (120 and 130 $/ton CO_2), important reductions in emissions throughout the Pareto front are obtained. Therefore, a maximum profit solution including low emissions can be attained. Notice that with the rest of the tax credits, only variations in the profit are observed. The impact of considering the penalizations and compensations in the integrated system can be observed by comparing the results with the generated emissions in a conventional system (without CLC or algae systems). This comparison is presented in Figure 5. The three solutions can be compared because their optimal configuration for the technologies in the power plant is equal and, thus, the net electricity is as well. Note that the highest compensation and the lowest penalization were considered in order to find the best economic and environmental solution (as described above, no further reduction of the emissions was found with greater taxes). The reduction in emissions for the considered tax and tax credit scenarios is similar: 70 and 71%, respectively. Regarding the economic objective, higher profits are attained with the carbon compensations. Therefore, it is concluded that involving tax credits for the avoided emissions gives better tradeoffs among the objective functions. Furthermore, the benefits of considering carbon policies as a strategy to reduce emissions and simultaneously attain a profitable system of power generation and biofuels production are identified. Through the different tradeoff solutions of the Pareto front, decision makers can select specific configurations depending on the power demand and on economic or environmental restrictions.

4. Carbon Policies in the Optimal Design of Water Distribution Networks Involving Power-Desalination Plants

The use of renewable resources, such as solar energy and biofuels, in energy systems has been proposed to reduce the environmental impact. Additionally, the simultaneous production of water and electricity in power-desalination plants (González-Bravo et al., 2015) and polygeneration plants (Rubio-Maya et al., 2011) has been presented as an alternative for the increasing demands of water and electric energy. In this regard, it is important to identify the optimum configurations and renewable energies that lead to efficient systems involving savings in power consumption and costs. Also, other aspects to consider are energy availability and consumption as well as economic and environmental impacts (Al-Karaghouli and Kazmerski, 2013). Particularly for desert regions, solar energy facilities can be more easily implemented because of the high solar radiation and the availability of large areas. González-Bravo et al. (2015) reported a case study for a problem related to water scarcity in the Sonoran Desert. This approach includes a system involving seawater desalination plants integrated with power plants using solar energy.

4.1 Model formulation

Recently, carbon taxes and tax credits have been evaluated in a macroscopic water distribution network integrated with power-desalination plants (Munguía-López et al., 2019). Taxes and tax credits are applied to water management as well. The optimization approach accounts for economic, environmental and social aspects to find a design that represents an alternative to reduce emissions and water scarcity. Solar energy, biofuels, water storage tanks, and the possibility of sending water to recharge aquifers are considered. The proposed model involves the selection of the optimum energy resource and the optimum configuration of the integrated system to satisfy demands for domestic, agricultural and industrial users. The objective function consists of maximizing the annual profit while, through the taxes and credits, a positive impact on the environmental (generated emissions, extracted water, and recharge of aquifers) and social (generation of jobs) functions is expected. Furthermore, the model formulation includes the electricity generation in existing and new dual-purpose power plants, availability restrictions for the renewable energy, existing and new water storage tanks, and variation in the demands. To represent the different alternatives for the design of the system under compensations and penalizations, the superstructures presented in Figure 6 are proposed and evaluated through an optimization model.

Several carbon taxes and carbon tax credits are evaluated in order to find their impact on the objective function and on the reduction in CO_2 emissions as well as on the generation of jobs. A decrease in emissions is achieved by satisfying part of the energy requirements with biofuels and solar collectors.

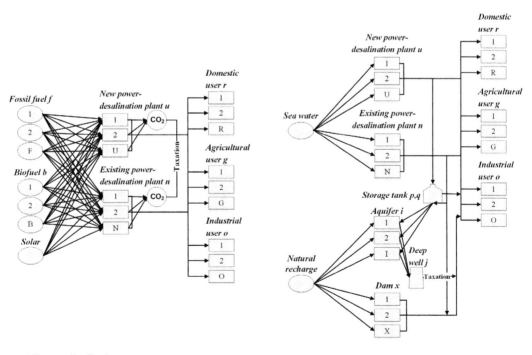

a) **Energy distribution**   b) **Water distribution**

Figure 6. (a) Superstructure for energy distribution in the integrated system. (b) Superstructure for water distribution in the integrated system.

The overall greenhouse gas emissions (*GHGE*) are estimated considering the emissions for fuels and biofuels, as presented in equation (4). The emissions for the solar collector are assumed to be zero. The economic penalty is function of the generated emissions and is computed as shown in equation (5). The parameter C^{tax} symbolizes the unitary carbon tax, which is a cost per ton of produced CO_2. In contrast, the economic compensation for avoiding emissions is presented in equation (6). The parameter C^{credit} refers to the unitary tax credit. The compensation depends on the variable amount of avoided emissions, which is given by the difference between the generated emissions in a conventional system (F^{max}) and the emissions in the proposed system, as indicated in equation (6).

$$GHGE = \sum_f \sum_u \sum_t \left[GHGE_f^{fossil} \cdot Q_{f,u,t}^{fossil} \right] + \sum_b \sum_u \sum_t \left[GHGE_b^{biofuel} \cdot Q_{b,u,t}^{biofuel} \right] \tag{4}$$

$$CT = H_Y \, C^{tax} \, GHGE \tag{5}$$

$$CTC = H_Y \, C^{credit} \left(F^{max} - GHGE \right) \tag{6}$$

The problem was solved considering the objective function of maximizing the annual profit. This economic function was estimated involving the sales of water and energy minus the total annual costs. When the carbon tax was evaluated, the corresponding penalization was included in the objective function as well as the compensation when the tax credits are analyzed. The variation on the reduction in emissions and on the generated jobs with the different taxation schemes was described as well.

4.2 Results

The presented model is general and applicable to any case study. In this approach, the results were obtained based on a water and energy management problem in Hermosillo, Sonora, Mexico. Hermosillo city is located in a desert region (Sonoran Desert) and it represents a potential place to implement the proposed system. It is an optimum location for solar collectors because of the high direct normal irradiation (around 3000 kWh/m²). In this case study, existing and new power-desalination plants are considered to satisfy the electricity demands. Also, the variation in the availability of biofuels depending on the seasons and the generation of agricultural wastes is involved. More details of the case study have been previously reported (González-Bravo et al., 2015).

The optimization results related to the carbon tax are presented in Figure 7. Besides the penalizations reported in the literature (10, 15, 25, 32, 41 and, 52 $/ton CO_2), a greater tax of 115 $/ton CO_2 was included in the analysis. This economic penalty represents the necessary tax to find better values for environmental and social functions. That means the minimum amount of greenhouse gas emissions (*GHGE*) and the maximum number of jobs. The solution with the highest tax shows that with the appropriate penalization further environmental and social benefits can be achieved, although the economic objective can be compromised. As shown in Figure 7, the solution with this tax gives the best values for the emissions and jobs (14,484,014 ton CO_2/year and 12,647 generated jobs) and the worst value of the Pareto front for the profit (−39 MM$/year). On the other hand, it was found that even with the lowest tax (10 $/ton CO_2), a reduction in the emissions with a positive value for the profit is obtained. However, this reduction is not maximum. Evaluating the rest of the taxes, lower values for the profit are achieved as the penalty increases. Therefore, tradeoffs among the economic, environmental and social functions can be identified through the Pareto front, as well as the cost of reducing emissions and generating jobs.

The sets of optimal solutions for the carbon tax credits analysis are presented in Figure 8. The Pareto sets show compromise solutions for the emissions, generated jobs and profit. Various economic compensations were evaluated (0.3, 1, 4, 7, 10, 80, 120 and, 130 $/ton CO_2 avoided) (Kossoy et al., 2015). The decrease in emissions and the rise in the generation of jobs start at a tax credit equal to 7 $/ton CO_2. As expected, with higher tax credits better values for the environmental and social functions are achieved as well as for the economic objective (as opposed to the carbon taxes). Specifically, with the greatest compensation, the reduction in emissions and the generated jobs are maximum as well as with

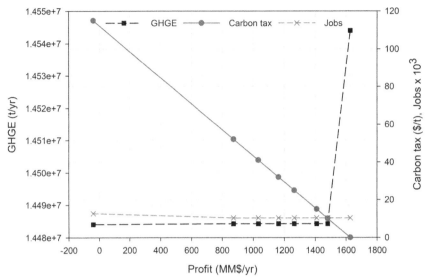

Figure 7. Pareto curves considering different carbon taxes for the integrated system.

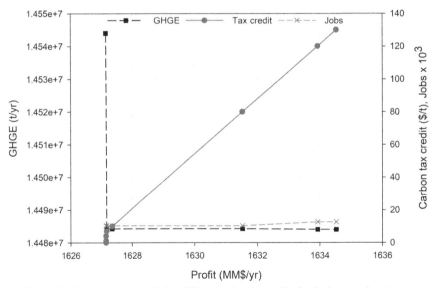

Figure 8. Pareto curves considering different carbon tax credits for the integrated system.

the highest penalization. However, involving the tax credits higher profits can be attained. Despite the fact that the profit is contrary to the other functions, it is possible to find its highest value (1635 MM$/year) along with the minimum amount of emissions and the maximum number of jobs due to the considered tax credit. Therefore, in this analysis, the tradeoff between the environmental and economic functions is represented by the required tax credit. Regarding the energy resources, the case of the tax credit equal to 130 $/ton CO_2 results in the solution with fewer fossil fuels consumption. Biogas and biomass are used to fulfill the total energy requirement.

The proposed methodology presents optimal configurations for water distribution networks involving power-desalination plants. Furthermore, the potential application of the model was illustrated through a case study. Results show the benefits of applying taxes and tax credits to the generated and avoided emissions, while through the Pareto front, compromises between the environmental and social functions are identified while maximizing the profit.

5. Conclusions

A current problem in the energy sector is the immoderate generation of emissions. In this regard, the importance of including carbon policies to reduce emissions in energy systems has been discussed and analyzed. Particularly from an optimization point of view, previous approaches have addressed this issue. Macroscopic systems including different processes have been proposed and evaluated through multi-objective models, involving economic, environmental and social aspects. Specifically, studies related to the analysis of carbon taxes and tax credits in the optimal design of alternative power generation systems have been proposed. The proposed models have been applied to case studies and the obtained results include Pareto sets where tradeoff solutions among the objectives are identified. When the economic compensations are considered, higher values for the profit, reductions in emissions and generation of jobs are attained. Therefore, it can be concluded that the use of carbon tax credits based on the avoided emissions gives better tradeoffs for the economic, environmental and social functions. On the other hand, only limited reductions in emissions were obtained with the evaluated penalizations. This occurs because the current carbon policies for the carbon tax are not sufficient to promote the use of processes that contribute to further reducing emissions as much as when considering the tax credits. Solutions with higher taxes were proposed to show that, with the appropriate penalization, further environmental and social benefits can be achieved, although the economic objective can be compromised. The proposed methodologies present alternative systems to consider for decision and policy makers. However, if the systems were to be applied, a detailed engineering study would be needed. These taxation schemes can be evaluated in other processes in order to analyze the difference of their impact and present other possibilities for sustainable energy systems.

References

Al-Karaghouli, A. and Kazmerski, L.L. 2013. Energy consumption and water production cost of conventional and renewable-energy-powered desalination processes. Renew Sustain Energy Rev. 24: 343–56.

Alton, T., Arndt, C., Davies, R., Hartley, F., Makrelov, K., Thurlow, J. and Ubogu, D. 2014. Introducing carbon taxes in South Africa. Appl. Energy. 116: 344–54.

Anheden, M. and Svedberg, G. 1998. Exergy analysis of chemical-looping combustion systems. Energy Convers. Manag. 39(16): 1967–1980.

Avi-Yonah, R.S. and Uhlmann, D.M. 2009. Combating global climate change: Why a carbon tax is a better response to global warming than cap and trade. Stan. Envtl. L.J. 28(1): 3–50.

Barbose, G., Bird, L., Heeter, J., Flores-Espino, F. and Wiser, R. 2015. Costs and benefits of renewables portfolio standards in the United States. Renew Sust. Energ. Rev. 52: 523–533.

Chen, Z.M., Liu, Y., Qin, P., Zhang, B., Lester, L., Chen, G., Guo, Y. and Zheng, X. 2015. Environmental externality of coal use in China: Welfare effect and tax regulation. Appl. Energy. 156: 16–31.

Clarkson, P.M., Li, Y., Pinnuck, M. and Richardson, G.D. 2015. The valuation relevance of greenhouse gas emissions under the European Union carbon emissions trading scheme. Eur. Account Rev. 24: 551–80.

Cristóbal, J., Guillén-Gosálbez, G., Jiménez, L. and Irabien, A. 2012. MINLP model for optimizing electricity production from coal-fired power plants considering carbon management. Energy Policy 51: 493–501.

Dickinson, S., Mientus, M., Frey, D., Amini-Hajibashi, A., Ozturk, S., Shaikh, F., Sengupta, D. and Halwagi, M.M. 2017. A review of biodiesel production from microalgae. Clean Technol. Environ. Policy 19(3): 637–668.

Diwekar, U. 2008. Introduction to Applied Optimization; Springer Science and Business Media: Clarendon Hills, IL.

Fuentes-Cortés, L.F., Serna-González, M. and Ponce-Ortega, J.M. 2017. Analysis of carbon policies in the optimal design of domestic cogeneration systems involving biogas consumption. ACS Sustain Chem. Eng. 5: 4429–42.

Galán-Martín, A., Pozo, C., Azapagic, A., Grossmann, I.E., Mac Dowell, N. and Guillén-Gosálbez, G. 2018. Time for global action: An optimized cooperative approach towards effective climate change mitigation. Energy Environ. Sci. 11(3): 572–581.

González-Bravo, R., Nápoles-Rivera, F., Ponce-Ortega, J.M. and El-Halwagi, M.M. 2015. Involving integrated seawater desalination-power plants in the optimal design of water distribution networks. Resour. Conserv. Recycl. 104: 181–93.

Graefe, S., Dufour, D., Giraldo, A., Muñoz, L.A., Mora, P., Solís, H., Garces, H. and Gonzalez, A. 2011. Energy and carbon footprints of ethanol production using banana and cooking banana discard: A case study from Costa Rica and Ecuador. Biomass Bioenergy 35(7): 2640–2649.

Gutiérrez-Arriaga, C.G., Serna-González, M., Ponce-Ortega, J.M. and El-Halwagi, M.M. 2014. Sustainable integration of algal biodiesel production with steam electric power plants for greenhouse gas mitigation. ACS Sustainable Chem. Eng. 2(6): 1388–1403.

Hoel, M. 1996. Should a carbon tax be differentiated across sectors? J. Public Econ. 59: 17–32.

IEA. 2017. CO_2 emissions from fuel combustion: Highlights. Paris.

IPCC. 2013. Climate Change 2013: The Physical Science Basis. Contribution of Working Group I to the Fifth Assessment Report of the Intergovernmental Panel on Climate Change. Stocker, T.F., Qin, D., Plattner, G.K., Tignor, M., Allen, S.K., Boschung, J., Nauels, A., Xia, Y., Bex, V. and Midgley, P.M. (eds.). Cambridge University Press, Cambridge, United Kingdom and New York, NY, USA: 1535 pp.

Ishida, M. and Jin, H. 1996. A novel chemical-looping combustor without NO_x formation. Ind. Eng. Chem. Res. 35(7): 2469–2472.

Ishida, M. and Jin, H. 1997. CO_2 recovery in a power plant with chemical looping combustion. Energy Convers. Manage. 38: S187–S192.

Ishida, M. and Jin, H. 2001. Fundamental study on a novel gas turbine cycle. J. Energy Resour. Technol. 123(1): 10–14.

Kaufman, N., Obeiter, M. and Krause, E. 2016. Putting a price on carbon: Reducing emissions. In World Resources Institute: Issue Brief; Washington, DC.

Kossoy, A. 2015. State and Trends of Carbon Pricing. World Bank Publications.

Liang, Q.M. and Wei, Y.M. 2012. Distributional impacts of taxing carbon in China: Results from the CEEPA model. Appl. Energy 92: 545–51.

Liu, Y. and Lu, Y. 2015. The economic impact of different carbon tax revenue recycling schemes in China: A model-based scenario analysis. Appl. Energy 141: 96–105.

Marron, D.B. and Toder, E.J. 2014. Tax policy issues in designing a carbon tax. Am. Econ. Rev. 104: 563–8.

Munguía-López, A.D.C., Rico-Ramírez, V. and Ponce-Ortega, J.M. 2018. Analysis of carbon policies in the optimal integration of power plants involving chemical looping combustion with algal cultivation systems. ACS Sustain Chem. Eng. 6: 5248–64.

Munguía-López, A.D.C., González-Bravo, R. and Ponce-Ortega, J.M. 2019. Evaluation of carbon and water policies in the optimization of water distribution networks involving power-desalination plants. Appl. Energy 236: 927–936.

Newell, R.G., Pizer, W.A. and Raimi, D. 2013. Carbon markets 15 years after Kyoto: Lessons learned, new challenges. J. Econ. Perspect. 27: 123–46.

NOAA. 2018. Trends in Atmospheric Carbon Dioxide. https://www.esrl.noaa.gov/gmd/ccgg/trends/mlo.html#mlo (accessed December 2018).

Pascual-González, J., Jiménez-Esteller, L., Guillén-Gosálbez, G., Siirola, J.J. and Grossmann, I.E. 2016. Macro-economic multi-objective input-output model for minimizing CO_2 emissions: Application to the U.S. economy. AIChE J. 62: 3639–56.

Petriz-Prieto, M.A., Rico-Ramirez, V., Gonzalez-Alatorre, G., Gómez-Castro, F.I. and Diwekar, U.M. 2016. A comparative simulation study of power generation plants involving chemical looping combustion systems. Comput. Chem. Eng. 84: 434–445.

Ren, H. and Gao, W. 2010. A MILP model for integrated plan and evaluation of distributed energy systems. Appl. Energy 87: 1001–14.

Rubio-Maya, C., Uche-Marcuello, J., Martínez-Gracia, A. and Bayod-Rújula, A.A. 2011. Design optimization of a polygeneration plant fueled by natural gas and renewable energy sources. Appl. Energy 88: 449–57.

Sánchez-Bautista, A.D.F., Santibañez-Aguilar, J.E., You, F. and Ponce-Ortega, J.M. 2017. Optimal design of energy systems involving pollution trading through forest plantations. ACS Sustain Chem. Eng. 5: 2585–604.

SEMARNAT and INECC. 2012. Fifth national communication to the united nations framework convention on climate change (accessed December 2018).

Schreifels, J. and Kruger, J. 2003. Tools of the Trade: A Guide to Designing and Operating a Cap and Trade Program for Pollution Control. EPA430-B-03-002, June.

The World Bank. 2014. Electricity production from oil, gas and coal sources. https://data.worldbank.org/indicator/EG.ELC.FOSL.ZS (accessed January 2019).

Vatn, A. 2015. Markets in environmental governance. From theory to practice. Ecol Econ. 117: 225–33.

Wong, S., Bhattacharya, K. and Fuller, J.D. 2010. Long-term effects of feed-in tariffs and carbon taxes on distribution systems. IEEE Trans. Power Syst. 25(3): 1241–1253.

Zhu, L., Jiang, P. and Fan, J. 2015. Comparison of carbon capture IGCC with chemical-looping combustion and with calcium-looping process driven by coal for power generation. Chem. Eng. Res. Des. 104: 110–124.

Suggested Further Reading

Fuentes-Cortés, L.F., Ma, Y., Ponce-Ortega, J.M., Ruiz-Mercado, G. and Zavala, V.M. 2018. Valuation of water and emissions in energy systems. Appl. Energy 210: 518–528.

González-Bravo, R., Nápoles-Rivera, F., Ponce-Ortega, J.M. and El-Halwagi, M.M. 2016. Multiobjective optimization of dual-purpose power plants and water distribution networks. ACS Sustain Chem. Eng. 4: 6852–66.

Hernández-Calderón, O.M., Ponce-Ortega, J.M., Ortiz-del-Castillo, J.R., Cervantes-Gaxiola, M.E., Milán-Carrillo, J., Serna-González, M. and Rubio-Castro, E. 2016. Optimal design of distributed algae-based biorefineries using CO_2 emissions from multiple industrial plants. Ind. Eng. Chem. Res. 55(8): 2345–2358.

Hernández-Martinez, J.F., Rubio-Castro, E., Serna-González, M., El-Halwagi, M.M. and Ponce-Ortega, J.M. 2016. Optimal design of integrated solar power plants accounting for the thermal storage system and CO_2 mitigation through an algae system. Ind. Eng. Chem. Res. 55(41): 11003–11011.

Jin, H. and Ishida, M. 2000. A novel gas turbine cycle with hydrogen-fueled chemical-looping combustion. Int. J. Hydrogen Energy. 25(12): 1209–1215.

Judd, S., van den Broeke, L.J., Shurair, M., Kuti, Y. and Znad, H. 2015. Algal remediation of CO_2 and nutrient discharges: A review. Water Res. 87: 356–366.

Lira-Barragán, L.F., Gutiérrez-Arriaga, C.G., Bamufleh, H.S., Abdelhady, F., Ponce-Ortega, J.M., Serna-González, M. and El-Halwagi, M.M. 2015. Reduction of greenhouse gas emissions from steam power plants through optimal integration with algae and cogeneration systems. Clean Technol. Environ. Policy 17(8): 2401–2415.

Martín, M. and Grossmann, I.E. 2017. Optimal integration of a self-sustained algae-based facility with solar and/or wind energy. J. Cleaner Prod. 145: 336–347.

CHAPTER 7

Carbon Mitigation in the Power Sector
Challenges and Opportunities

Bruce Rising

1. Introduction

The human connection to global scale weather and climate boils down to CO_2 emissions from three primary activities: Power generation, transportation, and industrial-commercial operations. While other pollutants (e.g., PM, sulfates, and tropospheric ozone) also exhibit the capacity to absorb or reflect incoming solar radiation, CO_2, an infrared absorbing poly-atomic molecule, has been assigned the bulk of the blame. Broad assessments of continental temperature shifts reveal noticeable changes taking place. Data presented in Figure 1 from the Energy Information Administration (EIA) on heating and cooling degree days for the United States suggests that summers do appear to be getting warmer, and winters milder (EIA, EIA Monthly Energy Review April 2018, 2018). With this data covering approximately 3,000,000 square miles, it represents a formidable set of data, suggesting that warming is taking place.

While massive amounts of CO_2 are released and absorbed through natural processes on a global scale, the carbon emissions from modern industrial society are easily identifiable in the atmosphere. Overall, the atmospheric burden of carbon related to consumption of fossil fuels appears to be about half the total amount. To counterbalance the increasing load, much of the scientific community has concluded

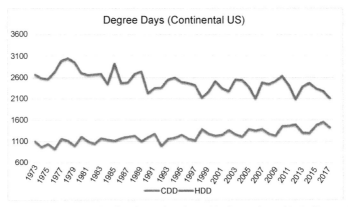

Figure 1. Cooling degree days (CDD) and heating degree days (HDD) in the continental U.S. (Source: EIA Monthly Energy Review Table 1.9 and Table 1.10 Heating and Cooling Degree Days by Census Division).

Executive Advisor, ADI-Analytics, Inc.
Email: Brising@ADI-Analytics.com

that regulation, specifically a reduction of CO_2 emissions, is a necessity. The concept is an extrapolation of air pollution regulation applied to the more toxic emissions (NO_x, CO, and SO_2), where strict emission regulations yielded technical evolutions ultimately producing visible improvements in air quality from innovative pollution control.

1.1 The fossil fuel dilemma

The challenge that we face is to find a way to use known fossil resources, while also mitigating the problem related to CO_2 emissions (in addition to problems of NO_x, SO_2, etc.). All fossil fuels are carbon bearing, although the abundance of coal and its wide use has made it a target for carbon control strategies directed at the power sector. Virtually every major economy that has become an economic powerhouse has done so using coal. This was a path charted by England, Germany, France, the United States, and now China and India. While early on coal was used in more distributed ways, like home heating, it ultimately evolved into the fuel of choice for producing power and refining metals.

Following coal's introduction as a fuel for heating and making iron, its use in illumination was explored in the 19th Century. Coal was the primary feedstock for gasifiers that provided "coal gas" for illumination up until the 20th century in the United States and Europe. Thousands of coal gasifiers were built around the country, producing a toxic mix of combustible carbon monoxide (CO), hydrogen, water vapor and carbon dioxide that was delivered to homes and businesses for indoor lighting. The first use for illumination in the U.S. was reported in Rhode Island (Melville's Gas Apparatus, 1876), although gasification had been in use in Europe as well. Eventually coal-gas illumination was replaced by kerosene lamps (creating a market to support the petroleum era), and kerosene lamps would be replaced by electricity. The bulk of this generation was accomplished with coal, but through complete combustion, not the partial oxidation experienced with gasification. At its peak, almost half of power generated was produced with coal used in the steam Rankine cycle, consuming nearly one billion tons of coal annually (MacIntyre, 2018). While none of those early gasifiers remain, gasification technology development has been continuously pursued, although not primarily for power generation. Attempts to adapt more advanced technologies, such as integrating gasified coal into a gas turbine, failed in markets where the primary product to be delivered was electricity. Ultimately, the number of Integrated Gasification Combined Cycle (IGCC) plants constructed globally is probably less than ten (Phillips, 2017).

2. Innovation Timeline

An almost irresistible solution to the problem of carbon finding its way to the atmosphere via combustion is to simply find a way to "decarbonize" the fuel, or possibly switch fuels entirely, or switch to a different technology. At the rate of change of technical innovation, it would seem that a new technology might be expected to achieve one or more of these goals. However, almost every new technology is preceded by an incubation period during its evolution. That period might include the first concept, multiple re-designs, field trials, and hopefully commercialization. To move innovations along, sometimes they are given protected status (designs are secret), or state subsidies, or market control (monopolies and protected markets). Yet even simple innovations, like the refrigerator or electric light, have experienced time lags of 30 to 40 years from initial exposure until adoptions of about 50% or more (Desjardins, 2018; Gordon, 2016).

Most novel technical concepts require near continuous innovation in order to achieve widespread adoption. For example, the steam cycle was essentially introduced in 1712, with technological modifications taking place in the intervening decades. However, it would be nearly 150 years before the cycle was adapted to power generation, while most of the previous applications were for transportation. It would take new developments using the steam turbine, and manufacturing infrastructure to produce equipment on a large enough scale in order to provide a significant amount of generating capacity. Before the mid 1950s, much of the power generation in the United States was based on renewables, and a significant portion of the hydropower was dependent upon major works programs undertaken by the Federal government during the worst economic crisis of the century. It wouldn't be until the collapse of nearly all major economies during World War II that the steam cycle, and all its variants, took a commanding role in power generation, a role it still maintains.

At first glance, solar energy would seem the most logical of the renewables. The sun is always available (although solar energy is not always accessible), and the output predictable, and it cannot be controlled by any regulatory body or commercial group. Furthermore, with appropriate technical innovation, there was hope that it could be harvested inexpensively, but the developmental timeline followed nowhere near the optimism put forth upon its first identification in the corporate research laboratory. Table 1 summarizes some of the events and the expectations in the development of photovoltaics. As Table 1 suggests, that process was much slower than the early optimism had promoted.

Figure 2 reveals just how long that induction period can last for two key renewables: Solar and wind. The first solar breakthrough was identified in 1931 (and expected to be competitive with hydropower at the time) and had only minimal impact on total installed capacity for 67 years. By 2016, total energy supplied by the solar energy was less than 2% of total generation (EIA, Form 2_3_4_5_6_7_2017_Early_ Release, 2017). For wind power generation, the induction period was about 54 years,[1] but nearly another 20 years before wind would begin to approach ten percent of the installed capacity (and less than 6% of total energy supplied).

Table 1. Technology development timeline for solar photovoltaic energy sources.

Year	Event
1931	Siemens and Westinghouse reveal the potential for a CdS cell that could make electricity from sunlight, potentially at costs less than hydroelectricity (NYTimes, Use of Solar Energy is a Near Solution, 1931).
1954	A solar battery is constructed by Bell Labs, Murray Hill, NJ (NYTimes, Vast Power of Sun is Tapped by Using Sand Ingredients, 1954).
1974	As the "energy crisis" makes its impact known, nearly a dozen homes are fitted with solar collectors to demonstrate alternative energy resources (Webster, 1974).
1976	By the year 2000, ERDA (DOE's predecessor agency) expects solar energy to provide 50 GW of capacity to the United States; actual reported value was less than 1% of the 1976 forecast.
2015	United States solar capacity 20% of the 1976 forecast.

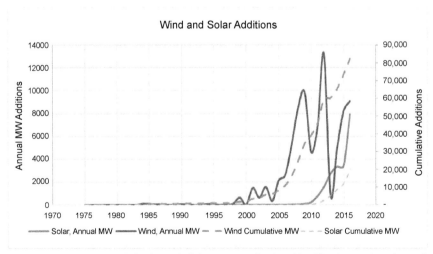

Figure 2. Example of the time delay (induction period) between a technology identification or adaptation and commercial application. Note that, for both these examples, heavy government support was required in order to reach commercialization (Source: U.S. Department of Energy, The Energy Information Administration (EIA), EIA-923 Monthly Generation and Fuel Consumption Time Series File, 2016 Final Revision).

---

[1] In the case of wind, the period is measured from the first application for power generation in 1943.

3. Technologies in Play

Technologies that can counter the growing CO_2 emissions in the power sector are already in hand, and with near-zero carbon emissions, these will serve as a stepping stone for progress toward reducing overall manmade CO_2 emissions. Currently, there are some notable successes to highlight, with some nations deriving over 90% of their electricity from hydro-electric power systems (e.g., Norway is over 90%, and Brazil over 60%) (EIA, Country Analysis Brief-Norway, 2018). Even the United States derived most of its power generation from hydroelectric systems up until mid-20th century. France and Japan both have significant supplies of low-carbon power generation in the form of nuclear power (WNA, Nuclear Power In Japan, 2019). France's substantial investment in nuclear power was designed to balance the lack of gas and oil, much like Japan's, but Japan is also a major importer of Liquid Natural Gas (LNG), a necessity to supply the nation's high demand for power while fulfilling the role of one of the world's largest economies.

One of the more remarkable successes can be found where carbon emissions have been reduced through high level coordinated planning coupled with grass-roots commitments to addressing the issue. Germany's *Energiewende* (energy transition) is an example. Intensive planning and pressure at the national level, as well as public support, resulted in the development of significant volumes of wind and solar to supply the nation (in 2017, generation from wind and solar was approximately 30% of all energy produced). A similar story could be told for Denmark, which some of the world's premiere wind turbine suppliers call home. But the technologies deployed are not necessarily new. Both wind and solar power had been well established, they just had not penetrated any markets dominated by conventional fossil power systems. What changed was a top-down committed policy to improve and expand on these technologies. Funding through government support, and direct subsidies, along with improvements in scale (e.g., wind turbine sizes moved from the kW range to the MW range in two decades) resulted in massive deployments throughout Germany. While Germany's CO_2 emissions have been substantially reduced, the massive effort expended has yet to result in any measurable reductions in the atmospheric CO_2 concentration. It does suggest that direct command-and-control may be required in order to solve the problem of the scale being faced. Wind and solar energy production continues to be dogged by the problem of intermittency. The need for reliable, dispatchable generation implies continued use of fossil fuels to supply those power systems already in operation, and expected to come online in the near future.

This leads to the application of technologies that directly extract CO_2 and prevent its release to the environment. In section 4, we consider those methods were CO_2 could be recovered at the point of generation. Extensive technical expertise on how to extract CO_2 from industrial operations, most notably in the oil and gas sector. Success in the petrochemical sector has led many to conclude that a similar solution could be adapted to the power sector. Perhaps it might be nothing more than an engineering and economic challenge that is the limiting agent. As it turns out, that adaptation has been incredibly difficult. While the technology to extract CO_2 from gases is available, we haven't been able to breach the gap where we can do so affordably, effectively making it an expensive off ramp for addressing the problem.

3.1 Technology innovations impacting carbon emissions

An unexpected event in the power sector has been the near disappearance of load growth in the United States. Historically, the power sector had planned new capacity additions based on a 1–2% increase in demand. In the first decade of the 21st century, this growth rate was projected to ignite a rebirth in the nuclear power industry. New nuclear capacity was planned in virtually every region of the country. But the economic turndown in 2008 put the brakes on much of this new capacity. Surprisingly, even with the following recovery, demand had not recovered, ultimately leading to the cancellation of nearly every nuclear project in the planning stages.

While this deceleration in demand was unexpected, part of the observed downshift can be attributed to expanded use of more efficient devices found in nearly every aspect of daily living. Perhaps the most significant impact came in the lighting sector where conventional incandensant lamps were replaced

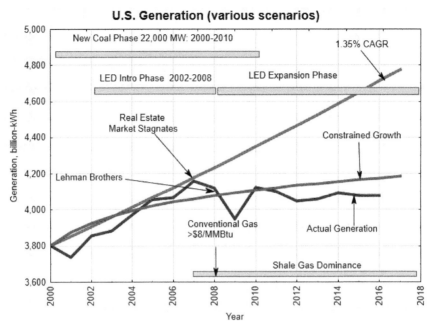

Figure 3. The slowdown of load growth in the United States (Source: *Blue line*: EIA Annual Generation by State Historical Tables 2017, *red line*: Extrapolation of historical growth from 2000, *green line*: Constrained growth reflecting impact of end-user efficiency upgrades).

carte blanche with Light-Emitting Diode bulbs (LED's). Figure 3 notes where U.S. generation was headed based on historical growth rates (red line), but the interruption of the great recession, coupled with the introduction of new, efficient lighting technologies, led to a significant departure from the simple extraction of historical generation. In addition to lighting, many other end user devices consumed much less power than units they replaced (the conventional CRT TV vs the LED flat screen). A modest 3–4% turnover rate in lamp replacement (incandescent to LED) alone can explain a significant component in the halt of load growth (green line noted as "constrained growth"). This combination of economic output changes, coupled with new technologies, seemed to put the United States on a new, and reduced, CO_2 emissions trajectory.

3.2 CO_2 reversals and economic output

Possibly except for the industrial revolution itself, there have been no innovations (or technologies) that correlate directly with any observed changes in atmospheric CO_2 concentrations. However, economic growth does appear to correlate with changes in atmospheric CO_2 levels, and severe economic shifts might be identifiable in those concentration changes.

To expand on that, temporal measurements of CO_2 are correlated with economic output in Figure 4. This figure summarizes the month-to-month changes of the measured atmospheric CO_2 concentration since 1973 (Keeling et al., 2005). The atmospheric data were compared with those periods of economic slowdown, classified as a "recession" by the Federal Reserve (FRED, 2018). In virtually every instance of a reported economic pullback, the growth in the atmospheric CO_2 concentration slows, and in one case it reverses. Moving from left to right, the impact of the U.S. economy (the largest in the world, and most likely to strongly correlate with the shift) is lessened as growing economies across the globe exert greater influence. Even by the 2000–2010 decade, the evidence of a general slowdown (the global financial crisis, which saw a dramatic reduction in industrial output and power generation) appears to be detectable by noting the shift in the CO_2 concentrations.

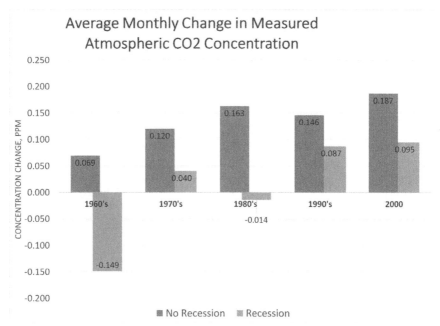

Figure 4. Monthly increase (decrease) of CO_2 burden compared to economic growth (Keeling, 2005).

4. The Current Low Carbon Power Portfolio

To address the atmospheric CO_2 burden, and the opportunities to successfully achieve a 2 °C temperature rise, the mix of components in the power system must be considered. The only widely available low-carbon energy sources operating in the power generation sector are limited to the following:

1. Nuclear power (principally light water reactors)
2. Hydroelectric, including pumped storage
3. Solar—primarily photovoltaic
4. Wind
5. Conversion of a coal fired facility to a combined cycle (achieving nearly 60% CO_2 reduction)

Biomass and geothermal also qualify as low-carbon alternatives, but are available only in limited markets, and have yet to penetrate the power supply markets with greater than single digits generation (DOE/EIA, Form EIA-923, 2015). Nuclear has great potential, but a hefty financial burden. In 2018, the United States operated 100,000 MWe operating nuclear, and 2,500 MWe under construction, but as much as a third of the operating nuclear could be expected to retire in the next 10–15 years. Beyond the United States, over 50,000 MWe of new nuclear is under construction, scheduled to be online over the next decade, much of it in Asia (WNA, World Nuclear Association, 2019), but that volume is dwarfed by the additions of fossil coal expected to enter service.

Fossil fuels, and coal specifically, remain the major contributor to the global CO_2 emissions. Since 2000, over 1,600,000 MWe of generating capacity had been ordered, although not all that capacity appears to be fully operational as of 2018. Looking forward, the immediate order backlog of steam turbines (i.e., coal fired) is reported to be near 100,000 MWe (McCoy, 2019). Considering the various sources available, the near term continued expansion of coal generation is perhaps no less than 100,000 MWe and possibly as high as 300,000 MWe (Endcoal.org, 2019).

This continued and rapid global buildup of even more CO_2 emissions from power sources seems almost irreversible just based on the near-term backlog of equipment. How the technology selection and fuel choice relate is highlighted in Figure 5. The graphic suggests that just by switching fuels, a reduction in CO_2 emissions occurs (based on CO_2 tonne/MWh as a benchmark). A typical coal plant in the U.S.,

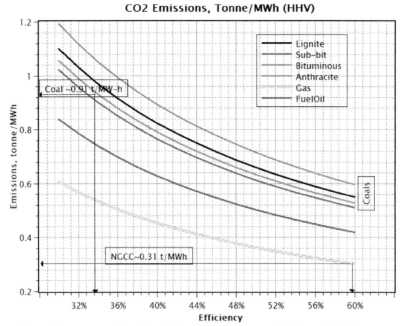

Figure 5. Comparison of CO_2 emissions (tonne/MWh) for various fuels and technology efficiencies (Based on nominal fuel heating values and carbon content and EIA Electric Power Annual Table 8.2).

operating at 33% cycle efficiency, would have a CO_2 emissions rate of approximately 0.90 tonne CO_2/MWh (depending upon the exact fuel selection) (DOE/EIA, Electric Power Annual, 2017). A combined cycle operating at a fleet average of 56% on natural gas would produce about 0.31 tonne CO_2/MWh, although the actual reported efficiency is much closer to 46%. Assuming newer plants could reach the higher performance standard, coupled with a switch to gas fuel, they might yield a CO_2 reduction fuel equivalent to approximately 66%. This without the addition of expensive carbon capture equipment, as well as the increased operating costs associated with maintaining the more complex carbon capture equipment. And almost as important, since the CO_2 release is minimized, there is no supplemental requirement to address the technical challenge of disposing of the CO_2. In the short term, this readily accessible simple solution is likely to be a top contender for a solution to reducing CO_2 emissions. It is also likely to establish the gas turbine as the *de facto* technology choice for carbon control for the next few decades. Medium term, the combined cycle may be amenable to carbon capture. Longer term, and more likely, the gas turbine cycle itself could be adapted into an oxy-fuel system, substantially reducing the complexity of the post combustion carbon capture.

4.1 Carbon capture and CO_2 reduction

An overview of power generation technologies applicable for carbon capture is summarized in Figure 6. The three primary approaches include pre-combustion (*in situ*) and post combustion capture, except that in the case of the Oxy-Fuel, CO_2 is not captured directly, but isolated by removal of the H_2O from a CO_2+steam mixture. Pre-combustion (IGCC) and Post Combustion carbon capture rely almost heavily on solvents (e.g., ethanolamine, di-ethanolamine, etc.), to extract the CO_2, but then the solvents must be regenerated and recycled. There are only a few IGCC facilities operating, thousands of conventional air and vapor power cycles, and, as will be discussed next, the Oxy-Fuel design. In every case, however, proper treatment of the CO_2 is required prior to compression (to minimize the risk of pipeline corrosion), and ultimately the disposal of CO_2 in a safe reservoir.

Some current projects on coal-fired boiler retrofits include Boundary Dam (in Canada), and W.A. Parish in the United States (the Petra Nova project). Integrated Gasification Combined Cycle (IGCC)

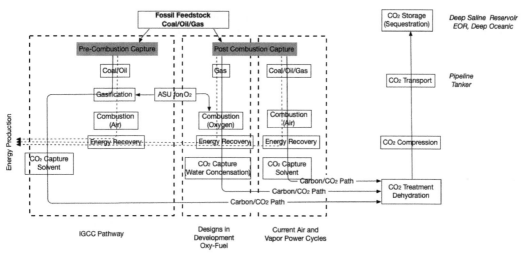

Figure 6. Carbon capture in the power sector. CO_2 recovery represents one of several critical steps required to address the CO_2 emission challenge. Initial fuel choice, energy conversion technology, and CO_2 capture method are tightly interconnected, impacting where the ultimate energy extraction occurs and the overall process efficiency.

was also expected to simplify the carbon capture process by allowing treatment of smaller volumes of carbon rich fuel prior to combustion (e.g., Kemper and North Dakota Gasification). As of 2017, there over 25 carbon capture projects globally (CCS, 2018), although that number drops dramatically for power generation carbon capture (Kapetakia, 2017), in contrast to some 800 coal thermal plants operating in the United States alone. For those successfully operating projects, a key factor in their commercial success has been a viable storage location for the recovered CO_2, as well as a revenue stream for the gas. In two projects, Petra Nova and North Dakota Gasification, CO_2 is sold for enhanced oil recovery. In addition, one of the original projects developed specifically to include carbon capture (Kemper) has since been converted to natural gas operation only, effectively scrapping the carbon capture elements, which contributed to much of the cost escalation (Kelly, 2018). A primary driver in this decision has been the excessive costs associated with both the gasification stages and the carbon capture components.

4.1.1 Post combustion capture

Despite so few examples in the power industry, carbon capture (CO_2 recovery) is widely practiced in the oil and gas industry. The extensive track record in oil and gas operations (O&G) led to the expectation that this technology could be easily adapted to power generation. To date, success in the power sector has been limited, with project costs representing a major hurdle (along with plant complexity, limited capabilities to cycle, and limited turn down capabilities). Nevertheless, costs are clearly not the only hurdle. Storage of carbon remains a major stumbling block. There are no locations in the U.S. that are permitted to store the three million tonnes of CO_2 that could delivered annually. All in all, CO_2 recovery at high concentrations (and pressures) is established practice, it just isn't well understood how to make it work cost-effectively at a power plant, of any type, without substantially degrading plant performance. Demonstrations currently operating have required substantial subsidies or external funding.

In spite of this, there are those examples worth considering, such as the recent retrofit at the W.A. Parrish coal plant in Texas (Patel, 2017). While the project is based on well-established chemical principles related to carbon chemical absorption (amine absorption), moving the project forward required investment by the Department of Energy to support the project. Plant performance was minimally impacted by providing a second power source to supply much of the required energy and heat to operate the carbon capture process. Since the supplemental power source is a combined heat and power (CHP) using a gas turbine, nominal project cost estimates would be in the range of $500–$700/kWe. Ultimately, the post combustion carbon capture on the coal unit is achieved using natural gas in a second power plant. The completed project captures roughly 33% of the carbon emissions from the retrofitted coal unit but none of the CO_2 from the gas fired CHP installed to support the carbon capture[2] (Dubin, 2017). The 240 MWe plant total project costs were estimated at $1 billion (Patel, 2017).

4.1.2 Carbon capture-natural gas

A study from 2017 goes into more detail on how the process might work on a clean gas stream in a combined cycle. The work was based on modelling of power plants using a conventional amine scrubber to recover CO_2 from the exhaust (Carapellucci, 2017). Their results underscore the crux of any post combustion capture: performance degradation is substantial; raising the cost of electricity, and potentially making the facility less competitive.

Results for three natural gas combined cycles (NGCC) noted in Table 2 are not wholly unexpected. The high pressure drops necessary to move the exhaust gases through the amine absorption tower, and the parasitic losses associated with recovering the amine solvent drive the plant economics into non-competitive status. Not evident in the study is the role of cycling. In the modern grids with large renewable footprints (wind and solar), power plant cycling has become the norm, possibly with starts and stops up to four times per day. The post combustion chemical processes necessary to make carbon capture work are much better suited for base load operation, where temperatures, pressures, and flow rates are steady, and unchanging. The evolving power markets encountered today appear to be in conflict with the demands of the basic chemistry of solvent extraction of CO_2.

Table 2. Model results for post combustion retrofit of three combined cycles using amine absorption, with targeted 90% carbon capture. Modeled results at the source only. Compression, pumping, and injection place even greater losses on total system available capacity.

Facility	Base plant, MW rating	Change in cost of electricity, COE	Change in capacity
NGCC 1	384	+35%	−11%
NGCC 2	153	+41%	−11%
NGCC 3	49.9	+57%	−14%

4.1.3 Carbon capture-coal

Coal's growth, and continued contribution to the atmospheric CO_2 burden, will likely increase substantially in the coming decades. The large number of fossil coal plants built, under construction, or planned will easily offset any coal retirements achieved in the United States and Europe. The implication here is that some measure of carbon control adapted to coal is necessary in order to realize the 2 °C temperature rise. However steep the challenge for carbon capture with natural gas, it is significantly greater for coal. A landmark study carried out by Alstom used American Electric Power's Conesville power plant as a basis. While numerous iterations were explored, none demonstrate any result where the plant performance was not substantially degraded (Nsakala, 2001).

In this study, three cases were considered. Two with amines and one as an Oxy-Fuel. Concept A (as describe in the paper reference) was a simple monoethanolamine post CO_2 capture. Concept B was essentially an oxy-fuel conventional steam system, where nitrogen is eliminated from the air supply, and the exhaust gas is a high concentration CO_2 product stream, effectively avoiding the need for chemical extraction of the carbon dioxide. Concept C is a modification of the first concept, where the amine is a modified version of that used in Concept A, an attempt to improve performance and efficiency.

In every case studied, performance was substantially degraded, even though the carbon capture potential was substantial. These results are typical of what has been found in virtually every post combustion capture study. For a load serving entity (LSE), this loss of performance and flexibility would have been unacceptable. By 2019, changing market dynamics resulted in the closure of this facility, essentially eliminating all CO_2 emissions, with power generation now made up by a mixture of natural gas combined cycles and renewables in the region.

---

[2] The process treats only a portion of the exhaust from unit No. 8, extracting 90% of the CO_2.

Table 3. Performance comparison of existing coal fired power plant for post combustion carbon capture. Based on AEP Conesville Unit No. 5.

Design configuration	Power, MW, net	Plant efficiency	Carbon capture
Base Plant	433.7	35%	94%
Concept A: MEA	260.7	21%	91%
Concept B: Oxy-Fuel, no amine	279.7	23%	91%
Concept C: Modified MEA solution	341.5	23%	87%

4.2 Alternatives: Oxy-fuel combustion

The primary difficulty with carbon capture is the need to selectively extract CO_2 from large volumes of relatively inert gases that also include large quantities of reactive pollutants (e.g., SO_2). The extraction of CO_2 becomes far more difficult if the treated exhaust stream includes particulates, sulfates, and nitrates, which might be the case if the fossil fuel used is coal. Hence, for a coal-fired power plant, extensive gas cleanup would be necessary in order to limit the contamination (and potential rapid degradation) of the solvents. A preferred approach to the problem would be to work with a relatively clean fuel, such as natural gas, which is primarily methane, with product streams that are also gaseous.

In the conventional gas turbine, the largest inert diluent is the atmospheric nitrogen. Nitrogen removal prior to combustion would leave carbon dioxide and water would make up the bulk of the remaining constituents, greatly reducing the volumes of gas to be treated. In fact, if the exhaust gases consisted only of a mixture of CO_2 and H_2O, there would be no need for chemical treatment to extract CO_2 at all, since the water can be easily condensed, isolating the CO_2. Eliminating the solvent for carbon capture greatly simplifies the concept of capture, not to mention making it easier to permit since a relatively toxic solvent can be a challenge to permit. Condensing the water vapor from the exhaust through a heat exchanger will also yield a relatively high purity CO_2 stream.

If the fuel and oxygen are mixed in a stoichiometric mixture, however, the combustion process results in extreme temperatures capable of distressing hot-gas-path components. The challenge is to find a way of moderating the peak combustion temperature. An obvious choice would be to recycle one or more of the exhaust products. Having CO_2 and H_2O in abundance in the exhaust makes them an obvious choice to moderate temperatures, as well as the serving the duty as the working fluid. Recycling part of the exhaust gases will moderate the combustor temperatures, but still leave the basic design unaffected— isolation of CO_2 by condensing water from the exhaust.

There are two novel designs exploring this concept. One approach uses H_2O (steam) as the recycled working fluid to moderate temperature conditions. Another approach recycles CO_2 to achieve a similar goal (Rathi, 2018). In each case, the energy released is recovered in a power turbine (or a heat exchanger if a boiler is used to raise steam). With the emphasis on efficiency, the focus here will be on energy recovery using a rotating apparatus such as a power turbine.

Both cycles make use of a uniquely designed expansion turbine element to recover power. Gas temperatures will have to be limited to a maximum value based on the materials of construction. The most critically stressed element is the first stage blade, where temperatures, pressures, and stresses are greatest. Another common requirement for both systems is the use of an air separation unit (ASU) to provide high purity oxygen as the oxidizer for the fuel. This step eliminates the need to separate CO_2 from inert nitrogen and argon in the exhaust gases, although it creates additional parasitic loads for the entire process.

4.2.1 Oxy-fuel H_2O cycle

The Oxy-Fuel (Clean Energy Systems, n.d.) cycle shares many features with a conventional gas turbine combined cycle, including an expansion turbine, a heat recovery section, and similar plant auxiliaries. The major differences are the lack of a compressor section (which nominally could absorb 60 to 70%

of the power turbine's output) and the inclusion of the air separation plant (ASU) for the supply of the oxygen for combustion (Anderson, 2008).

The working fluid in the power turbine is a mixture of H_2O and CO_2, with steam being the dominant working fluid. Because of the high concentration of steam (over 90%), it may function more like a steam turbine even though the original design basis was a gas turbine.

Several features of the turbine design are significant departures from conventional Brayton cycle. One that stands out is the performance improvement. A recent demonstration (based on a heavily modified W251 B12 gas turbine) is reported to be capable of 150 MWe output, despite the original turbine design limit from the manufacturer of 43 MWe (Clean Energy Systems, n.d.). Removal of the compressor yielded a three-fold increase in unit rating. This boost in output is roughly equivalent of taking a sub-scale D-Class turbine and modifying it to achieve F-Class performance.

4.2.2 Oxy-CO₂

As an alternative to water (steam) as moderator/working fluid, carbon dioxide could be used as well. Carbon dioxide is compressed and recycled to blend with the fuel (and oxygen) (Breaking ground for a groundbreaker: the first Allam Cycle power plant, 2016). Like the Oxy-H_2O, water is continuously extracted in order to prevent its buildup somewhere in the cycle. The information provides an estimate of the power turbine output for the noted mass flow, and a calculated turbine inlet temperature. The key point is that both the Oxy-CO_2 and Oxy-H_2O are capable of producing impressive amounts of power, while in the conventional gas turbine application 60% or more that power would be consumed by the compressor.

4.2.3 Other benefits

Oxy-Fuel systems are very nearly a closed loop system. The exhaust products (only two are expected) can be either condensed or compressed and recycled back to the combustor, where more fuel is added with fresh oxygen. This results in a cycle where there is minimal, or zero, release of pollutants such as NO_x, CO, and SO_2. On paper, this is clearly a win-win, CO_2 control achieved without use of chemical solvents and near zero emissions of priority pollutants.

Added to this emission benefit is one more feature. An Oxy-Fuel system can also be a net water producing cycle. Hydrogen from the fuel will add to the total material balance and will have to be removed continuously. Thus, the Oxy-Fuel product streams (power, CO_2, and water) all have some inherent value. Limited water access, which has become a problem for many new power projects, could be offset by the net production of water possible from an Oxy-Fuel system.

Commercializing the Oxy-Fuel design is likely to demand extensive financial support as well strong engineering capabilities, and the design features deviate so far from current Brayton cycle design trajectories that it could encourage new entrants into the power equipment (and environmental) market. This is not unlike the changes that occurred when the electronic market shifted from vacuum tubes to solid state devices creating an opening that established new industry leaders.

4.2.4 Alternatives that include coal

The world is still faced with several bedrock issues:

1) the demand for more power,
2) the widespread availability of coal as a fuel of choice, and
3) the extreme technical and economic challenges for post combustion CO_2 capture.

Not too many years ago, the direction seemed to be using coal (a low-cost fuel) with the combined cycle (a low capital cost technology choice), with the IGCC (sometimes referred to as "pre-combustion capture" or *in situ* capture) offering a unique technical solution. For many reasons, those plans failed, resulting in few operating IGCC units. Beyond the technical horizon was an unexplored approach combining the gasification technology with coal to produce a synthetic natural gas, and alternatively supplying this gas to a combined cycle fleet. This shifts the carbon extraction from the point of generation to the point of gas production. The carbon content delivered to the power generator is reduced because hydrogen has been added to the fuel, thus reducing the CO_2 emissions at that point (as noted in Figure 5, where both the efficiency and the fuel carbon content have shifted favorably). Additionally, the by-product gas streams at the point of syn-gas production can be more easily treated to extract additional CO_2 with effectively zero impact on the efficiency, performance or the cost of electricity at the point of generation. Potentially, only the cost of fuel may be affected, but alternatively, CO_2 has marketable value that could offset any fuel price changes, and while CH_4 (methane) may be the primary product for the power market, the gasification process can be adapted to produce a range of hydrocarbon products from C1 to C6 and higher, where the longer chain hydrocarbons are of increased value, making the conversion process economically viable.

While a wide range of chemicals can be produced from the synthesis gases generated in a gasifier, perhaps one of the most appealing is the production of hydrogen via:

$$CO + H_2O \rightarrow CO_2 + H_2$$

This expression is revealing since the product streams are now carbon dioxide, which can be easily separated, and hydrogen, which is representative of the ultimate low-carbon fuel. Hydrogen has applications that include power generation, transportation, and even energy storage. And the chemical separation of CO_2, as already noted, is well established. Likewise, the energy released by this reaction can also drive the chemical recovery of CO_2 from a solvent extraction process.

4.2.5 Yet to be solved

The Oxy-Fuel offers great opportunity, but it comes along with many new challenges, none of them trivial. Actual field demonstrations are likely to reveal many more issues than simple paper study, or even a Front-End Engineering Design (FEED). Only modest scale demonstrations, or short-term operating runs have been achieved so far. For commercialization, however, there are more pressing challenges to address, including:

- Ramping and transients: Most combustion systems today operate over a range of conditions, usually characterized as the turndown ratio. Typically, peak performance occurs at the baseload, but it is likely that any new cycle designs would require significant supplemental development in order to demonstrate the kind of turndown capacity markets need to function properly.
- Materials design: Combustor (and hot section) designs are not trivial. Every gas turbine OEM devotes extensive research and development effort into these components, often taking years to release a new commercial design. The Oxy-fuel combustor and nozzle designs represent radical departure from existing designs and will likely take several iterations in service before achieving original design objectives.
- Time lag in startup from an air separation plant. An ASU (air separation units) could take considerable time to reach effective operating conditions. This time delay, which could be days, makes it difficult to adapt known ASU designs into either of these cycles.

4.3 Ultimate challenge—The final fate of CO_2

The final hurdle yet to be overcome is the ultimate disposal of CO_2 extracted from any of the processes considered. A hallmark of these newest design concepts is that they can eliminate (or at least minimize) many of the parasitic load problems associated with conventional carbon recovery plants. The handling of solvents is reduced, if not eliminated, but the final disposition of the CO_2 is still indeterminate. CO_2 for enhanced oil recovery offers economic benefits, but it may ultimately only recycle some of the carbon dioxide, and potentially increase emissions. Underground storage may come in several forms, but an alternative to EOR is storage in saline aquifers. Deep saline storage capacity for CO_2 has been estimated to be in excess of 1,000 Gt of CO_2 (Aydin, 2010). Compare this with annual releases in the range of 35+ from all industrial sources. In different terms, one research team estimated that a 0.7 km^3 volume could store 80% of the CO_2 released from a 500 MW coal plant operating over 30 years (Jordan Eccles, 2009). Because of the physical dimensions of a large coal plant, that estimated storage volume would be roughly equivalent to the surface area of the facility.

5. Looking Forward

Given the long developmental timeline of 30–50 years, it's unlikely that any new and innovative technology to mitigate CO_2 emission growth will become available to end users prior to 2050. Few of the existing coal plants are likely to be retrofitted with carbon capture. The examples we have to date show both enormous costs, and significant loss of performance. These two hurdles by themselves would, in practical terms, disqualify them immediately based on any competing alternative (e.g., a combined cycle, or even a combined cycle with some measure of carbon capture). During this same period, we can expect very high growth rates of new coal thermal plants to come online, a fact that will likely accelerate the rapid growth in the atmospheric CO_2 burden.

We also know that some things that can be accomplished in the short term that would at mitigate some of this explosive growth, and potentially offset some of the predicted temperature increases.

- *CO_2 emissions mitigated by technology choice*. Maximizing the efficiency at the generation source has a significant impact on source emissions. CO_2 reductions have also been substantially lowered with improvements in end-user efficiency. Most notable here is the impact of the LED.
- *CO_2 emissions mitigated by fuel choice*: Using methane (natural gas) where available and using this in a combined cycle would effectively meet the previously stated requirement. A gas fired thermal plant produces about 50% less CO_2 than the same unit operating on coal.
- *Where coal is in abundance, gasification and production of syn-gas will yield a gaseous fuel product compatible with any gas turbine*. CO_2 production at the point of gasification can be dealt with as a separate challenge.
- *Additional renewables*. This is an obvious choice being effectively free of CO_2, but where renewables are dominant, retail power prices have shown a propensity to rise. Disconnecting the increase in power prices from choice of renewables would lower the barriers to increased usage, but energy storage, the ability to make energy available when it's needed, and store when generation is in excess, is a major technology and economic gap. To date, we have not solved the "energy density" problem, where a simple hydrocarbon like kerosene stores forty times as much energy as the best batteries available. That situation continues to evolve, but landmark breakthroughs have yet to be realized.
- *Modular nuclear*. Nuclear power's cost challenge has been a nearly insurmountable hurdle, especially in open, competitive power markets. Much of this cost is associated with on-site construction and extremely long build intervals. A modular reactor design has the potential to transfer large portions of that build cycle back to the factory, where cost controls and construction times may result in a cost competitive design.

Meanwhile, since a gas-fired combined cycle yields about one third of the carbon emissions of a thermal plant (see Figure 5), one solution to the vast buildup of thermal plants now happening would be the eventual phased conversion of many of the recent coal plants into natural gas combined cycles. Such a conversion process would likely take a decade, as a gas infrastructure would need to be co-developed in tandem.

5.1 Longer term

Longer term, new technologies, or improvements to existing ones, need to be introduced if we are to slow, or at least minimize the accelerating growth of greenhouse gases. These technologies need to address some of these critical issues:

- Commercialization of technologies that permit isolation of CO_2 without the use of chemical extraction methods, and without sacrificing efficiencies to insurmountable parasitic losses. Preliminary Oxy-Fuel designs are promising technology, but they are years away from commercialization and all new turbine orders for the next decade are either conventional gas turbines or steam turbines. The Oxy-Fuel design does have a head start in that existing turbine designs can be adapted to operate in this mode.

- Solutions to the disposition of CO_2. Certainly not an easy objective. Simply using CO_2 for EOR only replaces CO_2 from the power sector with CO_2 from the transport and industrial sectors where the fuel is used. There are alternatives. For example, the Sabatier Process (Sabatier and Senderens, 1902; Borman, 2017) can process CO_2 and H_2 into CH_4 and H_2O, using a catalyst, and at high pressure. With the CO_2 already at high pressure, a source of hydrogen and a catalyst could provide one solution to disposing some of the CO_2. The concept is being studied to develop colonies on, of all places, Mars. One might expect a more receptive development here on earth (Junaedi, 2011). Finally, the underground storage of CO_2 in gigatonne quantities; we have only limited experience with any successful storage reservoirs. Whether this experience can be extrapolated on the scale required remains to be seen.

- Improved air separation technology. ASU systems that produce oxygen exhibit very long startup periods (hours if not days), while today's gas turbines have startup periods measured in minutes. There is a total mismatch in timescales to place these two technologies in tandem. Chemical looping (Moghtaderi, 2012) is an example of a technology that solves the air separation challenge, while still retaining the capacity to isolate CO_2 in the exhaust like the Oxy-Fuel. Demonstration of this concept in a power generation example is, however, several decades away.

- Improved energy storage in the power sector. Mechanisms to better integrate intermittent supply from wind and solar generation, making the energy supply available as close to demand as possible. This is the ever elusive "better battery" somewhere over the horizon.

- Improved energy storage in the transport sector. Better battery storage technology will bring the mobile transport sector and the power sector closer together. Effectively increasing the efficiency of the end user, providing further mitigation.

- Integration of the transport and power generation sectors. Together, these two (power generation and transportation) comprise nearly 2/3rds of all CO_2 from anthropogenic sources. Current technology is promising, but energy storage capacity on the mobile side needs additional improvement, perhaps by as much as a factor of 2 or 3. In addition, storage needs to be achieved cost competitively, and with an emphasis on end-user safety. These are non-trivial leaps in technology and could represent a decade's worth of research, but the financial rewards could easily translate to $100 billion annually.

In total, these advances could have a substantial impact towards the goal of reducing CO_2 emissions from the power sector; but, as hinted early on, that developmental cycle must be navigated between a working concept and commercial deployment could be fifty years or more. So far, the only periods where CO_2 concentrations in the atmosphere have either slowed or reversed appear to be associated with economic downturns, not the application of innovative technologies. Whether current fuel and technology applications are sufficient to reverse any impact related to climate change remains unknown.

References

Anderson, R.E. 2008. Adapting gas turbines to zero emission Oxy-fuel power plants. Proceedings of ASME Turbo Expo: Power for land, sea and air, GT2008-51377, Berlin.

Aydin, G. 2010. Evaluation of geologic storage options of CO_2: Applicability, cost, storage capacity, and safety. Energy Policy 5072–5080.

Borman, S. 2017. Iron-based catalyst reduces CO_2 to CH_4 photochemically. C&EN Global Enterprise, p. 8.

Breaking ground for a groundbreaker: the first Allam Cycle power plant. (2016, May 15). Modern Power Systems.

Carapellucci, R. 2017. Application of an amine-based CO_2 capture system in retrofitting combined gas-steam power plants. Energy 808–826.

CCS. 2018. Global CCS Institute. Retrieved from Status Report: https://www.globalccsinstitute.com/resources/global-status-report/.

Clean Energy Systems. n.d. Retrieved from Oxy Fuel Turbines: http://www.cleanenergysystems.com/oxy-fuel-turbines/.

Desjardins, J. 2018, Feb 14. The Rising Speed of Technological Adoption. Retrieved from Markets: https://www.visualcapitalist.com/rising-speed-technological-adoption/.

DOE/EIA. 2015. Form EIA-923. Retrieved from Page 1 Generation and Fuel Data: https://www.eia.gov/electricity/data/eia923/.

DOE/EIA. 2017. Electric Power Annual. Retrieved from Table 8.2 Average Tested Heat Rates by Prime Mover: https://www.eia.gov/electricity/annual/html/epa_08_02.html.

Dubin, K. 2017. Today In Energy. Retrieved from U.S. Energy Information Administration: https://www.eia.gov/todayinenergy/detail.php?id=33552. Accessed 31 October.

Eccles, J. 2009. Physical and economic potential of geological CO_2 storage in Saline Aquifers. Environmental Science and Technology 1962–1969.

EIA. 2017. Form 2_3_4_5_6_7_2017_Early_Release. EIA.

EIA. 2018. Country Analysis Brief-Norway. Retrieved from https://www.eia.gov/beta/international/analysis_includes/countries_long/Norway/Norway.pdf.

EIA. 2018. EIA Monthly Energy Review April 2018.

Endcoal.org. 2019, January. Retrieved from Plant Tracker: https://endcoal.org/global-coal-plant-tracker/summary-statistics/.

FRED. 2018. Annual Report 2018. St. Louis: U.S. Federal Reserve.

Gordon, R. 2016. The Rise and Fall of American Growth. Princeton University Press.

Junaedi, C. 2011. Compact and Lightweight Sabatier Reactor for Carbon Dioxide Reduction. American Institute of Aeronautics and Astronautics, 1–10.

Kapetakia, Z. 2017. Overview of Carbon Capture and Storage (CCS) demonstration project business models: Risks and Enablers on the two sides of the Atlantic. Energy Procedia 6623–6630.

Keeling, C.D., Piper, S.C., Bacastow, R.B., Wahlen, M., Whorf, T.P., Heimann, M. and Meijer, H.A. 2005. Atmospheric CO_2 and $13CO_2$ exchange with the terrestrial biosphere and oceans from 1978 to 2000: observations and carbon cycle implications. *In:* Ehleringer, J.T. (ed.). A History of Atmospheric CO_2 and its Affect on Plants. Springer Verlag.

Kelly, S. 2018. How America's clean coal dream unravelled. Retrieved from Guardian: https://www.theguardian.com/environment/2018/mar/02/clean-coal-america-kemper-power-plant. March 2.

MacIntyre, S. 2018. Today In Energy. Retrieved from Monthly Energy Review: https://www.eia.gov/todayinenergy/detail.php?id=37692. December 4.

McCoy. 2019. McCoy Global Fossil Power Report. Private communication.

Melville's Gas Apparatus. 1876, March 2. American Gas Light Journal, p. 92.

Moghtaderi, B. 2012. Review of the recent chemical looping process developments for novel energy and fuel applications. Energy & Fuels 15–40.

Nsakala, J.M. 2001. Engineering feasibility of CO_2 capture on an existing U.S. coal-fired power plant. First National Conference on Carbon Sequestration. Washington, DC.

NYTimes. 1931, April 4. Use of Solar Energy is a Near Solution. New York Times.

NYTimes. 1954, April 26. Vast Power of Sun is Tapped by Using Sand Ingredients.

Patel, S. 2017. Power Magazine. Retrieved from https://www.powermag.com/capturing-carbon-and-seizing-innovation-petra-nova-is-powers-plant-of-the-year/. August 1.

Phillips, J.N. 2017. The history of integrated gasification combined-cycle power plants. ASME Proceedings Coal, Biomass and Alternative Fuels (p. V003T03A007). ASME.

Rathi, A. 2018. Quartz. Retrieved from Safety Net Power: https://qz.com/1292891/net-powers-has-successfully-fired-up-its-zero-emissions-fossil-fuel-power-plant/. May 3.

Sabatier, P. and Senderens, J.B. 1902. C.R. Acad. Sci. (Paris), 514.

Webster, B. 1974, July 5. As Energy Grows Scarcer Science Looks Toward the Sun. New York Times.

WNA. 2019, Jun. Nuclear Power in Japan. Retrieved from http://www.world-nuclear.org/Information-Library/Country-Profiles/Countries-G-N/Japan-Nuclear-Power.aspx.

WNA. 2019, April. World Nuclear Association. Retrieved from Plans for New Reactors World Wide: http://www.world-nuclear.org/information-library/current-and-future-generation/plans-for-new-reactors-worldwide.aspx.

CHAPTER 8

The Environmental Impact of Implementing CO$_2$ Capture Process in Power Plants
Effect of Type of Fuel and Energy Demand

Carolina Mora-Morales,[1] Juan Pablo Chargoy-Amador,[2]
Nelly Ramírez-Corona,[1] Eduardo Sánchez-Ramírez[3] and
Juan Gabriel Segovia-Hernández[3,]*

1. Introduction

Global warming is currently recognized as a major environmental problem that affects humans worldwide (McCarthy et al., 2002; O'Neill and Oppenheimer, 2003). The high CO$_2$ concentrations and continuous emissions tend to deteriorate the environment. Currently, CO$_2$ is produced by several industrial processes, such as the combustion of fossil fuels, to produce electricity and in the transport sector. Recently, the Energy Information Administration of the United States (EIA) reported that global CO$_2$ emissions will grow at an average rate of 0.6% from 2015 to 2040, compared to the 1.3% growth from 1990 to 2015. The total amount of CO$_2$ emissions continues to grow, especially in developing countries, including China and India. This is due to the growing economy which results in an increase in the country's energy demand. The general combustion reaction for hydrocarbons can be represented by equation (1), where it is the molar ratio of air required in excess of stoichiometric oxygen required.

$$C_mH_n + \theta(m + n/4)(O_2 + 3.77)\,N_2 \rightarrow mCO_2 + \frac{n}{2}\,H_2O + (\theta - 1)\,(m + \frac{n}{4})\,O_2 + \theta(m + \frac{n}{4})\,3.77\,N_2 \qquad (1)$$

Studies that have analyzed the environmental effect of greenhouse gases consider that CO$_2$ contributes 60% of the effects of global warming (Olajire, 2010). The intergovernmental panel on climate change

[1] Departamento de Ingeniería Química, Alimentos y Ambiental, Universidad de las Américas Puebla. ExHda. Santa Catarina Mártir s/n, San Andrés Cholula, Puebla, México, 72820.
Email: nelly.ramirez@udlap.mx
[2] Centro Análisis de Ciclo de Vida y Diseño Sustentable (CADIS), Bosques de Bohemia 2, No. 9, Bosques del Lago, Cuautitlán Izcalli, Estado de México, México, 54766.
Email: jpchargoy@centroacv.com.mx
[3] Departamento de Ingeniería Química, Universidad de Guanajuato, Noria Alta s/n, Guanajuato, Gto., 36050, México.
* Corresponding author: gsegovia@ugto.mx

(IPCC) has predicted that the atmosphere could contain up to 570 ppm of CO_2, causing a remarkable increase in the world temperature of approximately 1.9 °C and an increase in sea levels of 3.8 m by 2100 (Olajire, 2010). Moreover, other studies forecast rises in temperatures in a range of 2–4 °C (Godard, 2008), on the other hand, the IPCC also predicts CO_2 concentration of about 705 ppm by 2100 and over 900 ppm by 2200 in the A1B scenario (A future world of very rapid economic growth, low population growth and rapid introduction of new and more efficient technology).

A special report from the IPCC on CO_2 capture and storage reports global emissions in the year 2000 of 23.5 GT, attributing more than 60% to 4942 electricity production stations that issued about 10.5 GT/year of CO_2. The remaining 40% of the emissions were due to the transport sector and the rest from other sectors, such as buildings, industry and so on. For example, Figure 1 shows an average measure of global CO_2 emissions.

Considering the aforementioned, the production of greenhouse gases due to electric generation is of relevant importance.

Combined cycle thermoelectric plants are systems that jointly produce electricity and thermal energy from a single fuel. In combined cycles, the fuel, mainly natural gas, is injected into a mixture with air to a turbine where combustion takes place. The kinetic force of the combustion gases causes the turbine to turn, and taking advantage of the movement in an alternator, electricity is produced. The heat that prevails in the exhaust gases of the turbine is used to boil water through a heat exchanger and steam generator. The steam produced is used in a conventional thermal cycle to move another turbine and generate electricity by turning another alternator. The advantage of these over conventional plants is that they take better advantage of the energy produced in the boiler by burning the fuel, thus achieving greater efficiencies and, in turn, having lower CO_2 and NO_x emissions (Franco and Giannini, 2005).

Currently, electricity from power plants represents a main source of energy around the globe. For example, in the United States, the electricity generation from coal and natural gas is in the three major categories along with nuclear energy and renewable energy. Specifically, within the fossil fuels, natural gas is the largest source (about 32% in 2017). Natural gas is used directly to produce steam or even to operate a gas turbine to generate electricity. An example of this type of thermal power plant is the one of Iligan in the Philippines, which has an installed capacity of 1,251 MW and is the plant with the highest thermal efficiency in the world thanks to the use of type G gas turbines produced by Mitsubishi Heavy Industries, Ltd. (Tsutsumi et al., 2003). Table 1 shows some numerical advantages of using combined cycle.

On the other hand, in the USA, coal was the second energy source (about 30% in 2017), in the same sense as natural gas, coal is almost all used in coal-fired power plants which use a turbine to generate electricity. Finally, petroleum was in 2017 the source with less than 1% of use in the U.S. In the same way as previous examples, residual fuel and petroleum coke are used in steam turbines. Note in Figure 2, the electricity generation by major source from 1950 to 2017 in the U.S.

Currently, the options to reduce total CO_2 emissions can be summarized in three options: (1) enhance conversion efficiency; (2) use a low/carbon-free fuel, and (3) improve CO_2 capture. The first option requires the efficient use of energy. The second requires a radical change in current fuels, for example, hydrogen or renewable energies. The third option involves the development of technology to improve CO_2 capture. Note, a combination of these three options is probably the best choice, rather than a single one.

2. CO_2 Capture Technologies and Life Cycle Assessment

The CO_2 capture of a gaseous effluent is an essential parameter for the viability of coal and CO_2 capture plants. Freund (1996) showed that capture technologies for emissions generated by electricity generation can be categorized via these three options: Pre-combustion, oxyfuel combustion and post-combustion. The selection of the appropriate method depends on the concentration of CO_2 in the gas stream, the gas pressure in the stream and the type of fuel. Below is a brief analysis of each alternative.

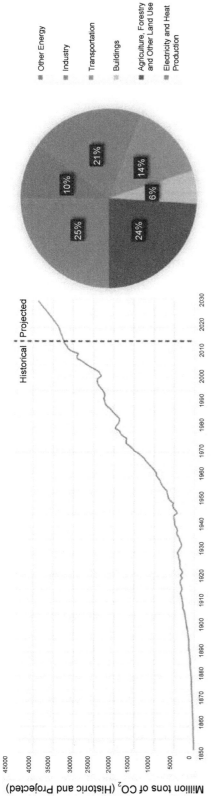

Figure 1. Global CO$_2$ emissions and Global Greenhouse Gas emissions by sector. Partially extracted from C2ES (2017).

Table 1. Advantages of using combined cycles (Tsutsumi et al., 2003).

	Combined cycles	Conventional process
Energetic Efficiency	55–57%	35–40%
CO_2 Emissions	360 g CO_2/kWh	850 g CO_2/kWh
Water Consumption	435 m³/h (400 MW)	875 m³/h (400 MW)

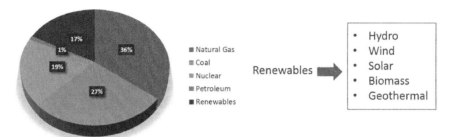

Figure 2. Electricity generation by several sources in the U.S.

2.1 Pre-combustion capture

In pre-combustion capture, the fuel reacts with oxygen or air, and in some cases with steam, to produce mainly carbon monoxide. This process is known as gasification, partial oxidation or reforming. The mixture of CO and H_2 is passed through a catalytic reactor where the CO reacts with steam resulting in CO_2 and more H_2. The CO_2 is separated and the H_2 is used as fuel in a gas turbine in a combined cycle plant. This technology is usually useful for coal gasification (IGCC), however, it can also be applied to liquid or gaseous fuels. Typically, the reaction for IGCC is shown in equations (2)–(4).

$$2C + O_2 + H_2O \rightarrow H_2 + CO + CO_2 \tag{2}$$

$$C + H_2O \rightarrow H_2 + CO \tag{3}$$

$$CO + H_2O \rightarrow H_2 + CO_2 \tag{4}$$

2.2 Oxyfuel combustion

Oxyfuel combustion is, in fact, a modified post-combustion method. The fuel is burned with almost pure oxygen instead of air, which results in high concentrations of CO_2 in the gas flow. Operationally speaking, if the fuel is burned in pure oxygen, the temperature of the flame is likely to be excessively high, so there is the possibility of recycling the CO_2 rich combustion gas to the burner to make the flame temperature similar to that obtained in a combustion chamber with conventional air.

The advantage of oxygen combustion is that the gas effluent can reach a concentration higher than 80%, thus, only a simple CO_2 purification is required. Additionally, NO_X formation is suppressed. On the other hand, the great disadvantage of oxyfuel combustion is the large amount of pure oxygen that is needed.

2.3 Post-combustion capture

The capture of CO_2 in post-combustion involves the separation of CO_2 from a gas stream produced by the combustion of some fuel. The post-combustion capture is a separation process and, in several aspects, has characteristics similar to gas desulfurization (FGD), which is widely used to capture the SO_2 in gas effluents from power plants where coal and oil are burned. The low concentration of CO_2 in the gas effluents of power plants (typically between 4% and 14%) means, operationally, that a large amount

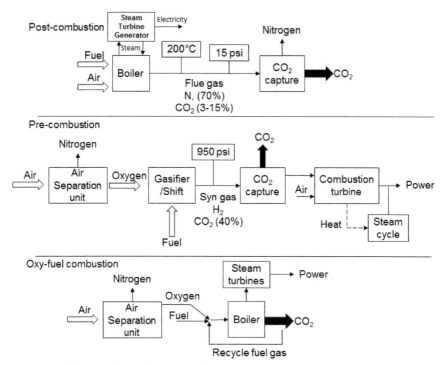

Figure 3. Block diagrams illustrating post-combustion, pre-combustion and oxyfuel combustion techniques.

of gas volume is processed, and consequently, the size of the equipment is large and comes with a high capital cost.

The capture of CO_2 in post-combustion is a significant challenge due to the low partial pressures of CO_2 in the gas stream. Additionally, the relatively high gas outlet temperature offers an additional challenge. On the other hand, a clear disadvantage of the low concentration of CO_2 is that, for its capture, powerful solvents must be used and, subsequently, the solvents must be regenerated by investing a certain amount of energy. Figure 3 shows the difference between the three alternatives already mentioned.

2.3.1 Chemical absorption

A typical chemical absorption process consists of an absorber and a separator where the absorbent is regenerated. In a chemical absorption process for CO_2 capture, the gas effluent enters the bottom of an absorber and comes into contact in countercurrent with a CO_2 absorber. After the absorption process, the gas effluent enters a separator for thermal regeneration. After regeneration, the CO_2 burner is returned to the absorber for reuse. The pure CO_2 is released from the separator to be subsequently compressed and transported. The operating pressure is about 1 bar and the operating temperatures of the absorber and separator are generally in the range of 40 °C to 60 °C and 120 °C to 140 °C, respectively. Theoretically, the minimum energy required for the recovery of CO_2 from a gas effluent and its subsequent compression is approximately 0.396 GJ/ton (Yu et al., 2012), considering compression of 150 bar. Therefore, there is a great opportunity to improve the absorption efficiencies, as well as the regeneration of the solvent. Currently, several authors have reported data associated with regeneration energy, for example Versteeg and Rubin (2011) and Jilvero et al. (2011) reported an energy regeneration between 2.2–2.8 GJ/ton$_{CO_2}$ using NH_3 (10 °C), in the same sense, Mirfendereski and Geuzebroek (2009), reported 2.33 GJ/ton$_{CO_2}$ using CANSOLV.

The advantage of a chemical absorption is that it is currently the technology with the highest maturity for the capture of CO_2 and has been commercialized for several years. Another advantage of this technology is that the retrofitting process is quite adequate for existing power plants.

Regarding absorbents, alkanolamines are widely used for CO_2 capture. The structure of the alkanolamines (including their primary, secondary and tertiary variants) is that they contain at least one OH group and one amino group, for example, monoethanolamine (MEA), diethanolamine (DEA) and N-methyldiethanolamine (MDEA). The reactivity of said amines to CO_2 follows the primary, secondary and tertiary order, for example, the reaction constants for CO_2 are 7000, 1200 and 3.5 m^3/s/kmol for MEA, DEA and MDEA at 25 °C, respectively (Sada et al., 1976). On the contrary, the loading capacity for a tertiary amine is approximately 1 mole of CO_2 per mole of amine, a greater number compared to primary and secondary amines that report load capacities between 0.5–1 mole of CO_2 per mole of amine. Table 2 shows the type of alkanolamines used in the capture of CO_2, highlighting monoethanolamine, 2-(2aminoethylaminp) ethanol (AEEA), Piperazine (PZ), N-methyldiethanolamine (MDEA), 2-amino-2-methyl-1-propanol (AMP) and diethylenetriamine (DETA).

In general, the absorption/separation process is a process with a certain energy cost. According to Rochelle et al. (2009), the price per ton of CO_2 is in the range of 52 to 77 $/T$_{CO_2}$. On the other hand, the energy needed to capture CO_2 in a conventional coal-burning power plant is between 3.24 and 4.2 GJ/ton (McCarthy et al., 2002). In this process, most of the energy consumed is associated with the regeneration of the absorbent, approximately 60%. In this way, a point of opportunity to reduce energy expenditure would be to improve the operation of the separator.

Table 2. Physicochemical properties of the common alkanolamines used as absorbents.

Property	MEA	AEEA	PZ	MDEA	AMP	DETA
MW (g/mol)	61.08	104.15	86.14	119.16	89.14	103.17
Density (g/cm³)	1.012	1.029	1.1	1.038	0.934	0.955
Boiling Point (K)	446	513	420	243	438	207
Vapor Pressure (393 K) (kPa)	0.0085	0.00015	0.1066	0.0013	0.1347	0.02
Solubility (293 K)	Freely soluble	Freely soluble	14 wt%	Freely soluble	Freely soluble	Freely soluble
CO_2 Absortion capacity (mol of CO_2/mol of absorbent)	0.5	1	1	1		1

2.4 Life cycle assessment

The capture and use of CO_2 (CCU) and its potential environmental benefits are now gaining attention (Cokoja et al., 2011; MacDowell et al., 2010; Plasseraud, 2010). The capture of CO_2 and the subsequent use of CO_2 as an alternative source of carbon promises a reduction both in greenhouse gases and in fossil fuel depletion (Peters et al., 2011; Quadrelli et al., 2011). However, both the capture and subsequent activation of CO_2 is an operation with considerable energy requirements, indirectly causing emissions of greenhouse gases. In this way, the intuitive benefit of CO_2 capture and reuse is apparently not so clear and analysis needs to be done in a critical and systematized manner. Moreover, as mentioned earlier, the regeneration of the solvent plays a main role in the energy requirements, so in the same sense, the solvent chosen plays the same role in the life cycle assessment. On the other hand, due to the greenhouse gas emissions, the fuel burned in the power plant also plays a significant role.

A systematic and standardized way to evaluate the environmental impact of both processes is a life cycle assessment (LCA), which will be described in the section below. LCA is the methodology used to measure the potential environmental impact of any product, process or system, from raw materials extraction to end of life stage (IHOBE, 2009).

According to ISO 14040, LCA has 4 phases: Goal and scope definition, inventory analysis, impact assessment, and interpretation. In the first one, the products to be studied shall be clearly defined in terms of the function that the product performs. The second, inventory analysis, involves data collection and calculation procedures to quantify the consumption of energy, raw material, air emissions, water discharges, and solid wastes. The next phase is impact assessment, where the results of the inventory analysis are added up into environmental impacts using common equivalent units; for example, burning a fuel in a given process can be associated with effects on the impact category of global warming, which are measured in kilograms of CO_2 equivalent. The fourth phase of an LCA is an interpretation, where the results obtained should be analyzed in order to establish understandable recommendations and decision arguments.

Next, the four stages (Figure 4) involved in LCA study are presented (IHOBE, 2009).

With all this background, in the present chapter, the environmental impact of different scenarios of thermoelectric plants coupled to CO_2 capture processes will be analyzed with a reactive absorption and desorption system using monoethanolamine (MEA) as a liquid solvent. The scenarios to be evaluated include the use of four different fuels: Biogas, coal, natural non associated gas and associated gas; as well as variants in the combined cycle to analyze the effect of the processes with constant fuel flow and constant energy demand.

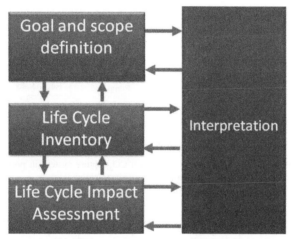

Figure 4. LCA steps (ISO 14040:2006).

3. Methodology

In order to analyze the environmental impact in different scenarios of electric power generation plants with the coupling of a CO_2 capture process in post-combustion, two cases were studied: Constant fuel flow and constant energy demand. For each of these cases, four different fuels were used: Biogas, coal, non-associated gas and associated gas. The simulation of the power plant and the CO_2 capture process was carried out in the Aspen Plus V8.8® process simulator.

According to the information reported by (Hasan et al., 2012), the design of the power generation plant was carried out using the Peng-Robinson method to estimate the thermodynamic properties. The simulation of the combustion chamber was carried out using the RGibbs reactor module, considering a molar ratio of air to fuel of 30:1 and a fuel flow of 1000 kmol/h for all the analyzed cases. The compositions in the mass percentage of the fuels used are shown in Table 3. Please note, associated gas refers to the natural gas found in association with oil within the reservoir. There are also reservoirs that contain only natural gas and no oil, this gas is termed non-associated gas.

The CO_2 capture process was designed by chemical absorption using an aqueous solution of monoethanolamine (MEA) at 30% weight of the solvent. RadFrac balance stage block was used for the simulation of the absorber and the regenerator (see Figure 5). An equilibrium stage model of a tower packed with Sulzer Mellapak 250 Y™ type packaging was used in the absorber and in the regenerator a non-equilibrium stage model of a tower packed with Sulzer Mellapak 150 Y™ packaging.

Table 3. Fuel composition in mass percent.

	CH_4	C_2H_6	C_3H_8	i-C_4H_{10}	N_2	CO_2
Natural Gas	96.00	1.80	0.40	0.15	0.70	0.95
Associated Gas	87.20	4.50	4.40	1.20	2.70	–
Biogas	60.00	–	–	–	2.00	38.00
	C	H	O	N		S
Coal	78.20	5.20	13.60	1.30		1.70

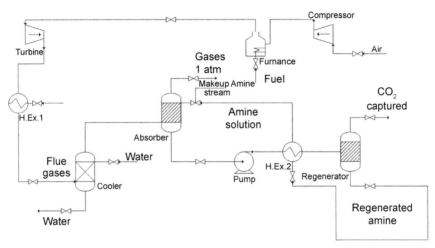

Figure 5. Flow diagram of a thermoelectric power plant with CO_2 capture system in post-combustion using chemical absorption with monoethanolamine.

Table 4. Kinetics of reactions (Zhang et al., 2018).

	Equation 8	Equation 9	Equation 10	Equation 11
Kinetic constant, k (kmol/m3 s)	1.33×10^{17}	6.63×10^{16}	3.02×10^{14}	5.52×10^{23}
Activation energy, E (kJ/mol)	55.38	107.24	41.2	69.5

The chemical reactions involved in the reactive absorption/desorption process are presented in equations (5)–(11). Table 4 shows the equilibrium constants.

$$MEAH^+ + H_2O \longleftrightarrow MEA + H_3O^+ \tag{5}$$

$$2H_2O \longleftrightarrow OH^- + H_3O^+ \tag{6}$$

$$HCO_3^- + H_2O \longleftrightarrow CO_3^{2-} + H_3O^+ \tag{7}$$

$$OH^- + CO_2 \xleftrightarrow{k_1} HCO_3^- \tag{8–9}$$

$$MEA + CO_2 + H_2O \xleftrightarrow{k_2} MEACOO^- + H_3O^+ \tag{10–11}$$

The power plant and capture process were simulated separately considering the combustion gases of the first process as a feed of the absorption tower. In order to carry out the LCA of the different scenarios, it was necessary to homogenize the processes so that they could be comparable to each other. In this case, the variables were adjusted in order to guarantee a 95% molar recovery in the CO_2 stream in the absorber. Because the components present in the absorption process dissociate, it is necessary to achieve a recovery of CO_2 in the gas output stream of the same equipment. In the case of the regenerator, the distillate flow and the reflux ratio were adjusted in order to capture the greatest amount of CO_2 from the combustion gas stream coming from the thermoelectric power plant and, thus, reduce the CO_2 emissions to the atmosphere and the environmental impact that they generate. For this reason, in all the analyzed cases, they were normalized to a purity of 99 mol% of CO_2.

To develop the LCA it is necessary to define the functional units. In the case of power plants, the functional unit is 1 MWh; for capture process, the functional unit is 1 kg of CO_2 captured. Impact Assessment was performed with SimaPro 8® software using ReciPe EndPoint (H) method. Scenarios evaluated for carbon capture at constant fuel flow and constant energy demand are presented in Figures 6 and 7.

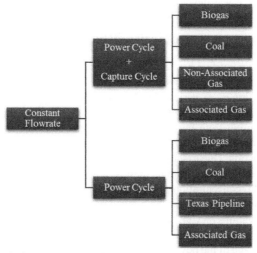

Figure 6. Carbon capture scenarios, at constant fuel flow, analyzed with LCA.

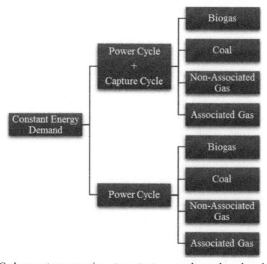

Figure 7. Carbon capture scenarios, at constant energy demand, analyzed with LCA.

4. Results

As discussed earlier, the characteristics of the flue gases depend on the type of combustion and fuel used within the power plant, particularly the volumetric flowrate and CO_2 content, which may affect the capture effectiveness during the CO_2 capture. In this work, the analysis of two different configurations, plant simulation with or without CO_2 capture system by chemical absorption, was conducted. The type and flowrate of flue gas have a significant effect on the capture effectiveness, because of the high variation in CO_2 content. Energy generation in power plants may vary depending on peaks of power demands and/ or variations in electricity prices, or even due to changes in the fuel characteristics. Therefore, capture plants should be able to capture different CO_2 loads, depending on those changes. Two different operating scenarios for the power plant were considered, the first one for a specified flowrate of fuel to be burned and the second one for specified energy production in the turbine of the power plant. This allows us to evaluate variations in the type of fuel and energy production. Four different types of fuels were selected. In order to evaluate the effect of the type of fuel and the capture plant implementation on the environmental impact of the process, an LCA was developed by means of the commercial software SimaPro.

4.1 Case study 1: carbon capture scenarios, at constant fuel flow

It has been stated that the type of fuel has not only an important role in energy production of the power plant but in the composition of the flue gases. As reported in several works (Nagy and Mizsey, 2013), the composition of combustion gases varies depending on the fuel (Table 5). The flue gas obtained when mineral coal is burned presented the lower CO_2 content, while the larger concentrations of CO_2 were observed for flue gases coming from burning natural gases. For all studied cases, the air flowrate was specified in a value of 33000 kmol/h, such that the oxygen to fuel ratio was around 3:1, i.e., enough to guarantee complete combustion. Due to the difference in the fuel compositions, there is a slight variation in the oxygen excess, as can be noticed in the compositions of N_2 and O_2 compositions in the flue gases (Table 5). Furthermore, CO_2 concentration also depends on water generation during combustion, which is larger during burning gases.

As expected, these variations on CO_2 concentration clearly influence the capture plant effectiveness, energy, and solvent requirements, as well as the CO_2 recovery. Table 6 summarizes the energy production-consumption among each stage of the power plant and the energy consumption due to the implementation of the capture process. When implementing the capture plant, there is a significant reduction in the efficiency of energy production, due to the energy consumption in the column for the amine regeneration. Such reductions range from 19.47% to 65.27%, wherein the larger efficiency reductions are observed for the process with mineral coal. This result can be explained in terms of lower energy production during electricity generation, in this case, 30% lower than that obtained by burning gases. Additionally, energy demand in the desorber column is also larger during the CO_2 capture; there is a direct relationship between CO_2 content and capture process efficiency.

Regarding impact assessment for the power plant working with the different fuels (Figure 8a), the use of mineral coal and non-associated gas present the major impact in most categories, followed by associated gas and biogas. This could be explained because, in these two cases, the fuels are directly obtained from as raw materials, while the associated gas is obtained as sub-product from oil wells

Table 5. Flue gas composition, reported in molar fraction, from the combustion of the different fuels, Case 1.

	Biogas	Coal	Non associated gas	Associated gas
N_2	0.767	0.786	0.767	0.766
O_2	0.168	0.189	0.144	0.139
CO_2	0.029	0.024	0.030	0.034
H_2O	0.035	0.002	0.059	0.062

Table 6. Case 1. Simulation of the power plant (CP) + capture process (CC), with a feed flowrate of fuels equal to 1000 kmol/h.

		Equipment	Biogas	Mineral coal	Non associated gas	Associated gas
Power plant (CP)						
Energy consumption	MW	Compressor	93.55	93.55	93.55	93.55
Energy production	MW	Turbine	109.65	100.07	138.58	145.26
Energy recovered	MW	Heat exchanger	89.49	72.68	151.21	165.93
Net energy	MW	CP	105.60	79.20	196.25	217.65
CO$_2$ generation	kgCO$_2$/MW	CP	408.22	433.86	228.12	230.81
Fuel flowrate	kg/h	CP	26910	14759	16820	18737
Power plant + Capture process (CP+CC)						
Energy consumption (Capture columns)	MW	Reboiler (CC)	43.80	51.69	44.38	42.39
Net energy (CP-CC)	MW	CP+CC	61.79	27.51	151.87	175.26
Energy efficiency reduction	%	CP+CC	41.48	65.27	22.61	19.47
Solvent requirement	kg amine/kg CO$_2$rec	CC	8.30	4.17	4.65	4.13

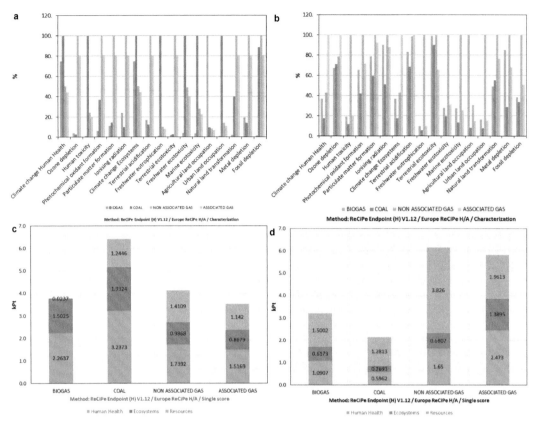

Figure 8. Life cycle impact assessment results, Case 1. (a) potential impact of power plant (CP), (b) potential impact of power plant + capture process (CP+CC), (c) Power plant single score (CP), (d) Power plant + Capture process single score (CP+CC).

and biogas is produced from residual biomass, such that all the impacts related with the ecosystems exploitation are reduced. After normalization and weighting to obtain a single score (Figure 8c), it is observed that the use of coal generates the greatest potential impact. This is due to the fact that climate change has a high weight within the single score calculation and coal has the biggest impact in this category.

Figure 8b shows the potential impact when the power plant is coupled to the capture process. It is noticed that, in this case, the impact of coal is reduced in several categories, while for the gases the impact is redistributed. The efficiency and solvent flowrate have an important role in these results (Table 6, Case 1). Figure 8d shows the single score results, where it is observed that implementing the CO_2 capture significantly reduces the environmental impact of the energy production for the biogas and coal systems, while for the natural gases this process does not represent a friendly environmental technology. As shown in Table 6, when equivalent flows of all fuels are considered, CO_2 produced during energy generation with biogas and mineral carbon are almost double that of natural gas, such that the capture process represents a larger benefit for those systems.

It is important to highlight that natural gases present a lower environmental impact during energy production, as these systems generate a lower amount of CO_2 to produce a kW in the power plant than the mineral coal or biogas. For the implementation of the capture process, however, those systems present a similar requirement of solvent and energy to recover a kg of CO_2, compared to the coal system, and, therefore, seem to be the less effective. Furthermore, for all gases, the impact associated with the resources is increased for the CO_2 capture implementation, due to the solvent requirement and energy demand.

4.2 Study case 2: carbon capture and scenarios, at specified energy production

The second scenario evaluated here considers a specified energy production within the turbine of the power plant. In this case, the feed flowrate of each fuel was adjusted such that the energy production goal is reached. As in the first case, the composition of combustion gases varies depending on the fuel. From these results, we can see that flue gases with larger CO_2 content are obtained from biogas and coal combustion (Table 7). Although the CO_2 concentration increases in this second scenario, the CO_2 generation per MW produced in the power plant for both these fuels is reduced in comparison to the first scenario (case 1), because of an increase in the energy production (Table 8).

Table 8 presents a comparison between the net energy of the power plant and the net energy of the same process when the capture process is coupled. In this scenario, the study cases with the lowest reduction in the energy efficiency were the associated and the non-associated gas, with 19.50% and 19.60%, respectively, and the biogas was the one with the highest reduction, since it is the system with the larger energy requirement during the capture process.

Regarding the LCA, the fuel with the greatest impact in most of the categories is mineral coal, to mention some categories, a noticeable impact is observed in climate change and human health, human toxicity, terrestrial acidification, terrestrial toxicity, and eutrophication, among others. After normalization and weighting to obtain a single score, results shown in Figure 9b seems to have a different trend than that observed for case 1. However, it is important to point out that, in both cases, natural gases present a single score close to 3, similar than that obtained in case 1. The real difference is observed for biogas and mineral coal, wherein the respective Eco points are significantly reduced as a result of the increase in energy production.

Table 7. Flue gas composition, reported in molar fraction, from the combustion of the different fuels, Case 2.

	Biogas	Coal	Non associated gas	Associated gas
N_2	0.749	0.781	0.765	0.766
O_2	0.135	0.162	0.139	0.139
CO_2	0.052	0.053	0.033	0.034
H_2O	0.064	0.004	0.064	0.062

Table 8. Case 2. Simulation of the Power plant (CP) + Capture process (CC), with energy production in turbine equal to 145 MW.

		Equipment	Biogas	Mineral coal	Non associated gas	Associated gas
Power plant (CP)						
Energy consumption	MW	Compressor	93.55	93.55	93.55	93.55
Energy production	MW	Turbine	145.00	145.00	145.00	145.00
Energy recovered	MW	Heat exchanger	164.99	164.52	164.99	165.93
Net Energy	MW	CP	216.44	215.97	216.44	217.38
CO_2 generation	kgCO₂/MW	CP	367.26	366.54	228.11	231.09
Fuel flowrate	kg/h	CP	49514	33502	18334	18737
Power plant + Capture process (CP+CC)						
Energy consumption (columns)	MW	Reboiler (CC)	78.15	62.26	42.42	42.39
Net Energy	MW	CP+CC	138.29	153.71	174.02	175.00
Energy efficiency reduction	%	CP+CC	36.11	28.83	19.60	19.50
Solvent requirement	kg amine/kg CO₂rec	CC	3.90	4.05	4.15	4.13

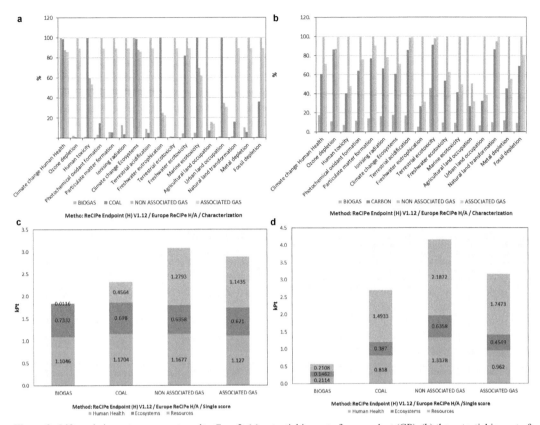

Figure 9. Life cycle impact assessment results, Case 2. (a) potential impact of power plant (CP), (b) the potential impact of power plant + capture process (CP+CC), (c) Power plant single score (CP), (d) Power plant + Capture process single score (CP+CC).

In this case, the single score (Figures 9c and 9d) indicates that implementing the capture process to the power plant working with biogas greatly reduces the environmental impact from 1.8494 kPt to 0.5684 kPt. This result can be attributed to: (i) the energy production was increased in the power plant so that the CO_2 generated per MW produced was diminished, (ii) A larger concentration of CO_2 in the flue gas that reduces the solvent requirement in the capture process.

On the other hand, the global impact of implementing this CO_2 capture technique to the process with the other fuels seems to be only redistributing among the different categories, and does not show a significant difference among them when the single scores are compared.

It is important to highlight that, even if the CO_2 capture implementation may reduce the environmental impact associated with the human health and/or ecosystems, the impact associated with the exploitation of the resources is always increased due to the energy and solvent requirements during the capture process. As discussed, the efficiency reduction observed during the capture implementation has an important effect on the environmental impact, so there is a clear incentive for optimizing the operating conditions of this post-combustion alternative in order to enhance its effectiveness.

5. Conclusions

The implementation of the CO_2 capture process in power plants has thus far been considered as the most mature technology to reduce the environmental impact associated with electricity production. Most research efforts in this field have been focused on performing techno-economic analysis and optimizing the energy efficiency of the capture process. However, it is essential to analyze the process from a holistic point of view, considering not only the CO$_2$ capture as a strategy to reduce the negative effects of the power plant but also by identifying new environmental effects due to the implementation of such capture process.

In this work, a Life Cycle Analysis was conducted in order to evaluate the environmental impact of different scenarios during the generation of electricity, as well as the energy and environmental implications of coupling a CO$_2$ capture process. For a specified feed flowrate of fuel to a power plant, the fuel with the lowest environmental impact is associated gas (natural gas found in association with oil within the reservoir), with a single score of 3.52 kEcopoints. When the CO$_2$ capture process is coupled to the power plant, the process with the greatest reductions in overall impact is coal, with 2.14 kEcopoints. For the scenario of specified energy production, the fuel with the lowest environmental impact is biogas, with a single score of 1.85 kEcopoints, and such impact is further reduced to as low as 0.57 kEcopoints, when the CO$_2$ capture plant is implemented.

Regarding energy efficiency, power plants suffer important energy penalties when the capture process is implemented, mainly due to the energy required for solvent regeneration. For the systems considered here, those processes working with associated gas and non-associated gas remain the most efficient in terms of net energy produced.

On the other hand, it is clear that the global demand for electricity is continuously growing and several efforts are centered on switching to less carbon-intensive systems by means of increasing the use of renewable energy sources. In this work, we evaluated the hypothetical case of using a biogas during electricity generation, bearing in mind that the environmental impact of this system should be minimal in comparison to the fossil fuels and that, for renewables sources, the implementation of a capture process should not be necessary. However, our findings indicate that even for this green fuel the environmental impact depend on the process parameters.

The environmental impact of the transport and storage of the captured CO$_2$ to depleted wells, mines or depth saline aquifers should be examined in a next stage, in order to considerer the long-term effect of these processes. However, in the light of the obtained results, we consider that, before developing a more comprehensive study, it is necessary to enhance the efficiency of coupling the power plant and CO$_2$ capture process. Our research group is undertaking this challenge by developing a model to solve a multiobjective optimization problem for each process variable and design parameter in both processes.

Further discussions should focus on exploring the benefits and weaknesses of CO$_2$ capture during real operation, considering combined technologies that are able to use different fuels, as well as variable energy demands.

References

Center for Climate and Energy Solutions. March, 2017. Global emissions. Retrieved from https://www.c2es.org/content/international-emissions/. Consulted on Consulted on September 10th.

Cokoja, M., Bruckmeier, C., Rieger, B., Herrmann, W.A. and Kühn, F.E. 2011. Transformation of carbon dioxide with homogeneous transition-metal catalysts: A molecular solution to a global challenge? Angewandte Chemie–International Edition 50: 8510–8537. https://doi.org/10.1002/anie.201102010.

Franco, A. and Giannini, N. 2005. Perspectives for the use of biomass as fuel in combined cycle power plants. International Journal of Thermal Sciences 44: 163–177. https://doi.org/10.1016/j.ijthermalsci.2004.07.005.

Freund, P. 1996. The IEA Greenhouse Gas R&D Programme. Energy Conversion and Management 37: 0–5.

Godard, O. 2008. The Stern Review on the Economics of Climate Change: Contents, insights and assessment of the critical debate. Surveys and Perspectives Integrating Environment and Society 1: 17–36. https://doi.org/10.5194/sapiens-1-17-2008.

Hasan, M.M.F., Baliban, R.C., Elia, J.A. and Floudas, C.A. 2012. Modeling, simulation, and optimization of postcombustion CO$_2$ capture for variable feed concentration and flow rate. 1. Chemical absorption and membrane processes. Industrial and Engineering Chemistry Research 51: 15642–15664. https://doi.org/10.1021/ie301571d.

IHOBE. 2009. Análisis de ciclo de vida y huella de carbono. Bilbao.

ISO 14040:2006. Environmental management–Life cycle assessment—Principles and framework.

Jilvero, H., Normann, F., Andersson, K. and Johnsson, F. 2011. Thermal integration and modelling of the chilled ammonia process. Energy Procedia 4: 1713–1720. https://doi.org/10.1016/j.egypro.2011.02.045.

MacDowell, N., Florin, N., Buchard, A., Hallett, J., Galindo, A., Jackson, G., Adjiman, C.S., Williams, C.K., Shah, N. and Fennell, P. 2010. An overview of CO_2 capture technologies. Energy and Environmental Science 3: 1645–1669. https://doi.org/10.1039/c004106h.

McCarthy, J.J., Canziani, O.F., Leary, N.A., Dokken, D.J. and White, K.S. 2002. Climate change 2001: Impacts, adaptation, and vulnerability, First. ed. Cambridge University Press, Cambridge, Uk. https://doi.org/10.5860/choice.39-3433.

Mirfendereski, Y. and Geuzebroek, F. 2009. Cansolv technologies: The value of integration. *In*: 12th Meeting of the International Post-Combustion CO_2 Capture Network.

Nagy, T. and Mizsey, P. 2013. Effect of fossil fuels on the parameters of CO_2 capture. Environmental Science and Technology 47: 8948–8954. https://doi.org/10.1021/es400306u.

O'Neill, B.C. and Oppenheimer, M. 2003. Dangerous Climate Impacts and the Kyoto Protocol. Science 302: 1718–1719.

Olajire, A.A. 2010. CO_2 capture and separation technologies for end-of-pipe applications—A review. Energy 35: 2610–2628. https://doi.org/10.1016/j.energy.2010.02.030.

Peters, M., Kohler, B., Kuckshinrichs, W., Leither, W., Markewitz, P. and Muller, T. 2011. Chemical Strategies for Exploiting and Recycling CO_2. First. ed. WILEY-VCH Verlag.

Plasseraud, L. 2010. Carbon dioxide as chemical feedstock. Aresta, M., Ed. ChemSusChem. 3: 631–632. https://doi.org/10.1002/cssc.201000097.

Quadrelli, E.A., Centi, G., Duplan, J.L. and Perathoner, S. 2011. Carbon dioxide recycling: Emerging large-scale technologies with industrial potential. ChemSusChem. 4: 1194–1215. https://doi.org/10.1002/cssc.201100473.

Rochelle, G.T. 2009. Amine scrubbing for CO_2 capture. Science 325: 1652–1654. https://doi.org/10.1126/science.1176731.

Sada, E., Kumazawa, H. and Butt, M.A. 1976. Gas absorption with consecutive chemical reaction: Absorption of carbon dioxide into aqueous amine solutions. The Canadian Journal of Chemical Engineering 54: 421–424. https://doi.org/10.1002/cjce.5450540507.

Tsutsumi, A., Musayuki, M., Yamanomoto, A., Nakamoto, Y. and Yuri, M. 2003. Description of the Latest Combined Cycle Power Plant with G type Gas Turbine Technology in the Philippines. Mitsubishi Heavy Industries, Ltd. Technical Review 40: 1–5.

Versteeg, P. and Rubin, E.S. 2011. A technical and economic assessment of ammonia-based post-combustion CO_2 capture at coal-fired power plants. International Journal of Greenhouse Gas Control 5: 1596–1605. https://doi.org/10.1016/j.ijggc.2011.09.006.

Yu, C.H., Huang, C.H. and Tan, C.S. 2012. A review of CO_2 capture by absorption and adsorption. Aerosol and Air Quality Research 12: 745–769. https://doi.org/10.4209/aaqr.2012.05.0132.

Zhang, Z., Li, Y., Zhang, W., Wang, J., Soltanian, M.R. and Olabi, A.G. 2018. Effectiveness of amino acid salt solutions in capturing CO_2: A review. Renewable and Sustainable Energy Reviews 98: 179–188. https://doi.org/10.1016/j.rser.2018.09.019.

CHAPTER 9

Systems Integration Approaches to Monetizing CO_2 via Integration of Shale Gas Processing and Industrial Waste Mineralization

Jared Enriquez and *Mahmoud M El-Halwagi**

1. Introduction

It is no surprise that there is a growing concern over atmospheric conditions and how greenhouse gases are (GHGs) affecting our environment. Carbon dioxide (CO_2) is one of the main components of growing GHG levels and accounts for nearly 77% of industrial GHG emissions (Rahman et al., 2017). In order to begin the process of restoring atmospheric GHG levels to acceptable conditions, sustainable practices should be implemented in order to reduce and minimize emissions created through industrial processes. This is easier said than done, as GHGs like CO_2 are typically considered as waste with no inherent chemical value due to their low-energy nature. Luckily, attention has been focused on finding ways of capturing, storing, or utilizing CO_2 in meaningful ways to help mitigate emissions and turn CO_2 into a useful commodity.

Generally speaking, CO_2 uses can be separated into two categories: Sequestration and utilization. Carbon capture and sequestration (CCS) is the method of capturing carbon dioxide from sources like emissions or even the atmosphere in order to separate and store the gas in an environmentally beneficial manner. One of the most popular forms of CCS currently is that of enhanced oil recovery (EOR). With this method, CO_2 is injected at well sites in order to increase pressure and flush out oil that may have remained after initial pumping. Typically, at a new drill site, only about 20–40% of the oil is initially obtained. EOR can help recovery around 5–30% of oil that would otherwise be missed (Abidin et al., 2012). Afterwards, the CO_2 can be geologically sealed off in the well to prevent its escape back into the atmosphere. While this makes good use of CO_2, it does not affect net levels or help to reduce its presence. There is some uncertainty towards using EOR or other geological sequestration techniques as a long-term solution due to the possibility of leaks and the need for constant monitoring of the sites (Meylan et al., 2015). Along with this, there is the issue of public acceptance when storing CO_2 underground.

Carbon dioxide utilization (CCU) can circumvent these issues, in that the CO_2 is chemically converted and can be considered as a more permanent "molecular" sequestration. As mentioned earlier, one of the main issues with CCU is that CO_2 is a very stable molecule and requires either highly reactive co-reagents

Department of Chemical Engineering, Texas A&M University, College Station, TX 77845, USA.
* Corresponding author: el-halwagi@tamu.edu

or large energy inputs in order to convert it into anything chemically useful. However, research has shown there are several viable routes for utilization, like polymer synthesis, fuel production, and biological conversion. While large scale implementations of these practices are not yet common, it has been argued that they have the potential to mitigate climate change and could help lead to a low-carbon economy (Barbato et al., 2014). With proper integration and optimization, more routes for CCU could become economically viable and turn CO_2 into a useful feedstock (Panu et al., 2019; Alsuhaibani et al., 2019; Tilak and El-Halwagi, 2018; Afzal et al., 2018; Pokoo-Aikins et al., 2010).

With the advancements in horizontal drilling and fracking techniques, the natural gas trapped inside shale formations has become much more accessible. Due to the rock's low permeability, normal drilling techniques couldn't adequately release the gas. Shale gas is quickly becoming a dominant source of energy and feedstocks (Elbashir et al., 2019; Al-Douri et al., 2017). Like natural gas, shale gas consists primarily of methane along with other heavier hydrocarbons. Small fractions of the gas are also composed of CO_2, nitrogen, and sometimes even hydrogen sulfide. The exact composition of the gas varies from well to well, which may also contain other trace chemicals. Before the gas can transported along pipelines, it must go through processing in order to remove impurities down to acceptable levels. One of the first steps of this process is the acid gas removal stage, where most of the CO_2 is separated. What is interesting about this stage is how pure the CO_2 stream exiting from this acid gas removal stage is. It is typical to have streams that are around 99% pure CO_2, yet most often this stream is vented directly into the atmosphere (Grobe, 2010). There is growing interest in using CO_2 for enhanced gas recovery (EGR), however this is nowhere near as prevalent or developed as EOR. While EGR and EOR are usually the main alternatives to venting the CO_2, not much focus is present on utilizing the stream for purposes of CCU and creating value-added products.

Mineralization is a CCU method where CO_2 is reacted with calcium- and magnesium-containing minerals to produce carbonates. What is unique about this, compared to other CCU processes, is that the conversion of CO_2 into carbonate is thermodynamically favorable and exothermic. With this transformation into a more stable molecule, the CO_2 is permanently stored in a solid matrix. Two main routes exist for CO_2 mineralization: *In situ* and *ex situ*. *In situ* mineralization involves injecting pressurized CO_2 underground to react with minerals and form carbonates. *Ex situ* instead uses mined minerals or alkaline materials and reacts them with CO_2 in controlled conditions. While there are abundant geological minerals that are capable of mineralizing CO_2, like olivine, serpentine, and wollastonite, the *in situ* method only serves as a permanent form of CO_2 storage and does not contribute to creating any valuable products (Gadikota and Park, 2015). One of the major downsides of mineralization as a form of CCU is that the reaction is kinetically limited. *In situ* mineralization is a slow process and one of the benefits of the *ex situ* method is the capability to carefully control and optimize the process to the point of reactions only taking hours rather than decades. However, in order to reduce reaction times, there is a large penalty to pay in terms of energy. One of the first steps in *ex situ* mineralization is to grind down alkaline minerals to increase surface area usually to particle sizes on the order of 10–100 μm (Gadikota and Park, 2015). This process, along with high reaction temperatures, is quite energy intensive and careful consideration must be taken to ensure this energy demand is not creating more CO_2 than it is utilizing. Table 1 summarizes key properties of common non-carbonated minerals.

Additional steps may be taken in order to help facilitate the carbonation process. Acid dissolution is one way of chemically preparing the minerals for easier reactivity. In a two-step process, the dissolved mineral ions are then subjected to basic aqueous conditions where CO_2 is bubbled through. Catalysts and

Table 1. Mineralization properties of common non-carbonated minerals.

Mineral	Carbonation products	Carbonation potential (kg CO_2/kg mineral)	ΔH_{rxn} (kJ/mol CO_2)
Olivine	$MgCO_3$(s), SiO_2(s)	0.56–0.63	–89
Serpentine	$MgCO_3$(s), SiO_2(s), H_2O(aq)	0.40–0.53	–64
Wollastonite	$CaCO_3$(s), SiO_2(s)	0.38	–90

Adapted from sources Gadikota and Park, 2015; Zhao et al., 2013.

chelating agents have also shown potential in further accelerating the process. When operated under the right conditions, relatively pure products, like calcium carbonate, magnesium carbonate, and silica, are some of the possible value-added materials to be obtained. Applications for these products range from paper, plastic, and construction fillers to glass and ceramic materials.

An interesting approach to avoid some of these complications is to utilize alkaline industrial waste sources as a substitution to these natural minerals. Common wastes, like fly ash and waste cement, are high in calcium content and can undergo carbonation to produce calcium carbonate products. As some of these sources already exist as particulate matter, there is little to no need for comminution.

2. Problem Statement and Approach

The main objective of this chapter is to investigate the prospects of utilizing and taking advantage of the high purity CO_2 stream produced from shale gas processing and integrating it with the CCU process of mineralization in order to create a sustainable and profitable system. Specifically, industrial waste will be investigated as a feedstock for mineralization due to its high carbonation potential and additional sustainability implications. Several waste sources will be considered and screened through an initial economic analysis in order to determine viability. Through literature analysis and reported experimental data, the integrated carbonation processes will be simulated using Aspen modelling as a method of detailed analysis. The optimized and finalized flow sheets will help evaluate mass and energy consumptions as well as estimations on equipment costs in order to fully evaluate the capital and operating costs of implementing the integrated processes. This will determine which processes are viable, profitable, and worth pursuing.

In order to get the most out of purchased and installed equipment for the carbonation process, it would be preferable to have a system that is adaptable to a variety of waste sources without the need for individualized reaction or separation vessels. Figure 1 shows this concept of a very simple and generalized process diagram which highlights the most important units. Here, the way in which the two processes of shale gas processing and waste mineralization are integrated with the connection of a carbon dioxide exchange stream is clearly visible. This CO_2 stream is fed into a reaction vessel along with the chosen industrial waste particles and brine solution where carbonation occurs. Next, the mixture is transferred to a separations unit, where reacted carbonate products are collected and unreacted materials and solvent are recycled back to the reaction vessel.

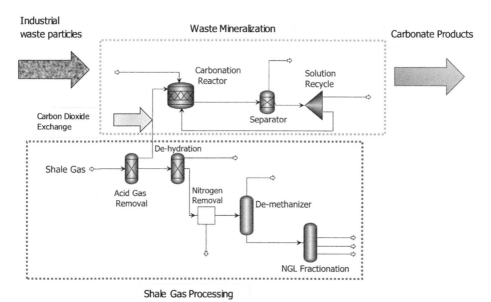

Figure 1. Generalized flowsheet of integrated gas and mineralization processes.

2.1 Waste source considerations

For the purposes of this work, industrial waste sources will be considered, based some of the following characteristics: Calcium content, availability, composition variability, and initial physical properties, like particle size. Obviously, the ideal waste source would have high calcium content, small particle sizes, and would be readily available in large quantities. However, industrial wastes vary greatly from one location to another and even within the same processing plant. In this section, some of the more viable waste sources will be considered and discussed.

2.1.1 Fly ash

Fly ash is generally produced as a byproduct of coal combustion but can be produced through other combustion processes, like municipal solid waste incineration (MWSI). The current production of fly ash is estimated to be around 500 million tonnes globally, with only around 16% of it being utilized in ways other than being disposed of in landfills (Ahmaruzzaman, 2010). Fly ash particles consist of toxic trace elements which can lead to environmental concerns when disposed of without treatment. While compositions vary, fly ash is broken into two classes: Class F and Class C. The main difference between the two is the calcium, silica, and iron content. Class F ash contains around 1–12% calcium while Class C contains around 30–40% calcium content. Fly ash exists as fine spherical particles, typically with sizes around 75 µm and surface areas as high as 1000 m^2/kg. As a hazardous byproduct, the cost of purchasing fly ash mostly consists of transportation costs which approximate to around $15/ton (Ahmaruzzaman, 2010). Compared to the criteria for what is considered a desirable waste source, fly ash is a highly viable option for mineralization due to its abundance and high calcium content.

2.1.2 Waste cement/cement kiln dust

As buildings are demolished and waste concrete is pulverized, powder byproducts formed as aggregates are recycled. This powder is known as waste cement powder, or simply waste cement. This waste cement can make up as high as a third of total waste concrete and currently is mostly used as roadbed material or is disposed of. Waste cement averages around 30% calcium content and has a typical particle size distribution of around 10–200 µm (Katsuyama et al., 2005). The source of cement kiln dust (CKD), a byproduct of the cement manufacturing process, is similar. Cement manufacturing produces millions of tons of CKD annually, the majority of which is disposed of in landfills. Calcium oxide content can range from 20–60% with compositions varying depending on where the CKD was obtained (Huntzinger et al., 2009). With around 15–20 tons of CKD produced for every 100 tons of cement, it is also a highly abundant waste source (Bobicki et al., 2012). Both sources are also potentially hazardous, but due to their abundance and small particle sizes, it is likely mineralization could be used to viably transform them into safer, useful products.

2.1.3 Steelmaking slag

Steel slag is a byproduct of the steel manufacturing process. Initially a molten liquid, steel slag cools into a mixture of oxide and silicate materials. This can refer to multiple steps of the process and the waste may have corresponding names, like furnace slag or ladle slag. Typically, steelmaking slag contains around 25–55 wt% calcium oxide and has been proven to effectively produce calcium carbonate at relatively low pressures and moderate temperatures with proper solution conditions (Romanov et al., 2015). Steelmaking slag is produced globally at about 200 Mt annually and is typically formed at a ratio of 0.2 tons of slag for every ton of steel (Said et al., 2013). Size distributions vary greatly from source to source, with ranges on the order of 1 mm^{-1} cm, and may require further comminution (Lekakh et al., 2008). As with the other sources, efficient CO_2 utilization will depend on the characteristics of the slag undergoing carbonation, but present work has shown promising results for a variety of steelmaking slag compositions and reaction conditions.

2.1.4 Other wastes

While the waste sources above are the most prevalent, there are a vast amount of other alkaline industrial waste sources, like red mud and other process waste. While these sources may also have high alkalinity and small particle sizes, the main limitation will be the availability for industrial scale operations.

2.2 Brine water as a solution

Saline wastewater is a common byproduct of the production process of oil and gas. About 20–30 billion barrels of this wastewater is produced annually in the U.S., where around 65% is reinjected into well sites as a means of pressure control (Soong et al., 2006). The remaining wastewater is typically either treated or discharged. Discharging saline water into the environment has obvious negative implications and treatment costs can range anywhere from a few cents to a few dollars per barrel. Part of this discrepancy is due to differences in local regulation standards, but another reason stems from the fact that brine has varying concentration of Ca, Mg, and Fe along with the standard Na and Cl ions. While adjustments need to be made to the brine water, its utilization as a medium for mineral carbonation could lead to a higher productivity due to the presence of these ions, specifically calcium.

Carbonate formation occurs under pH conditions of 7.8 or higher. Because the typical pH range of brine is about 3 to 5, it requires modification before carbonation can happen. The addition of the industrial waste material will help to increase the pH of the brine and can reach reaction conditions with enough waste. However, the pH can be more readily adjusted with the addition of bases like NaOH. While additional reaction reagents like NaOH could help to decrease process time, its environmental and economic impacts need to be carefully considered. Existing research on what conditions are optimal for carbonation in brine (pressure, temperature, etc.) is limited, but simulation optimization will hopefully help to supplement these missing parameters.

3. Analysis

It would be an inefficient use of time if every consideration for a waste source feedstock was simulated in order to determine if the choice is economically feasible or not. Luckily, there are screening methods that provide a rough but quick analysis as to whether or not a process may be profitable. For these methods, the only necessary information is the purchase price of feedstock, the selling price of products, and the stoichiometric relationships between the materials. One such method is the Metric for Inspecting Sales and Reactants (MISR) which is a ratio of product sales to reactant purchases (El-Halwagi, 2017). Any process that has an MISR greater than 1 has a chance of being profitable, while any value lower than 1 indicates the process is not economically feasible. The MISR equation is defined below as:

$$MISR = \frac{\sum_{P=1}^{N \, products} Annual \ Production \ Rate \ of \ P \ * Selling \ Price \ of \ P}{\sum_{R=1}^{N \, reactants} Annual \ Feed \ Rate \ of \ R \ * Purchase \ Price \ of \ R}$$

It should be noted that an MISR value greater than 1 does not guarantee a profitable process, as a full detailed analysis is necessary in order to determine profitability. However, a value greater than 1 does justify further investigation into the process. This method helps quickly weed out any processes that have no chance of viability rather than wasting effort on unnecessary analysis.

3.1 Waste evaluation

In order to perform an MISR analysis on previously mentioned waste materials, the purchasing prices and product selling prices must be realistically set first. This data was gathered through literature and will be used for this analysis. The more difficult aspect of the MISR evaluation is determining the stoichiometric relationship between the waste and the carbonate products due to the variation in composition. For the purposes of this analysis, it will be assumed that the wastes consist of 30 wt% calcium oxide, as this falls within the range of each of the considered waste sources. The CO_2 feed is obtained from the shale gas process, and brine is considered to be present within the existing infrastructure and, therefore, assumed to also be available as feed.

The representative reaction scheme for creating calcium carbonate from calcium oxide and carbon dioxide is as follows:

$$CaO + CO_2 \rightarrow CaO_3$$

If we assume the reaction goes to completion, a stoichiometric conversion of one mole of calcium oxide produces one mole of calcium carbonate. By incorporating molecular weights, we can describe the reaction in terms of mass as:

(56.08 g/mol) * 1 mole CaO + (44.01 g/mol) * 1 mole CO_2 → 100.09 g CaO_3

For the purposes of this analysis, it will be assumed that the wastes consist of 30 wt% calcium oxide, as this falls within the range of each of the considered waste sources. In order to get the desired amount of calcium oxide from the waste to satisfy the conversion scheme above, the following must be true:

30% * Waste Mass = 56.08 g CaO

Waste Mass = 186.92 g

Finally, the reaction scheme can now incorporate this theoretical waste as:

186.92 g Waste (30 wt% CaO) + (44.01 g/mol) * 1 mole CO_2 → 100.09 g CaO_3

The primary target product for this study will be precipitated calcium carbonate (PCC), a high purity material used in ceramics, fillers, and other chemical applications. From the reaction scheme above, a stoichiometric ratio of 1.87 kg waste/kg PCC is established, assuming all reactant material is converted. While ultrahigh purity calcium carbonate can reach market prices of USD 10,000 per metric ton, a more realistic quality (~ 98%) will be considered here with a typical market price around USD 400 per metric ton (Katsuyama et al., 2005). The CO_2 feed is obtained from the shale gas process, and brine is considered to be present within the existing infrastructure and, therefore, assumed to also be available as feed. Table 2 shows the results for the MISR calculations.

As indicated by the MISR values in Table 2, the waste carbonation reaction appears to be potentially profitable. Of course, this evaluation does not consider utility costs or other potential reactants, but this is saved for further analysis.

Table 2. MISR evaluations of waste carbonation.

Waste material	Approximate price (USD/metric ton)	MISR
Fly Ash	15	14.3
Cement Kiln Dust	17	12.6
Furnace Slag	13	16.5

3.2 The two-step approach

Two main strategies exist for mineralizing waste particles: The single-step approach and the two-step approach. A single step approach involves mixing waste particles in an aqueous solution where CO_2 is then also bubbled through in the same reactor. The two-step approach separates these processes, where the calcium ions are first leached into solution and the remaining solid waste is filtered out so CO_2 can be mixed to form pure calcium carbonate without other solids affecting purity. While the single-step approach is much more direct and easier to implement, the two-step approach will produce the purified product of interest and is more economically favorable in this case. The two-step approach also allows further process control and tuning. Leaching calcium ions out of solid particle waste is favored at lower pH ranges, while carbonation only occurs in more basic conditions. These competing reaction conditions make it difficult to be efficient in the single-step approach without the use of complicated pH swing techniques. All simulated processes in this research will implement the two-step approach in order to increase efficiency of producing high purity PCC.

3.3 Reaction modelling

Other than the weight percent of calcium content, the major difference between the three types of industrial waste being considered in this project is the form in which the calcium is stored. These waste particles contain various types of silicates, ores, metals, and hydrates which influence the effects and composition of the leaching solution. In order to account for these differences, the proper chemical systems of each waste type need to be considered in order to describe the processes. Before these systems are described, there are a few simplifications that can be made. First, it has been noted that with these types of industrial wastes, calcium, hydroxide, and sulfate ions are the main leachable components, followed by potassium and minor levels of sodium, aluminum, and magnesium. The sum of the three main leachable ions typically account for 90–95% of the electroneutrality condition. With this information, we can drastically reduce the amount of chemical equilibriums that need to be incorporated in the simulation and focus on the major components. While it has been stated that the dissociation of calcium sulfide (CaS) is also present, this can be neglected at lower mass fractions (< 0.1 wt%). Other than these leachable components, the rest of the waste can be assumed as an inert solid which will be filtered out following the leaching stage.

3.3.1 Leaching waste cement

Waste cement is typically composed of SiO_2, CaO, Al_2O_3, MgO, and Fe_2O_3. From this, we can infer that the free lime (CaO) component will be the main calcium source. When free lime is mixed within an aqueous solution, calcium hydroxide ($Ca(OH)_2$) is formed. This is a somewhat soluble precipitate which will dissociate a hydroxide group first, followed by the remaining hydroxide and calcium ions in a second equilibrium step. The equilibrium descriptions are stated below.

$$CaO + H_2O \rightarrow Ca(OH)_2$$

$$Ca(OH)_2 \rightarrow CaOH^+ + OH^-$$

$$CaOH^+ \rightarrow Ca^{2+} + OH^-$$

While the first step can be assumed to react to completion, the following two dissociations require equilibrium data to determine the resulting concentrations. Both equilibrium constants were found through literature and were input into Aspen to help model the system.

3.3.2 Leaching fly ash

Fly ash can contain a great variety of materials in varying amounts; however, the main components are typically SiO_2, CaO, Al_2O_3, MgO, K_2O, $CaSO_4$ and Fe_2O_3. While similar to the components found in waste cement, one key difference is the presence of calcium sulfate, which represents another viable calcium source. In the presence of water, calcium sulfate forms a hydrated complex known as gypsum. This solid hydrate is slightly soluble and dissociates to form calcium and sulfate ions along with the complexed water. While the free lime component can be assumed to follow the same equilibrium conditions as the waste cement, this gypsum component is described by the equations below.

$$CaSO_4 + 2H_2O \rightarrow CaSO_4 \bullet 2H_2O$$

$$CaSO_4 \bullet 2H_2O \rightarrow Ca^{2+} + SO4^{2-} + 2H_2O$$

The parameters for these equilibrium states are present in Aspen and were utilized in the simulation.

3.3.3 Leaching steelmaking slag

Steelmaking slag contains a complex mixture of silicates, srebrodolskite, and calcium/magnesium-wustite type phases. While free lime is present in small amounts, it is typically bound within the wustite phases and cannot react in the leaching process. In this case, the main leachable calcium content comes in the form of dicalcium silicate (Ca_2SiO_4). For this project, the dissociation equation will be assumed as stated below.

$$Ca_2SiO_4 \rightarrow 2Ca^{2+} + SiO_4^{4-}$$

The calcium in this phase is more difficult to extract than the calcium present in the previous waste types. While acids can improve extraction efficiency, they impose greater environmental concerns as well as a higher operating cost. An interesting solution that will be implemented in this project is the use of ammonium salts, as described in the next section.

3.3.4 Ammonium salts

As mentioned in the section on the two-step approach, the leaching process favors acidic conditions while the carbonation stage requires basic conditions. Using acid to improve extraction efficiencies would cause difficulties and the possible necessity of additional basic material to make the carbonation process possible. One suggestion is the addition of ammonium salts, like NH_4Cl, which, while less effective at leaching than acids, still manage to improve efficiencies without drastically impacting the pH level. In addition to this, the ammonium salts are regenerated after the carbonation step, which helps to improve process economics. The following reaction equations show an example of how the salts interact in the leaching and carbonation stages.

$$2CaO.SiO_2 + 4NH_4Cl \rightarrow 2CaCl_2 + 4NH_3 + 2H_2O + SiO_2$$

$$2CaCl_2 + 4NH_3 + 2CO_2 \rightarrow 2CaCO_3 + 4NH_4Cl$$

Different ammonium salts have different impacts on both the extraction efficiencies as well as the carbonation efficiencies of varying waste types. A high extraction efficiency doesn't correlate to a good carbonation efficiency. Additionally, the efficiency of some salts may vary greatly with concentration while others aren't impacted as heavily. For example, in the study of ammonium salt effects on waste cement leaching, NH_4NO_3 varied from an efficiency of 68.8% at 1 M to an efficiency of 60.1% at 0.5 M, while the salt CH_3COONH_4 varied from 69% at 1 M to 23.8% at 0.5 M.

3.3.5 Carbonation of leached calcium

While the extraction mechanisms differ between the waste types, the carbonation process is virtually the same between them all since it is only the leached calcium ions taking part in the reaction and none of the other ions. Because of this, the reactor can be modelled using the same set of equilibrium equations across all three waste types. In this reactor, CO_2 is bubbled through the solution to first produce bicarbonate, which in turn reacts to form the carbonate species. This carbonate ion reacts with the leached calcium ions to precipitate as the target calcium carbonate. The simplified set of equations is shown below.

$$CO_2 + H_2O \rightarrow HCO_3^- + H^+$$

$$HCO_3^- + OH^- \rightarrow CO_3^{2-} + H_2O$$

$$Ca^{2+} + CO_3^{2-} \rightarrow CaCO_3$$

As an example, literature has shown that with the process of leaching and carbonation of ash waste, a precipitated calcium carbonate product of ~ 99% purity is obtained.

3.4 Defined project efficiencies

While it is possible to simulate these systems in Aspen utilizing kinetic and equilibrium data, the results are not always accurate. Instead, this project will utilize present experimental data from literature to define reaction efficiencies. While this may limit the operating conditions to those defined in the literature, the resulting computational model should produce a more realistic simulation. With everything considered so far, the following three waste source scenarios will be explored, along with the determined efficiencies:

1. Fly ash (no ammonium salts)-extraction efficiency will be determined by Gibbs reactor; Carbonation efficiency is approximately 87%.
2. Waste Cement with NH_4NO_3 salt-extraction efficiency of 60% at salt concentration of 0.5 M; Carbonation efficiency is approximately 74%.
3. Steelmaking Slag with NH_4Cl salt-extraction efficiency of 35% at salt concentration of 2.0 M; Carbonation efficiency is approximately 84%.

All data was produced at ambient condition (25 °C, 1 bar) and will be reflected as such in simulation parameters.

4. Simulation Results

Three separate flowsheets were created for each of the waste sources being analyzed. While the feed streams and compositions may vary between them, the units and connections between them are the same in order to maximize the flexibility of the system when switching between types of industrial waste. An example of the process design can be seen below. First, the raw industrial waste is processed through a crushing unit in order to get the particles down to the designated size distribution. The refined waste is then sent to the leacher where it is mixed with water, brine, extraction salt if applicable to the system, and an aqueous recycle stream. After the particles have been leached of the calcium ions, the solution is sent through a filter to remove any remaining solid particles, as detailed by the two-step approach. The filtered solution is then passed into the carbonation reactor, where CO_2 is bubbled in to be mineralized with the alkaline solution. The stream, now present with precipitated carbonate material, is then passed through another filter, represented as a separator, in order to remove the PCC product. Finally, the remaining electrolytic solution is sent to a splitter in order to recycle a specified fraction. Figure 2 is a schematic representation of the flowsheet.

The scale for this project will be based on a case scenario of a 100 MMscfd shale gas processing plant. This equates to an approximate feed flow of 94200 kg/hr worth of shale gas. The mass composition of CO_2 in shale gas varies from well to well, usually in the range of 1 to 9 wt%. Taking a median value of 5 wt% CO_2 and assuming the acid gas removal process produces a pure CO_2 stream, this equates to a flow rate of about 4700 kg/hr. Based on this reasoning, the flowsheets in the project will utilize a 5000 kg/hr feed of CO_2.

In order to accurately simulate the intended processes, the chemistry needs to be properly defined in the program. The ELECNRTL property method was used in Aspen in order to describe the highly electrolytic solutions being modelled and to account for the numerous dissociation and precipitation reaction equilibriums. While some equilibrium data is available in Aspen for the present species, missing parameters must be filled in with external data. The process can be divided into two major reaction components: The leaching unit and the carbonation unit. Because the leaching unit only contains sets of dissociating and equilibrating species, a GIBBS reactor block was used to model this process. The carbonator, however, is the specified unit for containing the mineralization reaction and is modelled using a STOICH reactor block. Details on these two units are listed in the sections below.

4.1 The leacher

The GIBBS reactor block in Aspen takes the defined chemistry, thermodynamic, and equilibrium data to minimize the Gibbs free energy of the mixture input in order to calculate the thermodynamic equilibrium.

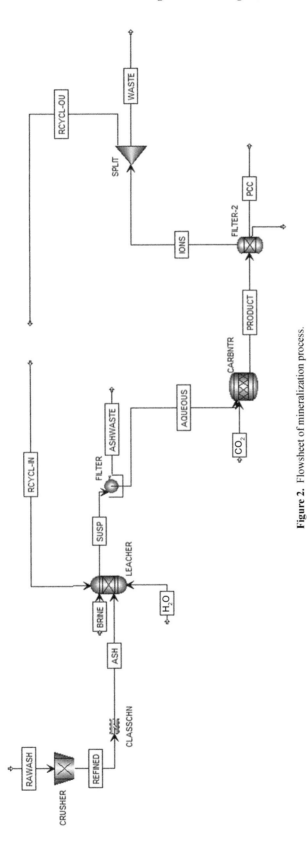

Figure 2. Flowsheet of mineralization process.

Data from literature indicates that, for leaching systems using waste particles of sizes smaller than 150 microns, steady-state is reached in well under two hours. This steady-state system can be approximated as an equilibrium and justifies the use of a GIBBS reactor unit. The specific components present in the reactor depends on the waste being utilized.

As specified in the defined efficiencies section, flow sheets involving the use of ammonium salts will utilize the extraction efficiencies reported in literature as a constraint in the Gibbs equilibrium calculations. This is done by placing restricted equilibrium definitions in the unit description parameters. By modifying the molar extent of the specific calcium leaching reaction in question, the proper extraction efficiency can be replicated and the remaining components in the system can reach equilibrium based on this constraint.

4.2 The carbonation vessel

The two input streams to this unit are the pure CO_2 coming from the theoretical shale gas plant and the filtered aqueous stream containing the leached ions. Because we are considering the calcium to be the only reacting component from this aqueous stream and we have the available carbonation efficiency data, we can model this reaction vessel using a STOICH reaction block. Here, the defined carbonation reaction mechanisms, as mentioned in the previous chapter, can be defined and restrained with a fractional conversion of calcium equivalent to the efficiency data.

4.3 The crusher

The first unit in the process design is the crusher, wherein unrefined ash is pulverized down to a predetermined size and fed into the leaching unit. While the cited literature typically uses particle sizes in the region of less than 150 microns, for this project, particles will be comminuted down to a distribution around 75 microns in order to assure proper equilibrium is reached in the leaching unit. Indeed, it was determined that between variations of particle size, vessel temperatures, and leaching times, the size distribution of the waste particles had the largest impact on how quickly the system reached steady-state (Hall et al., 2014).

The utility of the crushing unit is a function of the particle size distribution being input into the system. In order to evaluate and take into consideration the varying possible particle sizes of imported wastes, a sensitivity analysis will be performed to determine the utility usage and performance across a range of plausible waste sizes.

4.4 Filters and recycle

Two filters are implemented in the process design, one after the leaching unit and another after the carbonation step. The first filter unit is set to separate all the solid components of the suspension stream, which contain both inert components originally present in the waste particles as well as any additional precipitated materials, like gypsum and calcium hydroxide, depending on the specific flow sheet. The second filter is represented as a separations unit, where the precipitated calcium carbonate is separated as its own stream as well a stream of unreacted CO_2. This CO_2 in reality would be a product of the carbonation vessel, but was represented in the separations unit due to the limitations of the STOICH block.

A splitting block is used to represent the recycle of solvent back to the leaching unit and the remaining discarded waste stream. The recycle flow is described using two separated streams, a recycle-out of the splitter and a recycle-in to the leacher. While it is possible to represent a recycle stream in Aspen using a single connection, convergence issues can quickly arise if error tolerance standards are not met. This disconnected stream approach uses iteration as a means of approximate convergence. The recycle-out stream composition is copied into the recycle-in stream and the simulation is executed once more. This procedure is then repeated several times until mass compositions are in agreement with mass discrepancies well under 1%.

4.5 Solvent considerations

In this project, the influence of brine as both a solvent and a calcium source is explored alongside the use of waste particles. However, because brine is naturally acidic and the carbonation process requires basic conditions, the system needs to be supplemented with an additional water stream. In total there will be three streams acting a source of water: The brine stream, the water stream, and the recycled stream from the end of the process. A liquid to solid ratio of 50:1 was chosen as a compromise between efficiency and feed costs. As a consequence, the solid waste input will feed at a rate of 1000 kg/hr with the combined liquid streams equating to a rate 50,000 kg/hr.

The carbonation process naturally depletes the hydroxide concentration in the system and the output stream is highly concentrated with ions. Careful considerations must be taken to ensure that recycling solvent doesn't impact the pH levels of the next iteration of input streams. A recycle flow of 1000 kg/hr was chosen after calculations showed that regardless of the waste composition input the pH level of the aqueous stream was maintained above 8.5, as calcium carbonate begins to dissolve back into solution at pH levels below 8.3.

5. Results and Discussion

Across each of the three flowsheets, sensitivity analyses were performed by varying calcium weight fractions of the wastes. Subsequently, profit rates were determined, given the rate of product produced through each variation based on feed costs as well as waste disposal costs. Afterwards, the built-in economic analysis tools in ASPEN Plus were used to map and size equipment in order to calculate a rough estimate of the capital cost and evaluate rates of returns based on the varying production capacities. We will explore each waste individually and then will make recommendations on how to proceed forward based on the calculated results.

5.1 Utilities

One of the main interests going into this project was determining how comminution of particles would impact the economics of the carbonation process. While the final particle size was set to be 75 microns across every simulation and waste type, a range of plausible input waste sizes was examined to see its effect on the grinding utility. A key reason for choosing industrial waste as an alkaline mineral source was the fact that the particles generally are already at viable reaction sizes. After evaluation at a few interval ranges, it is clear that the impact on grinding utility is near identical across the whole considered region. Particles in the size interval 1000–1050 microns resulted in a grinder electrical usage of 3.54 kW, while particles in the interval of 1550–1600 drew a usage of 3.80 kW. This already pushes the boundaries of typical of waste particle sizes and, for our purpose, can be assumed as the theoretical limit. ASPEN Plus Process Economic Analyzer also gives a theoretical annual utility usage and estimates a value of approximately 470 MWh over an 8000-hour working year. This is equivalent to an approximate rate of 58.7 kW of power, and at an assumed electrical cost of 12 cents per kWh results in a utility rate of 7.04 dollars per hour.

5.2 Fly ash process evaluation

Fly ash is a special case in this analysis as there are two major calcium components typically found in this waste. Both calcium oxide and calcium sulfate are major leachable components that come at a variety of mass fractions. To account for this, a sensitivity analysis was performed individually for both components. A calcium composition range of 20–55 percent was analyzed as this represented a plausible range for this waste type. One of the resulting process flow sheets is shown by Figure 3 with labeled flow rates for each stream. Figure 4a–d represent the PCC production rate and the resulting profit flow incorporating all of the costs (e.g., utilities, feed, waste) and assuming the selling price of $400 per ton of product.

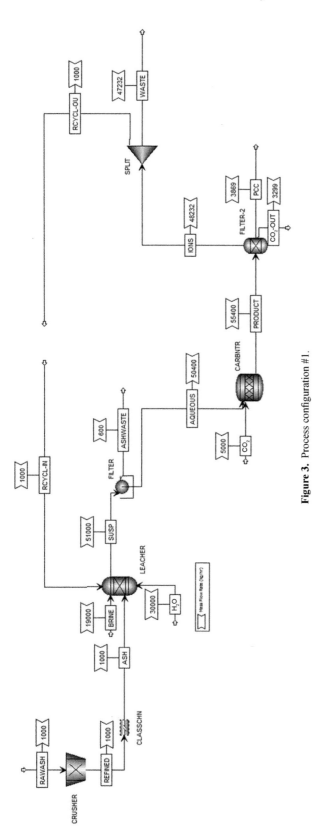

Figure 3. Process configuration #1.

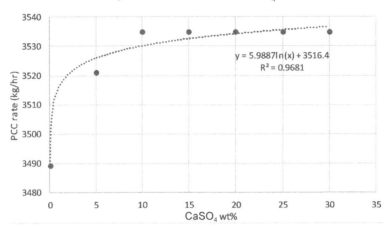

Figure 4a. PCC production rate vs. $CaSO_4$ wt%.

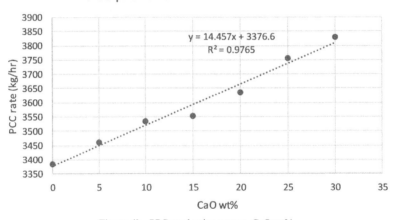

Figure 4b. PPC production rate vs. CaO wt%.

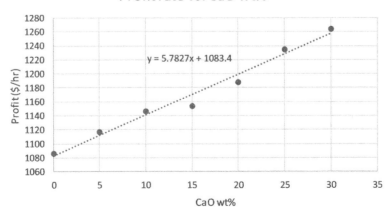

Figure 4c. Hourly profit vs. CaO wt%.

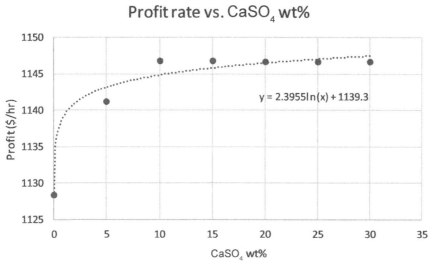

Figure 4d. Hourly profit vs. CaSO$_4$ wt%.

5.3 Waste cement evaluation

Similar to the fly ash evaluation, a sensitivity analysis was performed by varying the calcium oxide content from a weight percent range of 20–60. The process was performed with the addition of NH$_4$NO$_3$ in an amount that would result in a 0.5 M concentration in the leacher (2000 kg/hr). This feed cost was considered in the profit rate analysis, with ammonium nitrate assumed to be valued at \$300/ton. An example flowsheet from this simulation is shown by Figure 5, with stream flow rates displayed. Results of the analysis are shown by Figures 6a and b.

5.4 Steelmaking slag evaluation

As with the other wastes, the composition of the calcium source was varied; in this case, the source is dicalcium silicate. A weight percent range of 20–55 was evaluated and the process included the use of the ammonium salt NH$_4$Cl. In order to give a concentration of 2.0 M, a flow rate of 5349 kg/hr of salt is needed. This feed cost is considered in the profit analysis, where ammonium chloride is valued at \$130/ton. An example flow sheet is presented by Figure 7, with stream flows. The results are shown by Figures 8a and b.

5.5 Fixed capital and investment return

ASPEN Plus Process Economic Analyzer was used to produce an estimated cost for the capital investment of the theoretical process after specifying the type of equipment each unit represented. For example, the crushing unit was specified as a ball mill grinder used for real world mineral pulverizing purposes and the two reactor vessels were representing enclosed agitation units. After allowing ASPEN Plus to automatically size the units based on flow rates, the estimate for a fixed capital cost was approximately \$1.49 million. A sensitivity analysis was carried out in order to assess how the fluctuation in PCC pricing affects the profitability of the designed process. As by Figure 9, the range of PCC pricing from 200 to the original 400 dollars per ton is evaluated for all of the absolute minimum calcium mass percentages considered in the previous evaluation sections beforehand.

It is worth noting that the aforementioned case study is intended to illustrate the potential for mineralization as a strategy for monetizing CO$_2$ into value-added chemicals. Analogies can be made with other sources of CO$_2$ and other products. A key aspect to understand about shale gas processing is that CO$_2$

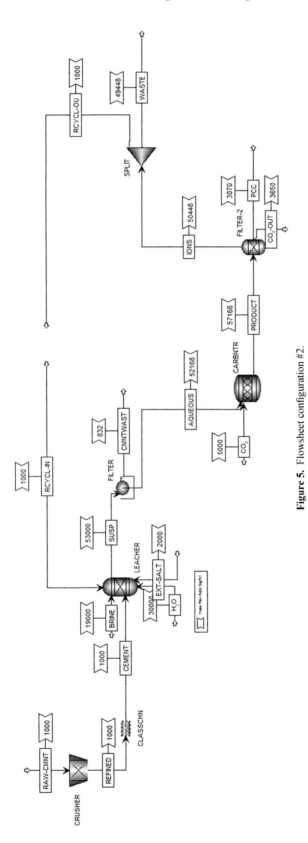

Figure 5. Flowsheet configuration #2.

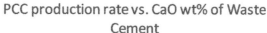

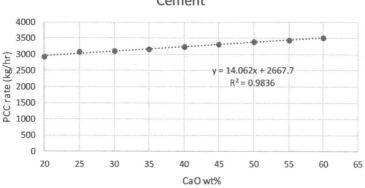

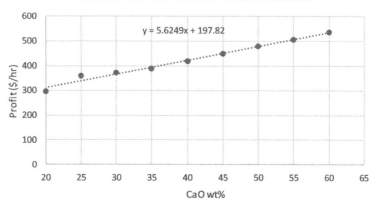

Figures 6a and b. Results for configuration #2.

is not the predominant GHG emission. Methane is emitted during multiple steps of the treatment of shale gas and is present in higher concentrations than conventional gas. More CO_2 is produced from combustion of the gas rather than its processing. What makes shale gas appealing for this type of integration process, as opposed to other facilities, is the purity of the CO_2 stream that occurs as a by-product of treatment. However, the applications of this mineralization technology are not limited to this specific instance of industrial development. Other common industrial sources of CO_2, like flue gas emissions, show promise of acting as a feed source for mineralization. A previous pilot scale study investigated the viability of mineralizing CO_2 emitted from a coal-fired power plant by reacting fly ash with flue gas emissions in a fluidized bed reactor. It was determined that an appreciable amount of CO_2, as well as SO_2 and Hg, was captured and mineralized in the process (Reddy et al., 2011). The exact purity of the calcium carbonate is unknown and additional separation steps would be necessary, but further research into this technology would be a worthwhile endeavor. While the economic viability of such a process would have to be evaluated on a case-by-case basis, the possibility of adapting this technology towards major industrial emitters of CO_2 is appealing.

Another important consideration is the demand and market size for carbonate products. A report from 2018 estimated that the global demand for calcium carbonate in 2016 was 113.7 million tons and is expected to grow to 180.1 million tons by 2025 (Market Research Reports & Consulting). Alkaline particle wastes are available in large enough quantities to supplement this carbonate market with transportation being the main obstacle. While this market growth is encouraging for mineralization technology, it has also been estimated that the amount of anthropogenic CO_2 produced is orders of magnitude greater than

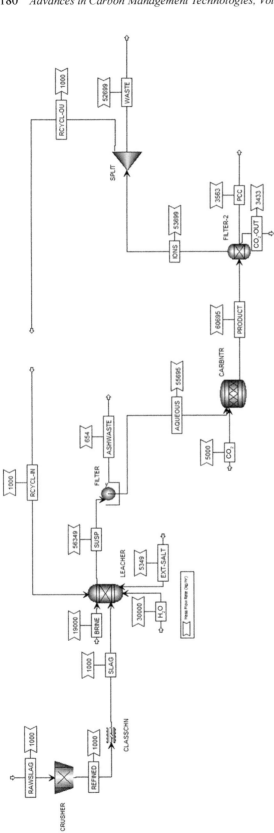

Figure 7. Flowsheet for configuration #3.

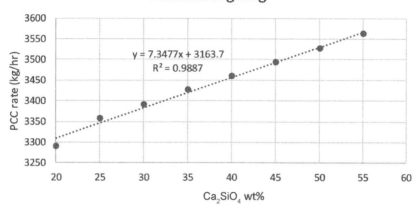

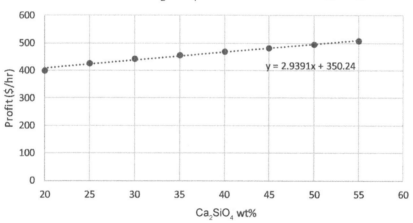

Figures 8a and b. Results for configuration #3.

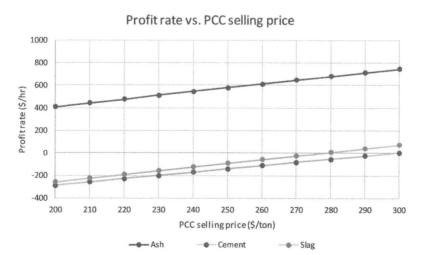

Figure 9. Sensitivity analysis of hourly profit as a function of selling prices.

the market size for carbonate products. Further investigation into the economics of other mineralization pathways, like the production of magnesium carbonate, could be one option to diversify and promote mineralization technology. More likely though, the solution to alleviating the global CO_2 burden will come from a combination of sequestration and utilization technologies.

6. Concluding Remarks

Mineralization is a relatively new consideration for carbon dioxide sequestration. While the benefits of long-term storage and value-added product potential have made it a compelling pathway for utilization, hurdles that have kept it from becoming a major solution to the challenges of CO_2 reduction still remain. It has been said that the current knowledge of mineralization is insufficient for us to be able to conclude if the process is energetically and economically feasible. However, no industrial scale process has been implemented or thoroughly investigated and this project aims at helping to determine its potential.

Multiple layers of sustainable opportunities have been merged together in order to create a unique and practical operation. Carbon dioxide utilization, hazardous waste fixation, and brine water application are independent processes that, with proper integration, could lead to possibilities for creating a greener but still profitable industrial system.

References

Abidin, A., Puspasari, T. and Nugroho, W. 2012. Polymers for enhanced oil recovery technology. Procedia Chemistry 4: 11–16.
Afzal, S., Sengupta, D., Sarkar, A., El-Halwagi, M.M. and Elbashir, N.O. 2018. An optimization approach to the reduction of CO_2 emissions for syngas production involving dry reforming. ACS Sustainable Chem. Eng. 6(6): 7532–7544.
Ahmaruzzaman, M. 2010. A review on the utilization of fly ash. Progress in Energy and Combustion Science 36(3): 327–363.
Al-Douri, A., Sengupta, D. and El-Halwagi, M.M. 2017. Shale gas monetization—A review of downstream processing to chemicals and fuels. Journal of Natural Gas Science and Engineering 45: 436–455.
Alsuhaibani, A.S., Afzal, S., Challiwala, M., Elbashir, N.O. and El-Halwagi, M.M. 2019. The impact of the development of catalyst and reaction system of the methanol synthesis stage on the overall profitability of the entire plant: a techno-economic study. Catalysis Today, https://doi.org/10.1016/j.cattod.2019.03.070.
Barbato, L., Centi, G., Iaquaniello, G., Mangiapane, A. and Perathoner, S. 2014. Trading renewable energy by using CO_2: An effective option to mitigate climate change and increase the use of renewable energy sources. Energy Technology 2(5): 453–461.
Bobicki, E.R., Liu, Q., Xu, Z. and Zeng, H. 2012. Carbon capture and storage using alkaline industrial wastes. Progress in Energy and Combustion Science 38(2): 302–320.
Elbashir, N.O., El-Halwagi, M.M., Hall, K.R. and Economou, I. 2019. Natural Gas Processing from Midstream to Downstream. Wiley.
El-Halwagi, M.M. 2017. Sustainable Design through Process Integration: Fundamentals and Applications to Industrial Pollution Prevention, Resource Conservation, and Profitability Enhancement. Second Edition, IChemE/Elsevier.
Gadikota, G. and Park, A.-H.A. 2015. Accelerated carbonation of Ca- and Mg-bearing minerals and industrial wastes using CO_2. Carbon Dioxide Utilisation, 115–137.
Grobe, M. 2010. Carbon Dioxide Sequestration in Geological Media: State of the Science; AAPG: Tulsa, OK.
Hall, C., Large, D., Adderley, B. and West, H. 2014. Calcium leaching from waste steelmaking slag: Significance of leachate chemistry and effects on slag grain mineralogy. Minerals Engineering 65: 156–162.
Huntzinger, D.N., Gierke, J.S., Sutter, L.L., Kawatra, S.K. and Eisele, T.C. 2009. Mineral carbonation for carbon sequestration in cement kiln dust from waste piles. Journal of Hazardous Materials 168(1): 31–37.
Katsuyama, Y., Yamasaki, A., Iizuka, A., Fujii, M., Kumagai, K. and Yanagisawa, Y. 2005. Development of a process for producing high-purity calcium carbonate ($CaCO_3$) from waste cement using pressurized CO_2. Environmental Progress 24(2): 162–170.
Lekakh, S., Rawlins, C., Robertson, D., Richards, V. and Peaslee, K. 2008. Kinetics of aqueous leaching and carbonization of steelmaking slag. Metallurgical and Materials Transactions B 39(1): 125–134.
Market Research Reports & Consulting, Calcium Carbonate Market Worth $34.28 Billion By 2025: CAGR: 5.7% [https://www.grandviewresearch.com/press-release/global-calcium-carbonate-market] (accessed on August 28, 2019).
Meylan, F.D., Moreau, V. and Erkman, S. 2015. CO_2 utilization in the perspective of industrial ecology, an overview. Journal of CO_2 Utilization 12: 101–108.

Panu, M., Topolski, K., Abrash, S. and El-Halwagi, M.M. 2019. CO$_2$ footprint reduction via the optimal design of carbon-hydrogen-oxygen SYmbiosis Networks (CHOSYNs). Chem. Eng. Sci. 203: 1–11. https://doi.org/10.1016/j. ces.2019.03.066.

Pokoo-Aikins, G., Nadim, A., Mahalec, V. and El-Halwagi, M.M. 2010. Design and analysis of biodiesel production from algae grown through carbon sequestration. Clean Technologies and Environmental Policy 12: 239–254. DOI: 10.1007/ s10098-009-0215-6.

Rahman, F.A., Aziz, M.M.A., Saidur, R., Bakar, W.A.W.A., Hainin, M., Putrajaya, R. and Hassan, N.A. 2017. Pollution to solution: capture and sequestration of carbon dioxide (CO$_2$) and its utilization as a renewable energy source for a sustainable future. Renewable and Sustainable Energy Reviews 71: 112–126.

Reddy, K.J., John, S., Weber, H., Argyle, M.D., Bhattacharyya, P., Taylor, D.T., Christensen, M., Foulke, T. and Fahlsing, P. 2011. Simultaneous capture and mineralization of coal combustion flue gas carbon dioxide (CO$_2$). Energy Procedia 4: 1574–1583. doi: 10.1016/j.egypro.2011.02.027.

Romanov, V., Soong, Y., Carney, C., Rush, G.E., Nielsen, B. and O'Connor, W. 2015. Mineralization of carbon dioxide: a literature review. ChemBioEng. Reviews 2(4): 231–256.

Said, A., Mattila, H.-P., Järvinen, M. and Zevenhoven, R. 2013. Production of precipitated calcium carbonate (PCC) from steelmaking slag for fixation of CO$_2$. Applied Energy 112: 765–771.

Soong, Y., Fauth, D., Howard, B., Jones, J., Harrison, D., Goodman, A., Gray, M. and Frommell, E. 2006. CO$_2$ sequestration with brine solution and fly ashes. Energy Conversion and Management 47(13-14): 1676–1685.

Tilak, P. and El-Halwagi, M.M. 2018. Process integration of calcium looping with industrial plants for monetizing CO$_2$ into value-added products. Carbon Resources Conversion 1: 191–199. DOI: 10.1016/j.crcon.2018.07.004.

Zhao, H., Park, Y., Lee, D.H. and Park, A.-H.A. 2013. Tuning the dissolution kinetics of wollastonite via chelating agents for CO$_2$ sequestration with integrated synthesis of precipitated calcium carbonates. Physical Chemistry Chemical Physics 15(36): 15185.

CHAPTER 10

Energy-Water-CO$_2$ Nexus of Fossil Fuel Based Power Generation

Kyuha Lee and *Bhavik R Bakshi**

1. Introduction

Thermoelectric power plants generate electricity and flue gas scrubbers mitigate emissions from the plants. However, at the same time, these plants withdraw large quantities of water from the watershed. In the United States, this withdrawal corresponds to 45% of the total 2010 water use (Maupin et al., 2014). In addition, NO$_x$ emissions from the scrubbers cause deposition of excessive nutrients in the watershed. Therefore, increasing electricity generation could increase water stress, deteriorate water quality, and contribute to climate change. Also, in the watershed where power plants are located, other activities, such as farming, interact with the power plants since these activities also require water from the watershed as well as energy from the power plants and release nutrients to the watershed. To prevent shifting of environmental impacts across multiple flows (Bakshi et al., 2018), the energy-water nexus between different activities in the watershed needs to be understood in assessing the impacts of power plants and sustainability of the watershed (Sanders, 2014).

Fossil fuel power plants not only require a huge amount of water, but also emit 28% of the 2016 U.S. greenhouse gas emissions (EPA, 2018a), 67% of the 2014 U.S. SO$_2$ emissions, and 12% of the 2014 U.S. NO$_x$ emissions (EPA, 2014). Ecosystem services, such as climate change regulation and air quality regulation, could play an important role in mitigating these pollutants and emissions. The ecosystem provides many essential goods and services to society and for our well-being. Ecosystem goods include water and fossil resources that our society has been extensively utilizing. Ecosystem services include air and water quality regulation and carbon sequestration by soil and vegetation. Therefore, in addition to the conventional energy-water nexus concept, we need to expand the system boundary to include the role of ecosystems. The framework of Techno-Ecological Synergy (TES) has been developed to account for the role of ecosystems in engineering and other human activities (Bakshi et al., 2015; Gopalakrishnan et al., 2016). In this framework, the demand for ecosystem services imposed by human activities, which correspond to the emissions and resource use, must not exceed the capacity of the corresponding ecosystem to supply the demanded goods and services. This condition needs to be satisfied in order to claim environmental sustainability of any activity. For example, for environmental sustainability of a power plant in a watershed, the amount of water consumed by the power plant should be smaller than the amount of renewable water available to the plant from the watershed. Otherwise, power plants will likely fail at some point because water resource will become scarce.

William G. Lowrie Department of Chemical and Biomolecular Engineering, The Ohio State University, Columbus, OH 43210.
* Corresponding author: bakshi.2@osu.edu

In the watershed where power plants are located, other activities, such as residential, industrial, and farming, also share the supply of ecosystem goods and services with electricity generation activity. Also, to address watershed-scale sustainability, the demand for ecosystem services from all activities in a watershed needs to be considered. This raises the need for a holistic assessment to investigate watershed sustainability. In such an analysis, the trade-offs between multiple objectives, such as water quality, water quantity, net electricity generation, climate change, and air quality objectives, could be identified as well.

In this work, we employ a holistic analysis approach to investigate the energy-water-CO$_2$ nexus for thermoelectric power plants, with specific focus on the Muskingum River Watershed (MRW) in Ohio in the United States. In 2014, two coal-fired power plants and three NG-fired power plants were located in the MRW. The year 2014 is selected because of data availability from a variety of online sources and reports. The holistic analysis boundary includes thermoelectric power generation, mining, residential, commercial, industrial, agricultural, transportation and wastewater treatment activities.

To suggest a better recommendation for sustainable watershed management, various alternatives for both technological and agroecological activities are considered. Technological alternatives include fuel mining and cooling technology options for power generation, CO$_2$ conversion options, and renewable power generation options. Agroecological alternatives include tillage practice options and land use change options. Scenarios for each alternative are analyzed in order to understand the trade-offs between multiple flows.

CO$_2$ flows are affected by both technological and agroecological alternatives in various ways. Technological alternatives mainly aim to reduce CO$_2$ emissions, while agroecological alternatives improve carbon sequestration in soil and vegetation. Broader implications in terms of CO$_2$ management are addressed through the holistic analysis of watershed activities.

The goal of this work consists of three parts. First, we identify ecological overshoots for activities in the MRW. Second, we investigate various alternative scenarios in order to understand the trade-offs between energy, water, and CO$_2$ flows. Third, we suggest better watershed management solutions that could be "win-win" in terms of multiple objectives for watershed sustainability.

2. Watershed Activities

In this section, we describe the characteristics of various watershed activities that are included in this holistic assessment study. Data collection is a challenging task for such a holistic assessment because we need to rely on multiple data sources that often have different spatial and temporal data resolutions. Thus, it is important to keep spatial and temporal consistency between data. In this study, watershed-scale data for the year 2014 is preferred because most data are available for this year. The watershed boundary is determined by the Hydrologic Unit Code (HUC) system that assigns a unique HUC code to each watershed (Seaber et al., 1987). Each HUC region is defined by distinct hydrologic features, such as rivers, lakes, and drainage basins. The Muskingum River Watershed (MRW) studied in this work corresponds to a region where 8-digits of HUC (HUC8) is assigned as 05040004. The MRW is located in southeast Ohio. Water flows from the MRW drains into the Muskingum River, which eventually flows into the Ohio River. Figure 1 shows land use and land cover features in the MRW.

If any data are not available for HUC8 spatial resolution, the data is allocated to HUC8 based on the ratio of population or area. For example, data for air pollutants is available for U.S. counties (EPA, 2014), which do not match HUC spatial resolution. The MRW includes eight counties in Ohio: Coshocton, Licking, Knox, Muskingum, Perry, Morgan, Washington, and Noble. The county-level data are then allocated to the MRW, based on the ratio of HUC8 area to eight counties' area. For other data, such as residential and commercial activity data, that are more dependent on population rather than area, the allocation is performed based on the population. Table 1 summarizes data inventories, data sources, and spatial data resolution for this work. When life cycle inventory databases, such as GREET (Wang, 2016) and USLCI (NREL, 2018), are used to obtain data, only on-site data are collected since the scope of this study is limited to the watershed scale. Figure 2 represents the scope of this study that includes various watershed activities and resource, waste, and ecosystem flows. In the following sections, we provide brief descriptions of each watershed activity.

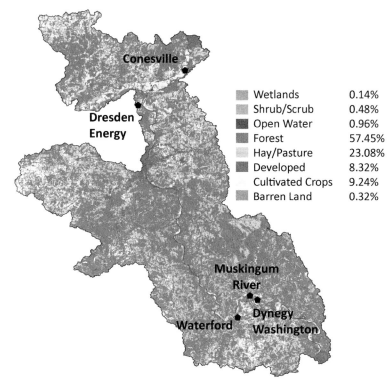

Figure 1. Land use and land cover features in the Muskingum River Watershed (HUC8: 05000405) in Ohio. Five thermoelectric power plants (◆) were located in the MRW in 2014.

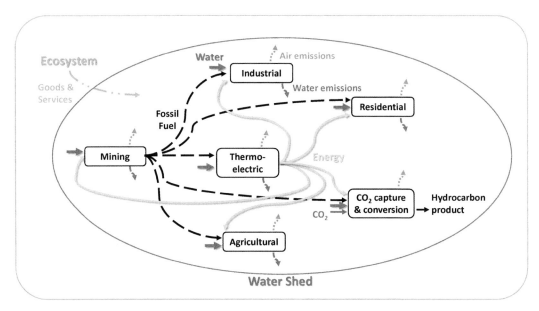

Figure 2. Scope of this study that includes various watershed activities and resource, waste, and ecosystem flows. Food production flow from agricultural activity is excluded from this study.

Table 1. Data sources for activities and environmental interventions in the MRW. If the spatial resolution of data is larger than HUC8 scale, the data is allocated to the HUC8 scale based on the ratio of population or area.

Activity	Environmental & material flow	Data source	Spatial resolution
Theremoelectric	GHG emissions	EPA eGRID (EPA, 2016)	Facility
	Air pollutants	EPA NEI (EPA, 2014)	County
	Water pollutants	EPA NPDES (Ohio EPA, 2018)	Facility
	Thermal water pollution	EIA-923 (EIA, 2018a)	Facility
	Water withdrawal	EIA-923 (EIA, 2018a)	Facility
	Water consumption	EIA-923 (EIA, 2018a)	Facility
	Natural gas use	EIA-923 (EIA, 2018a)	Facility
	Electricity use	EIA-923 (EIA, 2018a)	Facility
	Electricity generation	EIA-923 (EIA, 2018a)	Facility
Mining	GHG emissions	GREET (Wang, 2016)	U.S. average
	Air pollutants	EPA NEI (EPA, 2014)	County
	Water pollutants	NETL (Skone, 2016; Skone et al., 2016)	Appalachia average
	Water withdrawal	USGS (Solley et al., 1998)	Ohio
	Water consumption	GREET (Wang, 2016)	U.S. average
	Natural gas use	EIA (EIA, 2018b)	Ohio
	Electricity use	USLCI (NREL, 2018)	U.S. average
Agricultural & Other Activities (Residential, Commercial, Industrial, Transportation, Wastewater treatment)	GHG emissions	EPA GHGRP (EPA, 2016)	Facility
		EPA NEI (EPA, 2014)	County
	Air pollutants	EPA NEI (EPA, 2014)	County
	Water pollutants	(Agricultural) SWAT (Khanal et al., 2018)	HUC8
		(Other activities) EPA (Ohio EPA, 2016)	HUC4
	Water withdrawal	EnviroAtlas (EPA, 2018b)	HUC8
	Water consumption	USGS (Solley et al., 1998)	Ohio
	Natural gas use	EIA (EIA, 2018b)	Ohio
	Electricity use	EIA (EIA, 2018d)	Ohio
Ecosystem Supplies	Carbon sequestration	i-Tree Landscape (*i-Tree Landscape*, 2018)	HUC8
	Air quality regulation	i-Tree Landscape (*i-Tree Landscape*, 2018)	HUC8
	Water quality regulation	(Kadlec, 2008, 2016)	Average
	Water provision	AWARE (Boulay et al., 2018)	HUC2
	Natural gas provision	(Le Quéré et al., 2017)	Global

2.1 Thermoelectric power generation

In the MRW, there are two coal-fired steam turbine power plants and three natural gas-fired combined cycle (NGCC) power plants, as shown in Figure 1. The Muskingum River Power Plant was retired in 2015 due to environmental regulations (Gearino, 2013). However, since our study is for 2014, we assume that the Muskingum River Power Plant is still operating in order to keep the temporal consistency of data.

In terms of the power generation technology of thermoelectric power plants, 99% of coal-fired power plants in the U.S. employ steam turbine boilers, while 84% of NG-fired power plants in the U.S. employ combined cycle boilers (Wang, 2016). Since five power plants in the MRW are also operated by using these generation technologies, we only consider these two types of power generation technologies in this study.

Fossil power generation is responsible for about 45% of freshwater withdrawals in the U.S. (Maupin et al., 2014). Most water withdrawn is used for the cooling of boilers. Depending on which cooling methods are employed in the power plant, water and energy requirements are varied. Once-through cooling technology, also known as the open-loop cooling system, withdraws a massive amount of water but returns most of it at a warmer temperature to the watershed. The once-through cooling system has mainly been installed in power plants in the eastern U.S. On the other hand, recirculating cooling technology, also known as the closed-loop cooling system and cooling tower, withdraws only a fraction of the water that systems require, recirculates water, but consumes most of it through evaporation from the cooling tower. Therefore, recirculating cooling technology has higher water consumption than once-through cooling technology, even though its amount of water withdrawn is significantly lower. Also, it has lower electricity generation efficiency and is more expensive than once-through cooling technology (Tawney et al., 2005). The recirculating cooling system is widespread in the western U.S. In contrast to those wet cooling methods, the dry cooling technology uses no water, but requires more energy and higher capital and operation costs, and results in lower generation efficiency than wet cooling technologies. The lower generation efficiency means that more fossil resources are required to generate electricity, and thus, it will increase the impacts from upstream processes, such as mining and transportation of fossil fuels.

Among the five power plants in the MRW, only one coal-fired plant, the Muskingum River Power Plant, employs once-through cooling technology, the other four plants employ the recirculating cooling technology. In this study, three cooling technologies (once-through, recirculating, and dry cooling) are considered as technological alternatives for thermoelectric activity. It is reported that 0.3–1% and 2–4% reductions in generation efficiency are expected for thermoelectric power plants in Texas if the once-through cooling system is converted to recirculating and dry cooling systems, respectively (Loew et al., 2016). Also, 0.60–0.63 cents/kWh of cost is required for the plant operator to retrofit a recirculating cooling system to a dry cooling system.

2.2 Mining of fossil resources

In 2017, coal and natural gas accounted for 34% and 26% of energy sources that were used to generate electricity in the U.S., respectively (EPA, 2018c). Depending on which fossil resources are exploited, total environmental interventions from fossil power generation are hugely varied. This is mainly attributed to the fact that each fossil resource contains different chemical compositions. Furthermore, in the case of NG extraction, the hydraulic fracturing of shale gas requires substantial water resources relative to the conventional NG extraction (Sanders, 2014; Clark et al., 2013; Jenner and Lamadrid, 2013). According to the previous study, however, the amount of water withdrawn for the fracking is much smaller than the amount of cooling water for power generation (Clark et al., 2013). Fracking also contaminates water resources due to wastewater discharged from shale wells (Vengosh et al., 2013; Vidic et al., 2013).

The transportation of fossil fuels from mining sites to the power plants is considered as well. While coal is generally transported by truck and railroad, NG is primarily transported by pipeline. The leakage of gas from the pipeline transportation is also considered.

2.3 Farming

Figure 1 shows land use and land cover in the MRW. Total area for agricultural land use is approximately 3.8×10^8 m². The MRW is not a region where agricultural production dominates. However, nutrient runoff from farming activity can be varied depending on which farming practices are performed. Hence, agricultural practice management is crucial to prevent deteriorating water quality. According to the previous findings, farming activity in the river basin that includes MRW contributes significantly to total nitrogen (TN) and total phosphorus (TP) loads in the river basin (Khanal et al., 2018).

Tillage practices are performed to prepare the soil for crop production. However, these practices damage the soil structure and increase nutrient runoffs. Currently, four different tillage practices are employed in the MRW. 57.0% of the agricultural land area performs no-till practice. 22.9%, 20.0%, and 0.11% carry out conservation (chisel plow), reduced (tandem-disc plow), and intensive (moldboard plow) tillage practices, respectively. The conservation tillage practice is defined as tillage that has more than 30% of crop residues on the soil. The reduced and intensive tillage practices have 15–30% and less than 15% of crop residues that remain on the soil, respectively. The increased mixing efficiency leads to low crop residues, requires many fertilizer inputs, and increases the risk of soil erosion.

No-till practice helps reduce soil erosion and nutrient runoffs since more crop residue remains in the soil, and thus, fewer fertilizers need to be applied (Ohio EPA, 2013). Also, no-till practice is economically cheaper than tillage practices due to the reduced labor and fuel requirements. The long-term crop yield may be increased as well due to improved soil fertility. However, the food production flow is excluded from this study to focus on the nexus of energy and water flows. Differences in other interventions, such as air emissions and resource uses, between tillage practices are also taken into account, even though their contribution is relatively negligible in comparison to thermoelectric activity (NREL, 2018; Nemecek et al., 2007).

2.4 Miscellaneous activities

Although mining, thermoelectric, and agricultural activities account for most of the water use and pollutant emissions, other activities also consume water and energy and release emissions. Therefore, for the comprehensive analysis of the energy-water-CO_2 nexus in the MRW, various activities that include residential, commercial, industrial, transportation, and wastewater treatment, are included in the analysis. This comprehensiveness of the analysis boundary is particularly important when the TES framework is applied. Since ecosystem supplies, such as water provisioning, account for the entire amounts of supplies from the MRW, their associated ecosystem demands, such as water consumption, also need to be the total demands from all activities in the MRW.

2.5 Supply of ecosystem services

Traditional sustainability assessment approaches, such as conventional life cycle assessment (LCA), account for environmental impacts from activities. The impacts include emissions and resource use. These traditional methods quantify only relative sustainability in order to answer the question: Whose impacts are smaller than others? To claim absolute sustainability, however, we need to address how the surrounding ecosystem offsets those impacts. The TES framework accounts for the supply and demand of specific ecosystem goods and services and introduces a TES sustainability metric: $V_k = (S_k - D_k)/D_k$ (Bakshi et al., 2015; Gopalakrishnan et al., 2016). The ecosystem service demand (D) corresponds to environmental flows, such as air emissions and water consumption. The ecosystem service supply (S) corresponds to the provisioning of ecosystem goods and services, such as air quality regulation service and water resource provision. The TES metric V_k is calculated for each type of ecosystem good and service (k). A positive V_k indicates that impacts do not overshoot the capacity of ecosystem supplies, which means the system is sustainable in terms of k ecosystem goods and services.

Forest ecosystems and tree canopies provide carbon sequestration and air quality regulation services. Wetlands regulate water quality by removing excessive nutrient runoffs, although only 0.14% of land area in the MRW is wetlands as shown in Figure 1. With respect to ecosystem goods, such as water and fossil resource provisioning, only the renewable portion can be included as the supply of such ecosystem goods. To assess water provisioning in the MRW, for example, various factors about the water cycle, such as precipitation, evapotranspiration from canopy and soil, infiltration into the soil, and surface/subsurface runoffs, need to be considered. In this study, the Available Water Remaining (AWARE) model is used to estimate the available amount of water in the MRW (Boulay et al., 2018). This model has data for the HUC2 spatial scale, which represents a much larger area than the HUC8 scale. The value for the Ohio region (HUC2: 05) is allocated to the MRW region (HUC8: 05040004) based on the ratio of the land areas in HUC8 versus HUC2.

Fossil resources are formed through anaerobic decomposition of organic matter in the earth over very long periods of time. With respect to the natural gas supply from the ecosystem, only renewable NG should be considered as the ecosystem supply. However, since the formation rate of NG is significantly slower than the NG consumption rate, the TES metric for NG (V_{NG}) is very close to negative one (–1) regardless of any scenarios. In this work, therefore, we consider the other case where 2 °C of global warming since the pre-industrial period is allowed for exploiting accumulated NG (Le Quéré et al., 2017). According to the Intergovernmental Panel on Climate Change (IPCC), the rise in global temperature must be limited to 2 °C above pre-industrial levels in order to avoid disastrous consequences of climate change

(Le Quéré et al., 2017). The IPCC has calculated the carbon budget, which represents the amounts of global CO_2 emissions that must not be exceeded in order to maintain global warming under 2 °C. The remaining budget for greenhouse gas emissions is estimated to be 275 GtC (WRI, 2014). Since 22% of GHG emissions are attributed to the use of NG (EPA, 2018c), 60 GtC of the budget can be allocated to the GHG emissions from NG use. This budget is further allocated to the NG use in the MRW, based on the ratio of NG consumption in the MRW to global NG consumption (Dudley et al., 2018). The resulting budget is the amount of NG supply in the MRW that allows 2 °C of global warming.

2.6 Potential CO_2 conversion technologies

As a way of mitigating CO_2 emissions, extensive research is being conducted on technologies for converting CO_2 into a variety of hydrocarbon products (Frauzem et al., 2015; Artz et al., 2017; Wolf et al., 2016; Behr and Nowakowski, 2014). In this work, three CO_2 conversion technologies are selected as technological alternative options that could improve watershed sustainability. First, methane (CH_4) is produced from carbon dioxide and hydrogen through Sabatier exothermic reaction, as follows:

$$CO_2 + 4H_2 \rightarrow CH_4 + 2H_2O \qquad (\Delta_R H^O_{298} = -165 \text{ kJ/mol}).$$

CO_2 feedstocks are assumed to be captured from thermoelectric power plants. To provide hydrogen for hydrocarbon products, it is assumed that a water resource is utilized through an electrolysis process, as follows:

$$H_2O \rightarrow H_2 + 1/2O_2 \qquad (\Delta_R H^O_{298} = 286 \text{ kJ/mol}).$$

Therefore, the overall reaction is:

$$2H_2O + CO_2 \rightarrow CH_4 + 2O_2 \qquad (\Delta_R H^O_{298} = 979 \text{ kJ/mol}).$$

The converted CH_4 product is assumed to displace natural gas in the MRW since most of the NG composition is CH_4. Therefore, displacement credits of all kinds of avoided emissions and resource uses are given to CO_2 conversion technologies.

CO_2 is also used to produce synthetic gas through the reverse water-gas shift reaction, as shown below:

$$CO_2 + H_2 \rightarrow CO + H_2O \quad (\Delta_R H^O_{298} = 41 \text{ kJ/mol}).$$

If syngas that has a ratio of H_2 to CO of 2:1 is required, hydrogen is obtained from the electrolysis of water shown above. Also, formic acid is synthesized from the hydrogenation of CO_2 as follows:

$$CO_2 + H_2 \rightarrow HCOOH \qquad (\Delta_R H^O_{298} = -31.5 \text{ kJ/mol}).$$

Similarly with the displacement approach for CH_4, it is assumed that CO_2-converted syngas and formic acid displace each of the products from conventional processes. Syngas is produced from conventional coal gasification process (Dunn et al., 2015). 69% of formic acid is synthesized from methyl formate and the rest of it is produced from butane (Sutter, 2007).

Since the CO_2 conversion technologies described above still have not been fully developed for commercialization, stoichiometric reactions are assumed for all CO_2 conversion scenarios for the simplicity of analysis. The energy demand for CO_2 conversion is estimated using the standard enthalpy of each conversion reaction ($\Delta_R H^O_{298}$). Numerous experimental data are available from literature (Behr and Nowakowski, 2014; Jessop et al., 2004). However, those data vary substantially depending on process configurations, such as the use of a specific catalyst. For example, one report employed 30 bar of CO_2 pressure for converting CO_2 to formic acid (Lau and Chen, 1995), while the other one employed 120 bar of CO_2 pressure (Jessop et al., 1996). This makes the analysis challenging since we cannot just randomly select one technology for the comparison between different conversion scenarios (i.e., CO_2 conversion to methane, syngas, and formic acid). For a more accurate analysis, process data for CO_2 capture using monoethanolamine (MEA) is included in the analysis (Althaus et al., 2007) since the CO_2 capture process is common for all conversion technologies. It is assumed that MEA is produced from outside MRW. With respect to the energy demand for CO_2 compression, we assume that 30 bar of CO_2 pressure is required for all conversion technologies since many formic acid production technologies from CO_2 employed 30 bar of CO_2 pressure (Behr and Nowakowski, 2014; Jessop et al., 2004).

The overall CO_2 conversion that includes the electrolysis of water is not only water-intensive but also energy-intensive. Thus, significant amounts of water and energy resources are required. This could make CO_2 conversion options infeasible by shifting impacts of GHG emissions to the impacts of energy and water consumption since the interventions from power generation are also increased.

2.7 Potential renewable electricity sources

To lessen the interventions from the increased demand for electricity for CO_2 conversion technologies, renewable electricity sources are considered as means to provide electricity for CO_2 capture, compression, and conversion by displacing conventional, thermoelectric power generation (Artz et al., 2017). In this study, solar and wind power sources are considered as means to replace fossil fuel sources in the MRW. These renewable power generation technologies do not consume as much water as thermoelectric power generation (Lampert et al., 2015). Solar power generation includes two technologies: Photovoltaics (PVs) and concentrated solar panels (CSPs). 57% of solar power generation in the U.S. uses PVs and the rest of it employs CSPs (Mendelsohn et al., 2012). In this study, it is assumed that the national average solar power generation technologies are employed. While the water consumption for PVs is negligible, the solar power generation using CSPs requires similar water consumption as thermoelectric power generation due to the cooling of panels and steam turbines. In terms of wind power generation, its water consumption is insignificant since it does not require cooling.

In the same manner of the displacement approach for CO_2 conversion technologies, it is assumed that various environmental interventions that are associated with thermoelectric power generation and its upstream activities, such as mining, can be avoided since electricity is produced from renewable power generation. Also, data for solar and wind power generation potentials in the MRW are investigated and considered as constraints that represent the maximum amount of renewable electricity generated from the available land area in the MRW (EPA, 2018b).

2.8 Potential land use changes

The MRW has 0.32% of barren land area, as shown in Figure 1. This corresponds to approximately 13×10^6 m² of land area. In this study, three land use change scenarios for the available land are investigated as follows: Reforestation, wind farm installation, and wetland construction. If the available land is reforested, supplies of various ecosystem goods and services are increased.

The supplies of carbon sequestration and air quality regulation services are enhanced since these services are provided from vegetation in a forest (*i-Tree Canopy*, 2018). Nutrient loads in streams are reduced due to the reduction in runoff (*i-Tree Hydro*, 2018). With respect to water provision, there is a debate on whether reforestation helps improve water supply in watersheds. A majority of reports claim that water availability is reduced in a short period of time due to the reforestation and is recovered over a considerable amount time because of the improved soil infiltration capacity and the increased groundwater levels (Filoso et al., 2017).

In the MRW, wetlands occupy only a very small portion of the landscape, as shown in Figure 1. Newly-constructed wetlands can improve the supply of water quality regulation service. Previous reports have discussed how much nutrient runoff could be reduced by constructed wetlands (Kadlec, 2008, 2016). In this study, we calculate the increased amounts of nitrogen and phosphorus removals due to the constructed wetlands in the available land. The barren land can also be used to install new wind farms. Wind farm-generated electricity can displace electricity from thermoelectric power generation, as described in section 2.7.

3. Results and Discussion

3.1 Base case analysis

To investigate the base case condition of activities in the MRW, various environmental interventions from each activity are plotted in Figure 3. Thermoelectric activity shows the most dominant contribution to many environmental interventions. Thermoelectric power plants are responsible for 69.6% of water consumption, 82.4% of GHG emissions, 68.4% of NO_X emissions, and 97.1% of SO_2 emissions in the MRW. These results align well with the U.S. national average (Maupin et al., 2014; EPA, 2018a, 2014), although the contribution from thermoelectric activity in the MRW to those interventions is much larger than the national average. This is because the MRW is an intensive area in terms of thermoelectric activity. The MRW is a region where 17.4% of Ohio's electricity is generated (EIA, 2018a), although the population in the MRW is only 1.1% of the state's population. For some air emissions, such as PM_{10} and CO, transportation activity is the dominant contributor. Also, most of the water nutrient emissions, such as total N and P loads, are attributed to agricultural activity. For electricity consumption, industrial activity is the most dominant activity.

Conventional sustainability assessment approaches only account for environmental interventions, such as emissions and resource uses, which correspond to ecosystem demands. As described in section 2.5, however, all ecosystem demands need to be compared with their corresponding ecosystem supplies when assessing the sustainability of watershed activities. Figure 4 represents base case analysis results in terms of the demands and supplies of electricity and ecosystem goods and services. The TES sustainability metrics are calculated for each ecosystem good and service and are shown in Figure 5.

As shown in Figure 4, net electricity generation in the MRW is calculated to be 9.6 TJ/day. Most ecosystem supplies are smaller than their corresponding ecosystem demands. Only the water demand from activities in the MRW does not exceed the water supply from the ecosystem in the MRW. This indicates that the MRW is not a water-scarce region but has a scarcity of other ecosystem services. As shown in Figure 3, the amount of water withdrawn is significantly large, but it is not considered as the ecosystem demand for water provision since most of the withdrawn water returns to the water body. Rather, the amount of water consumption is considered as the ecosystem demand. In terms of the natural

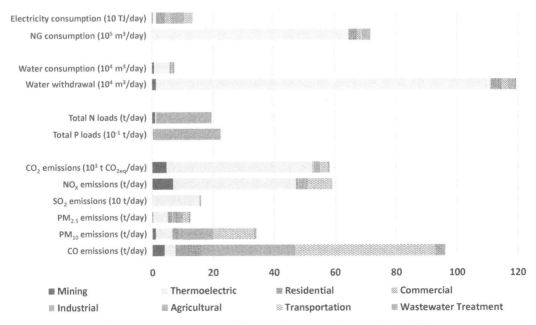

Figure 3. Various environmental interventions from each activity in the MRW.

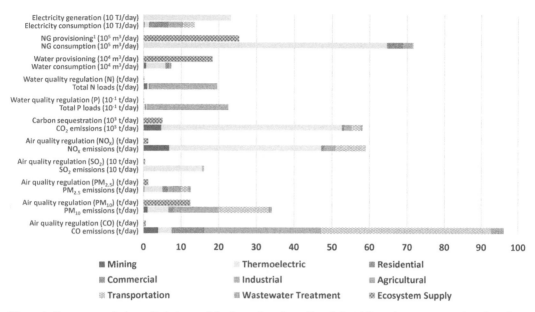

Figure 4. Base case analysis results in terms of the demands and supplies of electricity and ecosystem goods and services. ([1]NG provisioning includes the amount of accumulated NG by allowing for 2 °C global warming since the pre-industrial period.)

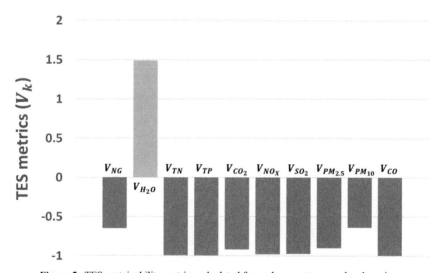

Figure 5. TES sustainability metrics calculated for each ecosystem good and service.

gas supply, approximately 2.5×10^6 m³/day of accumulated NG is considered as the supply of NG by allowing for 2 °C of global warming, as described in section 2.5. The TES metric for NG is then calculated for this adjusted NG supply.

To identify ecological overshoots for activities in the MRW, various TES metrics are calculated for the base case. As shown in Figure 5, most TES metrics are negative, which indicates that environmental interventions overshoot the capacity of ecosystems in the MRW. Most of those interventions are attributed to the thermoelectric activity, as shown in Figure 3. Activities in the MRW are environmentally sustainable only with respect to water consumption since the TES metric for water is positive.

3.2 Technological alternatives

To address the nexus of multiple flows in the analysis, multiple objectives that represent these flows need to be considered. In this work, the following objectives are included: Net electricity generation, TES metrics for greenhouse gases and air pollutants (V_{CO_2}, V_{NO_X}, V_{SO_2}, $V_{PM_{2.5}}$, $V_{PM_{10}}$, and V_{CO}), TES metrics for water nutrient runoffs (V_{TN} and V_{TP}), and TES metrics for ecosystem goods (V_{H_2O} and V_{NG}). Various scenarios about technological alternatives are examined in order to improve the sustainability of activities in the MRW. As a functional unit for the comparison between alternatives, power plants in the MRW are assumed to generate 230 TJ/day of electricity, regardless of which alternatives are adopted. This corresponds to the amount of electricity generated from the five fossil plants in the MRW in 2014. To identify the interactions between objectives for each scenario, all TES metric values (V_k) are converted to the positive values by adding one (+1) to the original TES metric values, then normalized by the maximum value ($V_{k,max}$) in each scenario, as shown below:

$$\overline{V_k} = \frac{V_k + 1}{V_{k,max} + 1} = \frac{S_k/D_k}{S_{k,max}/D_{k,max}} .$$

S_k/D_k may be interpreted as an inverse of scarcity index (Aitken et al., 2016). Overbar notation refers to the normalized value. Similarly, net electricity generation (E), which corresponds to total electricity generation minus total electricity consumption, is also normalized by the maximum value (E_{max}), as shown below:

$$\overline{E} = \frac{E}{E_{max}} .$$

The results are plotted in radar diagrams, as shown in Figure 6. The normalized values ($\overline{E}$ and $\overline{V}_k$) range from 0 to 1. Larger values are preferred for each objective.

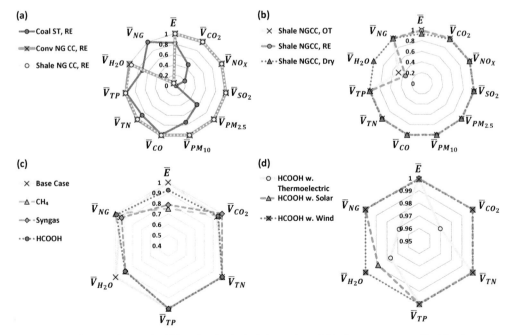

Figure 6. Radar diagrams for sustainability of technological alternatives. All indicators are normalized by the maximum value. Larger values represent better sustainability of corresponding ecosystem service. (a) Fuel and electricity generation technology options. (b) Cooling technology options. (c) CO₂ conversion technology options. (d) Renewable power generation technology options.

3.2.1 Technology options for fuel and power generation

Figure 6(a) shows eleven normalized indicators for the scenario where one type of fuel and electricity generation technology are adopted for all five fossil power plants in the MRW. Three kinds of fuel and generation technology options include coal-fired steam turbine power plants (Coal ST), conventional NG-fired combined cycle power plants (Conv NG CC), and shale NG-fired combined cycle power plants (Shale NG CC). All options are assumed to employ recirculating cooling technology (RE).

For most indicators, except $\bar{V}_{NG}$, the coal-fired steam turbine option is the worst of the fuel and generation technology options. This is not only because burning coal causes more air emissions than burning NG but also because coal-fired steam turbine plants have lower generation efficiency than the NGCC plants. In particular, $\bar{V}_{SO_2}$ indicator can be improved significantly by employing NGCC plants since NGCC plants have negligible SO_2 emissions compared to coal-fired plants. Also, coal-fired steam turbine plants consume more water than NGCC plants per kWh of electricity generated. Moreover, coal mining and coal-fired steam turbine plants require more electricity than NG extraction and NGCC plants. Only the $\bar{V}_{NG}$ indicator shows that the coal-fired steam turbine option is better than two NG options because coal is selected to be burned as fuel instead of NG. In terms of the comparison between conventional NG and shale NG options, there are very minor differences. For instance, $\bar{V}_{H_2O}$ value for the conventional NG option is only 4.5% larger than the shale gas option. This is because power generation technology is the same between these two NG options and the impacts of mining activity are negligible compared to the impacts of the thermoelectric activity, which is the most dominant activity for most interventions. Moreover, the slight decrease in $\bar{V}_{H_2O}$ indicator for the shale option compared to the conventional NG option does not imply a meaningful difference since water resource in the MRW is abundant. Rather, we should take account of the availability of NG reserves because the production rate of shale NG in Ohio is expected to keep increasing (EIA, 2018c).

3.2.2 Cooling technology options

Two types of wet cooling technologies, once-through cooling (OT) and recirculating cooling (RE), and dry cooling technology are compared for shale NG-fired combined cycle power plants, as shown in Figure 6(b). For $\bar{V}_{H_2O}$ indicator, the dry cooling option is better than wet cooling options since it does not require the use of water. However, since dry cooling is 1–3% less efficient than wet cooling methods (Loew et al., 2016), dry cooling requires more electricity and fuel inputs to produce the same amount of electricity. Accordingly, net electricity generation of dry cooling is smaller than that of the wet cooling options. Also, dry cooling technology is much more expensive than wet cooling technologies. It is estimated that the average cost to plant operator for converting the existing power plants in the MRW to the power plants with dry cooling is approximately $340,000/day (Loew et al., 2016).

The once-through cooling option shows a larger $\bar{V}_{H_2O}$ value than the recirculating cooling option. However, once-through cooling technology withdraws a huge amount of water and returns most of it at a warmer temperature, which results in significant thermal water pollution. According to the records for the existing power plants in the MRW, coal-fired steam turbine plant with once-through cooling system causes 19 times larger thermal water pollution than that with recirculating cooling system (EIA, 2018a). Therefore, once-through cooling technology is not advisable.

3.2.3 CO$_2$ conversion technology options

Although $\bar{V}_{CO_2}$ indicator can be improved by employing NG instead of coal, its TES metric value is still negative ($V_{CO_2} = -0.87$). Figure 6(c) exhibits the results for several CO$_2$ conversion technology scenarios to mitigate CO$_2$ emissions in the MRW. In this work, it is assumed that 1,000 t/day of CO$_2$ are converted to CH$_4$, synthetic gas, and formic acid. The results exhibit the increase in $\bar{V}_{CO_2}$ indicator and the reduction in $\bar{E}$ indicator for CO$_2$ conversion options compared to the base case.

Table 2 shows the interventions from CO$_2$ conversion processes and displacement credits from conventional processes. The CO$_2$ conversion processes include CO$_2$ capture, compression, and conversion to products. Total CO$_2$ emissions from the conversion processes are −745 t/day when 1,000 t/day of CO$_2$

Table 2. Interventions from CO_2 conversion processes and displacement credits from conventional processes. [1]CO_2 conversion includes CO_2 capture, compression, and conversion to products. Stoichiometric reactions are assumed for CO_2 conversion processes.

Conversion alternative	Conversion to CH$_4$			Conversion to synthetic gas			Conversion to formic acid		
	Electricity use (TJ/day)	Water use (m³/day)	CO_2 emissions (t/day)	Electricity use (TJ/day)	Water use (m³/day)	CO_2 emissions (t/day)	Electricity use (TJ/day)	Water use (m³/day)	CO_2 emissions (t/day)
Interventions from conversion process[1]	24.10	9,486	−745	22.28	9,486	−745	7.64	9,076	−745
Displacement credits from conventional process	−0.08	−199	−289	−1.73	0	−2,576	−0.54	−1	−13

is captured and converted to products since approximately 255 t/day of CO_2 is emitted from the CO_2 capture process. The conventional coal gasification process to produce syngas has high greenhouse gas emissions. Since these GHG emissions can be avoided as a displacement credit for CO_2 conversion to syngas, the syngas option shows the best $\bar{V}_{CO_2}$ value among the conversion options. On the other hand, in terms of $\bar{E}$ indicator, the formic acid option shows a higher value than the other two conversion options. This is because the total electricity requirement for CO_2 conversion process to formic acid is relatively smaller than that for other conversion processes.

In this work, hydrogen is assumed to be provided from water resource. As shown in Figure 6(c), $\bar{V}_{H_2O}$ values are decreased when CO_2 is captured and converted to the products. This is because a CO_2 capture system increases water consumption for cooling (Magneschi et al., 2017). Even though CO_2 conversion scenarios intensify water consumption, the sustainability of watershed is affected very little because the MRW is not a water-scarce region, as shown in Figure 5. V_{H_2O} values for CO_2 conversion scenarios are approximately 1.20, which is still positive.

As the results show, in choosing the best option between CO_2 conversion alternatives, it is crucial to consider the impacts that are avoided from conventional processes as well as technological advances of CO_2 conversion processes because the displacement credits from the conventional processes can be significant. Also, the results indicate that the increased water consumption from the CO_2 conversion options is not of much concern since the MRW has abundant renewable water to offset total water consumption. This emphasizes the needs for holistic assessment and TES assessment when addressing the sustainability of watershed.

Using CO_2 as a source of carbon is an energy-intensive process. In most cases, therefore, CO_2 conversion technologies are economically expensive. According to the study, the net present value of CO_2 conversion to formic acid is negative at least for 10 years due to the capital investment cost (Agarwal et al., 2011). However, the profitability of CO_2 conversion process to formic acid can be enhanced if the cost of consumable chemicals is reduced and the life time of catalyst is increased. Also, if the carbon price is included, the additional revenues can be earned through the CO_2 conversion process.

The limitation of market capacity for formic acid from CO_2 needs to be considered, especially if the cost for converting CO_2 to formic acid can be cheaper than that for conventional formic acid. Formic acid is generally used as a preservative and antibacterial agent in animal feed. The global production capacity of formic acid in 2009 was roughly 720,000 t/y (Sankaranarayanan and Srinivasan, 2012). 1,000 t/day of CO_2 conversion to formic acid in this study corresponds to more than half of the worldwide formic acid production. However, the demand for formic acid could be increased if its production cost is decreased significantly. Formic acid can also be used to remove impurities on the metal surface if its price is competitive enough to replace HCl and H_2SO_4 (Agarwal et al., 2011). Nonetheless, other CO_2

conversion pathways, such as syngas production, must be utilized as well to maximize the opportunity for converting CO_2 to valuable products.

3.2.4 Renewable power generation technology options

CO_2 conversion options have a considerable electricity requirement, as shown in Figure 6(c). This could make CO_2 conversion technologies infeasible to be implemented if electricity is provided from fossil fuel power plants which have significant environmental impacts. To offset the impacts of generating electricity for CO_2 conversion technologies, solar and wind power generation technology options are considered. Figure 6(d) shows the results for CO_2 conversion to formic acid with different power sources. If 1,000 t/ day of CO_2 is converted to formic acid, 7.64 TJ/day of electricity is required for the conversion process. Given the barren land area in the MRW, energy potentials for solar and wind power generation in the MRW are 158 and 56 TJ/day, respectively. Therefore, there is enough land area in the MRW for solar and wind power generation for CO_2 conversion to formic acid.

If renewable power generation technology is employed for CO_2 conversion to formic acid instead of fossil fuel-based technology, the impacts from thermoelectric power generation can be avoided. Figure 6(d) exhibits the increase in $\bar{V}_{CO_2}$ and $\bar{V}_{H_2O}$ indicators by employing renewable power generation options. Wind power option shows a higher $\bar{V}_{H_2O}$ value than the solar power option since some water resource is still required for solar power generation technology, as described in section 2.7. $\bar{V}_{NG}$ indicator can also be improved since renewable power generation technologies do not require the use of NG to generate electricity. The scale of those changes is very small because only a tiny portion of fossil power plants (7.64 TJ/day out of 230 TJ/day) are replaced by renewable power generation technologies for 1,000 t/day of CO_2 conversion.

For renewable power generation technologies to be economically feasible, they need to be cheaper than conventional fossil power generation. According to the U.S. EIA report, levelized costs of electricity (LCOE) for NGCC, solar (PV), and wind (onshore) power generation technologies, respectively, are estimated to be 42.8, 48.8, and 42.8 $/kW if new generation facilities are introduced in 2023 (EIA, 2019). The report also describes that the LCOE for solar and wind power generation technologies can be cheaper if federal tax credits are included for the renewable technologies: 37.6 $/kW for solar (PV) and 36.6 $/kW for wind (onshore). In this context, appropriate tax incentives can accelerate the use of renewable generation technologies. Economy models, such as general or partial equilibrium models, could address these tax incentives in the analysis (Golosov et al., 2014).

3.3 Agroecological alternatives

Although technological alternatives can improve many sustainability indicators, as described above, they only help reduce the environmental interventions from activities, which correspond to the ecosystem demands. According to the TES framework, watershed sustainability can also be enhanced by increasing the supply of ecosystem services (Bakshi et al., 2015). Also, none of the technological alternatives have significant impacts on water quality indicators, such as $\bar{V}_{TN}$ and $\bar{V}_{TP}$. As shown in Figure 3, most water nutrient runoffs (total N and P loads) are attributed to agricultural activity. In this section, we discuss potential agroecological alternative options that include various tillage practice and available land use change options.

3.3.1 Tillage practice options

Figure 7(a) shows the results of sustainability indicators for the base case and various tillage practice options. Unlike most technological alternatives addressed in section 3.2, tillage practice options affect water quality-related indicators. No-till option is most sustainable with respect to $\bar{V}_{TN}$ and $\bar{V}_{TP}$ indicators. Although no-till option does help improve $\bar{V}_{TN}$ and $\bar{V}_{TP}$ indicators, changes in these indicators are insignificant. In fact, the prime mover for nutrient runoffs is precipitation rather than the implementation of certain agricultural practices. In this static analysis, however, the amount of precipitation is fixed and only the impacts of agricultural practices on nutrient runoffs are examined.

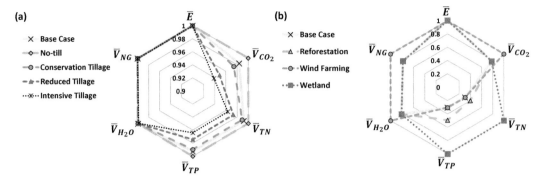

Figure 7. Radar diagrams for sustainability of agroecological alternatives. All indicators are normalized by the maximum value. Larger values represent better sustainability of corresponding ecosystem service. (a) Tillage practice options. (b) Available land use change options.

No-till farming also improves soil carbon sequestration. According to the previous study, the conversion of intensive tillage to no-till for the corn and soybean farming in Ohio can increase carbon sequestration rate by 82.5 $gCm^{-2}y^{-1}$ (West and Post, 2002). If reduced tillage practice is converted to no-till, additional 69.3 $gCm^{-2}y^{-1}$ can be sequestered in soil. As a result, the increase in $\bar{V}_{CO_2}$ is obtained from the no-till option. However, The no-till practice may reduce food production yield, which is not included in this analysis.

3.3.2 Available land use change options

Figure 7(b) represents the results for available land use change alternatives. If the barren land is converted to forest, $\bar{V}_{TN}$ and $\bar{V}_{TP}$ indicators are improved since tree cover helps reduce nutrient runoff through soil infiltration. The impacts of reforestation on other indicators, such as $\bar{V}_{CO_2}$ and $\bar{V}_{H_2O}$, are negligible although reforestation certainly increases the amount of carbon sequestration service. Since GHG emissions are much larger than the supply of carbon sequestration service, and therefore, the reforestation does not improve the TES metric for CO_2 considerably.

Constructed wetlands can be built in the available land as well. The supply of water regulation service is provided from the wetland. As shown in Figure 7(b), the positive impacts of constructed wetland on $\bar{V}_{TN}$ and $\bar{V}_{TP}$ indicators outweigh the reforestation option.

In addition to two ecological alternatives to land use change, wind farming option is considered as a technological land use change alternative. Contrary to ecological land use change options, the installation of wind farms improves $\bar{V}_{CO_2}$, $\bar{V}_{H_2O}$, and $\bar{V}_{NG}$ indicators, instead of water quality-related indicators. The increase in those indicators is attributed to the displacement credits from thermoelectric activity, which is replaced by wind power generation.

3.4 Solutions to improve watershed sustainability

In sections 3.2 and 3.3, various technological and agroecological alternatives are investigated as means to improve the overall sustainability of watershed. While technological alternatives mainly help reduce air emissions and water consumption, which correspond to the ecosystem demands, agroecological alternatives improve water quality indicators by enhancing the supply of water quality regulation service and reducing water nutrient runoffs. As best case scenarios, the following technological alternatives are selected as technological solutions: Shale NG-fired combined cycle power plants with recirculating cooling system and 1,000 t/day of CO_2 conversion to formic acid with wind power generation. Also, an agroecological solution is determined as follows: The implementation of no-till practice and the construction of wetlands on available land.

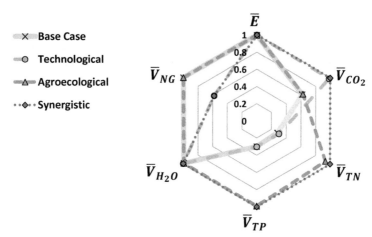

Figure 8. Radar diagram for sustainability of best case scenarios. Technological solutions include shale NG-fired combined cycle power plants with recirculating cooling system and 1,000 t/day of CO_2 conversion to formic acid with wind power generation. Agroecological solutions include the implementation of no-till practice and the construction of wetlands on available land. The synergistic solution combines both technological and agroecological solutions.

Figure 8 shows sustainability indicators for the two solutions described above. As compared to the base case results, $\bar{V}_{CO_2}$ indicator is greatly improved for the technological solution. The reduction in $\bar{V}_{NG}$ indicator for the technological solution is inevitable since NG is used instead of coal for thermoelectric power generation. On the other hand, the agroecological solution enhances $\bar{V}_{TN}$ and $\bar{V}_{TP}$ indicators. Since each solution recommends alternatives to different groups of activities, both solutions can be combined to obtain synergies between the solutions. Synergistic solution includes alternatives that are selected for both technological and agroecolgial solutions addressed above. As shown in Figure 8, the synergistic solution can give a "win-win" situation between objectives.

4. Conclusions

In this work, various alternatives in terms of fossil power generation are investigated in order to identify interactions between multiple flows while addressing the energy-water-CO_2 nexus in the watershed. To examine watershed sustainability, a holistic TES assessment approach is employed since total ecosystem demands from activities in the watershed need to be compared with total ecosystem supplies when investigating if activities in the watershed are sustainable or not. From the results, it is identified that the amount of water supply in the MRW is larger than the amount of water demand for any scenarios discussed in this work. This implies that the reduction in water quantity indicator may not be a huge concern since the TES metric for water supply is still positive. However, TES metrics for other ecosystem goods and services, such as NG, CO_2, and air and water pollutants, show negative values, which indicate unsustainable conditions of activities in the MRW.

Since most of the air emissions and NG consumption are attributed to thermoelectric power generation, various technological alternatives that include different fossil fuels, cooling technologies, CO_2 conversion technologies, and renewable power generation technologies are examined. Overall, it is identified that TES sustainability metrics for carbon sequestration and air quality regulation services can be improved by employing NGCC power plants with recirculating cooling system and CO_2 conversion to formic acid that uses electricity from wind power generation. Coal option is better in terms of NG indicator, however, it has deleterious impacts on other sustainability indicators. Also, dry cooling technology has the potential to improve water quantity indicator. However, its lower generation efficiency and higher cost than wet cooling technologies may pose constraints for its implementation. Moreover, improving water quantity sustainability may not be urgent for a water-affluent region such as the MRW.

With respect to CO_2 conversion technology alternatives, stoichiometric reactions are assumed for potential CO_2 conversion technologies since data for commercialized technologies are not available at this point. It is noted that displacement credits from conventional processes also need to be considered when examining which conversion technology is better than others. It remains as future work to perform a more realistic analysis for CO_2 conversion scenarios using detailed process data and technological constraints.

Agroecological alternatives that include farming practices and land use changes are in a better position to improve water quality sustainability rather than technological alternatives since water nutrient runoffs are mainly attributed to agricultural activity. No-till farming practice and constructed wetlands on barren land in the MRW can improve TES metrics for nitrogen and phosphorus runoffs. In conclusion, watershed sustainability can be improved by considering alternatives for multiple activities. The synergistic solution that includes both technological and agroecological alternatives could produce "win-win" outcomes in terms of multiple objectives.

Some alternatives addressed in this work may be superior in terms of environmental sustainability, but economically expensive. Thus, potential trade-offs between environmental and monetary objectives need to be identified to provide more insights for the sustainable management of the watershed. Specifically, the economic feasibility of renewable power generation and CO_2 conversion alternatives needs to be investigated more thoroughly. For the robust economic analysis, the use of sophisticated economy models is required. One example is to use partial or general equilibrium models to account for market changes due to technology choices (Golosov et al., 2014; Voll et al., 2012). Also, non-market monetary values of ecosystem goods and services need to be estimated and compared with market values of other technological and agroecological alternatives. Moreover, economic options based on nutrient trading schemes can be explored to avoid adverse impacts on water quality in the watershed since abating agricultural nutrient loadings is much cheaper than abatement at the regulated point sources (Ohio EPA, 2013).

The interventions from activities and the availability of ecosystem goods and services vary over space and time. Renewable energy potentials and nutrient runoffs from farming activities also vary with season. Additional energy storage systems may be needed to compensate for the intermittency of renewable energy resources. Therefore, regional and seasonal variations in those flows need to be considered. Moreover, the consequence of future climate change on watershed activities needs to be assessed in order to ensure watershed sustainability. Depending on future climate scenarios, water quantity indicator may not be positive anymore due to the increased risk of drought. Uncertainty analysis also needs to be performed in order to evaluate the robustness of results.

Considering the energy-water-CO_2 nexus in the analysis will avoid shifting of the environmental impacts across multiple flows by identifying interactions between flows. This work could be applied to any watershed to help decision-making of businesses and policymakers and improve the sustainability of watershed.

Acknowledgment

Partial financial support was provided by the Ohio Water Resources Center.

References

Agarwal, A.S., Zhai, Y., Hill, D. and Sridhar, N. 2011. The electrochemical reduction of carbon dioxide to formate/formic acid: Engineering and economic feasibility. Chem-SusChem. 4(9): 1301–1310.

Aitken, D., Rivera, D., Godoy-Faúndez, A. and Holzapfel, E. 2016. Water scarcity and the impact of the mining and agricultural sectors in chile. Sustainability 8(2): 128.

Althaus, H.-J., Chudacoff, M., Hischier, R., Jungbluth, N., Osses, M., Primas, A. et al. 2007. Life cycle inventories of chemicals. Final Report Ecoinvent Data V2.0: 8.

Artz, J., Müller, T.E., Thenert, K., Kleinekorte, J., Meys, R., Sternberg, A., Bardow, A. and Leitner, W. 2017. Sustainable conversion of carbon dioxide: An integrated review of catalysis and life cycle assessment. Chemical Reviews 118(2): 434–504.

Bakshi, B.R., Ziv, G. and Lepech, M.D. 2015. Techno-ecological synergy: A framework for sustainable engineering. Environmental Science & Technology 49(3): 1752–1760.

Bakshi, B.R., Gutowski, T.G. and Sekulic, D.P. 2018. Claiming sustainability: Requirements and challenges. ACS Sustainable Chemistry & Engineering 6(3): 3632–3639.

Behr, A. and Nowakowski, K. 2014. Catalytic hydrogenation of carbon dioxide to formic acid. In Advances in Inorganic Chemistry 66: 223–258. Elsevier.

Boulay, A.-M., Bare, J., Benini, L., Berger, M., Lathuillière, M.J., Manzardo, A., Margni, M., Motoshita, M., Núñez, M. and Pastor, A.V. 2018. The wulca consensus characterization model for water scarcity footprints: Assessing impacts of water consumption based on available water remaining (aware). The International Journal of Life Cycle Assessment 23(2): 368–378.

Clark, C.E., Horner, R.M. and Harto, C.B. 2013. Life cycle water consumption for shale gas and conventional natural gas. Environmental Science & Technology 47(20): 11829–11836.

Dudley, B. et al. 2018. Bp statistical review of world energy 2018. British Petroleum Co., London, UK.

Dunn, J.B., Adom, F., Sather, N., Han, J., Snyder, S., He, C., Gong, J., Yue, D. and You, F. 2015. Life-cycle analysis of bioproducts and their conventional counterparts in greet.

EIA. 2018a. Form EIA-923 detailed data. (Available at: https://www.eia.gov/ electricity/data/eia923/ Accessed Apr 2018.)

EIA. 2018b. Natural Gas Summary. (Available at: https://www.eia.gov/dnav/ng/ NG SUM LSUM DCU SOH A.htm Accessed Apr 2018.)

EIA. 2018c. Shale Gas. (Available at: https://www.eia.gov/dnav/ng/ng enr shalegas a EPG0 R5302 Bcf a.htm Accessed October 2018.)

EIA. 2018d. Total Energy Consumption, Price, and Expenditure Estimates, 2015. (Available at: https://www.eia.gov/state/seds/data.php?incfile=/state/seds/sep fuel/html/fuel te.html&sid=US&sid=OH Accessed Apr 2018.)

EIA. 2019. Levelized cost and levelized avoided cost of new generation resources in the annual energy outlook 2019.

EPA. 2014. 2014 National Emissions Inventory (NEI). (Available at: https://www.epa.gov/air-emissions-inventories Accessed Feb, 2018.)

EPA. 2016. Emissions & Generation Resource Integrated Database (eGRID). (Available at: https://www.epa.gov/energy/emissions-generation-resource-integrated-database-egrid Accessed Feb, 2018.)

EPA. 2018a. Inventory of U.S. greenhouse gas emissions and sinks: 1990–2016.

EPA. 2018b. EnviroAtlas Interactive Map. (Available at: https://www.epa.gov/ enviroatlas Accessed Jan 2018.)

EPA. 2018c. Where Greenhouse Gases Come From. (Available at: https://www.eia.gov/energyexplained/index. php?page=environment where ghg come from Accessed Apr 2018.)

Filoso, S., Bezerra, M.O., Weiss, K.C. and Palmer, M.A. 2017. Impacts of forest restoration on water yield: A systematic review. PloS One 12(8): e0183210.

Frauzem, R., Kongpanna, P., Roh, K., Lee, J.H., Pavarajarn, V., Assabumrungrat, S. and Gani, R. 2015. Sustainable process design: Sustainable process networks for carbon dioxide conversion. In Computer Aided Chemical Engineering 36: 175–195. Elsevier.

Gearino, D. 2013. AEP to close coal-fired power plant instead of converting. (Available at: http://www.dispatch.com/content/stories/business/2013/07/11/aep-to-close-coal-fired-power-plant-near-beverly.html Accessed May 2018.)

Golosov, M., Hassler, J., Krusell, P. and Tsyvinski, A. 2014. Optimal taxes on fossil fuel in general equilibrium. Econometrica 82(1): 41–88.

Gopalakrishnan, V., Bakshi, B.R. and Ziv, G. 2016. Assessing the capacity of local ecosystems to meet industrial demand for ecosystem services. AIChE Journal 62(9): 3319–3333.

Jenner, S. and Lamadrid, A.J. 2013. Shale gas vs. coal: Policy implications from environmental impact comparisons of shale gas, conventional gas, and coal on air, water, and land in the united states. Energy Policy 53: 442–453.

Jessop, P.G., Hsiao, Y., Ikariya, T. and Noyori, R. 1996. Homogeneous catalysis in supercritical fluids: Hydrogenation of supercritical carbon dioxide to formic acid, alkyl formates, and formamides. Journal of the American Chemical Society 118(2): 344–355.

Jessop, P.G., Joó, F. and Tai, C.-C. 2004. Recent advances in the homogeneous hydrogenation of carbon dioxide. Coordination Chemistry Reviews 248(21-24): 2425–2442.

Kadlec, R. 2008. The effects of wetland vegetation and morphology on nitrogen processing. Ecological Engineering 33(2): 126–141.

Kadlec, R. 2016. Large constructed wetlands for phosphorus control: A review. Water 8(6): 243.

Khanal, S., Lal, R., Kharel, G. and Fulton, J. 2018. Identification and classification of critical soil and water conservation areas in the muskingum river basin in ohio. Journal of Soil and Water Conservation 73(2): 213–226.

Lampert, D.J., Cai, H., Wang, Z., Keisman, J., Wu, M., Han, J., Dunn. J., Sullivan, J.L., Elgowainy, A. and Wang, M. 2015. Development of a life cycle inventory of water consumption associated with the production of transportation fuels.

Lau, C.P. and Chen, Y.Z. 1995. Hydrogenation of carbon dioxide to formic acid using a 6,6'-dichloro-2,2'-bipyridine complex of ruthenium, cis-[Ru(6,6'-Cl₂bpy)₂(H₂O)₂](CF₃SO₃)₂. Journal of Molecular Catalysis A: Chemical 101(1): 33–36.

Le Quéré, C., Andrew, R.M., Friedlingstein, P., Sitch, S., Pongratz, J., Manning, A.C., Korsbakken, J.I., Peters, G.P., Canadell, J.G. and Jackson, R.B. 2017. Global carbon budget 2017. Earth System Science Data Discussions 1–79.

Loew, A., Jaramillo, P. and Zhai, H. 2016. Marginal costs of water savings from cooling system retrofits: A case study for texas power plants. Environmental Research Letters 11(10): 104004.

Magneschi, G., Zhang, T. and Munson, R. 2017. The impact of CO_2 capture on water requirements of power plants. Energy Procedia 114: 6337–6347.

Maupin, M.A., Kenny, J.F., Hutson, S.S., Lovelace, J.K., Barber, N.L. and Linsey, K.S. 2014. Estimated use of water in the United States in 2010.

Mendelsohn, M., Lowder, T. and Canavan, B. 2012. Utility-scale concentrating solar power and photovoltaic projects: A technology and market overview.

Nemecek, T., Kägi, T. and Blaser, S. 2007. Life cycle inventories of agricultural production systems. Final Report Ecoinvent v2.0: 15.

NREL. 2018. U.S. Life Cycle Inventory Database. (Available at: https://www.nrel.gov/lci/ Accessed Feb 2018.)

Sanders, K.T. 2014. Critical review: Uncharted waters? The future of the electricity-water nexus. Environmental Science & Technology 49(1): 51–66.

Sankaranarayanan, S. and Srinivasan, K. 2012. Carbon dioxide—a potential raw material for the production of fuel, fuel additives and bio-derived chemicals.

Seaber, P.R., Kapinos, F.P. and Knapp, G.L. 1987. Hydrologic unit maps.

Skone, T.J. 2016. Life cycle analysis of coal exports from the powder river basin.

Skone, T.J., Littlefield, J., Marriott, J., Cooney, G., Jamieson, M., Jones, C., Demetrion, L., Mutchek, M., Shih, C. and Curtright, A.E. 2016. Life cycle analysis of natural gas extraction and power generation.

Solley, W.B., Pierce, R.R. and Perlman, H.A. 1998. Estimated use of water in the united states in 1995. U.S. Geological Survey (USGS).

Sutter, J. 2007. Life cycle inventories of petrochemical solvents. Final Report Ecoinvent Data v2.0 (22).

Tawney, R., Khan, Z. and Zachary, J. 2005. Economic and performance evaluation of heat sink options in combined cycle applications. Journal of Engineering for Gas Turbines and Power 127(2): 397–403.

i-Tree Canopy. 2018. (Available at: https://canopy.itreetools.org/ Accessed Sep 2018.)

i-Tree Hydro. 2018. (Available at: https://www.itreetools.org/hydro/ Accessed Sep 2018.)

i-Tree Landscape. 2018. (Available at: https://landscape.itreetools.org/ Accessed Feb 2018.)

Ohio EPA. 2013. Ohio Nutrient Reduction Strategy.

Ohio EPA. 2016. Nutrient Mass Balance Study for Ohio's Major Rivers.

Ohio EPA. 2018. Individual Wastewater Discharge Permit Information. (Available at: https://www.epa.ohio.gov/dsw/permits/npdes info Accessed Apr 2018.)

Vengosh, A., Warner, N., Jackson, R. and Darrah, T. 2013. The effects of shale gas exploration and hydraulic fracturing on the quality of water resources in the United States. Procedia Earth and Planetary Science 7: 863–866.

Vidic, R.D., Brantley, S.L., Vandenbossche, J.M., Yoxtheimer, D. and Abad, J.D. 2013. Impact of shale gas development on regional water quality. Science 340(6134): 1235009.

Voll, A., Sorda, G., Optehostert, F., Madlener, R. and Marquardt, W. 2012. Integration of market dynamics into the design of biofuel processes. Computer Aided Chemical Engineering 31: 850–854.

Wang, M. 2016. The Greenhouse gases, Regulated Emissions, and Energy use in Transportation (GREET) Model.

West, T.O. and Post, W.M. 2002. Soil organic carbon sequestration rates by tillage and crop rotation. Soil Science Society of America Journal 66(6): 1930–1946.

Wolf, A., Jess, A. and Kern, C. 2016. Syngas production via reverse water-gas shift reaction over a ni-AL_2O_3 catalyst: Catalyst stability, reaction kinetics, and modeling. Chemical Engineering & Technology 39(6): 1040–1048.

WRI. 2014. The Carbon Budget. (Available at: http://www.wri.org/ipcc-infographics Accessed Apr 2018.)

CHAPTER 11

Natural Gas Reforming to Industrial Gas and Chemicals Using Chemical Looping

Dawei Wang, Yitao Zhang, Fanhe Kong, L-S Fan and *Andrew Tong**

1. Introduction

Energy production via combustion of carbon-based fossil fuels is the major source for greenhouse gas emissions (International Energy Outlook, 2011). Compared to coal, natural gas emits 50 to 60% less carbon dioxide (CO_2) and, thus, has a much lower life cycle greenhouse gas emissions (NETL, 2010). The vast availability of natural gas resources in the United States have become economically accessible due to advances in production drilling technology. The U.S. Energy Information Administration has projected that the global natural gas consumption will rise by 50% between 2010 and 2035 (International Energy Outlook, 2011).

With the growing abundance of natural gas, research has accelerated to develop efficient methods to utilize natural gas to build long carbon chain chemical and liquid fuel products as opposed to the conventional approach of refining and cracking heavy hydrocarbons, such as crude oil. While technologies for directly converting methane (CH_4), the main component in natural gas, to chemicals are still under development, industrial chemical production processes rely on a two-step method, converting CH_4 to syngas, a mixture of mainly CO and H_2 with various ratios, and utilizing syngas for chemical synthesis and production. Producing syngas from CH_4 at an industrial scale is a mature technology, mainly through steam-methane reforming (SMR) and autothermal reforming (ATR). Conventional SMR plants operate at pressures between 14 and 40 atm with outlet temperatures in the range of 815 to 925 °C. Though commercially demonstrated, SMR is an energy intensive process due to the endothermic nature of its reaction and, thus, external heat sources are required for continuously stable operation, resulting in substantial carbon emissions. For conventional ATR, the energy intensive air separation unit greatly reduces the overall plant efficiency, increasing carbon emissions in the plant. For both processes, necessary down-stream processing steps are also required in order to adjust the quality of syngas, such as the H_2:CO molar ratio and syngas purity, in order to meet the requirements of various downstream processes. The capital cost of SMR plants is also prohibitive for small to medium size applications. Recently, catalytic partial oxidation of methane (CPOM), where CH_4 is reacted with O_2 only to produce syngas over a noble metal catalyst in a simple, one-step reaction scheme, has been treated as a promising alternative and received much attention (Hickman and Schmidt, 1993; Neumann et al., 2004; Ashcroft et al., 1990; Tsang et al., 1995). Owing to its operating temperature of over 1000 °C, the CPOM process has a high reaction rate, resulting in an extremely short residence time for the reactants (Hickman and Schmidt, 1993). CPOM offers several advantages over SMR, such as the exothermic nature of the reactions which allows for autothermal

William G. Lowrie Department of Chemical and Biomolecular Engineering, The Ohio State University, USA 43210.
* Corresponding author: tong.48@osu.edu

operation and, thus, lowers the carbon footprint of syngas production, and the compact reactor design with very high space-time yields owing to its very high reaction rate. However, several issues have limited its commercialization, including the safety concerns related to direct contact of fuel and oxygen at a temperature close to their upper flammability limit, the requirement of an air separation unit (ASU) to produce purified oxygen, and the high cost of producing catalysts from expensive noble metal.

The chemical looping concept provides possible solutions to directly address the issues for the CPOM processes. The spatial separation of the fuel conversion into two or more separate steps avoids any direct contact of feedstock and gaseous oxygen and, thus, alleviates safety concerns. In addition, using metal oxide as oxygen carrier eliminates the need for a cryogenic ASU, substantially reducing the plant's parasitic power demand.

Chemical looping was first practiced in the late nineteenth century when Franz Bergmann filed a German patent for producing calcium carbide (CaC_2) via reaction between manganese (IV) oxide in the presence of a hydrocarbon fuel and calcium oxide (CaO) (Bergmann, 1897). In 1903, the Lane hydrogen producer, or Lane process, was invented for producing hydrogen in a fixed bed reactor system where syngas was used to reduce iron oxide ores and steam was introduced to produce hydrogen from the reduced iron ore from the steam-iron reaction (Hurst, 1939; Gasior, 1961; Teed, 1919). The incomplete conversion of syngas and low recyclability of the iron ores prevented the Lane process from being economically competitive against the increased availability of oil and natural gas in the 1940s. The Lane process was eventually phased out at this time. In the 1950s, Warren Lewis and Edwin Gilliland patented a chemical looping process using iron and copper oxides as the metal oxide to produce CO_2 as the desired product in a countercurrent gas-solids contacting pattern for use in the beverage industry (Lewis and Edwin, 1954). In the 1960s and 1970s, the Consolidation Coal Company (now CONSOL Energy) developed a pilot-scaled CO_2 acceptor process to produce substitute natural gas (SNG) (Dobbyn et al., 1978). $CaS–CaSO_4$ redox cycle observed in the CO_2 acceptor process was different from the typical redox cycles based on metal/metal oxide. In the 1970s, the Institute of Gas Technology (IGT) developed the HYGAS process to produce SNG from coal via the methanation reaction. The required hydrogen for the methanation reaction came from the steam–iron reaction, where iron (Fe) was obtained from iron oxide reduction by syngas produced from coal gasification (U.S. Department of Energy, 1979). Two-stage countercurrent fluidized bed reactor design was used for both the reducer and the oxidizer in order to enhance the fuel gas and iron oxide conversions, as well as heat and mass transfer. The HYGAS process was demonstrated at pilot-scale but not commercialized. In the 1980s, Atlantic Richfield Company (ARCO) developed a gas to gasoline (GTG) process using a reactor configuration consisting of a two circulating fluidized beds reactor configuration (Jones et al., 1987; Sofranko et al., 1987). The process was designed to convert CH_4 to an ethylene-rich intermediate via oxidative coupling of methane for gasoline production using Mobil's olefin to gasoline and distillate (MOGD) process. Catalytic metal oxide for the oxidative coupling of methane (OCM) reaction was reduced in one reactor while reacting with CH_4, and were transported to the other reactor for regeneration. A pilot-scale demonstration of the process was operated. However, the process was discontinued towards the end of 1980s as the crude oil price decreased below the natural gas price. In the 1990s, DuPont developed a process, referred to as the DuPont process, to produce maleic anhydride through the selective oxidation of butane. A vanadium phosphorus oxide (VPO) catalytic metal oxide was used. The selective oxidation for butane conversion to maleic anhydride took place in a lean phase riser where a multifunctional VPO metal oxide provided catalytic activity as well as lattice oxygen. Meanwhile, the regeneration reaction of reduced VPO catalytic metal oxide by air took place in a fluidized bed reactor. This process was scaled to commercial demonstration but failed as the VPO catalytic metal oxide was not reactive enough to provide the oxygen and particle integrity was compromised (Dudukovic, 2009; Evanko et al., 2013). A fixed bed reactor system for partial oxidation of CH_4 to syngas using CeO_2 was also attempted in the 1990s. Using CeO_2 in a chemical looping process to convert CH_4 to syngas is thermodynamically favorable, but experimental results showed extensive carbon deposition under high temperature conditions, which negatively affected the CeO_2 reactivity, and reduced the reaction kinetics between CH_4 and CeO_2, resulting in a low CH_4 conversion (4% per pass) (Otsuka et al., 1998a). In the 2000s, a chemical looping process to provide the heat necessary to operate conventional steam methane reforming (SMR) reactions using metal oxide oxygen carriers while mitigating carbon emissions from the SMR furnace was developed (Rydén and Lyngfelt, 2006; Adanez et al., 2012; Pans et al., 2013). The SMR tubular reactor is placed inside either the reducer or combustor reactor to serve as the SMR furnace so as to allow the chemical looping reactors to provide the necessary heat for the endothermic SMR reaction. In this approach, the CO_2 produced from the furnace is captured without the need of a post combustion CO_2 capture unit to mitigate carbon emissions. This process requires the diversion of natural gas in an equal or greater quantity than the conventional approach to fulfill the endothermic heat requirements for the SMR reactions.

Chemical looping processes using a metal-based oxygen carrier to perform redox reaction with a carbon-based fuel can be categorized into 2 types of systems: Chemical looping combustion (CLC) for power generation with CO_2 capture and chemical looping reforming (CLR) for chemical and industrial gas production. The Lewis and Gilliland process represents a CLC system where the carbon fuel is fully oxidized to CO_2. CLR systems can be further divided based on the product from the CLR reactor. The Lane Producer and HYGAS processes represent CLR processes for H_2 production. The DuPont VPO and ARCO processes each represent CLR for selective oxidation systems where the metal oxide oxygen carrier serves to selectively oxidize the reactants to a desired product. CLR for syngas production systems represent chemical looping processes where a carbon-based gaseous fuel, such as natural gas, is partially oxidized to syngas (Fan et al., 2015; Luo et al., 2014; Ryden et al., 2008; Nalbandian et al., 2011; Dai et al., 2006). CLR for selective oxidation systems rely on a multifunctional metal oxide, which possesses catalytic and oxygen transfer properties, to selectively convert hydrocarbon feedstock to chemicals. In the reducer reactor, a catalytic metal oxide reacts with a hydrocarbon feedstock to selectively produce chemicals and reduced catalytic metal oxide. In the combustor reactor, the reduced catalytic metal oxide is regenerated by oxidation with air (Keller and Bhasin, 1982; Contractor, 1999). By directly producing chemicals in the reducer, the generation of syngas as an intermediate is not necessary. A simplified flow diagram of CLR is shown in Figure 1.

Metal oxide reaction engineering and particle science and technology are two key technical areas for the development of chemical looping concepts for combustion, gasification, or reforming applications. Understanding these metal oxide issues allows the metal oxide materials to be formulated effectively, synthesized, and used in a sustainable manner for desired chemical looping reaction applications.

A variety of metal oxides have been investigated for chemical looping applications (Fan, 2010; Messerschmitt et al., 1915; Lane, 1913; Ishida et al., 1987). The preliminary screening of the oxygen carrier is based on its thermodynamic properties as illustrated using the modified Ellingham diagram, as shown in Figure 2(left). The modified Ellingham diagram is based on the Gibbs free energy of reactions and its variation with temperature for metal oxides. The diagram can be divided into four sections, namely, combustion section (A), syngas production section (B), carbon deposition section (D) and inert section (C), based on the four fundamental reactions for carbon and hydrogen conversions, as is shown in Figure 2(right). Between reaction lines 1 and 2, a very small section (E) exists, which also produces syngas, however, the syngas will contain a significant amount of H_2O byproduct. Using Figure 2, one can identify metal oxides ideal for CLC systems, i.e., metals that lie in region A, and metal oxides ideal for

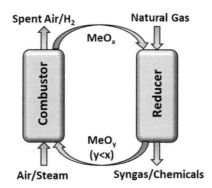

Figure 1. Process flow diagram for chemical looping reforming.

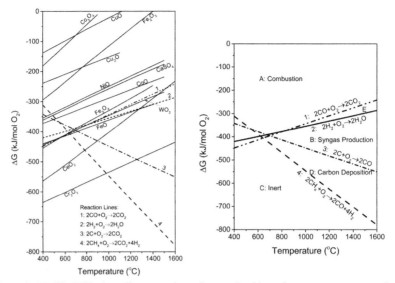

Figure 2. Modified Ellingham diagram to determine metal oxide performance as oxygen carriers.

CLR for syngas production systems, i.e., metals that lie in region B. Based on these sections, the metal oxides are identified as potential oxygen carriers for different chemical looping processes.

Metal oxides used in chemical looping combustion applications (i.e., full conversion of carbon fuels) include NiO, CoO, CuO, Fe_2O_3. Metal oxides lying in region B for CLR for syngas production, such as CeO_2, have mild oxidation properties to prevent over-conversion of the carbon-based fuel to CO_2 and H_2O. Metal oxides in carbon deposition and inert sections, such as Cr_2O_3 and SiO_2, lack the potential to be used as oxygen carriers and are generally considered to be inert and unfavorable. However, they are good candidates for support materials for active oxygen carrier materials, to enhance some physical properties. For example, when TiO_2, which lies in the inert region C in Figure 2, is combined with FeO, it will form a $FeTiO_3$ complex, which will generate a higher quality of syngas than FeO alone (Li et al., 2011). Thus, the addition of a support material in this case improves the oxygen carrier performance. On the other hand, when considering CLC systems, the addition of supports may hinder the oxidation properties of certain metals alone. Examples are CuO-based oxygen carriers for CLC which lose oxidizing potential when Al_2O_3 is used as a support material due to the formation of $CuAl_2O_4$ (Arjmand et al., 2011). It becomes essential to understand how different phases behave in the presence of each other during the development of high performing oxygen carriers. CH_4 is thermodynamically unstable at temperatures higher than 750 °C and spontaneously decomposes to form C and H_2 in the absence of an oxygen source.

It should be noted that the modified Ellingham diagram represents only a thermodynamic analysis of metal oxides to be used as potential oxygen carrier materials. In addition to thermodynamics, many other factors come into play when selecting oxygen carrier materials. With the base metal oxides identified, their physical and chemical properties can then be further characterized. For commercial applicability, the oxygen carrier should possess several properties, including redox reactivity, long-term stability, physical strength, toxicity, and appropriate production cost (Luo et al., 2015). Many oxygen carrier materials have been studied for CLR syngas production applications (Fan, 2015; Li et al., 2009; Luo et al., 2015). During the early development of these applications, single metal oxides or sulfates were considered to be the active components in oxygen carrier materials. Recent research has been directed towards the use of binary and ternary metal composite materials for improved process performance (Luo et al., 2015).

2. Chemical Looping Reforming for Syngas Production

Syngas, as an important feedstock for the production of many valuable industrial chemicals such as methanol, synthetic liquid fuels, ammonia, and hydrogen, significantly affects the operation and the economics of downstream chemical processes. With the development of the technology, the syngas production process has evolved from simply passing steam through hot coke to using large-scale solids circulating systems which process a variety of hydrocarbon fuels.

CLR for reforming natural gas to syngas represents a potentially efficient means of producing electricity, hydrogen, syngas, and/or liquid fuels with minimal carbon emissions compared to traditional natural gas reforming schemes (Fan et al., 2015). There is a distinct difference in operating strategies between CLR and CLC processes. CLC processes can achieve complete carbon fuel oxidation to CO_2 and H_2O with a low reactant gas to oxygen carrier flow ratio. For CLR processes, partial fuel oxidation requires careful control of the reactant gas to oxygen carrier flow ratio in order to maximize natural gas conversion to syngas.

Multiple CLR processes for syngas production from natural gas have been tested with multiple types of oxygen carrier materials and reactor configurations. The present section summarizes several of the CLR processes that have been and are under development to date and is organized based on the type of gas-solid flow regime used for the reducer reactor design.

2.1 Fluidized bed chemical looping reforming

CLR processes were typically designed as fluidized bed systems, similar to those of CLC processes where fluidized beds of various forms, i.e., bubbling, turbulent, or fast fluidized beds, are used for the reducer or combustor. The system is configured as a circulating fluidized bed (CFB) system, where metal oxide oxygen carriers continuously circulate among the different reactors of the system while undergoing reduction and oxidation reactions in the reducer and the combustor, respectively. Several major fluidized bed CLR process configurations are introduced below.

2.1.1 Welty process

The first chemical looping reforming approach for syngas generation in fluidized bed reactors from methane was the Welty process, disclosed in a patent by Welty et al. in 1951 (Welty, 1951). The Welty process consisted of two separate fluidized bed reactors, as shown in Figure 3.

In the process, methane was partially oxidized by the oxygen provided by oxygen carriers, such as Fe_2O_3 or CuO, to produce a mixture of H_2 and CO in the gas converter reactor (reducer). The reduced

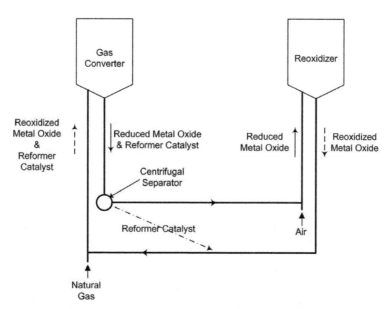

Figure 3. Welty process for syngas generation from methane (Welty, 1951).

oxygen carrier was re-oxidized with air in the reoxidizer (combustor). In addition to the oxygen carriers, nickel-based reforming catalysts with Al_2O_3 or MgO support were also used to selectively convert methane into syngas through steam methane reforming and dry reforming reactions. The addition of a nickel-based reforming catalyst was to inhibit the full oxidation of methane to CO_2 and H_2O in the reducer. In the initial design, the catalyst was transported to the combustor together with the reduced oxygen carrier where it was regenerated with air. However, this regeneration step also oxidized the reforming catalyst from Ni to NiO, causing the loss of the catalytic activity of the reforming catalyst. A solids–solids centrifugal separator, which is based on the density difference between the oxygen carrier and the nickel-based catalyst, was added to the modified version of the process in order to overcome the drawback. The low-density nickel-based reforming catalyst exited the top of the separator, which was directed back into the reducer, while the high-density oxygen carrier exited at the bottom of the separator and was directed to the combustor for regeneration.

The system had operational issues due to non-optimal heat integration as there were difficulties in maintaining an operating temperature difference between the reducer at 816–927 °C and the combustor at 871–982 °C. In addition to heat integration difficulties, the Welty process was challenging to operate, as careful control of the fluidized gas was required in order to maintain both proper oxygen carrier circulation and separation of the reforming catalyst to prevent its transfer to the combustor. The requirements of both these conditions on the fluidizing gas flow rate substantially reduced the window of operability limits for the Welty process. Between the work of Welty et al. and the early 2000s, little progress was made in further developing the CLR process, until a similar concept was applied by Mattisson et al. for methane reforming, using fluidized beds (Mattisson and Lyngfelt, 2001).

2.1.2 Steam methane reforming-chemical looping combustion (SMR-CLC) systems

The conventional SMR process, as briefly described in the introduction to this chapter, can be coupled with CLC for its energy integration, as the external heat generated by the CLC process can be used to compensate for the heat requirement of the endothermic methane reforming reaction while inherently capturing the CO_2 generated for the combustion of the natural gas. The tubular, packed bed SMR reactors can be embedded in either the reducer or the combustor of the chemical looping combustor system to maintain its operational temperature for continuous operation. This kind of process is referred to as SMR–CLC process. The two different configurations with SMR embedded in the reducer and the combustor, respectively, are shown in Figure 4.

When hydrogen is the desired product, water-gas shift (WGS) reactor and PSA unit are integrated into the SMR-CLC system. The WGS reactor is located after the SMR reactor to maximize the hydrogen yield, with the PSA unit being used to separate the hydrogen from the gas product to obtain purified hydrogen. The off-gas stream from the PSA unit, consisting of CO_2 produced in the SMR and WGS reactors, unconverted CO and CH_4, and a small amount of H_2, is sent back to the reducer of the CLC system together with necessary natural gas to reduce the oxygen carrier particles and ensure continuous operation of the CLC system. Since the CLC unit generates pure CO_2 stream from the reducer outlet, the integrated process provides a unique advantage of near 100% CO_2 capture without any further processing steps.

In the SMR-CLC systems, the SMR reactor can be embedded either in the reducer (Figure 4a) or in the combustor (Figure 4b). The two types of configurations have significantly different process schemes influencing the energy balance and performance of the SMR–CLC process. In the CLC reactor systems, the reaction taking place in the reducer can be either endothermic or slightly exothermic, depending on the fuel and oxygen carrier used, while the reaction in the combustor is strictly exothermic. As a result, more heat in the combustor can be used for the SMR process, and the temperature in the combustor is typically higher than that in the reducer for CLC systems (Fan, 2010; Lyngfelt, 2015).

As in the type a SMR–CLC process, the heat in the CLC reducer needs to support the endothermic SMR reaction and maintain the normally endothermic or slightly exothermic reaction between the metal oxides and fuel, the temperature of the CLC reducer is lower than that of stand-alone CLC systems under equivalent CLC operating conditions. The increase in heat requirement of the CLC reducer in

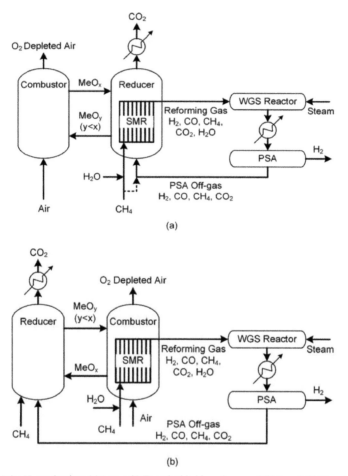

Figure 4. SMR–CLC for H_2 production: (a) type a SMR embedded in reducer and (b) type b SMR embedded in combustor (Rydén and Lyngfelt, 2006; Adanez et al., 2012).

the SMR-CLC process can be managed by increasing the recirculating temperature of oxygen carriers from the combustor to the reducer or increasing the oxygen carrier circulation rate of the CLC system. However, increasing the recirculating temperature of the oxygen carrier particles generally means increasing the operating temperature of the combustor, which puts forward harsher requirements on the physical properties of the oxygen carriers. When the oxygen carrier circulation rate of the CLC system is increased, the reactor volumes need to be increased, and the attrition rate of the oxygen carrier particles will also increase.

In the type b SMR–CLC process, extensive heat is released in the combustor during the regeneration process of the reduced oxygen carrier reaction, which can provide enough heat for SMR reaction. Thus, the reducer can be configured the same as that of stand-alone CLC systems. There is no need to increase the operating temperature or the oxygen carrier circulation rate of the CLC system.

Pans et al. (Pans et al., 2013) studied the detailed energy balance of both type a and type b SMR–CLC processes with two different iron-based oxygen carriers, namely pure iron oxide (Fe_2O_3) and alumina supported iron oxide (Fe_2O_3–Al_2O_3). The effects of swinging between different oxidation states, Fe_2O_3/Fe_3O_4 scheme for pure iron oxide and Fe_2O_3/FeO (Fe_2O_3-Al_2O_3/FeAl$_2O_4$) for alumina supported iron oxide, were analyzed for the H_2 yield from the SMR–CLC process. The different swinging schemes of iron oxide oxidation states alter the overall energy balance of the system as the reduction reaction from Fe_2O_3 to Fe_3O_4 by natural gas is endothermic; while it is exothermic from Fe_2O_3-Al_2O_3 to FeAl$_2O_4$. The performance of type a and type b SMR–CLC processes under an autothermal operating condition was compared. The operating temperature of the reactor where the SMR reactor was embedded was set at 900 °C. The type b SMR–CLC system produced a higher hydrogen yield than type a for both oxygen carriers. However, as shown in Figure 5, the difference between the two schemes decreases with the increase of the oxygen carrier circulation rate of the system. For the Fe_2O_3/Fe_3O_4 scheme, the temperature in the combustor needs to be higher than the reducer for both types of SMR–CLC systems. However, when Fe_2O_3-$Al_2O_3/FeAl_2O_4$ scheme is used, only in the type a SMR-CLC does the combustor temperature need to be higher than the reducer temperature; while in the type b SMR-CLC system, the reducer temperature is higher than the combustor temperature due to the exothermic characteristics of the reaction in the reducer. This is beneficial for the reactor design as the kinetics of the reaction between the fuel and the oxygen carriers can be enhanced under a higher operating temperature.

Fan et al. conducted an exergy analysis to investigate the benefits of SMR-CLC as compared to conventional SMR process (Fan et al., 2016). The conventional SMR and SMR-CLC processes used in the study were shown in Figure 6. The SMR-CLC system was a type b system where the SMR reactor was embedded in the combustor. Nickel-based oxygen carrier particles were used in the CLC system. The temperature of the combustor was set at 1000 °C.

The exergy distributions in the SMR and the SMR-CLC processes were listed in Table 1, generated using Aspen Plus®. The process simulation results showed that the overall exergy efficiency of the SMR-CLC system had an efficiency of 71.4%, an approximately 9.5% advantage over SMR, whose efficiency was 65.2%. The overall exergy destruction of methane in the SMR-CLC process was 217.87 kJ per mole of CH_4, while it was 299.66 kJ per mole of CH_4 for SMR process. The main exergy destruction

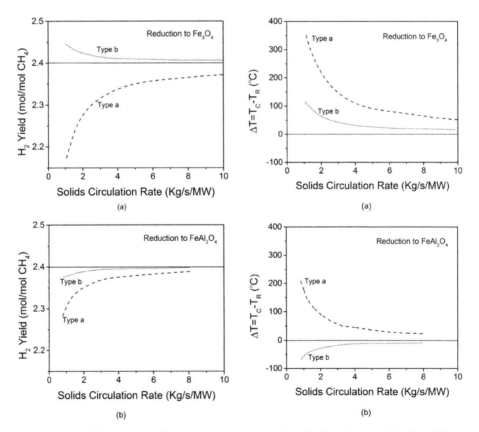

Figure 5. Performance of different SMR–CLC systems (type a and type b) with (a) pure iron oxide, Fe_2O_3 and (b) an alumina supported iron oxide, Fe_2O_3–Al_2O_3 oxygen carriers, where T_R and T_C are the temperatures for the reducer and the combustor, respectively.

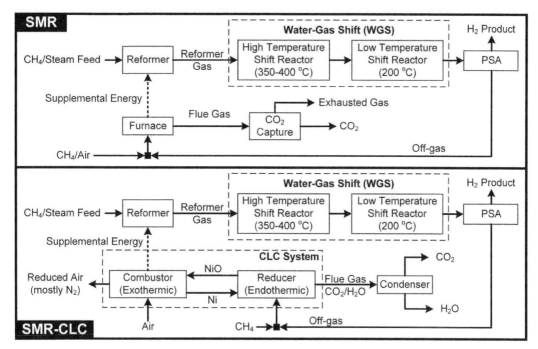

Figure 6. Schematic diagram of conventional SMR and SMR–CLC processes.

Table 1. Exergy of each component of SMR and SMR-CLC processes.

	SMR		SMR-CLC	
	Exergy (kJ/mol CH$_4$)	% of total Ex_{in}	Exergy (kJ/mol CH$_4$)	% of total Ex_{in}
Exergy in	940.55	100.0	866.17	100.0
Methane	830.19	88.27	830.19	95.85
$W_{compressors}$	30.86	3.28	30.86	3.56
W_{pump}	0.15	0.01	0.15	0.02
$W_{membrane}$	4.97	0.53	4.97	0.57
CO$_2$ capture	74.38	7.91	–	–
Exergy out	640.89	68.14	648.30	74.85
Hydrogen	613.59	65.24	618.70	71.43
Exhausted gas	27.30	2.90	29.60	3.42
Exergy destroyed	299.66	31.86	217.87	25.15
Exergy un-used	326.96	34.76	247.47	28.57
Exergy efficiency	65.24		71.43	

in the SMR process was the combustor and the CO$_2$ capture unit, which contributed 28.5% and 24.9% of the total destroyed exergy, respectively. While in the SMR-CLC process, the exergy destruction for combustion was reduced from 85.3 kJ per mole of CH$_4$ in SMR process to 79.1 kJ per mole of CH$_4$.

The economic feasibility of the SMR-CLC process was also examined by means of financial analysis. The capital cost estimation of both SMR and SMR-CLC processes were listed in Table 2. When compared to the SMR process, the SMR-CLC process can save about 0.02 million euro (M€) total investment costs, mainly due to the reduced total equipment costs which decreased from 15.22 M€ for the conventional SMR process to 15.05 M€ for the SMR-CLC process. This reduction revealed that the SMR-CLC process

Table 2. Capital cost estimation of SMR and SMR-CLC processes (Fan et al., 2016).

	Unit	SMR	SMR-CLC
Capital cost			
Total equipment costs	M€	15.22	15.05
Total install costs (excluding contingency)	M€	29.43	29.1
Total investment costs	M€	36.79	36.37
Fixed O&M costs	M€/year	4.69	4.5
Variable O&M cost			
Fuel cost	M€/year	8.79	8.79
Non-fuel (including water, power consumption and solvent loss)	M€/year	1.84	1.82
Total fixed and variable O&M costs	M€/year	15.32	15.11
Net Hydrogen Production	kg/year	5472000	5472000
Levelized cost of hydrogen	€/kg	3.28	3.24

is economically feasible and attractive, in addition to technically simplifying the overall process. The total fixed and variable O&M costs for SMR-CLC were 15.11 M€ per year, lower than the 15.32 M€ per year for the SMR process. As a result of reduction on both capital cost and O&M cost, the levelized cost of hydrogen for the SMR-CLC process is 3.24 €/kg, lower than the 3.28 €/kg for the SMR process.

Viktor et al. compared simulation results of a type b SMR-CLC process and several other novel SMR integration methods against the conventional SMR process for hydrogen production (Stenberg et al., 2018). Fe-based and Mn-based oxygen carriers were used in the simulation, where the temperatures of the reducer and the combustor of the CLC system were assumed to be the same. The simulation results showed that the SMR-CLC process is beneficial as its efficiency is much higher than that of conventional SMR process, almost as high as that of the SMR integrated with oxygen carrier aided combustion process, and it has a unique advantage of near 100% CO_2 capture without an additional energy penalty.

Evaluations of SMR-CLC processes were mainly performed using simulations, with minimal experimental tests conducted (Adanez et al., 2012). As CLC was merely used for externally heating the SMR reactor, while the experimental researches on the CLC systems were still in relatively small scales which cannot reach autothermal operating conditions yet. The large surface area to volume ratio of the small scaled CLC reactor makes the heat loss of the system significantly higher than the net energy produced by the system. Therefore, they require external heaters or burners to compensate for the heat loss. In addition, SMR-CLC does not adequately manifest a novel chemical looping technology-based reforming process as CLC is merely applied in order to provide external heat for the endothermic SMR process. This reforming technology is fundamentally based on the already existing SMR technology.

2.1.3 Reactor systems for fluidized bed chemical looping reforming

Unlike the SMR–CLC process, which requires the integration of conventional SMR process with the CLC process, the CLR processes produce syngas directly from partial oxidation of natural gas, while using an oxygen carrier to transport oxygen for the reaction. For natural gas, the overall reaction of the CLR process is highly exothermic. This enables the autothermal operation of the CLR processes. A widely applied design for the CLR process is based on the CLC system using a fluidized bed reducer.

The economic analysis of a fluidized bed CLR process was conducted in order to evaluate the natural gas to syngas (NGTS) process of producing syngas for liquid fuel production via Fischer-Tropsch (F-T) gas-to-liquid (GTL) technology. The NGTS process is one of the major CLR processes for syngas production that uses an iron-titanium composite metal oxide (ITCMO) as the oxygen carrier to convert natural gas to syngas (Li et al., 2011).

The liquid fuel production process is based on a reference GTL process with 50,000 barrel per day (bpd) liquid fuel production using autothermal reforming (ATR) to convert natural gas to syngas

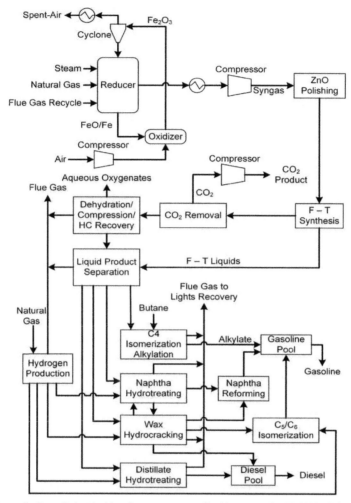

Figure 7. Overall chemical looping system gas to liquids process for 50,000 bpd plant.

(Gollener et al., 2013). As shown in Figure 7, the fluidized bed reducer replaces the reference ATR reactor for syngas generation. The NGTS-GTL process is designed to achieve syngas quality and quantity to match the amount of liquid fuel production of the baseline ATR-GTL process. Stoichiometric number (S#) is one of the key parameters to describe the quality of syngas, which is defined as

$$S\# = \frac{y_{H_2} - y_{CO_2}}{y_{CO} + y_{CO_2}}$$

Here, y_A represents the molar concentration of component A in syngas. The Fe_2O_3:C ratio and H_2O:C ratio are adjusted to reach specific syngas quality for liquid fuel production and low solids circulation rate to minimize the amount of metal oxide particles in the process. The recycle fuel gas stream is assumed to be completely converted. These assumptions represent the most optimal performance possible for a fluidized bed reactor. In a fluidized bed reactor, solids phase backmixing and heterogeneity of gas-solids fluidization lead to a wide residence time distribution and gas channeling. Experimental results obtained for methane conversion to syngas in a fluidized bed reactor showed that syngas with a H_2:CO ratio of 1.8 and a CO:CO_2 ratio of 12 can be produced from methane in a fluidized bed reactor of CLR process using ITCMO oxygen carrier. The methane conversion is limited to 70%, based on the testing results.

Table 3. Overall performance of the NGTS process in a 50,000 bpd GTL plant (Gollener et al., 2013).

Component	Base case	NGTS (10 atm) fluidized bed
Natural gas flow, kg/hr	354,365	452,992
Natural gas flow, kmol/hr	20,451	26,143
H_2O/C_1	0.68	0.25
H_2/CO	2.19	2.18
S#	1.59	~ 1.90
O_2/C_1	0.73	0.48
Butane feed flow, kg/hr	18,843	18,843
Diesel fuel, bbl/day	34,543	34,543
Gasoline, bbl/day	15,460	15,460
Total liquids, bbl/day	50,003	50,003
Electrical Load (kWe)		
Total Gross Power	303,700	303,700
Net Plant Power	40,800	150,480

These experimental results from fluidized bed CLR were used to develop the models of the fluidized bed NGTS-GTL process for its economic analysis.

Table 3 listed the performance results for the NGTS-GTL process using fluidized bed CLR, as compared to the ATR-GTL process with a 50,000 bpd liquid fuel output. The carbon efficiency of the process is lower than that of the baseline case, even though the net plant power output increases as the fluidized bed CLR eliminated the needs of energy intensive ASU. This is mainly because the fluidized bed CLR consumes 28% more natural gas than the conventional ATR.

300 W_{th} CLR Unit at Chalmers University of Technology

Chalmers University of Technology (Chalmers) in Sweden investigated a 300 W_{th} two-compartment fluidized bed reactor system for CLR, as shown in Figure 8, using Ni-based oxygen carriers (Rydén et al., 2006; Rydén et al., 2009; Johansson et al., 2006; Rydén et al., 2008). The unit was initially designed for CLC experiments and was later adopted for CLR for syngas tests (Rydén et al., 2008).

The design of the CLR for syngas system is derived from the shale oil reforming system proposed by Chong et al. in 1986, featured with solids exchange between two adjacent fluidized bed reactors without gas mixing (Chong et al., 1986). The 300 W_{th} unit consists of two adjacent chambers, the reducer reactor

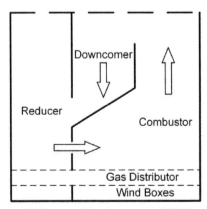

Figure 8. Chalmers 300 W_{th} fluidized bed CLR system (Kronberger et al., 2004) (Arrows in the figure denotes the direction of solids flow).

chamber and the combustor reactor chamber. The chambers are divided by a vertical wall with two slots, as shown in Figure 8 (Kronberger et al., 2004). The reducer chamber had a square cross-sectional area of 25 mm × 25 mm. The combustor had a rectangular cross-section at the bottom with a dimension of 25 mm × 40 mm. The total height of the reactors was 200 mm, with an enlarged section on top for solids disengagement from gas flow. A downcomer with a width of 12 mm was located between the reducer and the combustor for solids transportation from the combustor to the reducer. The two slots on the vertical wall connected the bottom of the two reactors. The slots consisted of two walls, one in each reactor, to minimize the gas leakage between the two reactors. To maintain the constant operating temperature of the reactor system, the 300 W_{th} unit was placed inside an electrically heated furnace to compensate the heat loss. A water seal with adjustable water column height was added to the exit pipe of the reducer to control the pressure of the reducer so as to minimize the air leakage from the combustor to the reducer. The system used natural gas, with a composition equivalent to $C_{1.14}H_{4.25}O_{0.01}N_{0.005}$, as the fuel for the CLR experiments.

The two chambers are operated under fluidized bed conditions using different fluidization regimes in the two chambers in order to control the direction of oxygen carrier circulation. The combustor reactor chamber is operated at a higher gas velocity than the reducer in order to increase the void fraction in combustor, allowing the oxygen carriers in the reducer to overflow into the combustor. The combustor reactor chamber is tapered at the top, which increases the gas velocity, causing the oxygen carrier to be entrained to the top and pass travel into the downcomer to enter the reducer. The reducer is operated with a lower gas velocity in order to form a dense phase fluidized bed which generated a larger pressure drop than the combustor chamber and in turn causes a pressure difference between the bottoms of the reducer chamber and the combustor chamber. This pressure difference drives the particles to move from the reducer to the combustor through an overflow slot between the two reactors, thus forming a continuous solids circulation loop.

Three types of Ni-based oxygen carriers, whose properties are given in Table 4, were tested in the Chalmers CLR experiments. Experiments were performed with a reducer temperature of between 800 °C and 950 °C, and various air-to-fuel ratios for all the oxygen carriers. Steady state of 1–3 hours were maintained for each experiment condition. In the CLR tests, a high fuel flow rate (0.8–1.5 L/min) and a moderate to high air flow rate (3.8–10 L/min) were used. In addition to the tests with pure CH_4, CH_4-CO_2 and CH_4-steam, co-feed CLR experiments were also tested, with feeds containing 30% CO_2 and/or steam in the fuel gas.

Typical results for Chalmers CLR experiments are given in Figures 9–11. Syngas generation by CLR was achieved by adjusting the air-to-fuel ratio. Operating conditions that prevented carbon deposition, a phenomenon that not only reduces the syngas yield from CH_4 and produces CO_2 emission in the combustor, but also weakens the mechanical strength of oxygen carriers and deactivates the particles, were investigated. As shown in Figure 9, carbon deposition can be eliminated when the operating temperature of the reducer is over 930 °C. Cho et al. reported that solid carbon formation depends strongly on the available oxygen of the Ni-based oxygen carrier (Cho et al., 2005). When 80% of NiO is reduced to Ni, carbon deposition occurs rapidly. However, the Chalmers CLR tests showed that carbon deposition became significant even when there were instances where only 33–44% of NiO was reduced to Ni.

The CLR tests showed that, for a dry natural gas feed case with an operating temperature greater than 930 °C, the syngas quality is improved when the oxygen-to-fuel ratio is reduced, as shown in Figures 10–11. However, there is a lower boundary for the reduction of oxygen-to-fuel ratio, as carbon deposition might be significant. The maximum syngas purity, defined as the molar percentage of CO and H_2 in the gas product of the reducer, was around 75%, with a H_2:CO ratio of 1.7.

Table 4. Ni-based oxygen carrier properties used in 300 W_{th} Chalmers's CLR unit (Rydén et al., 2008).

Oxygen carrier	Chemical composition	Production method	Size (μm)	Porosity (%)	Solids inventory (g)
N2AM1400	10% NiO on MgAl$_2$O$_4$	Freeze granulation	90–212	35	250
Ni18-αAl	18% NiO on α-Al$_2$O$_3$	Impregnation	90–212	53	180–250
Ni21-γAl	21% NiO on γ-Al$_2$O$_3$	Impregnation	90–250	66	170

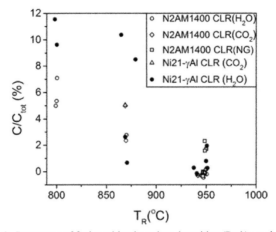

Figure 9. Percentage of fuel resulting in carbon deposition (Rydén et al., 2008).

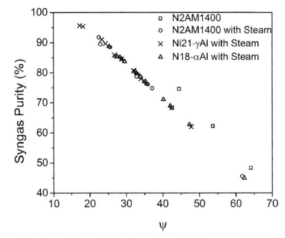

Figure 10. Syngas purity as a function of air-to-fuel ratio, with and without steam injection for different oxygen carriers (Rydén et al., 2008).

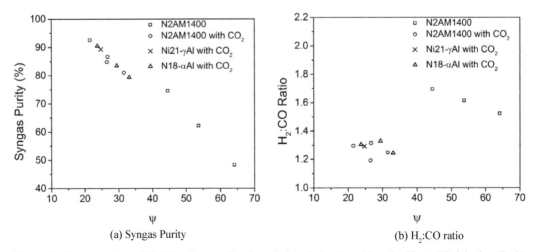

(a) Syngas Purity

(b) H_2:CO ratio

Figure 11. Syngas purity and H_2:CO ratio as a function of air-to-fuel ratio, with and without CO_2 injection (Rydén et al., 2008).

To overcome the limitation on oxygen-to-fuel ratio, steam or CO_2 co-feeding with natural gas were introduced in order to reduce or eliminate carbon deposition in the reducer. When 30% steam or CO_2 was co-fed with natural gas into the reducer with an operating temperature of 950 °C, a high syngas purity was obtained with a low air-to-fuel ratio. The co-feeding of steam with natural gas feed assisted the reactions in two aspects. The CH_4-steam co-feeding could yield a higher H_2 concentration compared to CH_4 only feedstock in the gas product due to the SMR reaction with Ni as a SMR catalyst. Furthermore, carbon deposition was reduced as the solid carbon was gasified by steam. As shown in Figure 10, the CH_4-steam co-feeding with an operating temperature of over 930 °C showed promising results for producing syngas with a high purity as the reducer was operated nearly under a condition of a steam methane reformer. The performance for 30% CO_2 co-feeding with CH_4 was given in Figure 11. The CH_4-CO_2 co-feeding also decreased carbon deposition, as the solid carbon was gasified by the CO_2. The maximum syngas concentration achieved was around 90%, as shown in Figure 11(a). However, the maximum H_2:CO ratio was only 1.3, as shown in Figure 11(b).

140 kW$_{th}$ DCFB CLR Unit at Vienna University of Technology

Vienna University of Technology (VUT) in Austria developed a dual circulating fluidized bed (DCFB) reactor configuration for the CLR process to generate syngas from natural gas (Pro et al., 2010; Pröll et al., 2010). The system, as shown in Figure 12(a), uses Ni-based oxygen carriers and has a thermal capacity of 140 kW (Pröll et al., 2005; Pröll et al., 2009). The reducer of the VUT DCFB system was operated under a turbulent fluidized bed regime, and the combustor was under a fast fluidized bed regime. The reactor system consisted of two interconnected circulation loops with two reactors, three loop seals, and two cyclones. The primary loop of the system consisted of the combustor, the reducer, the primary cyclone and two loop seals, one connecting between the bottoms of the combustor and the reducer, and the other connecting between the bottom of the primary cyclone and the middle of the reducer. The lower loop seal that connects the bottoms of the reducer and the combustor maintains a stable solids distribution between the two reactors. The internal loop, consisting of the reducer, an internal loop seal and an internal cyclone, enabled local solids circulation within the reducer. This design ensured a long average residence time of

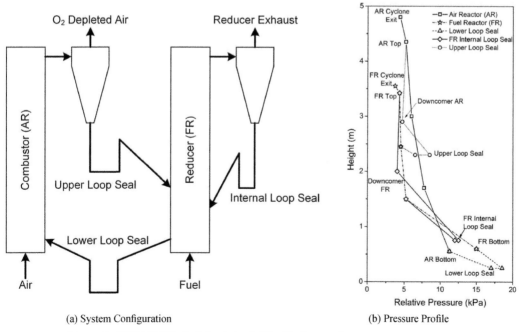

(a) System Configuration   (b) Pressure Profile

Figure 12. VUT 140 kW$_{th}$ dual circulating fluidized bed reactor for CLR (Pröll et al., 2009).

the oxygen carrier particles in the reducer, allowing for the reactions in the reducer to reach equilibrium state. This internal loop also enabled a local solids circulation rate of the reducer, independent of the global solids circulation rate of the system. Steam was used as the sealing gas in the loop seals.

Pressure profiles through the DCFB system, with a sample as shown in Figure 12(b), were obtained for several operating conditions in order to analyze the solids distribution in the system under different conditions. The sample pressure profile was obtained when the system was operated at a full capacity of 140 kW_{th} with a total solids inventory of 65 kg, with around 30 kg in the DCFB reactors and the balance being distributed in the cyclones and loop seals (Pröll et al., 2010). The global air-to-fuel ratio of the system was 1.1, with the reducer temperature maintained at 900 °C (Pröll et al., 2009). The pressure profile along the height of the combustor represented a typical fast fluidized bed operation. In the reducer, the pressure drop in the lower section was sharper, representing a typical dense-phase fluidized bed, while the top section had a much smaller pressure drop, reflecting the solids disengagement zone of the fluidized bed. Due to the different fluidization regimes of the reducer and the combustor, the overall pressure drop in the reducer was larger than the combustor. Correspondingly, more solids were present in the reducer compared to the combustor. For all the loop seals, the pressures at the gas inlets were higher than the both gas outlets, showing that the inert gas split and flew to the both ends, and good gas sealing between the connected reactors was achieved.

The oxygen carrier particles used in the system, whose production methods were described in detail by Linderholm et al. and Jerndal et al. (Linderholm et al., 2009; Jerndal et al., 2009), were made by mixing two different types of oxygen carrier materials with a 50:50 weight ratio. One oxygen carrier material is a sintered mixture of NiO/NiAl$_2$O$_4$ (N-VITO) synthesized from NiO and Al$_2$O$_3$. The other was a sintered mixture of NiO/MgAl$_2$O$_4$–NiAl$_2$O$_4$ (N-VITOMg) synthesized from NiO, Al$_2$O$_3$, and MgO. In total, there was about 40 wt% of active NiO material in the oxygen carrier particles. The mean diameter of the oxygen carrier particles was 135 μm.

Three different operation temperatures, 750 °C, 800 °C and 900 °C, were tested. Natural gas from the Viennese grid (98.7% CH$_4$) was introduced into the reducer and maintained at 140 kW_{th} capacity during stable operation. The global stoichiometric air-to-fuel ratio was decreased stepwise from 1.1 to 0.5, with an interval of 0.1, while the cooling duty of the reactor was adjusted accordingly to maintain the constant temperature.

The operational results of the VUT CLR system as a function of different air-to-fuel ratios under different operating temperatures of the systems were shown in Figure 13 (Pröll et al., 2010). As shown in Figure 13(a), under a given operating temperature of the system, the syngas purity increased with the decrease of the air-to-fuel ratio. Increasing the operating temperature of the reducer decreased the syngas purity under a given air-to-fuel ratio.

The combustor operating temperature of the VUT CLR system was designed to be higher than the reducer, such that the temperature difference is sufficient to obtain autothermal operation. The operating

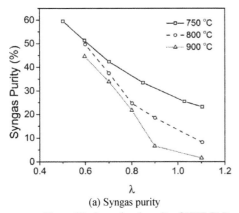

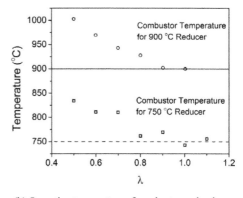

(a) Syngas purity   (b) Operating temperature of combustor and reducer

Figure 13. Operational results of VUT CLR system as a function of air-to-fuel ratio (Pröll et al., 2010).

temperatures of the combustor under different reducer operating temperature conditions as a function of different air-to-fuel ratios were given in Figure 13(b) (Pröll et al., 2010). Under a given reducer temperature, a higher air-to-fuel ratio translated to a lower operational temperature of the combustor. When the air-to-fuel ratio was above 1.1, the system was essentially operated in the CLC mode. The heat generated from combustion was sufficient to provide the heat required for full oxidization of fuel in the reducer. On the other hand, decreasing the air-to-fuel ratio resulted in a larger temperature difference requirement between the two reactors in order to achieve heat balance. The minimum air-to-fuel ratio was around 0.5.

During the operation of the VUT CLR system, gas leakage between the reducer and the combustor was observed. This may lead to a hazardous condition as the reducing gas in the reducer was mixed with the oxidizing gas from the combustor at high temperatures. During the operation, this gas leakage rate was controlled to be no more than 0.5% volume of the combustor gas, as deemed acceptable for safety and process performance. However, due to the gas leakage issue, a large amount of sealing gas in the loop seals flowed to the reactors. The majority of the gas entered the combustor and exited in the combustor exhaust gas. As much as 5% volume of steam was detected in the gas at the combustor outlet. The sealing gas leakage to the reducer could have a slight effect on decreasing the syngas concentration in the reducer product gas. During the experiments, there was no CO_2 or CO detected in the combustor, indicating minimal carbon deposition in the reducer and no gas leakage from reducer to combustor. The study is one of a few cases of fluidized bed CLR processes where no obvious carbon deposition was observed in the reducer without steam or CO_2 co-feeding with natural gas. However, the syngas purity under these conditions was relatively low, at less than 60%.

In summary of experimental studies on fluidized bed CLRs, the fluidized bed CLR processes are capable of syngas production from methane reforming. The syngas purity from these reducers can achieve from 0% (CLC case) to $\sim$ 70% by adjusting the air-to-fuel ratio from above 1 (for combustion) to about 0.4 (for reforming) without steam or CO_2 co-feeding. Further lowering the air-to-fuel ratio will result in significant carbon deposition, which lowers the syngas purity, affects the chemical and physical properties of the oxygen carrier particles and increases the CO_2 footprint of the process. Avoiding carbon deposition by limiting the oxygen carrier reduction to a low level leads to a high oxygen carrier circulation rate for the CLR system. However, steam and/or CO_2 co-feeding with natural gas can alleviate the carbon deposition issue. With the addition of steam and/or CO_2, the air-to-fuel ratio can be further lowered with a higher syngas purity achieved. Feedstock conversion may not be complete due to the inherent solids backmixing and channeling of gaseous hydrocarbons produced in a fluidized bed reducer.

2.2 *Moving bed chemical looping reforming for syngas production*

Some commonly used metal oxide oxygen carriers in CLC systems, such as iron oxide and copper oxide, pose challenges when used for fluidized bed CLRs, as they are thermodynamically favorable of fully oxidizing methane to CO_2 and H_2O instead of CO and H_2. To overcome this problem, the CLR systems have to be operated under conditions with a fuel-to-oxygen carrier ratio higher than the value of full combustion by stoichiometry in order to increase syngas generation. However, other problems, such as low fuel conversion, low syngas purity, and carbon deposition, will occur under such operating conditions. There is another metal oxide, NiO, that is suitable for both CLC processes and fluidized bed CLR, owing to the high syngas yield resulting from its excellent reactivity. However, the toxicity and price of NiO may limit the feasibility of its use in large-scale CLR processes.

Alternatively, when the reducer in a CLR for syngas process is designed in a moving bed mode, it is possible to have a high performance in both fuel conversion and syngas purity when an iron-based oxygen carrier is used (Fan, 2010; Zeng et al., 2012; Fan et al., 2015). This makes the moving bed reducer design an attractive configuration option for CLR processes from both economical and operational standpoints.

Similar to that of the fluidized bed CLR process, as described in section 2.1.3, the economic analysis of a moving bed CLR process using ITCMO as oxygen carrier particles was also conducted via an evaluation of a NGTS-GTL process (Fan, 2017). The overall process diagram is the same as that shown in Figure 7, except that instead of the fluidized bed CLR, the NGTS system uses a cocurrent moving bed CLR process to convert natural gas to syngas. The NGTS system uses steam-natural gas co-feeding scheme.

Since the flow in a moving bed reactor can be considered as a plug flow with a constant residence time (Barelli et al., 2008), a single stage RGibbs reactor block in Aspen Plus® can be used to simulate its fluid flow behavior. To ensure that the syngas production of the packed moving bed NGTS process is equivalent to the baseline syngas production from the ATR (Gollener et al., 2013), an optimized NGTS reducer configuration was developed, with the adjustment on the parameters consisting of H_2O:C molar ratio, Fe_2O_3:C molar ratio, temperature swing along the reducer reactor, natural gas pre-heat temperature, and steam pre-heat temperature.

Table 5 listed the performance results for the NGTS-GTL process using moving bed CLR, as compared to the baseline GTL process with a 50,000 bpd liquid fuel output. The natural gas feed required by the process with moving bed CLR is 11% less than that of conventional process with ATR for generating an equivalent amount of liquid fuel. This benefit, combined with the elimination of the energy intensive ASU, decreases the parasitic energy requirement for syngas production by 60%, resulting in doubled net power output compared to a conventional plant. The moving bed CLR can achieve a high syngas selectivity with less than 3% CO_2 produced and no carbon deposition.

The results of economic analysis of the NGTS-GTL process using moving bed CLR are listed in Table 6. The NGTS process shows a 72.5% reduction on the capital cost for syngas production, leading to a 31.6% reduction on the total plant cost. As a result, the NGTS-GTL process using moving bed CLR reduces the liquid fuel production capital cost to \$65,000/(bbl/day), a 25% decrease compared with the value of approximately \$86,000/(bbl/day) of the baseline GTL process. The substantial reduction on the capital cost enables the NGTS-GTL process using moving bed CLR to be competitive when the crude oil price is over \$48/bbl, a significant advantage over the baseline GTL process and the NGTS-GTL process using fluidized bed CLR which was described in section 2.1.3.

2.2.1 Iron-based oxygen carriers

Iron-based oxygen carriers are particularly attractive in chemical looping systems as they are abundantly available and low in cost. Certain types of iron ores, like ilmenite, can be used directly as oxygen carriers in chemical looping processes (Deshpande et al., 2015); they are also non-toxic.

There are three possible redox pairs for iron oxide when used as oxygen carriers, Fe_2O_3-Fe_3O_4, Fe_3O_4-FeO, and FeO-Fe. The Fe_2O_3-Fe_3O_4 redox pair is thermodynamically the most oxidative as Fe_2O_3 can fully convert the hydrocarbon fuel into CO_2 and H_2O, while being reduced to Fe_3O_4. As a result, the amount of unconverted CO/H_2 is significantly lower when compared to the case using NiO-Ni as the oxygen carrier redox pair. The Fe_3O_4-FeO and FeO-Fe redox pairs are comparatively less oxidative. As a

Table 5. Overall performance of the NGTS process using moving bed CLR in a 50,000 bpd GTL plant (Gollener et al., 2013).

Component	Base case	NGTS (10 atm) moving bed
Natural gas flow, kg/hr	354,365	317,094
H_2O/C in natural gas	0.68	0.25
H_2:CO	2.19	2.178
Stoichiometric number (S#)	1.59	1.96
Butane feed flow, kg/hr	18,843	18,843
Diesel fuel, bbl/day	34,543	34,543
Gasoline, bbl/day	15,460	15,460
Total liquids, bbl/day	50,003	50,003
Electrical Load (kWe)		
Total Gross Power	303,700	303,700
Net Plant Power	40,800	179,050

Table 6. Cost summary for NGTS-GTL plant producing 50,000 bpd liquid fuel (Gollener et al., 2013).

Component (2011 $)	ATR	NGTS
Total plant cost ($×1000)	2,750,000	1,880,000
Total as-spent cost ($×1000)	4,310,000	3,250,000
Capital costs ($/(bbl/day))	86,000	65,000
Capital costs ($/bbl)	51.5	41.8
O&M costs ($/bbl)	27.29	27.23
Fuel costs ($/bbl)	16.9	12.6
Electricity costs ($/bbl)	−1.15	−5.1
Required selling price* ($/bbl)	94.54	76.53
West Texas intermediate crude oil competitive price ($/bbl)	60.5	48.9

*: calculated when the required return on equity equals the internal rate of return for 30 years of operation with assumed financial structure and escalations.

result, CLC processes using Fe-based oxygen carriers in a fluidized bed reducer use only the Fe_2O_3-Fe_3O_4 redox pair with no further reduction of the oxygen carrier. For CLR applications, however, Fe-based metal oxide needs to be reduced to the FeO-Fe redox pair, where it falls in the "syngas production" region, as shown in the modified Ellingham diagram of Figure 2(b). This indicates that the FeO-Fe redox pair thermodynamically inhibits the full oxidation of syngas into CO_2/H_2O. In addition, adding support materials, such as TiO_2 and Al_2O_3, to an Fe-based oxygen carrier may result in the formation of complex materials ($FeTiO_3$ and $FeAl_2O_4$) that are even less oxidative than FeO, resulting in an even higher syngas purity when compared to a purely iron oxide oxygen carrier. Therefore, if the Fe-based oxygen carrier is sufficiently reduced to form a mixture of FeO and Fe in the CLR reducer reactor, a high syngas purity with low CO_2/H_2O concentration can be obtained. When iron-based metal oxide is used as oxygen carrier, the reducer reactor suitable for such a CLR reducer reaction scheme has to be designed in a moving bed mode, such that a high performance in both fuel conversion and syngas purity can be obtained.

A composite oxygen carrier consisting of 60% Fe_2O_3/40% Al_2O_3 by weight was tested in a thermogravimetric analyzer (TGA) at 900 °C using pure methane. The oxygen carrier had a particle diameter of 1.5 mm in order to accommodate the moving bed operating conditions. The weight change of the oxygen carrier during methane reduction process was shown in Figure 14. The process can be divided into two stages of Fe_2O_3 reduction, the stage where weight loss was observed and the carbon deposition stage where weight gain was observed. In the Fe_2O_3 reduction stage, the weight of the oxygen carrier decreased rapidly, owing to the loss of oxygen from oxygen carrier. The weight loss was up to approximately 11% of its original value. Then, the weight eventually increases sharply, when carbon deposition is catalyzed by reduced metallic iron.

TGA studies showed that two oxidation states of oxygen carrier particles are undesirable for syngas production, under-reduced states of Fe_2O_3/Fe_3O_4 and Fe_3O_4/FeO, and over-reduced state of Fe. The

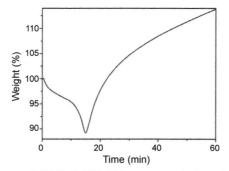

Figure 14. Weight change of 60% Fe_2O_3/40% Al_2O_3 oxygen carrier in methane reduction process.

particles with under-reduced states fully oxidize the reactants to CO_2 and H_2O, while those with over-reduced states cause carbon deposition. In a moving bed reducer, as the particles are in a plug flow mode, the residence time of the oxygen carrier particles can be precisely controlled. By selecting a suitable residence time of the oxygen carrier particles, both over-reduction and under-reduction of oxygen carrier particles can be avoided, and only FeO/Fe particles can present at the exit of the reducer. Additionally, with a cocurrent configuration where both syngas and oxygen carrier particles are fed from the top of the reactor and flow downwards together, the syngas is in contact with only reduced oxygen carrier particles consisting of FeO/Fe, which thermodynamically inhibits full oxidation of H_2 and CO. Therefore, producing syngas with a high purity can be achieved.

2.2.2 Lab-scale fixed bed tests

The feasibility of moving bed CLR using an iron-based oxygen carrier was first investigated with fixed bed experiments (Luo et al., 2014). A fixed bed reactor with a height of 38.1 cm and an internal diameter of 1.27 cm was used to investigate the performance of the cocurrent moving bed reactor. To mimic the oxidation state of the cocurrent moving bed reducer, two layers of iron-based oxygen carriers with different oxidation states were used to pack the fixed bed reactor with reduced oxygen carrier particles at the lower section and fully oxidized particles at the top. The mass ratio of the two different oxygen carrier particles was 2.78:1. The ratio was determined to represent the specific residence time for various oxidation states of iron obtained from TGA experiments for the kinetic study of an oxygen carrier. N_2 diluted CH_4 was injected as the reductive gas and N_2 diluted CO_2 was injected as the oxidative gas into the reactor from the top using digital mass flow controllers. The fixed bed reactor was put in an electrical furnace and was operated at 990 °C. The composition of the product at the gas outlet of the reactor was analyzed by both a non-dispersive infrared (NDIR) gas analyzer and a gas chromatograph (GC).

The composition of the gas stream from the outlet of the fixed bed reactor is shown in Figure 15(a). In the first 2500 seconds, 50 ml/min CH_4 and 50 ml/min N_2 were fed together to the reactor. The methane concentration increased at the beginning, maintained near constant levels in the middle, and then decreased at the end; while the CO concentration decreased at the beginning, maintained constant levels in the middle, and increased in the end. It indicated that the gas composition was regulated by different oxidation states of the ITCMO particles, including Fe, $FeTiO_3$, and/or Fe_2TiO_4. The intermediate oxidation state of Fe_3O_4 has a slower reaction rate with CH_4, compared to Fe_2TiO_5 and Fe_2TiO_4/ $FeTiO_3$, which is shown with the TGA reactivity test results. When the gas feeding rate was reduced to 60 ml/min (50% CH_4 balanced with N_2) at time point (1), as marked in Figure 15(a), an immediate decrease in the methane concentration was observed. In the meantime, the $CO:CO_2$ ratio increased to approximately 10, resulting in a very high syngas selectivity. This is mainly due to the longer gas

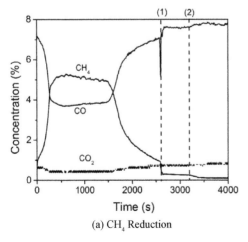

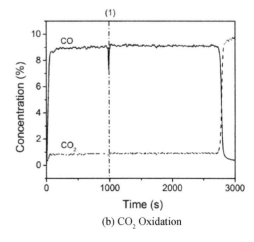

(a) CH_4 Reduction

(b) CO_2 Oxidation

Figure 15. Product gas composition of fixed bed experiment with ITCMO particles (Luo et al., 2014).

residence time as gas velocity is reduced. Slight improvements on the methane conversion and syngas selectivity were observed again when the reactor temperature was increased to 1050 °C at time point (2). The thermodynamic analysis using the modified Ellingham diagram (Figure 2) indicates that methane could be fully converted under all the experimental conditions. However, the incomplete methane conversion and its improvement towards long residence time and higher reactor temperature indicated that there is a kinetic restriction for the chemical reaction between ITCMO particles and methane.

The kinetic limitation of syngas production was further explored by oxidizing the above reduced ITCMO particles in the fixed bed with CO_2 at a flow rate of 30 ml/min and diluted by 30 ml/min N_2 for about 1000s, subsequently with a pure CO_2 stream at 60 ml/min. As shown in Figure 15(b), the reduced ITCMO particles were oxidized by the CO_2 stream which was converted to CO. The $CO:CO_2$ ratio reached the thermodynamic equilibrium shown in the modified Ellingham diagram (Figure 2) and had no change when CO_2 concentration in the feed increased at time point (1), marked in Figure 15(b), indicating the product gas composition was dictated by thermodynamics rather than kinetics. A sharp change in the product composition, where CO and CO_2 concentration flipped over near the end of the CO_2 oxidation experiment, was observed. The change illustrated the dependence of this reaction on oxygen carrier composition. When the oxygen carrier is in the oxidation state of Fe/FeO, the CO concentration is greater than CO_2 and a high purity syngas stream is generated when methane is fed, as shown in Figure 15(a). However, when the oxygen carrier is oxidized to a state above FeO, the $CO:CO_2$ ratio in the gas steam immediately drops to a low value.

The experiments involving CH_4 reduction and CO_2 oxidation indicated that the reaction of methane with iron-based oxygen carriers was relatively slow and, hence, rate limiting. The product gas composition, the syngas purity, was determined by thermodynamics rather than kinetics (Luo et al., 2014).

2.2.3 Bench-scale moving bed reducer tests

A bench-scale moving bed reactor system with an internal reactor diameter of 0.05 m and a height of 0.9 m was used to further investigate the performance of a moving bed reducer for CLR applications using ITCMO particles. The moving bed was operated in a cocurrent mode with all the feedstock and ITCMO particles introduced from the top of the reactor, moved downwards and exited the reactor from the bottom. The solids flow rate was controlled by a screw feeder at the bottom of the reactor. Gaseous products were sampled along the reactor to quantify the concentrations of CO, CH_4, CO_2, O_2, and H_2 at different stages of the reaction. The reactor was operated under ambient pressure.

The feedstock with a flow rate of 2 l/min and a gas composition of 90% CH_4 and 10% N_2 was tested at a temperature of 1000 °C, with a Fe_2O_3:CH_4 molar ratio of 0.8. As shown in Figure 16, stable syngas generation was achieved with a $CO:CO_2$ molar ratio higher than 9 and H_2:CO ratio of around 2. The CH_4 conversion was around 95%, with a syngas purity higher than 85%.

CH_4 can be co-fed with coal to the moving bed chemical looping systems as a H_2-rich feedstock for gasification purpose. Syngas with higher H_2 concentrations can be obtained, as compared with CLG process with coal only. As shown in Figure 17(a), when methane was co-fed with coal, the H_2:CO ratio in the syngas increased to 1, compared to around 0.65 for the coal only case. The syngas purity is above 95% on a dry basis with the $CO:CO_2$ ratio greater than 10 and minimal unconverted CH_4. When steam was also added together with coal and methane, the H_2:CO ratio in the syngas was further increased, as shown in Figure 17(b). When the CH_4:steam:coal mass ratio was about 1:0.9:1, the produced syngas had a H_2:CO ratio of around 1.8, a $CO:CO_2$ ratio of around 6, and a syngas purity of greater than 85% on a dry basis.

2.2.4 Sub-pilot moving bed CLR system

A 25 kW$_{th}$ sub-pilot scale moving bed CLR system using ITCMO as the oxygen carrier and CH_4 as feedstock, as shown in Figure 18, was tested (Fan et al., 2013). The reactor had a height of 1.52 m and an inner diameter of 0.1 m. The solids flow rate was controlled by a rotary solids feeder at the bottom of the reactor system. External heaters with a PID control program for maintaining the reactor temperature were used to heat the reactor system. A solids reservoir section was placed above the reducer such that the solid level was maintained above the reaction section and solids were at the required operating temperature

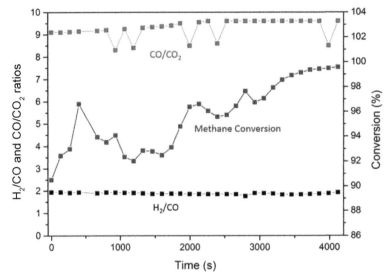

Figure 16. Syngas product distribution for CLR of methane (Luo et al., 2014).

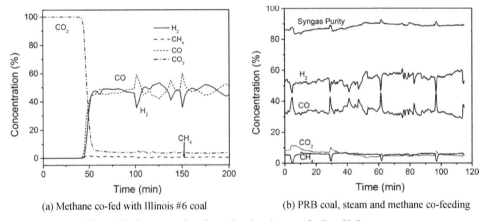

(a) Methane co-fed with Illinois #6 coal   (b) PRB coal, steam and methane co-feeding

Figure 17. Syngas product for coal and methane co-feeding CLG processes.

when entering the reaction section. Various parameters, including Fe_2O_3:CH_4 molar ratios from 0.5 to 1.4, temperatures from 900 to 1050 °C, and steam:CH_4 molar ratios from 0 to 0.4, were tested.

A typical product gas composition from the sub-pilot CLR unit, with an operating temperature of 975 °C, a Fe_2O_3:CH_4 molar ratio of 0.73 and a CH_4 flow rate of 10 l/min, was shown in the Figure 18 (Fan et al., 2013). When changing the operating conditions, such as reactant residence time or temperature, achieving a higher CH_4 conversion and a higher syngas purity is possible. The methane conversion rate was over 99.9%, with a H_2:CO ratio of 1.97 and a syngas purity of 91.3%. Comparing to the bench-scale unit described in section 2.2.3, the sub-pilot unit achieved a higher methane conversion and a higher syngas purity, even though the sub-pilot unit was operated under a condition with a slightly lower operating temperature and a lower Fe_2O_3:CH_4 molar ratio. This was mainly due to the higher reactant residence time (20%) compared to the bench-scale unit. Thus, it can be concluded that the reactant residence time is a key factor for optimization of a moving bed CLR process design. This can be achieved by using a non-mechanic valve, like an L-valve, which can control the solids flow rate by adjusting the injected aeration gas flow rate (Wang and Fan, 2015), as the reducer is operated in a packed bed mode and the oxygen carrier particles have a large diameter.

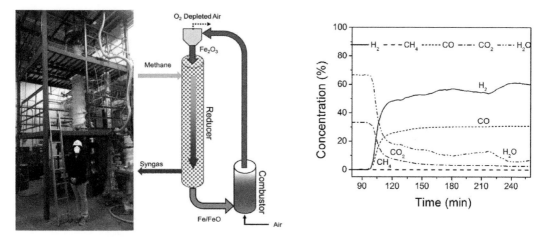

Figure 18. 25 kW$_{th}$ sub-pilot CLR process: Experiment unit (left); flow schematic (middle); product gas composition (right) (Fan et al., 2013).

2.2.5 Other natural gas reforming schemes using moving bed CLR

The moving bed CLR process can have various configurations applied for different feedstocks and applications. In addition to natural gas/steam co-feeding for the STS process, as described at the beginning of section 2.2, co-feeding natural gas with other source of carbon, such as solid fuels like coal and biomass, as well as gaseous CO_2 is possible. When CO_2 is used as a feedstock, co-feeding with natural gas, the relationship between the CO_2 flow rate and H_2:CO ratio in the product is nonlinear, which enables a modular approach to the reducer design of the moving bed CLR process for product yield enhancement. Cases with several different feedstock and modular design are described below.

Coal and natural gas co-feed

Natural gas is a good feedstock for supplemental hydrogen and carbon in chemical production processes relying on coal-derived syngas. Natural gas has a hydrogen-to-carbon ratio of about 4, a value much higher than that of coal, which is around 1. When the natural gas fraction in the co-feed feedstock increases, the carbon efficiency of the syngas generation step and the overall process increases. The economic feasibility of a coal and natural gas co-feeding moving bed coal to syngas (CTS) process was investigated via an evaluation of the coal and natural gas co-feeding CTS process for syngas generation to produce methanol (Li et al., 2011). 50% HHV ratio between natural gas and coal feeding rate, which allows both high utilization rate of coal and carbon efficiency improvement, was used in the process. Optimization analysis on the co-feed CTS process was conducted for identifying suitable Fe_2O_3:C and H_2O:C ratios to achieve autothermal operation and specific syngas compositions for downstream methanol synthesis.

The overall block diagram for the coal and natural gas co-feed CTS for a methanol production plant is shown in Figure 19, as based on a baseline case for a net plant output of 10,000 tonnes of methanol per day, described in a DOE/NETL report (Summers, 2014). The co-feed CTS process is integrated into the methanol production plant with a second sub-model developed to balance steam generation against the steam loads in the system, which includes the steam for co-feed CTS process usage and power generation.

Table 7 listed the performance of the integrated process with coal and natural gas co-feed CTS process, as compared to the baseline cases with and without CO_2 capture as well as CTS process without natural gas co-feed. By co-feed natural gas with coal, the overall process was improved with reduced consumption of water, steam, and air, as well as reduced sulfur and carbon emissions. The co-feed case requires less parasitic energy for coal and ash handling, with coal consumption also reduced. An improvement in carbon efficiency with natural gas co-feed leads to a higher syngas quality from the syngas generation process resulting in a lower CO_2 concentration in the syngas and correspondingly

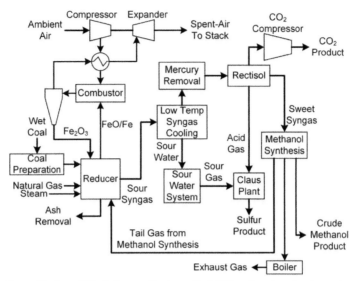

Figure 19. Overall block diagram for the co-feed CTS methanol production system.

Table 7. Overall performance of coal and natural gas co-feed CTS process in a 10,000 tonne/day methanol plant (Summers, 2014).

Case	MB-A	MB-B	CTS w/o co-feed	Co-feed CTS
Mass Flows (lb/hr)				
As Received Coal	1,618,190	1,618,190	1,395,457	718,631
Oxygen from ASU containing 95% O_2	1,010,968	1,010,968	NA	NA
Steam to Gasifier, Quench, Shift reactors, CTS	1,533,584	1.533,584	1,624,318	693,587
Air to Direct-fired boiler	121,518	121,518	181,009	606,106
Clean syngas for methanol production	1,183,080	1,183,080	1,025,106	1,039,864
Tail gas from Claus unit	61,476	61,476	50,089	25,589
Captured CO_2 (no capture for MB-A)	0	1,569,410	1,302,138	663,393
Electrical Load (kWe)				
Total Gross Power	320,680	390,170	20,830	31,491
Total Net Power**	12,280	21,480	–323,504	–245,692

Notes ** Negative value indicates power purchase required.

a smaller acid gas removal (AGR) unit, CO_2 compressor, syngas compressors, and syngas cooling loads. The results from the process efficiency standpoint alone show significant benefits associated with co-feeding natural gas to the reducer.

Table 8 listed the results of economic analysis for the cases of coal and natural gas co-feed CTS, baselines with and without CO_2 capture, and CTS without natural gas co-feed. The co-feed CTS case has a lower capital cost investment than that of the CTS without natural gas co-feed, owing to a higher carbon efficiency. However, the co-feed CTS case has a required selling price of methanol 5% higher than the CTS case, mainly due to the high natural gas price.

CO_2 and natural gas co-feed

Experimental results showed that co-feeding CO_2 with natural gas can effectively reduce or eliminate carbon deposition on the oxygen carriers (Rydén et al., 2008). In addition, adding CO_2 in the feedstock

Table 8. Comparative summary of capital and operating costs of methanol production.

Case (2011 $)	MB-A	MB-B	CTS w/o co-feed	Co-feed CTS
Total plant costs (Million $)	4,586	4,775	3,497	2,996
Total as-spent costs (Million $)	6,580	6,852	5,003	4,291
Capital costs ($/gal)	1.18	1.23	0.89	0.81
Fuel costs ($/gal)	0.24	0.26	0.18	0.39
O&M costs ($/gal)	0.23	0.23	0.16	0.14
CO_2 TS&M costs ($/gal)	0	0.06	0.05	0.03
Electricity costs ($/gal)	0	0	0.14	0.11
Required selling price* ($/gal)	1.64	1.78	1.41	1.48

*: calculated when the required return on equity equals the internal rate of return for 30 years of operation with assumed financial structure and escalations.

changes the syngas purity and the H_2:CO ratio in the gas product by affecting the reaction equilibrium in the CLR reducer. This co-feed scheme enables CLR to be CO_2 neutral, i.e., the CO_2 input to a process is equal to the CO_2 output from the process under a steady state condition, or even CO_2 negative, i.e., the CO_2 input to a process is greater than the CO_2 output from the process. A CO_2 neutral or negative process may involve a CO_2 recycle in the process system.

Process simulation on a CLR process with CO_2 co-fed with natural gas, steam and recycled fuel gas from downstream processes to produce syngas for liquid fuels production, whose schematic diagram was shown in Figure 20, was conducted (Kathe et al., 2017). The CLR process used the moving bed CLR process with ITCMO particles, as described above. Like the beginning of section 2.2, the moving bed reducer of the CLR process was simulated using single stage RGibbs block from Aspen Plus®. The reducer was set to a temperature of 900 °C and a pressure of 1 atm. The equilibrium condition at the outlet of the reducer was simulated by the reducer model. The performance of the process was compared to the baseline case in section 2.1.3, which uses the conventional ATR for syngas production (Gollener et al., 2013).

CO_2 separated from the outlet stream of the Fischer-Tropsch synthesis was recycled as CO_2 co-feed source to reduce CO_2 emission. The targets of syngas generation by the CLR process were to match the baseline performance with H_2:CO ratio of 2.19 and S# greater than 1.58 at a H_2 flow rate of at least 45,285 kmol/hr. 90% CO_2 recycling was assumed in the process simulation (Gollener et al., 2013).

The process simulation showed that an optimally designed CO_2 and natural gas co-feed CLR process can reduce the natural gas consumption by 22% and steam consumption by 27%, as compared to the ATR baseline case, as shown in Figure 21, with a H_2:CO ratio matching the ATR baseline case and H_2 flow

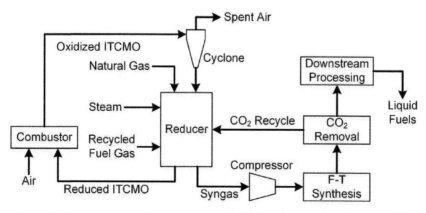

Figure 20. Chemical looping CO_2 recycle scheme for liquid fuels production (Fan et al., 2017).

rate and S# greater than the baseline case. At a natural gas price of $2/MMBtu, these benefits represented a $7,507.94/hr cost reduction for natural gas consumption, leading to an annual saving of $59.2 Million for a CLR plant, given an annual operation time of 90% of its 8,760 h/a designed operation time (U.S. Energy Information Administration, 2016; Gas-to-Liquids Conversion, 2012; Quality Guidelines for Energy System Studies, 2012).

When the CO_2 co-fed to the system, such as that shown in Figure 20, is higher than the available CO_2 recycled from downstream processes and, thus, external CO_2 sources are required, the process utilizes more CO_2 than it generates and is, thus, CO_2 negative (Fan et al., 2017). The simulated relationships among feedstock parameters of CO_2:CH_4 ratio and H_2O:CH_4 ratio, as well as syngas product parameters of H_2:CO ratio, CO_2 concentration, and carbon recycle parameter (CRP) defined by the ratio of CO_2 entering the reducer to CO_2 exiting the reducer, are shown in Figure 22. From the figure, the expected syngas quality can be obtained if CO_2, steam and methane co-feed ratios are given. In addition, the conclusion as to whether the process can be operated under CO_2 neutral or negative conditions can be obtained. The ability of the syngas production section to consume CO_2 is valuable to offset carbon emissions from other sections of the entire process or other processes and is able to transform the CO_2 market.

CO_2 and natural gas co-feed with reducer modularization

Further analysis on the above CO_2 co-feed scheme found that the amount of CO_2 co-fed to the CLR reducer posed a nonlinearity effect on the quality of syngas produced from the CLR process, as

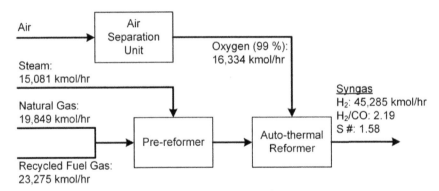

Figure 21. ATR baseline case for liquid fuels production using syngas from autothermal reforming (Gollener et al., 2013).

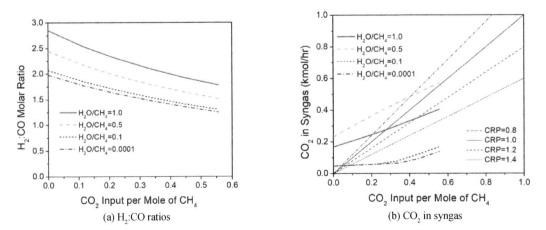

(a) H_2:CO ratios

(b) CO_2 in syngas

Figure 22. Simulation of the effect of CO_2 input rates at various CRP and H_2O:CH_4 ratios (Fan et al., 2017).

exemplified in Figure 23, where 1 mole of CH_4 is co-fed with steam and CO_2 to a moving bed CLR reducer at a temperature of 900 °C, a pressure of 1 atm, and an effective Fe_2O_3:CH_4 ratio of 0.40 (Fan et al., 2017). Such a nonlinearity relationship induced by CO_2 co-feed can be beneficial as it can be used to optimize the system configuration and enhance product yield using a reactor modularization approach, as exemplified in Figure 24, where a two reducer modularized system is shown (Fan et al., 2017).

When the system uses two reducers, one producing syngas product with a H_2:CO ratio of 2.50 and the other with a ratio of 1.27, the system can provide the same syngas with a combined methane consumption of only 0.83 mol/s, a 17% reduction. The details of this operating condition are shown in Table 9. The decreased carbon and hydrogen supply from methane are offset by the increase of steam and CO_2 co-feed rates. The economic benefits of the reduction in natural gas flow with a two-reducer system will outweigh the increased cost of higher steam and CO_2 input.

When the two reducer moving bed CLR system is integrated into the 50,000 bpd GTL plant, as shown in Figure 7, with one reducer processing 8850 kmol/hr of natural gas and the other processing 6000 kmol/hr, a liquid fuel production equivalent to that shown in Table 6 can be obtained. Compared to the single reducer system with CO_2 co-feed which consumes 15,500 kmol/hr of natural gas, the two reducer system reduces the natural gas consumption by 650 kmol/hr when 50,000 bpd of liquid fuel is produced. Considering this reduction is in addition to the reduction of 4349 kmol/hr (or 22%) over the

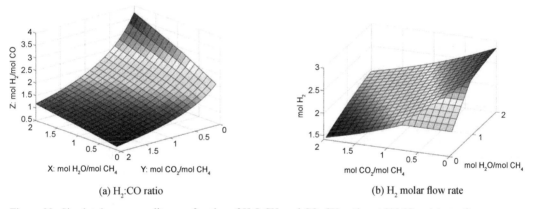

(a) H_2:CO ratio   (b) H_2 molar flow rate

Figure 23. Simulated syngas quality as a function of H_2O:CH_4 and CO_2:CH_4 ratios at 900 °C and 1 atm for a cocurrent moving bed reactor and an effective Fe_2O_3:CH_4 ratio of 0.8 (Fan et al., 2017).

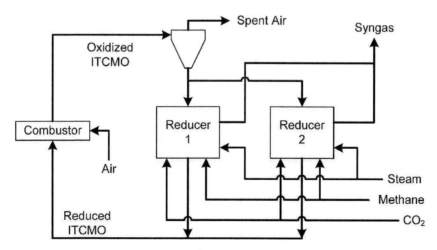

Figure 24. Chemical looping system operated with two reducers in parallel and a single combustor reactor with CO_2 input (Fan et al., 2017).

Table 9. Comparison of the syngas yields for the 2-reducer and the 1-reducer system.

	2-Reducer modularized system			1-Reducer system
	Reducer 1	Reducer 2	combined	
Input conditions				
$CH_{4\ in}$ (mol/s)	0.68	0.15	0.83	1
H_2O_{in} (mol/s)	0.54	0.03	0.57	0.23
$CO_{2\ in}$ (mol/s)	0.07	0.10	0.17	0
Output conditions				
$H_{2\ out}$ (mol/s)	1.74	0.30	2.04	2.04
CO_{out} (mol/s)	0.70	0.23	0.93	0.93
H_2/CO Ratio	2.50	1.27	2.19	2.19

ATR baseline case by using the CO_2 co-feed scheme, the overall benefit of CO_2 co-feed with reducer modularization configuration is, therefore, significant.

It should be noted that reducer modularization is not only limited to two reducers, the concept can be expanded to 'n' reducers (Fan et al., 2017). While each reducer may perform differently, their configurations can be optimized such that the combined performance can be optimal for product yield, considering temperature, pressure, reactor reaction time, H_2:CO ratio, and Fe_2O_3:CH_4 ratio. The optimization of these reducers can target to reduce overall cost and maximize the efficiency under given operating conditions (Fan et al., 2017).

Summarizing the moving bed CLR of methane process, it can efficiently convert feedstock into a high-quality syngas using an iron-based oxygen carrier. Methane can also be used as a co-fed feedstock with solid feedstock, such as coal or biomass, to effectively adjust the H_2:CO ratio of the syngas and improve the quality of the syngas product. Owing to the uniform and controllable residence time of oxygen carrier particles and avoided solids backmixing typically presented in fluidized bed reactors, the oxidation state of the oxygen carrier at the reducer outlet is controlled to those thermodynamically favorable for syngas generation. The cocurrent flow pattern in the moving bed reducer ensures that the syngas is only in contact with the desired oxidation state of the oxygen carrier, to ensure a favorable composition of the syngas. The oxygen carrier particles can donate more oxygen than those used in fluidized bed reactors as the oxygen carrier conversion from a moving bed reducer is higher. As a result, the moving bed CLR process requires a lower solids circulating rate than the fluidized bed CLR process operated at the same feedstock processing capacity. In addition, by limiting the reduction degree of oxygen carrier in order to avoid the formation of metallic iron, carbon deposition is effectively inhibited. With steam or CO_2 co-feed options, the H_2:CO ratio in the syngas can be adjusted to meet the downstream processing requirement, for example, a H_2:CO ratio of 2:1 for Fischer–Tropsch synthesis or methanol production. With the source of co-fed CO_2 coming from the recycled CO_2 of downstream processes, the moving bed CLR process can be CO_2 neutral or negative. As the amount of CO_2 co-fed to the CLR reducer poses a nonlinearity effect on the quality of syngas, CLR reducer modularization can further reduce the overall cost and maximize the efficiency under given operating conditions.

3. Solar Thermal Chemical Looping Reforming (SoCLR)

As stated above, the conventional processes of SMR and ATR processes for syngas and hydrogen generation are very energy intensive. The parasitic energy consumption associated with ASU operation for oxygen preparation can account for up to 40% of the operating cost of a syngas production plant (Rostrup-Nielsen, 2002). Using the CLR process can eliminate the need for ASU. However, the endothermic reforming reaction of methane consumes a large amount of heat and creates a significant challenge for the CLR reactor design. On the other hand, solar radiation is the most abundant source of energy that can be utilized for industrial applications. Combining the solar thermal process with CLR eliminates the need for ASU, provides a much better heat management and effectively utilizes solar energy for chemical production and electricity generation. From a thermal energy storage (TES) perspective, the solar reforming combined cycle can be viewed as a heat conversion process that stores intermittent solar thermal energy as latent chemical energy. Under this design, the steam reforming catalyst can serve as the heat transfer fluid to absorb and transport the thermal energy throughout the process.

SoCLR for syngas and hydrogen generation combines the reaction engineering of metal oxide lattice oxygen transfer with the process design of concentrating solar power systems. It has the potential to reduce the carbonaceous fuel consumption associated with conventional syngas/hydrogen production. The key technical challenge of this technology lies in the selection of a suitable metal oxide whose required properties are similar to other chemical looping processes of combustion, gasification and reforming.

The solar thermochemical process for methane reforming to produce syngas replaces fossil fuels with solar energy to supply the energy for the endothermic heat of reaction. This design can potentially reduce the fossil fuel consumption in a reforming process by 20–35% (He and Li, 2014). In addition, using a metal oxide-based reforming catalyst as the heat transfer fluid has the potential to simplify the process flow and improve the economics. Since reforming processes and solar thermal processes are separately available on a commercial scale, the key engineering challenge exists in the process integration and scale-up of this combined cycle design.

In the SoCLR of methane process, methane and steam are the feedstock for the two redox reactors. In the reducer, methane is oxidized by the metal oxides to generate syngas, with the reaction

$$M_xO_y + yCH_4 = xM + y(2H_2 + CO)$$

The reduced metal oxides are then transported to the combustor (oxidizer) to react with steam for H_2 generation,

$$xM + yH_2O = M_xO_y + yH_2$$

Compared to the SMR process, the SoCLR process produces syngas ready for production of methanol or liquid fuel, as well as pure H_2, with decreased CO_2 emissions and simpler syngas processing steps. CO_2 can also be used to replace steam in the combustor for CO generation with the reaction

$$xM + yCO_2 = M_xO_y + yCO$$

He et al. (He and Li, 2014) evaluated the performance of the SoCLR process for syngas production to produce liquid fuel while co-producing H_2. The simulation results were compared with conventional SMR and solar SMR processes. The process diagram of the SoCLR for liquid fuel and H_2 production is shown in Figure 25. Iron-based oxide with perovskite crystal structure support material of lanthanum strontium ferrite ($La_{0.8}Sr_{0.2}FeO_{3-\delta}$) cycled between the oxidation states of Fe_3O_4; Fe/FeO was used as the oxygen carrier in the process. The heat required in the reducer was compensated by solar energy. The syngas produced in the reducer was used to produce a hydrocarbon mixture with a Fischer-Tropsch (F–T) reactor and was then upgraded to naphtha and diesel in refining units. A pressure swing adsorption (PSA) system was used to purify hydrogen produced in the combustor after being compressed. Part of the hydrogen was used in the upgrader to upgrade fuels. Byproducts from the different units of F–T reactor, upgrader, and PSA were combusted to generate steam. Process heat was recovered by the heat recovery steam generator (HRSG) and steam turbines in order to produce enough electric power to satisfy the parasitic energy requirements of the system. As a large amount of solar energy is integrated into this system, the life cycle CO_2 emission from the SoCLR process is expected to be much smaller than traditional methane-reforming processes.

The methane processing capacity of the process was assumed to be 8 t/h, which required a solar input of approximately 60 MW$_{th}$ and allowed for the integration of existing concentrated solar thermal systems into the process. Three different cases of SoCLR, whose parameters were listed in Table 10, were simulated. Case 1 assumed the system was operated under the conditions of the reducer temperature being at 900 °C and system pressure at 1 atm. Case 2 had a system pressure of 10 atm and reducer temperature of 950 °C. Both case 1 and case 2 assumed the reducer and the combustor reached thermodynamic equilibrium. Case 3 had the same operational conditions as Case 1, while the products used the experimentally obtained results from a fixed-bed reducer whose data was kinetically limited by the reactor design and

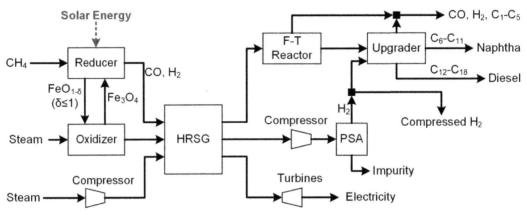

Figure 25. Simplified schematic of the SoCLR process.

Table 10. Operating conditions of simulation cases for SoCLR process.

	Case 1	Case 2	Case 3
Fe_3O_4:CH_4 molar ratio	0.26	0.26	0.64
Reducer temperature and pressure	900 °C, 1 atm	950 °C, 10 atm	900 °C, 1 atm
Oxidizer temperature and pressure	750 °C, 1 atm	750 °C, 10 atm	750 °C, 1 atm
CH_4 conversion	97.6%	95.1%	95%
Syngas yield	96.5%	93.2%	59%
Steam to H_2 conversion	68%	68%	60%
Hydrogen purity	100%	100%	97%

oxygen carrier performance. As shown in Table 10, for case 3, even with a much higher Fe_3O_4:CH_4 molar ratio, the purity of hydrogen is still less than the thermodynamic equilibrium, as shown in cases 1 and 2.

The SoCLR process has garnered significant research interest in recent years, with major researches on process thermodynamics and oxygen carrier materials. Steinfeld et al. conducted an appreciable amount of the pioneering work in this solar thermochemical scheme (Steinfeld et al., 1993; Steinfeld et al., 1995).

Steinfeld et al. proposed the iron/syngas solar co-production process based on an Fe_3O_4 and CH_4 system in 1993 (Steinfeld et al., 1993). A thermodynamic analysis of the reaction system suggests that at 1 atm and temperatures above 1300 K, the equilibrium reaction components consist of solid metallic iron and a mixture of gaseous H_2 and CO in a 2:1 ratio.

The direct contact of carbon with iron oxide can form intermediate products, such as cementite (Fe_3C), which deactivates the metallic iron. Therefore, injection of CO_2 to induce the Boudouard reaction and minimize carbon deposition is critical to maintaining iron reactivity.

A two-step cyclic process based on iron oxide for hydrogen and syngas production from water and methane can be illustrated in a chemical looping diagram, as given in Figure 26. In the first step of this process, the highly energy intensive reduction of Fe_3O_4 by methane to form syngas is driven by solar energy. In the second, regeneration step, metallic iron is oxidized by water to form hydrogen and Fe_3O_4. This reaction is exothermic and occurs at a lower temperature.

To examine this process, a fluidized bed reactor with a solar concentrator was constructed (Steinfeld et al., 1993). The reactor was a quartz tube of diameter 2 cm operated in a fluidized bed mode. The solar receiver, a 10 cm ID steel cylinder, was installed perpendicular to the reactor on its outer surface. A layer of specular reflective gold was electroplated onto the inner wall of the solar receiver in order to reflect infrared diffuse radiation, thus minimizing energy loss. A solar concentrator collected solar energy, and the reactor was located at the focus of the solar concentrator. A water-cooled steel plate with a circular

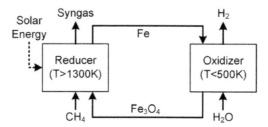

Figure 26. Schematic diagram for syngas and hydrogen co-production using Fe_3O_4–Fe redox.

aperture of 6 cm diameter was also attached to the solar receiver to prevent the radiation from spilling. The capacity of the solar heating was estimated to be 1.1 kW for the reactor. This is a typical fluidized bed experiment set-up that utilizes solar thermal energy to provide heat for the endothermic reaction. The experiment demonstrated that fluidized particles were effective in absorbing solar radiation. The reaction between Fe_3O_4 and methane occurred in two stages. In the first stage, Fe_3O_4 was reduced to FeO, and more CO_2/H_2O than CO/H_2 was generated. In the second stage, FeO was reduced to Fe, and more CO than CO_2 was generated. The first stage had a higher reactivity than the second stage. The sintering and recrystallization of metallic iron contributed to the slower kinetics in the second stage. The conversion of the methane was ~ 20%.

Steinfeld et al. also proposed a SoCLR process for a zinc and syngas co-production process which used a ZnO/Zn redox cycle (Steinfeld et al., 1995). The concept and advantages in the zinc process are the same as those in the iron process. The major difference is that zinc has a lower boiling point, 907 °C, at which metallic zinc in the product stream vaporizes. The zinc vapor is easily re-oxidized, lowering the overall process efficiency due to irreversible energy loss (Steinfeld et al., 1995). Even though the reactor system was not optimized for obtaining the best performance of the process, the feasibility of producing zinc and syngas simultaneously with solar power was demonstrated. Steinfeld et al.'s intention was to produce zinc, which was considered to be more valuable than syngas, and thus, the operational condition of the reactor was not intended to be used for high methane conversion and syngas generation. The reactor was operated with a CH_4:ZnO molar ratio larger than 10. Under this condition, the yield of zinc can reach over 90%, however, methane conversion was very low.

Kodama et al. experimentally compared the performance of different metal oxides, including Fe_3O_4, ZnO, In_2O_3, SnO_2, V_2O_5, MoO_2, and WO_3, for syngas and H_2 or CO co-production with SoCLR of methane a reactor temperature of 1000 °C (Kodama, 2003). WO_3 and V_2O_5 were found to be reactive and selective for the process. The reduced metallic tungsten from the reducer can split water for hydrogen generation at a temperature as low as 800 °C. Supported tungsten oxides were examined in the temperature range of 800–1000 °C in a bid to improve the reactivity of WO_3. Experimental results showed that the ZrO_2-supported WO_3 improved methane conversion from 40% to 70% and steam conversion from 7% to 30% with a H_2 selectivity from 69% to 97%. However, the CO selectivity dropped from 97% to 86% when CO_2 was used in the combustor. The improvements were partially attributed to the interaction between WO_3 and the support material. Shimizu et al. suggested that tungsten carbide (WC) was formed when ZrO_2-supported WO_3 was reduced above 850 °C (Shimizu et al., 2001). In a subsequent study at a temperature of 1077 °C, the solar simulator with 50 wt.% ZrO_2-supported WO_3 reached to a methane conversion of up to 93% with the selectivities of H_2 and CO of 46% and 71%, respectively (Kodama et al., 2002). However, the solar-to-chemical efficiency was low because H_2O conversion was low (less than 30%).

Kodama et al. also investigated the iron-based oxygen carriers with mixing of other bivalent metal oxides, including nickel, cobalt, and zinc oxide (Kodama et al., 2002). The particles were synthesized with a constant dopant to iron molar ratio of 0.15, giving a general chemical formula of $M_{0.39}Fe_{2.61}O_4$ (M = Ni, Co, Zn). A fixed bed reactor of SoCLR was used to examine the particles at an operating temperature of 900 °C. The nickel ferrite ($Ni_{0.39}Fe_{2.61}O_4$) particles provided a higher CO yield and selectivity (22% and 72%, respectively) than pure Fe_3O_4 (8% and 63%, respectively). The addition of highly reactive Ni attributed to the enhanced performance. However, severe sintering under the operating temperature significantly hindered the following water splitting reaction. To suppress the sintering effect,

ZrO_2 was then added to support the nickel ferrite particle (Kodama et al., 2008). Five redox cycles were tested with the ZrO_2 supported nickel ferrite with good recyclability reported. Methane conversions of around 46–58% with a CO selectivity of around 47% and a H_2:CO mole ratio of around 2.5 were obtained from the reduction reaction, while a steam conversion of around 20% was obtained with the oxidation reaction. Cu-ferrites with ZrO_2 support material were investigated by Cha et al. (Cha et al., 2009; Cha et al., 2010). It was reported that, in the syngas production step, increasing Cu content in the $Cu_xFe_{3-x}O_4$/ZrO_2 medium suppressed carbon deposition and enhanced the reaction rate, which reached its highest point when $x = 0.7$. As a result of less carbon deposition as compared to the particle without Cu, CO selectivity was higher. In the oxidation step, the addition of Cu promoted the gasification of the deposited carbon. In addition, adding Ce as a binder also improved reactivity in the syngas production step and yielded the highest reactivity when the Ce:Zr molar ratio was 3:1. The $Cu_{0.7}Fe_{2.3}O_4$/Ce–ZrO_2 oxygen carrier showed high durability and recyclability in ten repeated cycles.

Ceria has been studied widely as a redox material for SoCLR processes, mainly due to its stable fluorite structure up to a nonstoichiometry of 0.25 and the extremely fast kinetics of oxygen diffusion (William and Sossina, 2009; Chueh and Haile, 2010). Otsuka et al. tested a series of CeO_2–ZrO_2 composite oxides ($Ce_{1-x}Zr_xO_2$) with Zr content below 50% (Otsuka et al., 1998; Otsuka et al., 2007; Otsuka et al., 1998; Otsuka et al., 1999). It was reported that the syngas formation rates were increased and the activation energy was remarkably decreased due to the addition of ZrO_2 into CeO_2 (Otsuka et al., 1998a, 2007, 1998b; Otsuka et al., 1999). By adding Pt catalyst, the reaction was further accelerated. When using $Ce_{0.8}Zr_{0.2}O_2$ in the presence of Pt, the operating temperature for converting CH_4 to H_2 and CO can be as low as 500 °C. Krenzke et al. explored the performance of fibrous ceria oxide particles with a low surface area of 0.143 m^2/g under different temperature and methane flow rate conditions (Krenzke et al., 2016). It was reported that methane and steam conversion increased when temperature increased from 900 °C to 1000 °C, with modest improvements in syngas selectivity. When the temperature was increased to 1100 °C, carbon deposition was observed. There was a tradeoff between high methane conversion, high syngas selectivity and high steam conversion. The methane conversion increased when a slower methane flow rate was used, with reduced syngas selectivity and steam conversion observed.

Morphologies with higher specific surface area can improve syngas production rates if the surface area is stable over many redox cycles. Gao et al. demonstrated that nanostructured ceria with a high surface area and porosity can significantly enhance the initial and long-term syngas production performance (Gao et al., 2016). Among the three types of ceria morphologies they synthesized, flame-made CeO_2 nanopowders with a surface area of 77 m_2/g had up to 191%, 167% and 99% higher initial average production rates than the flower-like, commercial (8 m^2/g of surface area) and sol–gel ceria powders, respectively, given a reaction temperature of 900 °C. High H_2 and CO production rates as well as nonstoichiometry up to 0.25 were reported. However, after 10 redox cycles, 89% of the specific surface area and 96% of the pore volume of the particle were lost, resulting in syngas production dropping to 57%.

Tests of ceria oxide in lab-scale reactors were conducted. Warren et al. conducted fixed bed experiments in a solar simulator (Warren et al., 2017). The fixed bed held 1130 g of ceria which was then loaded into the reactor, corresponding to a 1/14th capacity of the reactor. SoCLR with carbon dioxide was operated at a temperature of 1120 °C. The reduction gas was argon diluted methane with a CH_4:Ar ratio of 1:9. The reported methane conversion was 52% with H_2 and CO selectivities of 83% and 59%, respectively. Significant carbon deposition was observed. Low efficiency of the system was obtained, partially due to diluted methane supply, high thermal losses in a prototype, and probably the selection of reactant flow rates and cycling times. Welte et al. investigated the reduction of ceria with diluted CH_4 in a solar particle-transport reactor (Welte et al., 2017). Ceria particles with an mean diameter of 40 mm were dosed downward through a vertically oriented alumina tube enclosed in a solar cavity-receiver, with a flow rate of 0.1–0.6 g/s. Experiments with both co-current and counter-current operations were conducted. Argon diluted CH_4 with a concentration of 2.5–10% and with total flow rates of 0.67–2.69 mmol/s was used in the test, which resulted in gas residence times less than 1 s. A peak thermal efficiency of 12% was projected for operation at 1300 °C with co-current flows of 0.13 g/s of ceria and 2.02 mmol/s of 10% CH_4 in Ar. The projected efficiencies are lower than thermodynamic predictions, partially due to high sensible heating requirements incurred by feeding the ceria particles and heavily diluted CH_4 into the reactor at near ambient temperature. Chuayboon et al. conducted SoCLR with reticulated porous ceria foam that was directly irradiated in a solar concentrator (Chuayboon et al., 2019). Ceria material with a porosity of 91.8% and a surface area of less than 1 m²/g was used. The nominal temperature of the ceria foam was varied between 950 and 1050 °C. Reduction was performed with 33–67% CH_4 in Ar at total flow rates from 11.0 to 20.9 mmol/s per gram of CeO_2. Oxidation was carried out with 55% H_2O in Ar at 16.3 mmol/s per gram of CeO_2. The energetic upgrade factors were 0.97–1.10 and thermal efficiencies were 2.73–5.22%. Severe carbon deposition was observed with a H_2:CO ratio of up to 3.5 in the syngas product. The efficiency was low because of the combined effects of high sensible heating requirements, long reduction and oxidation durations, and the loss of carbon particles. Fosheim et al. operated of a prototype high-flux fixed bed solar reactor with ceria at thermal steady state in a solar simulator for more than 20 redox cycles with CH_4 and CO_2 (Fosheim et al., 2019). First, incremental changes were made in the operating conditions in order to eliminate carbon accumulation and maximize efficiency (Fosheim et al., 2019). Then, the reactor was operated with a CH_4 concentration of 75% at 955 °C and 1000 °C for ten cycles at thermal steady state. Higher temperature promotes better performance of the reactor. At the operating temperature of 1000 °C, CH_4 conversion is 0.36, with H_2 and CO selectivities of 0.90 and 0.82, respectively. The oxidization step had a CO_2 conversion of 0.69. The energetic upgrade factor under this condition was 1.10, with a heat recovery effectiveness of over 95%. Reported solar-to-fuel efficiency and thermal efficiency were 7% and 25%, respectively, with projected 31% and 67% for the full-scale reactor, and 56% and 85% for a commercial reactor with lower thermal losses. The projected scaled-up efficiencies suggest SoCLR could be a competitive approach for the production of solar fuels.

4. Chemical Looping Reform for Selective Oxidation of Methane

There are two major routes to convert natural gas into higher value products: Indirect oxidation and direct oxidation. The indirect oxidation approach converts natural gas to an intermediate, syngas, which is then further converted to desired chemical products. Applying chemical looping in this indirect oxidation approach has been described in section 2. The direct CLR selective oxidation approach, however, converts natural gas directly into the desired value-added chemical products. By removing the intermediate syngas production process, the direct oxidation approach simplifies the overall process operations and, thus, can reduce the capital and operating costs. The major direct oxidation processes of methane to chemical products include selective oxidation to methanol, to formaldehyde (HCHO), and to ethylene and other olefins (Sinev et al., 2009; Wang et al., 1995; Anders, 2009). ARCO and DuPont have developed pilot scale CLR selective oxidation pilot plants. However, there are no commercial-scale CLR selective oxidation systems that exist to date. The highly stable methane molecule poses a challenge for industrial sized direct conversion methods for natural gas to chemicals production. For instance, the Oxidative Coupling of Methane (OCM) reaction must take place at a temperature of around 500–1000 °C due to the stability of the methane molecule (Gesser et al., 1985; Lunsford, 2000; Alvarez-Galvan et al., 2011; Ruiz-Martinez et al., 2016; Fleischer et al., 2016). With this high temperature, the commercialization of such a reaction system is also hindered by the high CH_4 conversion with low selectivity issue.

4.1 Chemical Looping Oxidative Coupling of Methane (CLOCM)

The principle of the chemical looping scheme has been widely used in OCM, especially in pioneering OCM work. Historically, the CLOCM process has been referred to as the reducible catalytic metal oxide OCM approach or the OCM redox approach. The redox cycles and reducible catalytic metal oxides of CLOCM processes are conceptually analogous to redox cycles and oxygen carriers used in CLR processes. However, the controlling factors for the two differ, as the chemical reactions in the CLR processes are dictated by thermodynamics while the reactions in CLOCM are dictated by reaction kinetics. In this sense, the catalytic metal oxides in CLOCM behave more similarly to the functionality and mechanism of catalytic metal oxides in conventional OCM processes.

In typical OCM processes, methane and molecular oxygen are co-fed to a reactor filled with catalyst particles (Armor, 2014). Generally, the co-feed OCM takes place through a heterogeneous–homogeneous reaction pathway (Zavyalova et al., 2011). The desired homogeneous gas-phase reactions compete with heterogeneous reactions which generate thermodynamically stable products of CO and CO_2. As a result, hydrocarbon selectivity is normally inversely proportional to the methane conversion. The highly reactive C_2 intermediate may further take part in the unwanted yet thermodynamically favorable oxidation reactions (Tiemersma et al., 2012). The key parameter that controls this relationship of co-feed OCM performance is the partial pressure of gaseous oxygen (Stangland, 2015). Several micro-kinetic models for co-feed OCM suggested that reactions responsible for CO/CO_2 generation are more dependent on oxygen partial pressure than those for producing C_2 hydrocarbons (Jenkins, 2012; Krylov, 2005).

Chemical looping uses the lattice oxygen from an oxygen carrier that is obtained and transported from the gaseous O_2 from air (Fan et al., 2015; Fan, 2010). This avoids direct contact between gaseous O_2 and the methane, thereby eliminating the need for an ASU, reducing the risk of highly flammable methane–oxygen mixtures, and allowing for inherent separation of CO_2. The chemical looping concept with various oxygen carrier particles and reactor designs has been extensively studied for combustion applications, as well as reforming applications, as described in section 2 (Fan et al., 2015; Fan, 2010).

In 1982, to minimize non-catalyzed gas-phase methane oxidation reactions in studying catalytic oxidative coupling, Keller and Bhasin fed methane and air cyclically over catalytic metal oxides, with a short time of purging with inert gas purge in between (Keller and Bhasin, 1982). Essentially, the catalytic metal oxides used in this OCM mode act as oxygen carriers that separate oxygen from air for methane oxidation by a way of redox reactions. The overall reaction scheme is similar to the catalytic process, except that lattice oxygen from an oxygen carrier is used instead of gaseous oxygen from the air. Significant efforts to commercialize this technology were made by companies like Union Carbide and ARCO, with various reducible, catalytic metal oxides tested for producing higher hydrocarbons up to C_7, such as toluene (Jones et al., 1987; Jones et al., 1984). However, due to the oil price crash in the 1980s, these technologies were abandoned before being scaled up to commercial olefin production or commercial liquid fuel synthesis (Keller and Bhasin, 1982).

4.1.1 CLOCM oxygen carriers

As with other chemical looping processes, the key for success and the efficiency of the CLOCM process is the oxygen carrier material (or so called "reducible catalytic metal oxides"), as they determine the selectivity to the desired products and reactant conversions. For the selective oxidation of methane, many possible reactions are competing with the desired methane to C_2 partial oxidation reaction. In addition to this, the desirable products generated from methane to C_2 partial oxidation reaction can take part in further oxidation reaction and be converted to undesired CO or CO_2.

Major efforts have been made to identify and screen the oxygen carrier candidates from a series of metal oxides (Jones et al., 1987). The screening results in these early tests are given in Table 11. Manganese-based catalytic metal oxides demonstrate high methane conversion, which has provided the basis for further material optimization. When pure Mn_2O_3 was tested, C_2 selectivity did not go beyond 20%. When support materials, such as silica or alumina, were used, the selectivity could be improved by three times. It was observed that braunite, Mn_7SiO_{12}, forms in the silica-supported manganese oxides, and this may be one of the reasons for the improved reaction performance. Furthermore, the addition of alkali components, such as sodium pyrophosphate, could further improve selectivity by about 10%. It was hypothesized that sodium promotes the formation of braunite, thus producing more selective and more active catalytic metal oxides. The metal oxide catalyst $Mn/Na_2WO_4/SiO_2$ has been observed, with methane conversions from 12–31% and C_{2+} selectivities from 60–80% (Wang et al., 1995; Salehoun et al., 2008; Li, 2001; Talebizadeh et al., 2009). Also, there is an indication that the support material should be low in acidity in order to obtain high selectivity. Silica is often used as the support material for this reason (Jones et al., 1987).

However, at a reaction temperature of 800 °C, even the best oxygen carrier with the most promising results, manganese oxides with support, is limited to a 20% of C_2 yield in a fluidized bed reactor, commonly

Table 11. Sample of CLOCM results with various catalytic metal oxides at 800 °C and 860 hr^{-1} GHSV (Jones et al., 1987).

Catalyst	(%) CH$_4$ conversion	Selectivity (%)								
		C$_2$H$_4$	C$_2$H$_6$	C$_3$	C$_4$	C$_5$	Benzene	Toluene	CO	CO$_2$
5% Mn/Al$_2$O$_3$	11		45.0^a							55
5% Mn/SiO$_2$	13.3	21.8	15.7	2.2	0.8	0.1	0.6	0.1	16.5	42.2
15% Mn/SiO$_2$	26	36.3	14.3	3.8	2.3	0.2	1.8	0.3	14.9	26.1
5% Bi/SiO$_2$	5.1	23.1	30.7	1.9	1.2	0.2	0.8	0.3	24.5	17.3
5% Ge/SiO$_2$	1.4	38.5	24.6	3.5	2.3	0.2	1.7	0.2	27	2
5% In/SiO$_2$	7.3	26.7	19.8	2.9	1.1	0.1	1.4	0.2	17.3	30.5
5% Pb/SiO$_2$	2.5	32.6	5.4	3.5	2	0.2	1.9	0.2	26.9	27.3
5% Sb/SiO$_2$	1.2	31.6	27.6	1.1	0.5	0.1	0.9	0.1	38.1	0
5% Sn/SiO$_2$	3.2	26.8	21.2	1.7	1.2	0.1	0.4	0.1	18.2	30.3
α-Al$_2$O$_3$	0.3	50.2	42.2	4.7	1.7	0.1	1	0.1	0	0

^{a} Combined C$_2$H$_4$ and C$_2$H$_6$.

used in the industry (Mleczko et al., 1996; Jašo, 2011). This yield is short of the minimum required yield to make commercial OCM economically feasible (Zavyalova et al., 2011; Keller and Bhasin, 1982; Su et al., 2003). Thus, the redox mode may require innovative reactor design in order to maintain high yields (Mleczko et al., 1996; Jašo, 2011; Zavyalova et al., 2011; Keller and Bhasin, 1982; Su et al., 2003). Previous research has compared the co-feed and chemical looping modes of manganese oxides, and demonstrated that, for the same conversion, higher C$_2$ selectivity occurs in the chemical looping redox mode than in the co-feed mode, as is shown in Figure 27 (Sofranko et al., 1988).

Under a CLOCM scheme, research has been focused on the selection of a suitable reducible metal oxide (Jones et al., 1987; Sofranko et al., 1987; Sofranko et al., 1988). In addition, under a redox operating mode, because of the low oxygen carrying capacity of the reducible catalytic metal oxides, the lattice oxygen is quickly depleted during methane oxidation, rendering it difficult to improve yields. In general, methane conversion decreases and C$_2$ selectivity increases over time, and available lattice oxygen is depleted within 30 minutes. The CLOCM differs from the methane and oxygen co-feed approach since oxygen diffusion within the catalytic metal oxide is critical to its performance. Oxygen diffusion in the catalytic metal oxide could take place via various mechanisms, such as diffusion along the interstices, exchange of vacancies and ions, or simultaneous cyclic replacement of atoms (Kofstad, 1972). If the oxygen diffusion rate is too slow, the oxygen in the bulk phase will have a minimal effect on the reaction. Various chemical composition and defective structures have been tested for their capacity to control the

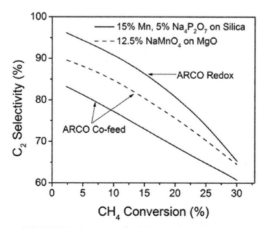

Figure 27. Comparison of CLOCM (redox) to co-feed for Mn-based metal oxides (Sofranko et al., 1988).

change in oxygen mobility, and hence, the reaction rate of the OCM reaction (Greish et al., 2010; Mestl et al., 2001; Sung et al., 2010).

Chung et al. investigated the performance of Manganese-based reducible metal oxides in the co-feed and the chemical looping schemes. The CLOCM experiments were conducted in a fixed bed at 840 °C with alternating methane and oxygen flows at a gas hourly space velocity (GHSV) of 2400 hr^{-1} of methane. The formation of higher hydrocarbons in the experiment was confirmed. A C$_{2+}$ yield of 23.2% with C$_{2+}$ selectivity of 63.24% at a methane conversion of 36.7% was reached (Chung et al., 2015, 2016). Hydrocarbons up to C$_7$ were observed but the C$_5$ and C$_7$ compounds could not be analytically identified.

Il'chenko et al. investigated a series of modified perovskite catalysts based on SrCoO$_3$, where metals of Li, Na and K were added as solid oxidants. It was found that K$_{0.125}$Na$_{0.125}$Sr$_{0.75}$CoO$_{3-x}$ had the highest activity and selectivity in addition to the catalytic stability on OCM without the presence of gaseous oxygen. The oxygen required for the reactions originated from the adopted oxygen on the surface of the catalyst and the lattice oxygen of the particles. The decrease of surface oxygen with time is partly compensated as a result of its diffusion from the volume of the catalyst (Il'chenko et al., 2000).

Novel OCM catalysts, in terms of favoring kinetics of C$_2$ production reactions and catalyst composition optimization, have been developed and studied. Hedrzak and Michorczyk developed a Mn–Na$_2$WO$_4$ catalyst using acrylic templates coupled with different methods of loading the active phases (Michorczyk and Hedrzak, 2017). Hou et al. investigated La-based oxides and proposed a La$_2$O$_2$CO$_3$ catalyst by way of hydrothermal and precipitation method with different morphologies (Hou et al., 2015). Elkins et al. investigated rare-earth oxides and discovered that using Li–TbO$_x$/MgO as a catalyst can achieve a high C$_2$ selectivity with a reaction temperature above 600 °C, however, the activity of the catalyst was not sustained (Elkins et al., 2016). Chung et al. identified a Mg–Mn composite oxygen carrier with a composition of Mg$_6$MnO$_8$ for the CLOCM process. The oxygen carrier showed both stable reactivity and high oxygen carrying capacity during the test of multiple redox cycles (Chung et al., 2016).

Cheng et al. investigated the enhanced C$_2$ selectivity of Mg$_6$MnO$_8$ using a Li dopant for the CLOCM process (Cheng et al., 2018). Li was selected as the dopant because of its high catalytic function in OCM and its similar ionic radius to Mg (Ito and Lunsford, 1985). However, a high Li-dopant concentration may modify the crystal phases of the oxygen carrier which, in turn, leads to a decrease of the oxygen carrying capacity of the oxygen carrier (Qin et al., 2017). Thus, the Li dopant concentration was controlled at a low value of around 1%. The study concluded that the Li-doped oxygen carrier in CLOCM universally has a higher C$_2$ selectivity than the undoped Mg$_6$MnO$_8$ oxygen carrier, with a maximum selectivity improvement of about 50%. Density functional theory calculations and redox experiments were conducted in order to reveal the reaction enhancement mechanism of the oxygen carrier. The way in which the doping-induced oxygen vacancy affects the selectivity of the oxygen carrier was revealed, which, in turn, provides a dopant-screening strategy for identifying a high-performance catalytic oxygen carrier for CLOCM.

4.1.2 CLOCM reaction kinetics

Oxygen diffusion in the catalytic oxygen carriers is an additional factor that increases the complexity of the CLOCM reaction kinetics, since diffusion of the lattice oxygen from the bulk to the surface must be considered. Reshetnikov et al. derived a relationship between oxygen diffusion and the OCM reaction for a K$_{0.125}$Na$_{0.125}$Sr$_{0.75}$CoO$_{3-x}$ perovskite (Reshetnikov et al., 2011). The oxygen carrier was identified as having the best performance among a series of SrCoO$_3$-based perovskite catalysts by Il'chenko et al. (Il'chenko et al., 2000). The model theoretically explained the effect of the mobility of the oxygen in a perovskite catalyst on the dynamics of the catalytic reaction of the OCM.

During the OCM reaction process, the catalytic metal oxide perovskite was assumed to be comprised of catalytic metal oxide centers with adsorbed oxygen (ZO), catalytically active metal oxide centers (Z), and reduced inactive catalytic metal oxide centers (Z$_R$). The rates of the reactions of the adsorbed oxygen on the catalytic metal oxide surface and methane, active catalytic metal oxide center and methane, and re-oxidation of the reduced catalytic metal oxide, are given as, $r_{ZO} = k_{ZO}C\theta_{ZO}$, $r_Z = k_ZC\theta_Z$, and $r_{ZR} = k_{ZR}\theta_{ZR}\sigma_s$, respectively. In the expressions, r is the rate of reaction, k is the rate constant of the rate equation, C is the mole fraction of methane, θ is the fraction of the ZO, Z and ZR, respectively, with $\theta_{ZO} + \theta_Z + \theta_{ZR} = 1$, and σ_s is the oxidizability of the catalytic metal oxide surface.

To simplify the analysis, it was assumed that the reactor is a continuous stirred tank reactor (CSTR). At the surface of the catalytic metal oxide, the changing rates of methane concentration and fractions of the three metal oxide centers as a function of time can be expressed as,

$$\frac{dC}{dt} = \frac{1}{\tau_g}(C^f - C) + a\sum_{k=1}^{N_r} v_{ki}(C,\theta)$$

$$\frac{d\theta_j}{dt} = \sum_{k=1}^{N_r} v_{ki} r_k(C,\theta)$$

$$C = C^0, \theta_j = \theta_j^0 \quad at \quad t = 0$$

The oxidizability at the surface of the catalytic metal oxide is a function of oxygen diffusion from the crystal lattice to the particle surface and the surface reduction rate due to reaction with methane, and is calculated by,

$$\frac{\partial \sigma}{\partial \tau} = \frac{1}{\varphi^2}\frac{\partial^2 \sigma}{\partial \xi^2}$$

The left-hand side of the equation represents the changing rate of oxidizability on the catalytic metal oxide surface, and the right-hand side represents the oxygen diffusion rate within the lattice. The boundary conditions of the equation are

$$\begin{cases} \dfrac{d\sigma}{d\xi_0} = 0 & at & \xi = 0 \\[2mm] \dfrac{d\sigma}{d\xi_1} = -\varphi^2 \theta_{ZR}\sigma_S & at & \xi = 1 \end{cases}$$

where $\varphi^2 = L^2 k_s/D$ is the analog Thiele parameter, D is the effective coefficient of volumetric oxygen diffusion in the catalytic metal oxide, L is the characteristic dimension of the oxide crystallite, $\tau = k_{ZR}t$ is the dimensionless time, and $\xi = l/L$ is a dimensionless coordinate in the crystallite.

The model was solved by numerical integration using the Runge-Kutta method with solution of the diffusion equation by the marching technique at each step of integration. The calculated coefficients obtained from the experiments with $K_{0.125}Na_{0.125}Sr_{0.75}CoO_{3-x}$ perovskite as catalytic metal oxide were $k_{zo} = 1$ s^{-1}, $k_z = 1.1e^{-2}$ s^{-1}, and $k_{ZR} = 1.2e^{-3}$ s^{-1} (Reshetnikov et al., 2011).

It was found that, in this model, the Thiele parameter was the most important factor to evaluate the effect of oxygen diffusivity in the catalytic metal oxide on the dynamics of the OCM reactions. When the Thiele parameter is small, the diffusion rate of lattice oxygen is high, hence, the rate of surface oxygen regeneration is fast. With a small value of φ, the lattice oxygen concentration will be uniform within the crystallite. However, when the value of φ is large, a lattice oxygen gradient will occur in the crystallite since the oxygen diffusivity is insufficient to re-oxidize all the reduced catalytic metal oxide surface in time. Hence, the methane conversion decreases with time due to a decrease in oxidizability of the surface. Reshetnikov et al. found that φ has to be less than 7, a value coinciding with the one obtained from the dynamics of the reactions occurring in the presence of gaseous oxygen (Ostrovskii and Reshetnikov, 2005) in order to let the surface reaction not be limited by oxygen diffusion in the crystallite. Also, as the characteristic size L for mixed metal oxide crystallites falls in the range 2–30 nm, the possible value for D was found to be in the range of 10^{-18} to 10^{-16} cm^2/s, a range for typical value of the effective coefficient of oxygen diffusion in metal oxides (Kofstad, 1972).

4.1.3 CLOCM reactor design

A typical continuous CLOCM system consists of two reactors, where one reactor performs the OCM reaction (reducer) and the other regenerates the catalytic oxygen carrier (combustor), with the catalytic oxygen carrier particles circulating between them. The reactors can be operated in either fluidized bed or moving bed modes. Another possible CLOCM can be made using a single reactor with a periodically operated mode. Instead of circulating between different reactors, the catalytic oxygen carrier particles stay in the same reactor, while undergoing subsequent reduction and oxidation reactions at different times by means of either feeding methane or air to the reactor.

ARCO dual circulating fluidized beds

Circulating fluidized beds possess excellent heat and mass transfer properties, which is crucial to the success of the OCM redox approach. The hydrodynamics of circulating fluidized beds, which allow the gas and solid flows in the circulating systems to be well controlled, are also well known.

In 1985, ARCO patented a double circulating fluidized bed configuration for its CLOCM process (Jones et al., 1985). It was claimed that improved results were obtained by employing a process wherein solids are continuously recirculated between two physically separated zones, a methane contact zone (reducer reactor) and an oxygen contact zone (combustor reactor). By maintaining fluidized beds of solids in the two reactors, the average solids residence time in each reactor was controlled. The mixing of the oxygen carrier particles and methane was also ensured. One possible layout of ARCO's process is exemplified in Figure 28 (Jones et al., 1987). Using this system, ARCO tested its catalytic metal oxide for 30,000 redox cycles over a period of six months. During the operation, the catalytic metal oxide maintained its reactivity, selectivity and fluidization properties (Sofranko and Jubin, 1989).

Fixed bed reactor design

The Lab-scaled reactors for CLOCM processes normally use the cyclic gas feeding to a single fixed bed reactor, as shown in Figure 29 (Fleischer et al., 2016; Parishan et al., 2018). It is based on the idea of dynamic experiments, where the feed is switched between different reactants. In the first step, oxygen is sent to the reactor to serve as the catalyst oxidation process. In the second step, methane is sent to the reactor for OCM reaction and reduction of the catalyst. Between the two steps, purging the reactor with inert gas is conducted so as to prevent oxygen and methane from mixing. A proper dosing strategy enables a continuous operation of the process.

Technische Universität Berlin realized such a system with two independently operated six-port pulse valves on the fixed bed reactor and tested $Na_2WO_4/Mn/SiO_2$ catalyst material. 2 grams of catalyst were loaded to the reactor. Repetitive continuous simulated CLOCM was carried out for a total time of 300 minutes and a total of 100 redox cycles, 50 cycles at 775 °C followed by another 50 cycles at 800 °C, with a CH_4 flow rate of 25 ml/min.

Moving bed reactor design

A moving bed reactor has been applied for the experimental study on physical separation of hydrocarbon products and methane for co-feed OCM operation. Kruglow et al. demonstrated the operation of a countercurrent moving-bed chromatographic reactor (CMBCR) (Kruglov et al., 1996). In the operation

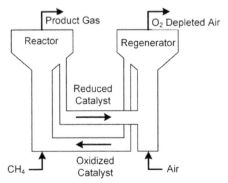

Figure 28. ARCO chemical looping dual circulating fluidized bed OCM process (Jones et al., 1987).

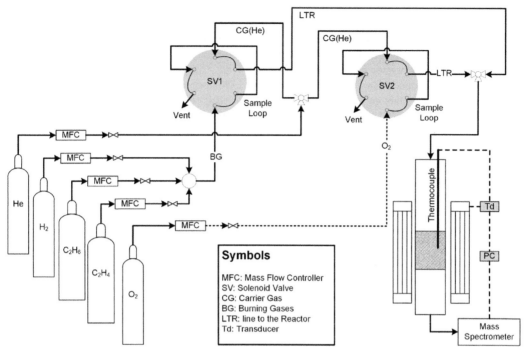

Figure 29. Schematic of the CLOCM experimental setup at Technische Universität Berlin.

of the unit, C_2 products were separated from unconverted reactants at low conversions and recycled back to the reactor. A 0.55 C_2 yield was reached in the experiment.

The moving bed reactor design can also be applied in a CLOCM process. It has been suggested that a moving bed reactor design would be beneficial for CLOCM applications (Iglesia, 2002; Sofranko et al., 1987). However, a countercurrent moving bed reducer as used in CMBCR will put the desired products at the gas outlet in contact with the highest oxidation state of the catalytic metal oxide, resulting in undesired oxidation reactions to CO_x, thus leading to a lower selectivity of chemicals. Therefore, for a CLOCM process, a cocurrent moving bed is more feasible. Such a cocurrent moving bed CLOCM reactor system, whose design shall be very similar to that of a moving bed CLR, as discussed in section 2, would have many advantages over fluidized bed and fixed CLOCM reactors.

4.1.4 Process engineering of CLOCM

As of now, research into OCM processes is being focused on catalyst development, with little focus on reactor design. The development of the overall process, which considers all the steps, from feedstock to purified end products, is rarely investigated. However, the overall process development is a step that cannot be overlooked in the commercialization of the OCM process because heat integration, product separation and process design also play significant role on the final design of the system and the economic feasibilities of the process.

From a process standpoint, the CLOCM approach possesses several inherent advantages over the methane-oxygen co-feed approach. With the use of catalytic metal oxide instead of gaseous oxygen, the undesired gas-phase oxidative reactions can be effectively inhibited, the reactor can be operated beyond the flammability limit of methane, and the heat generated from the OCM reactions can be effectively managed. As there is a large amount of heat released from the OCM reactions, which, if not removed properly, will, in turn, increase the reaction temperature of the reactor and negatively impact the C_2 yields as the combustion reaction of methane and intermediate products to CO_x are thermodynamically favorable under high temperature. The heat management is very important for the continuous and reliable operation of the OCM reactor. Contrary to co-feed OCM processes, which use a single fixed bed reactor with the inherent defect of heat removal, the CLOCM process circulates catalytic metal oxide solids between different reactors, which provides multiple heat integration and heat removal options. When the two reactors are operated under fluidized bed or moving bed mode, an in-bed heat exchanger can be easily placed inside the reactor for heat removal.

Chung et al. conducted a simulation study of the CLOCM process with Aspen Plus®. The process flow diagram of the product generation section of the process is shown in Figure 30. Two reactors, two gas-solid separators, heat exchangers, a water pump, and an air compressor are used in the process. The reduced oxygen carrier was pneumatically transported between the reducer and combustor while undergoing redox reactions. Gas-solids separators were placed after each reactor in order to continually cycle the solids between the two reactors. Heat exchangers are used to minimize the use of utilities, both hot and cold, through heat exchange with existing process streams and utilization of the heat of reaction. In the simulation, a CLOCM process that consumes 1,000 kmol/hr methane in the reducer was considered. Figure 30 shows the general process flow diagram and provides the temperatures of equipment and streams. While downstream processing, either separations or upgrading, which requires a high pressure OCM product, the gaseous stream from the reducer is always cooled to 40 °C. The depleted air is also cooled to 40 °C in order to maintain the temperature differential necessary for heat exchange. Based on a heat balance calculation, the OCM product generation section can generate excess heat for electricity generation or heating requirement from other sections of the OCM process.

From Figure 30, the simplest heat integration option would be to use heat of the O_2-depleted air from the combustor outlet to heat the natural gas feedstock before it is sent to the reducer and use the heat of the OCM products from the reducer gas outlet to heat compressed air for the combustor operation. In addition, the heat exchanger can be placed in the two reactors to remove the excess heat from the reactions, and the steam generated from the heat exchanger can produce 9.7 MWe of electricity. The electricity can be used to operate auxiliary equipment with excess electricity. Other heat integration options for this CLOCM process can also be developed. The reactants can be partially heated either directly or indirectly using the exothermic heat of reaction. Combinations of reactant–product and reactant–heat of reaction can also be performed.

Figure 31 shows the general process flow diagram for ethylene separation from OCM. The numerous stages, temperature differences between inlet gas and purified products, and overall complexity of the separations process causes the separation section to account for approximately half of the capital cost

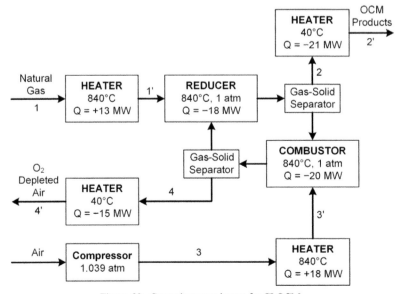

Figure 30. General process layout for CLOCM.

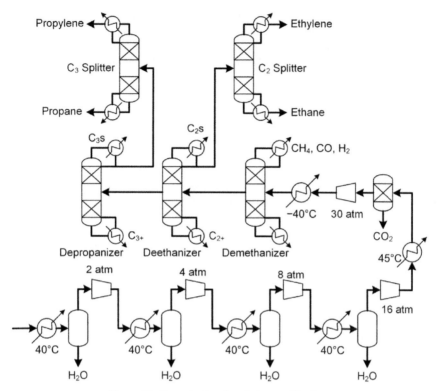

Figure 31. General set-up for ethylene purification.

of the entire OCM process. The product stream from the OCM reactor is cooled, compressed, rid of acid gases, dried, further compressed, and finally separated into individual components via cryogenic distillation.

The complexity of the traditional ethylene separation process, combined with the unique composition exiting the OCM reactor, provides an opportunity to develop alternative separation schemes. The use of silver complexes to selectively bind ethylene is one such promising idea. Complexation is used at industrial scale, but so far, no large-scale process for ethylene separation is in operation. Membranes have been researched for separation of a species from a gas mixture and can be applied for ethylene separation, either from typical pyrolysis gas or from OCM product gas. Issues with cost, durability, and separation efficiency have limited the industrial practice of membranes for such separation tasks. One final idea is to directly consume or convert the hydrocarbons in the product stream as they are a rich blend of various hydrocarbons.

From a process design standpoint, both co-feed and CLOCM provide numerous advantages over traditional steam cracking for ethylene production. First, methane is directly upgraded to a high value product, ethylene, without intermediates. Second, the OCM reactions are highly exothermic, whereas steam cracking is endothermic, reducing energy requirements. CLOCM is advantageous over co-feed OCM in this aspect since heat integration options are more flexible and easily applied. Finally, separation equipment is reduced using OCM since the separations require only a demethanizer, deethanizer, and C_2 splitter. Additional pathways for direct product upgrading are also possible without the need for separations.

4.2 Selective partial oxidation of CH_4 (SPOM) to formaldehyde (HCHO)

Formaldehyde (HCHO) is an important industrial chemical that is a precursor for the production of several resins used in the automobile and textile industries, in addition to its uses in disinfectants and biocides. Initially, CH_4 was converted to HCHO by a three-step process, through CH_4 to syngas, syngas to methanol, followed by air oxidation of methanol over a silver or iron molybdate catalyst to HCHO. The importance of HCHO as an industrial precursor led to research on direct oxidation of CH_4 to HCHO with investigations of various oxidants, catalyst formulations, and operating conditions. One-step HCHO production from methane has been the most commonly studied over supported catalytic metal oxides, which include SiO_2, MgO, MoO_3, V_2O_5, WO_3, and ZnO. Their catalytic activity and performance as support materials were studied (Hall et al., 1995).

Catalytic metal oxides utilize lattice oxygen, but have the added participation of the catalytic surface, which promotes formation of the desired products by incorporation of lattice oxygen, while suppressing further oxidation. This is similar to the CLOCM process. The main difference is that, in the CLOCM scheme, the catalytic metal oxides are re-oxidized in a separate reactor by air, whereas here, the catalytic metal oxides are simultaneously regenerated in the same reactor using molecular oxygen. On the other hand, according to the reaction mechanism of the SPOM to HCHO with catalytic metal oxides, the process could also be operated in the chemical looping mode, with cyclic reduction and regeneration of the catalytic metal oxides, and an effective reactor design. Thus, researches on the properties of some popular catalytic metal oxides that lead to high HCHO selectivity and the mechanisms are introduced here.

Recently, MoO_3 and V_2O_5-based catalysts are the most widely studied as they are considered to be the most effective for the selective partial oxidation of methane to HCHO (Hall et al., 1995). They have been extensively studied for the effect of their different physicochemical properties, such as their structure, dispersion of the metal oxide species, density of active sites. Their reducibility on the selective oxidation of CH_4 SPOM to HCHO were detailed investigated.

Faraldos et al. (Faraldos et al., 1996) investigated SiO_2 supported V_2O_5 and MoO_3 and compared their activity towards SPOM to HCHO. It was found that SiO_2 supported V_2O_5 exhibits a higher CH_4 conversion as compared to supported MoO_3. To achieve the same methane conversion, V_2O_5-based catalysts require a lower onset reaction temperature as compared to MoO_3, which is also an indication of the higher reactivity of V_2O_5. Similar observations were reported from other studies (Spencer and Pereira, 1989; Parmaliana et al., 1991; Miceli et al., 1993). It was reported that with SiO_2 supported MoO_3, CO_2 is produced even at low CH_4 conversions, and the CO_2 selectivity remained constant over the entire range of CH_4 conversions studied. SiO_2 supported V_2O_5, on the other hand, formed no CO_2 at low CH_4 conversions, with increasing CO_2 selectivity when the CH_4 conversion is increased. This suggests that, for SiO_2 supported MoO_3 catalyst, CO_2 is the primary product, whereas for SiO_2 supported V_2O_5 catalyst, it is formed by further oxidation of CO.

As for the HCHO selectivity of the two catalysts with similar metal oxide surface concentrations, MoO_3 catalyst always exhibits a higher HCHO selectivity than V_2O_5 under all methane conversion conditions. Faraldos et al.'s study suggested that there is no real correlation between HCHO selectivity and the reducibility of the two supported catalytic metal oxides, which is in agreement with other similar studies (Miguel A. Bañares et al., 1994).

It has been reported that SiO_2 supported MoO_3 does not adsorb oxygen but some degree of adsorption occurs on SiO_2 supported V_2O_5 (Miguel A. Bañares et al., 1994; Haber, 1979; Kartheuser et al., 1993). Thus, in the case of vanadium oxide, there is a higher interaction between the gaseous oxygen and the active sites. Isotopic studies have shown that selective oxidation of CH_4 to HCHO involves the incorporation of lattice oxygen from the catalytic metal oxides (Bañares et al., 1993; Kartheuser et al., 1993; Koranne et al., 1994; Mauti and Mims, 1993). The vacancies created on the metal oxide surface are filled by gaseous oxygen. In the case of SiO_2 supported V_2O_5, in addition to the lattice oxygen, surface-adsorbed oxygen species are also present, which makes the vanadium oxide more reactive and promotes further oxidation of HCHO to carbon oxides, resulting in a lower HCHO selectivity. Thus, while HCHO is the primary product for both SiO_2 supported MoO_3 and V_2O_5, the absence of adsorbed oxygen in molybdates suppresses the further oxidation of HCHO to CO and CO_2, leading to higher selectivity.

The catalytic conversion of a hydrocarbon molecule to a desired chemical is often highly sensitive to its structure. Specifically, selectivity of the product is dependent on the distribution of the crystal planes (Andersson and Hansen, 1988). Catalysts that exhibit this kind of behavior fall under the category of catalytic anisotropy (Volta et al., 1979), where the difference between crystal planes depends on the ratio of various active sites on each plane. The selectivity of a catalytic metal oxide can be improved by selecting suitable bond strengths and active sites. Establishing the relation between product selectivity and crystal planes/active sites is the first step in exploring the potential of catalytic metal oxides. MoO_3 exhibits such catalytic anisotropy.

Smith and Ozkan (Smith and Ozkan, 1993a) investigated the effect of structural specificity of unsupported MoO_3 on its reactivity and selectivity by preparing MoO_3 samples using temperature programmed techniques to preferentially expose different crystal planes. The selectivity and production rates of methane oxidation on MoO_3–C and MoO_3–R, which were synthesized using two different techniques to expose different crystal planes of MoO_3, were investigated. It was revealed that MoO_3–C had a significantly higher selectivity and production rate toward HCHO than MoO_3–R. The much higher concentration of (100) side planes for MoO_3–C was considered to contribute to the higher selectivity of MoO_3–C towards HCHO. While the (010) basal planes which tend to form carbon oxide products are in larger concentrations in MoO_3–R. The higher HCHO production rate for MoO_3–C also confirmed the importance of the (100) side plane in HCHO formation. Laser Raman spectroscopy, used to identify the type of Mo and O sites in the two different MoO_3 structures, suggests that MoO_3–C has more exposed terminal Mo=O sites. Hence, the Mo=O sites are concluded to be the main surface species that promotes partial oxidation of CH_4. On the other hand, the bridged Mo–O–Mo sites, which have a higher density in MoO_3–R surface, promote the formation of complete oxidation products. Consequently, oxidation of HCHO over the two different MoO_3 catalytic metal oxides revealed that MoO_3–R is more active towards complete oxidation.

The reducibility of both types of site was tested with temperature programmed reduction (TPR) experiments by subjecting both MoO_3–C and MoO_3–R to H_2. The experimental results suggested that MoO_3–R is more readily reducible than MoO_3–C, meaning that the Mo–O–Mo sites are easier to reduce than the Mo=O sites. The differences in reducibility of these sites are reflected in the selectivity towards HCHO. Metal oxides that are more difficult to reduce have a higher energy barrier for the formation of complete and partial oxidation products (CO_x) and are more selective towards formation of HCHO. The difference, with respect to selectivity, between these two sites reveals the preferred structure of MoO_3 for selective oxidation of methane and provides insight when designing a MoO_3 catalytic metal oxide.

With the preferred planes and sites for partial oxidation of CH_4 identified, isotopic labeling experiments using $^{16}O_2$ and $^{18}O_2$ as reactive gases on both MoO_3–C and MoO_3–R samples revealed the reaction pathway (Smith and Ozkan, 1993b). The isotopic experiments investigated the interaction between oxygen and the metal oxide surface in the absence of CH_4, as well as the source of O atoms in the products during CH_4 oxidation. The experiment results suggested that MoO_3–R is more susceptible to exchange of O atoms than MoO_3–C. It was found that MoO_3–R contributed ~ 50% more ^{16}O in the gas phase than the contribution of MoO_3–C and at more than twice the rate. The high rate of desorption and exchange of lattice oxygen in the metal oxide surface, combined with the comparatively easier reducibility, allow for the surface Mo–O–Mo site to be more active to donate lattice oxygen than the surface Mo=O site. Such high activity promotes the formation of complete oxidation products with methane as the reactant.

Isotopic oxygen switching experiments with CH_4 were conducted in order to determine the source of oxygen in the products. Normalized CO_2 isotope concentrations suggested that oxygen from MoO_3–R depletes faster than that from MoO_3–C. Isotopes of H_2O also showed similar trends. HCHO isotope concentrations suggested that the reaction path is highly dependent on the oxygen atoms from the metal oxide and not the gaseous O_2. However, the gaseous oxygen plays a role in replenishing the oxygen atoms on the surface.

The reduction and oxidation mechanism on the catalytic metal oxide surface is dependent on the type of the terminal metal–oxygen bond. Laser Raman spectroscopy on the reduced and re-oxidized MoO_3 sample revealed that the bridged site with Mo–O–Mo termination is more readily oxidized by gaseous oxygen. Mo=O site replenishment, on the other hand, is preferred via oxygen diffusion through the catalytic metal oxide lattice. Since HCHO is mainly formed from ^{16}O, it is concluded that the Mo=O site is key to the selectivity of HCHO. The differences between the re-oxidation mechanisms and the relative population of each bond on MoO_3–C and MoO_3–R distinguish the HCHO selectivity.

In practice, the structural advantage of MoO_3 can be used to promote the activity of silica. Arena et al. (Arena et al., 1997) extensively studied the conversion of CH_4 to HCHO on SiO_2 supported MoO_3 by varying the amount of MoO_3 on silica and evaluating their interaction with CH_4. The effect of MoO_3 loading has been studied using H_2 and CH_4 TPR and high temperature oxygen chemisorption (HTOC) experiments. H_2-TPR profiles of catalytic metal oxides of MoO_3 on precipitated silica with concentrations of 2 wt%, 4 wt% and 7 wt% were compared with pure precipitated silica and bulk MoO_3. For all MoO_3 on precipitated silica catalytic metal oxides, the onset temperature and peak locations are lower than pure precipitated silica and bulk MoO_3. This suggests a positive interaction that results in different phases of MoO_3 on silica (Arena and Parmaliana, 1996).

Methane reactivity with each catalytic metal oxide was studied using CH_4-TPR experiments. The rate of oxygen consumption was calculated by performing an oxygen balance through the total summation of oxygen atoms in product molecules such as HCHO, CO, CO_2, and H_2O. CH_4-TPR further confirmed the findings of the H_2-TPR studies. 7 wt% MoO_3 on precipitated silica exhibited the highest lattice oxygen consumption rate, while 2 wt% MoO_3 on precipitated silica had the lowest among all the MoO_3 on precipitated silica metal oxides. Pure precipitated silica exhibited low reactivity toward CH_4 and only increased slightly with temperature. HCHO and carbon are the only products formed in the absence of oxygen. For MoO_3 on precipitated silica catalytic metal oxides, reactivity with methane is slow, at 650 °C, but increases drastically at higher temperatures. The CH_4 conversion rate is inversely proportional to Mo loading. Oxygenated product formation also increased at beyond 650 °C. At the same time, high HCHO selectivity is only observed at low temperatures.

5. Conclusion

Owing to its low life cycle greenhouse emission, the abundantly available natural gas is a prime fuel choice for most countries to meet their growing need for energy with low greenhouse gas emissions. Researchers are investigating novel methods to efficiently utilize natural gas for industrial gas and chemical production. There are two major routes to convert natural gas into higher value products: Indirect oxidation and direct oxidation. The indirect oxidation approach converts natural gas to an intermediate, syngas, which is then further converted to desired chemical products. The direct selective oxidation approach, instead, converts natural gas directly into the desired value-added chemical products.

Chemical looping reforming represents a potentially efficient way for reforming natural gas to syngas as compared to traditional natural gas reforming schemes. The chemical looping reforming concept refers to the use of a metal oxide to partially oxidize natural gas to syngas while regenerating the reduced metal with air and/or steam. The advantage of CLR is the ability to produce syngas without the need for external heating or molecular oxygen, resulting in reduced unit operations, greater syngas yield, and minimal CO_2 emissions. Research and development work on advancing CLR processes to commercially relevant testing is ongoing around the world. Much of this research is directed at determining the oxygen carrier material and reactor design that can maximize syngas yield and natural gas conversion.

Fluidized bed CLR processes consist of reducer reactors operating as dense fluidized beds. Testing at pilot and lab scales show that up to ~ 70% syngas purity can be generated from this system by adjusting the air-to-fuel ratio from above 1 (for combustion) to about 0.4 (for reforming) without steam or CO_2 co-feed. Co-feeding of steam and/or CO_2 has been investigated to mitigate carbon deposition issues and to increase overall syngas yield. However, the wide oxidation state distribution in a fluidized bed reducer reactor design may result in some carbon deposition on the over-reduced oxygen carrier. The moving bed CLR process can efficiently convert feedstock into a high-quality syngas using an iron based oxygen carrier. The uniform and controllable residence time of oxygen carrier particles allows the oxidation state of the oxygen carrier at the reducer outlet to be controlled to a thermodynamically favorable condition for maximum syngas generation. In addition, by limiting the reduction degree of oxygen carrier to avoid the formation of metallic iron, carbon deposition is effectively inhibited. With steam or CO_2 co-feed options, the H_2:CO ratio in the syngas can be adjusted to meet the downstream processing requirement, for example, a H_2:CO ratio of 2:1 for Fischer–Tropsch synthesis or methanol production. Other options include co-feeding solid feedstocks, such as coal or biomass, in order to effectively adjust the H_2:CO ratio of the syngas. By recycling all the CO_2 produced from the downstream catalytic syngas conversion units back to the CLR moving bed reducer reactor, the CLR process can achieve CO_2 neutral or negative operations at the plant site. As the amount of CO_2 fed into the CLR reducer produces a nonlinear effect on the amount and purity of syngas produced, the CLR reducer can be operated as a series of parallel modules that can reduce overall plant costs and maximize the plant efficiency under the given operating conditions.

SoCLR processes combine the solar thermal process with CLR, thus eliminating the need for ASU, also providing much better heat management and effectively utilizing solar energy for chemical production and electricity generation. From a thermal energy storage (TES) perspective, the solar reforming combined cycle can be viewed as a heat conversion process that stores intermittent solar thermal energy as latent chemical energy. Under this design, the steam reforming catalyst can serve as the heat transfer fluid to absorb and transport the thermal energy throughout the process. SoCLR for syngas and hydrogen generation combines the reaction engineering of metal oxide lattice oxygen transfer with the process design of concentrating solar power systems. It has the potential to reduce carbonaceous fuel consumption associated with conventional syngas/hydrogen production. The key technical challenge of this technology is the selection of a suitable metal oxide whose required properties are similar to other chemical looping processes of combustion, gasification and reforming. The solar thermochemical process for methane reforming to produce syngas replaces fossil fuels with solar energy to supply the energy for the endothermic heat of reaction. This design can potentially reduce the fossil fuel consumption in a reforming process by 20–35% (He and Li, 2014). In addition, using a metal oxide-based reforming catalyst as the heat transfer fluid has the potential to simplify the process flow and improve the economics.

Direct chemical synthesis methods, including OCM and selective oxidation of methnol, offer operation options with fewer units for producing the desired chemical products. ARCO and DuPont have developed pilot scale CLR selective oxidation plants to produce olefins via OCM. Though the highly stable methane molecule poses challenges in maximizing methane conversion while maintaining high selectivity of the desired product, much research has been directed toward developing oxygen carrier particles for the selective catalytic oxidation of methane. Chemical looping direct chemical synthesis has many advantages. First, methane is directly upgraded to a high value product without intermediates. Second, the reactions are exothermic, reducing energy requirements, and thus having flexible and easily applied heat integration options. Finally, separation equipment is reduced. Additional pathways for direct product upgrading are also possible without the need for separations. CLOCM and methane to formaldehyde represent two example chemical looping schemes for selective oxidation with a metal oxygen carrier.

References

Adanez, J., Abad, A., Garcia-Labiano, F., Gayan, P. and De Diego, L.F. 2012. Progress in chemical-looping combustion and reforming technologies. Prog. Energy Combust. Sci. 38: 215–282.

Alvarez-Galvan, M.C. et al. 2011. Direct methane conversion routes to chemicals and fuels. Catal. Today 171: 15–23.

Anders, H. 2009. Direct conversion of methane to fuels and chemicals. Catal. Today 142: 2–8.

Andersson, A. and Hansen, S. 1988. Catalytic anisotropy of MoO_3 in the oxidative ammonolysis of toluene. J. Catal. 114: 332–346.

Arena, F. and Parmaliana, A. 1996. Silica-supported molybdena catalysts. Surface structures, reduction pattern, and oxygen chemisorption. J. Phys. Chem. 100: 19994–20005.

Arena, F., Giordano, N. and Parmaliana, A. 1997. Working mechanism of oxide catalysts in the partial oxidation of methane to formaldehyde: II. Redox properties and reactivity of SiO_2, MoO_3/SiO_2, V_2O_5/SiO_2, TiO_2, and V_2O_5/TiO_2 systems. J. Catal. 167: 66–76.

Arjmand, M., Azad, A.-M., Leion, H., Lyngfelt, A. and Mattisson, T. 2011. Prospects of Al_2O_3 and $MgAl_2O_4$-supported CuO oxygen carriers in chemical-looping combustion (CLC) and chemical-looping with oxygen uncoupling (CLOU). Energy and Fuels 25: 5493–5502.

Armor, J.N. 2014. Key questions, approaches, and challenges to energy today. Catal. Today 236: 171–181.

Ashcroft, A.T. et al. 1990. Selective oxidation of methane to synthesis gas using transition metal catalysts. Nature 344: 319–321.

Bañares, M.A., Fierro, J.L.G and Moffat, J.B. 1993. The partial oxidation of methane on MoO_3/SiO_2 catalysts: Influence of the molybdenum content and type of oxidant. J. Catal. 142: 406–417.

Bañares, M.A., Spencer, N.D., Jones, M.D. and Wachs, I.E. 1994. Effect of alkali metal cations on the structure of Mo(VI) SiO_2 catalysts and its relevance to the selective oxidation of methane and methanol. J. Catal. 146: 204–210.

Barelli, L., Bidini, G., Gallorini, F. and Servili, S. 2008. Hydrogen production through sorption-enhanced steam methane reforming and membrane technology: A review. Energy 33: 554–570.

Bergmann, F.J. 1897. Process for the Production of Calcium Carbide in Blast Furnaces. German Patent.

Cha, K.S. et al. 2009. Reaction characteristics of two-step methane reforming over a Cu-ferrite/Ce-ZrO_2 medium. Int. J. Hydrogen Energy 34: 1801–1808.

Cha, K.S. et al. 2010. A study on improving reactivity of Cu-ferrite/ZrO_2 medium for syngas and hydrogen production from two-step thermochemical methane reforming. Int. J. Energy Res. 34: 422–430.

Cheng, Z. et al. 2018. C_2 selectivity enhancement in chemical looping oxidative coupling of methane over a Mg-Mn composite oxygen carrier by Li-doping-induced oxygen vacancies. ACS Energy Lett. 3: 1730–1736.

Cho, P., Mattisson, T. and Lyngfelt, A. 2005. Carbon Formation on Nickel and Iron Oxide-Containing Oxygen Carriers for Chemical-Looping Combustion. 668–676.

Chong, Y.-O., Nicklin, D.J. and Tait, P.J. 1986. Solids exchange between adjacent fluid beds without gas mixing. Powder Technol. 47: 151–156.

Chuayboon, S., Abanades, S. and Rodat, S. 2019. Syngas production via solar-driven chemical looping methane reforming from redox cycling of ceria porous foam in a volumetric solar reactor. Chem. Eng. J. 356: 756–770.

Chueh, W.C. and Haile, S.M. 2009. Ceria as a thermochemical reaction medium for selectively generating syngas or methane from H_2O and CO_2. ChemSusChem. 2: 735–739.

Chueh, W.C. and Haile, S.M. 2010. A thermochemical study of ceria: Exploiting an old material for new modes of energy conversion and CO_2 mitigation. Philos. Trans. R. Soc. A Math. Phys. Eng. Sci. 368: 3269–3294.

Chung, E.Y. et al. 2015. Examination of Oxidative Coupling of Methane by Traditional Catalysis and Chemical Looping with Manganese-Based Oxides. Presented at 2015 AIChE Fall Meeting. Salt Lake City. UT. November 8–13.

Chung, E.Y. et al. 2015. Process development of manganese-based oxygen carriers for oxidative coupling of methane in a pressurized chemical looping system. Presented at 2015 AIChE Spring Meeting and 11th Global Congress on Progress Safety. Austin. TX. April 26–30.

Chung, E.Y. 2016. Investigation of Chemical Looping Oxygen Carriers and Processes for Hydrocarbon Oxidation and Selective Alkane Oxidation to Chemicals. Ph.D. Dissertation. The Ohio State University. Columbus. OH.

Chung, E.Y. et al. 2016. Catalytic oxygen carriers and process systems for oxidative coupling of methane using the chemical looping technology. Ind. Eng. Chem. Res. 55: 12750–12764.

Contractor, R.M. 1999. Dupont's CFB technology for maleic anhydride. Chem. Eng. Sci. 54: 5627–5632.

Dai, X.P. et al. 2006. Hydrogen production from a combination of the water-gas shift and redox cycle process of methane partial oxidation via lattice oxygen over $LaFeO_3$ perovskite catalyst. J. Phys. Chem. B 110: 25856–25862.

Deshpande, N., Majumder, A., Qin, L. and Fan, L.-S. 2015. High-pressure redox behavior of iron-oxide-based oxygen carriers for syngas generation from methane. Energy and Fuels 29: 1469–1478.

Dobbyn, R.C. et al. 1978. Evaluation of the Performance of Materials and Components Used in the CO_2 Acceptor Process Gasification Pilot Plant. Final report No. DOE/ET/10253-T1. Continental Oil Co., Library, PA (USA).

Dudukovic, M.P. 2009. Frontiers in reactor engineering. Science 325: 698–701.

Elkins, T.W., Roberts, S.J. and Hagelin-Weaver, H.E. 2016. Effects of alkali and alkaline-earth metal dopants on magnesium oxide supported rare-earth oxide catalysts in the oxidative coupling of methane. Appl. Catal. A Gen. 528: 175–190.

Energy Information Administration. International Energy Outlook. 2011. Natural Gas: World Natural Gas Consumption by region, Reference Case.

Evanko, B.W. et al. 2013. Efficient generation of H_2 by splitting water with an isothermal redox cycle. Science 341: 540–542.

Fan, J., Zhu, L., Jiang, P., Li, L. and Liu, H. 2016. Comparative exergy analysis of chemical looping combustion thermally coupled and conventional steam methane reforming for hydrogen production. J. Clean. Prod. 131: 247–258.

Fan, L.-S. 2010. Chemical Looping Systems for Fossil Energy Conversions. John Wiley & Sons, Hoboken, NJ.

Fan, L.-S., Luo, S. and Zeng, L. 2013. Methods for Fuel Conversion. U.S. Patent Application PCT/US2014/014877.

Fan, L.-S. 2015. Metal Oxide Reaction Engineering and Particle Technology Science: A Gateway to Novel Energy Conversion Systems. 2015 AIChE Annual Meeting.

Fan, L.-S., Zeng, L. and Luo, S. 2015. Chemical-looping technology platform. AIChE J. 61: 2–22.

Fan, L.-S. 2017. Chemical Looping Partial Oxidation: Gasification, Reforming, and Chemical Syntheses. Cambridge University Press.

Fan, L.-S., Empfield, A., Kathe, M. and Blair, E. 2017. Chemical Looping Syngas Production from Carbonaceous Fuels. U.S. Patent PCT/US2017/027241.

Faraldos, M. et al. 1996. Comparison of silica-supported MoO_3 and V_2O_5 catalysts in the selective partial oxidation of methane. J. Catal. 160: 214–221.

Fleischer, V., Littlewood, P. Parishan, S. and Schomäcker, R. 2016. Chemical looping as reactor concept for the oxidative coupling of methane over a $Na_2WO_4/Mn/SiO_2$ catalyst. Chem. Eng. J. 306: 646–654.

Fosheim, J.R., Hathaway, B.J. and Davidson, J.H. 2019. High efficiency solar chemical-looping methane reforming with ceria in a fixed-bed reactor. Energy 169: 597–612.

Gao, X. et al. 2016. Efficient ceria nanostructures for enhanced solar fuel production: Via high-temperature thermochemical redox cycles. J. Mater. Chem. A 4: 9614–9624.

Gasior, S.J. 1961. Production of Synthesis Gas and hydrogen by the Steam-Iron Process: Pilot Plant Study of Fluidized and Free-Falling Beds No. BM-RI-5911. Bur. Mines, Washington, DC (USA).

Gas-to-Liquids Conversion, Presented at Natural Gas Conversion Technologies Workshop of ARPA-E.

Gesser, H.D., Hunter, N.R. and Prakash, C.B. 1985. The direct conversion of methane to methanol by controlled oxidation. Chem. Rev. 85: 235–244.

Gollener, J.F. et al. 2013. Analysis of natural gas-to liquid transportation fuels via fischer-tropsch, DOE report, DOE/NETL-2013/1597.

Greish, A.A. et al. 2010. Oxidative coupling of methane in the redox cyclic mode over the catalysts on the basis of CeO_2 and La_2O_3. Mendeleev Commun. 20: 28–30.

Haber, J. 1979. Oxygen in catalysis on transition metal oxides. Catal. Rev. 19: 1–41.

Hall, T.J., Hargreaves, J.S.J., Hutchings, G.J., Joyner, R.W. and Taylor, S.H. 1995. Catalytic synthesis of methanol and formaldehyde by partial oxidation of methane. Fuel Process. Technol. 42: 151–178.

He, F. and Li, F. 2014. Hydrogen production from methane and solar energy—Process evaluations and comparison studies. Int. J. Hydrogen Energy 39: 18092–18102.

Hedrzak, E. and Michorczyk, P. 2017. The application of 3d printing in the designing of channel structures in monolithic catalysts dedicated to the oxidative coupling of methane. Technical Transactions 3: 31–40.

Hickman, D.A. and Schmidt, L.D. 1993. Production of syngas by direct catalytic oxidation of methane. Science 259: 343–346.

Hou, Y.-H., Han, W.-C., Xia, W.-S. and Wan, H.-L. 2015. Structure sensitivity of $La_2O_2CO_3$ catalysts in the oxidative coupling of methane. ACS Catal. 5: 1663–1674.

Hurst, S. 1939. Production of hydrogen by the steam-iron method. Oil Soap 16: 29–35.

Iglesia, E. 2002. Challenges and progress in the conversion of natural gas to fuels and chemicals. ACS Div. Fuel Chem. Prepr. 47: 128–131.

Il'chenko, N.I., Pavlenko, N.V., Raevskaya, L.N. and Bostan, A.I. 2000. Influence of the composition of co-containing perovskites on their catalytic properties in the conversion of methane into higher hydrocarbons in non-stationary conditions. Theor. Exp. Chem. 36: 48–53.

Ishida, M., Zheng, D. and Akehata, T. 1987. Evaluation of a chemical-looping-combustion power-generation system by graphic exergy analysis. Energy 12: 147–154.

Ito, T. and Lunsford, J.H. 1985. Synthesis of ethylene and ethane by partial oxidation of methane over lithium-doped magnesium oxide. Nature 314: 721–722.

Jašo, S. 2011. Modeling and Design of the Fluidized Bed Reactor for the Oxidative Coupling of Methane. Ph.D. Dissertation. Technische Universität Berlin. Berlin. Germany. 1–162.

Jenkins, S. 2012. Shale gas ushers in ethylene feed shifts. Chem. Eng. 119: 17–19.

Jerndal, E., Mattisson, T., Thijs, I., Snijkers, F. and Lyngfelt, A. 2009. NiO particles with Ca and Mg based additives produced by spray-drying as oxygen carriers for chemical-looping combustion. Energy Procedia 1: 479–486.

Johansson, E., Mattisson, T., Lyngfelt, A. and Thunman, H. 2006. A 300 W laboratory reactor system for chemical-looping combustion with particle circulation. Fuel 85: 1428–1438.

Jones, C.A., Leonard, J.J. and Sofranko, J.A. 1984. Methane Conversion. U.S. Patent 4,443,645.

Jones, C.A., Leonard, J.J. and Sofranko, J.A. 1984. Methane Conversion. U.S. Patent 4,443,646.

Jones, C.A., Leonard, J.J. and Sofranko, J.A. 1984. Methane Conversion. U.S. Patent 4,443,647.

Jones, C.A., Leonard, J.J. and Sofranko, J.A. 1984. Methane Conversion. U.S. Patent 4,443,648.

Jones, C.A., Leonard, J.J. and Sofranko, J.A. 1984. Methane Conversion. U.S. Patent 4,443,649.

Jones, C.A., Leonard, J.J. and Sofranko, J.A. 1984. Methane Conversion. U.S. Patent 4,444,984.

Jones, C.A., Leonard, J.J. and Sofranko, J.A. 1984. Methane Conversion. U.S. Patent 4,443,644.

Jones, C.A., Leonard, J.J. and Sofranko, J.A. 1985. Methane Conversion. U.S. Patent 4,560,821.

Jones, C.A., Leonard, J.J. and Sofranko, J.A. 1987. Fuels for the future: Remote gas conversion. Energy and Fuels 1: 12–16.

Kartheuser, B., Hodnett, B.K., Zanthoff, H. and Baerns, M. 1993. Transient experiments on the selective oxidation of methane to formaldehyde over V_2O_5/SiO_2 studied in the temporal-analysis-of-products reactor. Catal. Letters 21: 209–214.

Kathe, M. et al. 2017. Modularization strategy for syngas generation in chemical looping methane reforming systems with CO_2 as feedstock. AIChE J. 63: 3343–3360.

Keller, G.E. and Bhasin, M.M. 1982. Synthesis of Ethylene via Oxidative Coupling of Methane. 73: 9–19.

Kodama, T., Shimizu, T., Satoh, T., Nakata, M. and Shimizu, K.I. 2002. Stepwise production of CO-rich syngas and hydrogen via solar methane reforming by using a Ni(II)-ferrite redox system. Sol. Energy 73: 363–374.

Kodama, T. 2003. Thermochemical methane reforming using a reactive WO_3/W redox system. Energy 41: 304.

Kodama, T., Gokon, N. and Yamamoto, R. 2008. Thermochemical two-step water splitting by ZrO_2-supported $NixFe_{3-x}O_4$ for solar hydrogen production. Sol. Energy 82: 73–79.

Kofstad, P. 1972. Diffusion and electrical conductivity in binary metal oxides. Nonstoichiometry. 289.

Kofstad, P. 1972. Nonstoichiometry, Diffusion and Electrical Conductivity in Binary Metal Oxides. John Wiley & Sons, New York, NY.

Koranne, M.M., Goodwin, J.G. and Marcelin, G. 1994. Oxygen involvement in the partial oxidation of methane on supported and unsupported V_2O_5. J. Catal. 148: 378–387.

Krenzke, P.T., Fosheim, J.R., Zheng, J. and Davidson, J.H. 2016. Synthesis gas production via the solar partial oxidation of methane-ceria redox cycle: Conversion, selectivity, and efficiency. Int. J. Hydrogen Energy 41: 12799–12811.

Kronberger, B. et al. 2004. A two-compartment fluidized bed reactor for CO_2 capture by chemical-looping combustion. Chem. Eng. Technol. 27: 1318–1326.

Kruglov, A.V., Bjorklund, M.C. and Carr, R.W. 1996. Optimization of the simulated countercurrent moving-bed chromatographic reactor for the oxidative coupling of methane. Chem. Eng. Sci. 51: 2945–2950.

Krylov, O.V. 2005. Methods for increasing the efficiency of catalysts for the oxidative condensation of methane. Russ. Chem. Rev. 61: 851–858.

Lane, H. 1913. Process of producing hydrogen. U.S. Patent 1,078,686

Lewis, W. and Edwin, G. 1954. Production of pure carbon dioxide. U.S. Patent 2,665,971.

Li, F. et al. 2009. Syngas chemical looping gasification process: Oxygen carrier particle selection and performance. Energy and Fuels 23: 4182–4189.

Li, F., Luo, S., Sun, Z., Bao, X. and Fan, L.-S. 2011. Role of metal oxide support in redox reactions of iron oxide for chemical looping applications: Experiments and density functional theory calculations. Energy Environ. Sci. 4: 3661–3667.

Li, S.-B. 2001. Oxidative coupling of methane over W-Mn/SiO_2 catalyst. Chinese J. Chem. 19: 16–21.

Linderholm, C., Mattisson, T. and Lyngfelt, A. 2009. Long-term integrity testing of spray-dried particles in a 10-kW chemical-looping combustor using natural gas as fuel. Fuel 88: 2083–2096.

Lunsford, J.H. 2000. Catalytic conversion of methane to more useful chemicals and fuels: A challenge for the 21st century. Catal. Today 63: 165–174.

Luo, S. et al. 2014. Shale gas-to-syngas chemical looping process for stable shale gas conversion to high purity syngas with a H_2:CO ratio of 2:1. Energy Environ. Sci. 7: 4104–4117.

Luo, S., Zeng, L. and Fan, L.-S. 2015. Chemical looping technology: oxygen carrier characteristics. Annu. Rev. Chem. Biomol. Eng. 6: 53–75.

Lyngfelt, A. 2015. Oxygen carriers for chemical-looping combustion. Calcium Chem. Looping Technol. Power Gener. Carbon Dioxide Capture. Woodhead Publishing. 221–254.

Matherne, J.L. and Culp, G.L. 1992. Direct conversion of methane to C2's and liquid fuels: process economics. Chapter 14. pp. 463–482. *In*: Wolf, E.E. (ed.). Methane Conversion by Oxidative Processes: Fundamental and Engineering Aspects. Van Nostrand Reinhold. New York. NY.

Mattisson, T. and Lyngfelt, A. 2001. Applications of chemical-looping combustion with capture of CO_2. Second Nordic Minisymposium on CO_2 Capture and Storage, Göteborg, Sweden.

Mauti, R. and Mims, C.A. 1993. Oxygen pathways in methane selective oxidation over silica-supported molybdena. Catal. Letters 21: 201–207.

Messerschmitt, A. 1915. Process of producing hydrogen. U.S. Patent 971,206.

Mestl, G., Nagy, A.J. and Schlögl, R. 2001. The role of subsurface oxygen in the silver-catalyzed, oxidative coupling of methane. J. Catal. 188: 58–68.

Miceli, D., Arena, F., Parmaliana, A., Scurrell, M.S. and Sokolovskii, V. 1993. Effect of the metal oxide loading on the activity of silica supported MoO_3 and V_2O_5 catalysts in the selective partial oxidation of methane. Catal. Letters 18: 283–288.

Mleczko, L., Pannek, U., Niemi, V.M. and Hiltunen, J. 1996. Oxidative coupling of methane in a fluidized-bed reactor over a highly active and selective catalyst. Ind. Eng. Chem. Res. 35: 54–61.

Nalbandian, L., Evdou, A. and Zaspalis, V. 2011. $La_{1-x}Sr_xM_yFe_{1-y}O_{3-\delta}$ Perovskites as oxygen-carrier materials for chemical-looping reforming. Int. J. Hydrogen Energy 36: 6657–6670.

National Energy Technology Laboratory (NETL). 2010. Cost and performance baseline for fossil energy plants, Volume 1: Bituminous coal and natural gas to electricity. Revision 2. November. DOE/NETL-2010/1397. United States Department of Energy.

Neumann, D., Kirchhoff, M. and Veser, G. 2004. Towards an efficient process for small-scale, decentralized conversion of methane to synthesis gas: Combined reactor engineering and catalyst synthesis. Catal. Today 98: 565–574.

Ostrovskii, N.M. and Reshetnikov, S.I. 2005. The influence of oxygen mobility in solid catalyst on transient regimes of catalytic reaction. Chem. Eng. J. 107: 141–146.

Otsuka, K., Wang, Y., Sunada, E. and Yamanaka, I. 1998. Direct partial oxidation of methane to synthesis gas by cerium oxide. J. Catal. 175: 152–160.

Otsuka, K., Wang, Y. and Nakamura, M. 1999. Direct conversion of methane to synthesis gas through gas±solid reaction using CeO_2-ZrO_2 solid solution at moderate temperature. Appl. Catal. A Gen. 183: 317–324.

Otsuka, K., Sunada, E., Ushiyama, T. and Yamanaka, I. 2007. The production of synthesis gas by the redox of cerium oxide. 107: 531–536.

Pans, M.A. et al. 2013. Optimization of H_2 production with CO_2 capture by steam reforming of methane integrated with a chemical-looping combustion system. Int. J. Hydrogen Energy 38: 11878–11892.

Parishan, S., Littlewood, P., Arinchtein, A., Fleischer, V. and Schomäcker, R. 2018. Chemical looping as a reactor concept for the oxidative coupling of methane over the Mn_xO_y-Na_2WO_4/SiO_2 catalyst, benefits and limitation. Catal. Today 311: 40–47.

Parmaliana, A. et al. 1991. Factors controlling the reactivity of the silica surface in methane partial oxidation. Appl. Catal. 78: L7–L12.

Pröll, T., Rupanovits, K., Kolbitsch, P. and Hofbauer, H. 2005. Cold flow model study on a dual circulating fluidized bed (DCFB) system for chemical looping processes. 32: 418–424.

Pröll, T., Kolbitsch, P., Bolhàr-Nordenkampf, J. and Hofbauer, H. 2009. A novel dual circulating fluidized bed system for chemical looping processes. AIChE J. 55: 3255–3266.

Pröll, T., Hofbauer, H., Kolbitsch, P. and Bolha, P. 2010. Operating experience with chemical looping combustion in a 120 kW dual circulating fluidized bed (DCFB) unit. Int. J. Greenh. Gas Control 4: 180–185.

Pröll, T., Bolhàr-Nordenkampf, J., Kolbitsch, P. and Hofbauer, H. 2010. Syngas and a separate nitrogen/argon stream via chemical looping reforming—A 140 kW pilot plant study. Fuel 89: 1249–1256.

Qin, L. et al. 2017. Improved cyclic redox reactivity of lanthanum modified iron-based oxygen carriers in carbon monoxide chemical looping combustion. J. Mater. Chem. A 5: 20153–20160.

Quality Guidelines for Energy System Studies; Specification for Selected Feedstocks. 2012. DOE/NETL-341/011812.

Reshetnikov, S.I., Pyatnitskii, Y.I. and Dolgikh, L.Y. 2011. Effect of the mobility of oxygen in perovskite catalyst on the dynamics of oxidative coupling of methane. Theor. Exp. Chem. 47: 49–54.

Rostrup-Nielsen, J.R., Sehested, J. and Nørskov, J.K. 2002. Hydrogen and synthesis gas by steam-and CO_2 reforming. Adv. Catal. 47: 65–139.

Ruiz-Martinez, J. et al. 2016. Strategies for the direct catalytic valorization of methane using heterogeneous catalysis: challenges and opportunities. ACS Catal. 6: 2965–2981.

Rydén, M. and Lyngfelt, A. 2006. Using steam reforming to produce hydrogen with carbon dioxide capture by chemical-looping combustion. Int. J. Hydrogen Energy 31: 1271–1283.

Rydén, M., Lyngfelt, A. and Mattisson, T. 2006. Synthesis gas generation by chemical-looping reforming in a continuously operating laboratory reactor. Fuel 85: 1631–1641.

Rydén, M., Lyngfelt, A. and Mattisson, T. 2008. Chemical-looping combustion and chemical-looping reforming in a circulating fluidized-bed reactor using Ni-based oxygen carriers. Energy and Fuels 22: 2585–2597.

Rydén, M., Johansson, M., Lyngfelt, A. and Mattisson, T. 2009. NiO supported on Mg-ZrO_2 as oxygen carrier for chemical-looping combustion and chemical-looping reforming. Energy Environ. Sci. 2: 970–981.

Salehoun, V., Khodadadi, A., Mortazavi, Y. and Talebizadeh, A. 2008. Dynamics of Mn/Na_2WO_4/SiO_2 catalyst in oxidative coupling of methane. Chemical Engineering Science 63: 4910–4916.

Shimizu, T., Shimizu, K., Kitayama, Y. and Kodama, T. 2001. Thermochemical methane reforming using WO_3 as an oxidant below 1173 K by a solar furnace simulator. Sol. Energy 71: 315–324.

Sinev, M.Y., Fattakhova, Z.T., Lomonosov, V.I. and Gordienko, Y.A. 2009. Kinetics of oxidative coupling of methane: Bridging the gap between comprehension and description. J. Nat. Gas Chem. 18: 273–287.

Smith, M.R. and Ozkan, U.S. 1993. The partial oxidation of methane to formaldehyde: role of different crystal planes of MoO_3. J. Catal. 141: 124–139.

Smith, M.R. and Ozkan, U.S. 1993. Transient isotopic labeling studies under steady-state conditions in partial oxidation of methane to formaldehyde over MoO_3 catalysts. J. Catal. 142: 226–236.

Sofranko, J.A., Leonard, J.J. and Jones, C.A. 1987. The oxidative conversion of methane to higher hydrocarbons. J. Catal. 103: 302–310.

Sofranko, J.A., Leonard, J.J., Jones, C.A., Gaffney, A.M. and Withers, H.P. 1988. Catalytic oxidative coupling of methane over sodium-promoted Mn/SiO_2, and Mn/MgO. Catal. Today 3: 127–135.

Sofranko, J.A. and Jubin, J.C. 1989. Natural Gas to Gasoline: the ARCO GTG Process. Symposium on Methane Activation, Conversion, and Utilization. International Congress of Pacific Basin Societies. Honolulu. HI. December 17–19.

Spencer, N.D. and Pereira, C.J. 1989. $V_2O_5$$SiO_2$-catalyzed methane partial oxidation with molecular oxygen. J. Catal. 116: 399–406.

Stangland, E.E. 2015. The Shale Gas Revolution: A Methane-to-Organic Chemicals Renaissance? Dow Chemical Company.

Steinfeld, A., Kuhn, P. and Karni, J. 1993. High-temperature solar thermochemistry: Production of iron and synthesis gas by Fe_3O_4-reduction with methane. Energy 18: 239–249.

Steinfeld, A., Frei, A., Kuhn, P. and Wuillemin, D. 1995. Solar thermal production of zinc and syngas via combined ZnO-reduction and CH_4-reforming processes. Int. J. Hydrogen Energy 20: 793–804.

Stenberg, V., Rydén, M., Mattisson, T. and Lyngfelt, A. 2018. Exploring novel hydrogen production processes by integration of steam methane reforming with chemical-looping combustion (CLC-SMR) and oxygen carrier aided combustion (OCAC-SMR). Int. J. Greenh. Gas Control 74: 28–39.

Su, Y.S., Ying, J.Y. and Green, W.H. 2003. Upper bound on the yield for oxidative coupling of methane. J. Catal. 218: 321–333.

Summers, W. 2014. Baseline Analysis of Crude Methanol Production from Coal and Natural Gas. Natl. Energy Technol. Lab. DOE/NETL-341/101514. 1–83.

Sung, J.S. et al. 2010. Peculiarities of oxidative coupling of methane in redox cyclic mode over Ag-La$_2$O$_3$/SiO$_2$ catalysts. Appl. Catal. A Gen. 380: 28–32.

Talebizadeh, A., Mortazavi, Y. and Khodadadi, A.A. 2009. Comparative study of the two-zone fluidized-bed reactor and the fluidized-bed reactor for oxidative coupling of methane over Mn/Na$_2$WO$_4$/SiO$_2$ catalyst. Fuel Process. Technol. 90: 1319–1325.

Teed, P.L. 1919. The Chemistry and Manufacture of Hydrogen. Nature 103: 442–444.

Tiemersma, T.P., Tuinier, M.J., Gallucci, F., Kuipers, J.A.M. and van Sint Annaland, M. 2012. A kinetics study for the oxidative coupling of methane on a Mn/Na$_2$WO$_4$/SiO$_2$ catalyst. Appl. Catal. A Gen. 433–434: 96–108.

Tsang, S.C., Claridge, J.B. and Green, M.LH. 1995. Recent advances in the conversion of methane to synthesis gas. Catal. Today 23: 3–15.

U.S. Department of Energy. 1979. Development of the Steam–Iron Process for Hydrogen Production. Publication Number EF-77-C-01-2435. Pittsburgh. PA.

U.S. Energy Information Administration, Natural Gas. http://www.eia.gov/naturalgas/weekly/ (Accessed on 17th August 2016).

Volta, J.C., Desquesnes, W., Moraweck, B. and Coudurier, G. 1979. A new method to obtain supported oriented oxides: MoO$_3$ graphite catalyst in propylene oxidation to acrolein. React. Kinet. Catal. Lett. 12: 241–246.

Wang, D. and Fan, L.-S. 2015. L-valve behavior in circulating fluidized beds at high temperatures for group D particles. Ind. Eng. Chem. Res. 54: 4468–4473.

Wang, D.J., Rosynek, M.P. and Lunsford, J.H. 1995. Oxidative coupling of methane over oxide-supported sodium-manganese catalysts. J. Catal. 155: 390–402.

Warren, K. et al. 2017. Theoretical and experimental investigation of solar methane reforming through the nonstoichiometric ceria redox cycle. Energy Technol. 5: 2138–2149.

Welte, M., Warren, K., Scheffe, J.R. and Steinfeld, A. 2017. Combined ceria reduction and methane reforming in a solar-driven particle-transport reactor. Ind. Eng. Chem. Res. 56: 10300–10308.

Welty, J.A.B. 1951. Apparatus for conversion of hydrocarbons. U.S. Patent 2,550,741.

Zavyalova, U., Holena, M., Schlögl, R. and Baerns, M. 2011. Statistical analysis of past catalytic data on oxidative methane coupling for new insights into the composition of high-performance catalysts. ChemCatChem. 3: 1935–1947.

Zeng, L., Kathe, M.V., Chung, E.Y. and Fan, L.-S. 2012. Some remarks on direct solid fuel combustion using chemical looping processes. Curr. Opin. Chem. Eng. 1: 290–295.

CHAPTER 12

Alternative Pathways for CO_2 Utilization via Dry Reforming of Methane

Mohamed S Challiwala,[1,2,3] *Shaik Afzal*,[2] *Hanif A Choudhury*,[1,3]
Debalina Sengupta,[2,3] *Mahmoud M El-Halwagi*[2,3] *and Nimir O Elbashir*[1,3,*]

1. Introduction

The reforming of methane is an important step in the production of a valuable chemical precursor "Syngas" or synthesis gas. Syngas is a mixture of carbon monoxide and hydrogen and serves as an important intermediate to produce a large range of value-added chemicals (i.e., methanol, acetic acid, dimethyl ether, etc.) and ultra clean fuels via Fischer Tropsch (FT) technology (Abusrafa et al., 2019; Alsuhaibani et al., 2019; Choudhury et al., 2019). Commercially, the reforming of methane is done via three well-known technologies: Steam Reforming of Methane (SRM), Partial Oxidation of Methane (POX) and Autothermal Reforming (ATR). All the aforementioned reforming technologies utilize oxidants to chemically convert methane to syngas. For instance, SRM utilizes steam at high pressure, while partial oxidation utilizes oxygen to produce syngas. ATR utilizes a combination of both steam and oxygen in a specific ratio that yields a syngas ratio compatible with a downstream FT reaction, while allowing the reaction to occur under auto-thermal condition. The auto-thermal condition signifies a point in temperature wherein no external heat is required to drive the reaction, in other words, the reaction produces enough energy *in situ* to drive itself spontaneously. Equations (1)–(5) summarize the stoichiometric reactions and associated energy requirements for SRM, POX and ATR reactions:

SRM:

$$CH_4 + H_2O \rightarrow 2H_2 + CO \qquad \Delta H_{298K} = 206 \text{ kJ/mol} \qquad (1)$$

POX:

$$CH_4 + \frac{1}{2}O_2 \rightarrow 2H_2 + CO \qquad \Delta H_{298K} = -36 \text{ kJ/mol} \qquad (2)$$

ATR:

$$CH_4 + 1\frac{1}{2}O_2 \rightarrow CO + 2H_2O \qquad \Delta H_{298K} = -520 \text{ kJ/mol} \qquad (3)$$

[1] Petroleum Engineering Program & Chemical Engineering Program, Texas A&M University at Qatar, Education City, Doha 23874, Qatar.
[2] Texas A&M University, College Station 77840, TX, USA.
[3] Gas and Fuels Research Center, Texas A&M Engineering Experiment Station, College Station, TX, 77843, USA.
* Corresponding author: nelbashir@tamu.edu

$$CH_4 + H_2O \rightarrow CO + 3H_2 \qquad\qquad \Delta H_{298K} = 206 \text{ kJ/mol} \qquad\qquad (4)$$

$$CO + H_2O \rightarrow CO_2 + H_2 \qquad\qquad \Delta H_{298K} = -41 \text{ kJ/mol} \qquad\qquad (5)$$

First introduced in the 1930s (Van Hook, 1980), SRM technology was implemented at a very small scale and was limited to a few locations in the United States with easy access to natural gas. The major development of this technology took place in the 1960s, when ICI started two reformer plants using tubular reactors at high pressure condition using naphtha as the primary feedstock (Rostrup-Nielsen, 2004). However, almost a decade before ICI setup their facility, Haldar Topsoe had designed their first reformer and hydrogen plant based on the SRM technology at 40 bar pressure (Rostrup-Nielsen, 2004). The primary utilization of syngas from SRM was for hydrogen production for the Ammonia plants, which is a starting material for urea fertilizer. The implementation of Topsoe technology had significantly reduced the energy demands of the ammonia plants and resulted in significant cost savings (Dybkjaer, 1995). Further energy reductions were realized when M.W. Kelloggs built integrated reformer plants combining SRM with steam turbines. As naphtha was the primary feedstock in Europe, the steam reforming of naphtha became a key technology in building up the town gas industry of the United Kingdom, which consequently replaced erstwhile low-pressure gasification processes. Apart from the technologies of ICI and Topsoe (Rostrup-Nielsen, 2004), SRM at low temperatures under adiabatic conditions, which was pioneered by British Gas, served as an attractive option to produce methane rich gas for heating and other utility purposes (Appl and Gössling, 1972). Later, the focus of SRM was also shifted to methanol, acetic acid synthesis and other oxo-alcohols apart from Ammonia production (Rostrup-Nielsen, 1984). The low-temperature reforming process developed by British Gas had served to be an important pathway for conversion of naphtha to methane during the energy crisis of the 1970s. The application of SRM was not only limited to the production of hydrogen, methanol and other petrochemicals, but could also be used in the iron and steel industry for direct reduction of iron-ore (Appl and Gössling, 1972; Mondal Kartick et al., 2016). The trend in 1990s was mostly towards the hydrocracking of heavy hydrocarbons for production of gasoline fuels in refineries (Rostrup-Nielsen and Rostrup-Nielsen, 2002). The later part of the 1990s and the early 2000s saw a great demand for hydrogen production from SRM mainly due to two reasons: (a) the availability of cheap natural/shale gas, and (b) the boom in fuel cell technology and refineries around the globe. Although renewable technologies, such as wind, solar, biofuels and other alternative technologies, drew great attention in 2000s, the dependency on SRM for hydrogen production has never declined (Rostrup-Nielsen and Rostrup-Nielsen, 2002).

The main challenge associated with SRM from the time of its inception was its inherent ability to produce solid coke at low steam to carbon ratios. This challenge was addressed by operation at extremely high temperatures (> 1000 °C), and in the presence of large quantities of steam. The presence of enormous quantities of steam however lead to lower conversions and resulted in a significant rise in energy costs. Nevertheless, the utilization of SRM was never less popular throughout its age. Another less commonly adopted method of carbon reduction was to passivate with Sulphur (Rostrup-Nielsen, 1984), however, this resulted in the contamination of products, leading to issues in downstream FT and other synthesis processes. Noble metals, such as Pt and Rh, also provided an alternative catalytic system to Ni, but economically less viable options for reduction of coking tendency of the SRM reaction (Trimm, 1997).

In later 1970s, CO_2 utilization as a part of SRM provided an attractive pathway of stabilizing the syngas ratio of the SRM product (Rostrup-Nielsen et al., 1988; Rostrup-Nielsen, 2002). However, like SRM, the possibility of coke formation in the combined reaction was extremely high. Formation of coke from this reaction had always been a crucial factor limiting the application of this reaction on standalone basis. However, the presence of CO_2 in SRM was still necessary for adjusting the quality of syngas for downstream applications such as FT and methanol synthesis reactions that require a H_2:CO ratio of 2:1. The introduction of ATR technology also provided an attractive, albeit expensive, pathway to alleviate the coking tendency in SRM since its inception in the early 1980s (Rostrup-Nielsen et al., 1988; Rostrup-Nielsen, 2002). Apart from the SRM process, POX has also been a well-known and established technology since the 1940s (Prettre et al., 1946). POX is relatively simple reaction compared to SRM and DRM, as it typically represents the partial combustion of methane. The challenge with this reaction is the cost associated with the production of pure oxygen. However, the benefit is that it could be conducted under homogenous conditions at high temperature without the need for a catalyst. Low temperature operation requires a catalyst, which would result in lower energy requirements than high temperature homogeneous reactions. Additionally, this reaction could also be beneficial in generating high pressure steam as the combustion process is highly exothermic. In 2007, Shell built the world's largest gas-to-liquids (GTL) plant, the Pearl GTL Plant. The plant uses POX reforming technology to produce syngas and has the world's largest air separation units (ASU) for oxygen generation. Due to the benefit of economy of scale, the syngas production by POX at a large scale (140,000 bbl/day syncrude capacity plant) is much cheaper than smaller scale plants (Wood et al., 2012). The combination of POX and SRM as ATR has also gained significant traction recently. A joint venture between SASOL and Qatar Petroleum (QP) built a state-of-the-art GTL facility, the ORYX GTL plant, in Ras Laffan, Qatar using ATR technology. The capacity of this plant is about 35,000 bbl/day of syncrude products, mainly diesel, and naphtha. In this backdrop, the Dry Reforming of Methane (DRM) process becomes important to consider since it can produce syngas by utilizing CO_2 as the oxidant, as shown in equation (6). Since CO_2 is an abundantly available in the flue gas streams (albeit at lower concentrations), it is anticipated that using it will reduce the operating costs and carbon footprint of syngas production.

DRM:

$$CH_4 + CO_2 \rightarrow 2H_2 + 2CO \qquad\qquad \Delta H_{298K} = 247 \text{ kJ/mol} \qquad\qquad (6)$$

The major challenges associated with DRM are the high coking tendency, high endothermic nature and low quality of syngas ratio. Syngas ratio is particularly important as only a specific quality of syngas is desirable for downstream application. For instance, in FT reaction and methanol synthesis, a syngas ratio of 2:1 is acceptable. However, as evident from equation (6) below for DRM, the H_2:CO ratio is 1:1 only. Also, due to the reverse Water Gas Shift reaction activity at high temperatures, the H_2/CO ratio is lowered further.

The energy demand to drive the reaction is also very high, it requires about 1.2 times more energy than the endothermic SRM process. The CO_2 that results from combusting methane or other fossil fuels to drive this reaction can significantly increase the carbon footprint of the DRM process. Nevertheless, if the reaction is coupled with other renewable sources of energy like solar or from excess heat from other parts of the plant, the CO_2 fixation from this process would be meaningful.

Carbon formation in the DRM process generally takes place via following pathways:

Methane Decomposition:

$$CH_4 \rightarrow 2H_2 + C \qquad\qquad\qquad\qquad (7)$$

Boudouard Reaction:

$$2CO \rightarrow C + CO_2 \qquad\qquad\qquad\qquad (8)$$

The aforementioned reactions (Methane decomposition and Boudouard reaction) are highly active at low temperature conditions (400 °C to 650 °C), while at high temperatures (beyond 800 °C), the rate of primary reforming reactions is much higher, resulting in significantly lower coke formation. The other problem with coke formation is that it physically damages the reactor tube that would either require serious maintenance or complete replacement of the tube. Due to these three limitations, DRM is still considered to be a grey area that requires further investigations.

Based on extensive modelling and experimental work, it was observed that DRM can be implemented either in combination with other reforming technologies (Challiwala et al., 2017; Challiwala et al., 2017; Elbashir et al., 2018), or by implementing a separate scheme that could either handle the carbon formed or adjust the syngas ratio as per downstream requirements. The two new pathways developed for DRM reaction based on the concept of reaction segregation are as follows:

i) CARGEN (Or **CAR**bon **GEN**erator) Technology—a combination of low temperature carbon forming process and high temperature syngas reformer using a two-reactor setup (Elbashir et al., 2018).

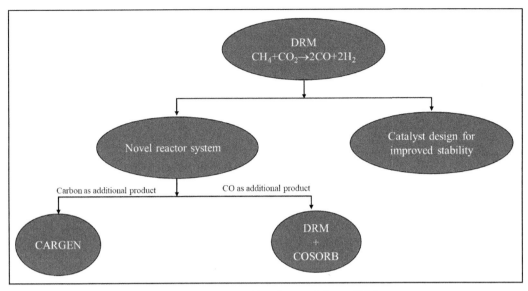

Figure 1. Strategy to address issues relating to DRM.

ii) DRM+COSORB Technology—a combination of DRM and a separate CO absorption unit (COSORB—**CO** Ab**SORB**tion) that removes CO to adjust the syngas ratio that meets downstream specifications. Figure 1 below provides a flowchart describing the options available to manage and implement DRM technology on a larger scale (Afzal et al., 2018).

In this chapter, these two process configurations are discussed in terms of the concept development, the merits and the challenges for implementation of these technologies.

2. Methodology Followed

2.1 Estimation of equilibrium composition

Reforming reactions generally take place at high temperature and pressure conditions, typically in the range of 900–1100 °C and 15–25 bar, in order to achieve the maximum allowable equilibrium conversion. The equilibrium composition of gas phase reaction is estimated via the Gibbs free energy minimization method. Gibbs free energy (or the chemical potential of the system) is the total free energy available in the system to do useful or external work, and is a function of temperature, pressure and composition of the system.

The Gibbs free energy of a multicomponent system is given by:

$$G^{total} = \sum n_i \mu_i \tag{9}$$

where,

'i' represents the chemical component present in the system
n_i represents the number of moles of component 'i'
μ_i represents chemical potential of component 'i'

The chemical potential of each component is calculated using the following equation:

$$\mu_i = \Delta G_{fi}^0 + RT \ln \left(\frac{y_i \widehat{\Phi_i} P}{P_0} \right) \tag{10}$$

where,

ΔG_{fi}^0 = Standard Gibbs free energy of formation for the component 'i'

R = universal gas constant $\left(\dfrac{Pa \times m^3}{mol \times K}\right)$

Since the reforming reaction produces solid carbon, the presence of solid phase in the Gibbs free energy equation is accounted as follows:

$$G^{total} = \sum n_i \left(\Delta G_{fi}^0 + RT\ln(y_i) + RT\ln(\widehat{\Phi}_1) + RT\ln(\frac{P}{P_0}) \right) + n_c \Delta G_{fc(s)}^0 \tag{11}$$

where,

y_i = Equilibrium composition of the component 'i' in the system
$\widehat{\Phi}_1$ = Fugacity coefficient of component 'i'
n_c = Moles of carbon produced
ΔG_{fc}^0 = Standard Gibbs free energy of formation of solid carbon
P = Total pressure of the system
P_0 = Standard pressure of 1 atm
T = System temperature
T_0 = Standard temperature of 298.15 K

To estimate gas phase composition, Peng Robinson (PR) equation of state was used, due to its relevance to reforming system (Challiwala et al., 2016; Challiwala et al., 2017; Challiwala et al., 2015).

Cubic equation model for compressibility factor calculation:

$$z = \beta + (z + \epsilon\beta)(z + \sigma\beta)\left(\frac{1+\beta-z}{q \times \beta}\right) \tag{12}$$

For each equation of state, different values of parameters β, ϵ, σ and q are available, and their incorporation in equation (12) will simplify the generic model to the pertinent equation of state. The parameters for the PR equation of state adopted from Chemical Engineering Thermodynamics textbook (Smith, 1950) are presented below:

$$\log(\phi_i) = z - 1 - \log(z - \beta) - \frac{\left(q\log\left(\frac{z+\sigma\beta}{z+\epsilon\beta}\right)\right)}{\sigma - \epsilon} \tag{13}$$

$$\alpha_{PR}(T_R; \omega) = \left[1 + (0.37464 + 1.54226\omega - 0.26992\omega^2)\left(1 - T_R^{\frac{1}{2}}\right) \right]^2 \tag{14}$$

$$\sigma = 1 + \sqrt{2} \tag{15}$$

$$\epsilon = 1 - \sqrt{2} \tag{16}$$

$$\Omega = 0.07780 \tag{17}$$

$$\psi = 0.45724 \tag{18}$$

$$Z_c = 0.30740 \tag{19}$$

The solution of the Peng Robinson equation of state enables the calculation of the fugacity coefficient based on the 'vapor' like root of the compressibility factor. The fugacity coefficient is then incorporated into the Gibbs free energy equation as a correction factor for the pressure term.

There are two mathematical approaches to minimize the Gibbs free energy function: (1) Lagrange's Undetermined Multipliers method and (2) the Direct Minimization method.

2.1.1 *Lagrange's undetermined multipliers method*

The first method of Lagrange's Undetermined Multiplier method is a technique to solve a non-linear equation by incorporating additional variables into the equation that artificially make the system of equations solvable. Details on the exact steps involved are provided below:

Step 1: Defining the Gibbs free energy equation

Let the total Gibbs free energy equation having "n" number of components be recognized as follows:

$$G^{total} = g(n_1, n_2 \dots . n_N) \tag{20}$$

G^{total} is fixed at a specific temperature T and pressure P of the system. However, the problem is to identify the equilibrium moles of all the components, i.e., $n_1, n_2 \dots . n_N$.

Step 2: Defining constraints:

As the total number of atoms remain constant for any reaction system, an atom balance provides element wise constraints that cannot be violated in search for equilibrium. These constraints can be written as follows:

$$\Sigma n_i a_{im} = A_m \qquad \text{for all elements "m" in the system.} \tag{21}$$

Or,

$$\Sigma n_i a_{im} - A_m = 0 \qquad \text{for all elements "m" in the system.} \tag{22}$$

where, "A_m" = Total number of atoms of element "m"
  "a_{im}" = Number of atoms of element "m" in component "i"

If the equation above is multiplied with a constant, its overall solution remains unchanged, this mathematical manipulation, therefore, becomes:

$$\lambda_m(\Sigma n_i a_{im} - A_m) = 0 \qquad \text{for all elements "m" in the system.} \tag{23}$$

Summation of the above equation for all the elements, therefore, becomes:

$$\Sigma_m \lambda_m(\Sigma n_i a_{im} - A_m) = 0 \tag{24}$$

Step 3: Incorporation of constraints in G^{total}:

Mathematically, if a new function "F" is formulated, which is a summation of the total Gibbs free energy "G^{total}" and equation (24) of atom balance constraint, then the new function would theoretically be equal to "G^{total}", since the total value of the atom balance constraint above is zero. This is shown below:

$$F = G^{total} + \Sigma_m \lambda_m(\Sigma_i n_i a_{im} - A_m) \tag{25}$$

Although the value of the above equation is the same as "G^{total}", its partial derivatives with respect to each component mole "n_i" are not. This is because function F now also incorporates the atom balance constraints.

Step 4: Gibbs free energy minimization

The minimum value of the modified Gibbs free energy equation "F" provided in equation (25), which is now also inclusive of atom balance, could be obtained by solving for partial derivative of function "F" with respect to molar composition "n_i" at any constant system temperature T and pressure P, as follows:

$$\left(\frac{\partial F}{\partial n_i}\right)_{T,P,n_k} = \left(\frac{\partial G^{total}}{\partial n_i}\right)_{T,P,n_k} + \Sigma \lambda_m a_{im} = 0 \qquad \text{for } k \in i \text{ system component} \tag{26}$$

Since, $\left(\dfrac{\partial G^{total}}{\partial n_i}\right)_{T,P,n_k} = \mu_i$, or the chemical potential of component "i", then the overall equation above could be rewritten in terms of chemical potential, as follows:

$$\left(\frac{\partial F}{\partial n_i}\right)_{T,P,n_k} = \mu_i + \Sigma\lambda_m a_{im} = 0 \qquad\qquad \text{for } k \in i \text{ system component} \qquad (27)$$

Incorporation of equation (17) of chemical potential given earlier, the equation above becomes:

$$\left(\frac{\partial F}{\partial n_i}\right)_{T,P,n_k} = \Delta G_{fi}^0 + RT\ln\left(\frac{y_i \widehat{\Phi}_i P}{P_0}\right) + \Sigma\lambda_m a_{im} = 0 \qquad\qquad \text{for } k \in i \text{ system component} \qquad (28)$$

The equation above is for all the components "N" in the system, and therefore furnishes "N" equations. "N" equations, along with constraint equations for "m" number of elements as per the atom balance, furnishes "N+m" number of total equations. This will be used to solve for "N" number of unknown equilibrium molar composition.

2.1.2 Direct gibbs free energy minimization

The direct method of Gibbs free energy minimization is a relatively simple technique and does not involve the mathematical manipulation provided in step 2 of the Lagrange's Undetermined multiplier method. However, the constraints of atom balance and mass balance remain the same.

The implementation of this method requires utilization of a built-in minimization tool box provided in MATLAB® called "fmincon", which directly searches for minima of any given function subjected to a set of external constraints that bounds the solution within a desirable range of conditions.

Although "fmincon" provides a simple and robust technique, Lagrange's undetermined multiplier's method is a general procedure that needs to be followed if any other mathematical solver is desired to be used.

2.2 Energy balance calculations

The total energy required or supplied from an endo/exothermic reforming reaction is estimated using energy balance calculations across the reformer block. This is done by using a simple equation that computes the difference between the enthalpies of the components that enter and leave the system and the energy liberated/consumed by the chemical reaction. The temperature and pressure of the components before entering and after leaving is of importance in this calculation, as per the following block diagram in Figure 2.

The following is the general equation to calculate the energy requirements in this process:

$$\text{Energy} = \Sigma n_{i,exit} H_{i,exit} - \Sigma n_{i,feed} H_{i,feed} + E_{rxn} \qquad (29)$$

where,

$n_{i,exit}$ = Molar flow of the exit gases
$n_{i,feed}$ = Molar flow of the entering gases
$H_{i,feed}$ = Enthalpy of gases in the feed

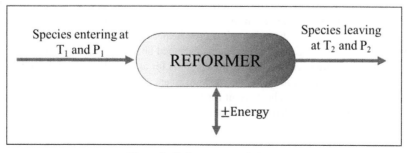

Figure 2. Reformer energy balance.

H$_{i,exit}$ = Enthalpy of gases in the exit
E$_{rxn}$ = Energy evolved/absorbed during chemical change
Energy = Total energy difference between the inlet and the outlet streams

The contrasting features between the three reforming technologies are apparent in terms of their net energy requirements, hydrogen to carbon monoxide ratio (or Syngas ratio) and the amount of carbon formed in each process.

Based on the thermodynamic analysis using the procedure mentioned in the methodology section, the overall equilibrium product distribution of the different species was estimated, including their net energy requirements, and the syngas ratio for all the three reforming technologies (i.e., SRM, POX and DRM).

2.3 LCA approach

2.3.1 Sources of CO$_2$ in a natural gas processing

To ensure that the comparison with existing technologies is fair, all major CO$_2$ emissions across different pathways should be considered. In particular, the LCA calculations include the following sources of CO$_2$ in the processes.

2.3.1.1 Emissions in natural gas upstream plants

Production of natural gas at the field and its subsequent transportation to the downstream contribute significantly to the CO$_2$ emissions. In the present work, the GREET (Greenhouse Regulated Emissions and Energy use in Transportation) model was used. This software was developed by Argonne National Laboratory, and is considered to be a reliable benchmark calculator for taking into account NO$_X$, CH$_4$ and CO$_2$ from the pertinent source of emission under consideration. The NO$_X$ and CH$_4$ emissions are calculated as equivalents of CO$_2$ emissions by using a multiplication factor, which is a function of the relative greenhouse gas potential of CH$_4$ and NO$_X$ in comparison to CO$_2$. More details on this are given in Afzal et al. (Afzal et al., 2018). For all calculations, a typical natural gas composition of 95% methane with a molecular weight of 16.81 kg/kmol was assumed. After incorporation of these values, the total CO$_2$ equivalent emission in the upstream is estimated to be about 620 g/kg of natural gas delivered at the GTL plant. This number, in other words, is an equivalent of 0.25 mole CO$_2$ produced per mole of methane processed.

2.3.1.2 Energy requirements in the reformer

An approach to calculate the energy requirements in the reformer is provided in section 2.2 of this chapter. All the conventional pathways to produce syngas (SRM, POX and ATR) lead to indirect CO$_2$ emissions resulting from heat duty for operating these units. For instance, in SRM, the contribution is both from the endothermic heat to boil the water at room temperature for production of steam of sufficient quality to feed the reformer in addition to the endothermic heat required to drive the reforming reaction. As for POX, even though it is a net energy producer, a huge energy footprint is associated with the preheater that heats the feed to approximately 1000 °C before it is allowed to enter the reactor. A DRM reaction, on the other hand, does not require a boiler unit like SRM, nor a pre-heater like POX for high temperature feed, but it requires extremely high heat duty via a fired heater to drive the highly endothermic reaction. Due to the large capacity and high-energy demands, as stated above, the reformer block in a GTL plant is one of the most energy intensive processes, and accounts for at least 40% of total plant energy. As per the literature available for industrial standards, 85% efficiency of the reformer block was used for all the calculations. In addition to this, a heat credit for the exhaust gases leaving at high temperatures from the reformer reactor was also included, as they may be used to heat another unit or produce high quality steam for power generation.

2.3.1.3 Oxidant production

The oxidants used in the SRM and POX reforming reactions are steam and oxygen, respectively. The role of an oxidant is to provide oxygen for the conversion of methane to carbon monoxide and hydrogen. There are carbon footprints associated with the production of steam in a steam boiler and oxygen in an air separation unit, therefore, they have to be accounted for in the overall LCA of the pertinent reaction. The main contributor to the carbon footprint of CO_2 in the DRM process is the energy required for its separation from flue gases and other major sources of CO_2 in the plant. Table 1 below provides carbon footprints for SRM, POX and DRM processes.

Table 1. Footprints associated with oxidant production.

Oxidant	Carbon footprint in terms of CO_2
Steam	471 [g/kg steam]
Oxygen	273 [g/kg oxygen]
CO_2	63 [g/kg CO_2]

The data provided in Table 1 for steam and oxygen is from GREET model, while for CO_2, it is adopted from a separate study done by David and Herzog (David and Herzog, 2000). Assuming a 3% concentration of CO_2 in flue gas and 90% capture efficiency, the study by David and Herzog reported a CO_2 equivalent emission of 63 g/kg of CO_2. The prediction from the GREET model for O_2 and H_2O, however, only accounts for the energy-related CO_2 emission resulting from the production of pure O_2 and H_2O. The approach implies a very conservative estimation, as a very dilute CO_2 concentration of flue gas is considered. However, a concentrated flue gas with a higher amount of CO_2 can safely enable exclusion of separation related emission from the overall calculation due to its negligible quantity.

2.3.1.4 Catalyst regeneration

Although the industrial conditions of the reformer are optimized to operate convincingly under the carbon free zone, the geometrical considerations of the reactor tube could still result in carbon formation. A typical industrial reformer tube is up to 13 m long and 0.1 m wide in a cylindrical form. The tubes are designed in such a way that maximum contact time is provided in a tortuous path that enables the highest utilization of the packed catalyst. In addition to this, the tubes operate at 20 bar pressure to overcome pressure drop, and to meet target production rates. Due to these conditions, there is a possibility of coke formation in localized regions of the reactor tube. This could not only disrupt the favorable coke-free equilibrium condition, but also lead to slow progression of deactivation throughout the reactor bed. Due to this reason, the catalyst bed needs to be regenerated frequently in order to avoid consequent issues of pressure drop and deteriorating syngas quality owing to coke formation. A typical procedure to regenerate the catalyst bed involves oxidation with air or pure oxygen to form carbon dioxide, which leads to CO_2 emissions. This has been incorporated in the calculations where 1 mole of carbon produces 1 mole of CO_2 during regeneration.

2.3.1.5 CO_2 from reforming reactions

Some of the side reactions in reforming processes produce CO_2. Since these reactions are coupled with primary reactions, they cannot be completely eliminated. However, the selectivity of these reactions could be reduced by using efficient catalyst materials and with the selection of proper operational conditions. The CO_2 forming reactions in particular are the Boudouard reaction ($2CO \rightarrow C + CO_2$) and the water-gas shift reaction ($CO + H_2O \rightarrow CO_2 + H_2$). These reactions contribute directly to the CO_2 emission as CO_2 is a part of the products. The LCA model accounts for such reactions by applying negative credits for these reactions in overall CO_2 emission calculation.

2.3.1.6 Optimization model formulation

The target for optimization in the present study is to maximize the number of moles of syngas produced while maintaining a fixed syngas ratio. The approach used in this work is adopted from a detailed methodology illustrated in El-Halwagi et al. (El-Halwagi et al., 2017). A systematic procedure adopted from the original work of El-Halwagi et al. (El-Halwagi et al., 2017) and Afzal et al. (Afzal et al., 2018) is presented below in context of the subject of reforming processes:

Objective function:

    Maximize syngas production.

Constraints:

- Carbon footprint reduction $\leq \epsilon_n$
  where, ϵ_n is increased iteratively from target for minimum carbon footprint
- Carbon footprint calculation = Σ(Carbon footprint in terms of CO_2 equivalents)
- Reforming equilibrium calculation using Gibbs free energy minimization
- H_2:CO ratio to be changed iteratively from a low value of 1 to a high value of 4 so as to assess all the possible case scenarios and combinations of reforming processes

The equations for calculation of CO_2 emissions of each reforming process provided in section 2.3.1 are given below:

Carbon footprint of SRM=

    {[CO_2 equivalent in reformer outlet]
    + [CO_2 equivalent from SRM furnace duty supplied by Natural Gas]
    + [CO_2 equivalent of upstream emissions of Natural gas and feed]
    + [CO_2 equivalent due to steam production]
    + [CO_2 equivalent of catalyst regeneration-burning of coke]
    − [CO_2 equivalent credit due to reformer outlet heat integration and reduction in upstream emissions due to reduced use of natural gas]}

Carbon footprint of POX=

    {[CO_2 equivalent in reformer outlet]
    + [CO_2 equivalent from POX furnace duty supplied by Natural Gas]
    + [CO_2 equivalent of upstream emissions of Natural gas and feed]
    + [CO_2 equivalent due to oxygen production]
    + [CO_2 equivalent of catalyst regeneration-burning of coke]
    − [CO_2 equivalent credit due to reformer outlet heat integration and reduction in upstream emissions due to reduced use of natural gas]}

Carbon footprint of DRM=

    {[CO_2 equivalent in reformer outlet]
    + [CO_2 equivalent from DRM furnace duty supplied by Natural Gas]
    + [CO_2 equivalent of upstream emissions of Natural gas and feed]
    + [CO_2 equivalent due to CO_2 separation]
    + [CO_2 equivalent of catalyst regeneration-burning of coke]
    − [CO_2 equivalent credit due to reformer outlet heat integration and reduction in upstream emissions due to reduced use of natural gas]
    − [CO_2 in reformer feed]}

The last term in the carbon footprint equation for DRM represents carbon fixation achieved by DRM. If the value of this quantity is greater than the total CO_2 footprint, then a net CO_2 fixation is realized and DRM is a CO_2 sink. Otherwise, the process is still a net CO_2 producer.

Additional constraints applied to the reformer in lieu of fair comparison between the three reforming processes are provided below:

- Syngas ratio $= \dfrac{\sum_i n_{out,H_2}}{\sum_i n_{out,CO}}$

- Operational temperatures of the reformers:

  SRM: 850 °C to 1000 °C

  POX: 1300 °C to 1400 °C

  ATR: 950 °C to 1100 °C

- Total methane fed to each reactor:

  $\sum n_{in,CH_4} = 1$

- Oxygen flow to POX or ATR reactor:

  n_{in,O_2}: 0 to 2

- CO₂ flow for DRM:

  n_{in,CO_2}: 0 to 2

- Steam to Carbon (S/C) ratio: $\dfrac{n_{in,H_2O}}{n_{in,CH_4} + n_{in,CO_2}}$

- For SRM,

  n_{in,H_2O}: 2 to 4

- For POX,

  n_{in,H_2O}: $0.1 \times n_{in,CH_4}$

- For DRM,

  n_{in,H_2O}: $0.1 \times (n_{in,CH_4} + n_{in,CO_2})$

The reported optimization algorithm was implemented in LINGO® and solutions were computed for various case scenarios pertaining to a syngas ratio ranging from 1 to 4.

3. Performance Results

In this section, a comparison is made between SRM, POX and the DRM processes in terms of syngas ratio, energy requirements and carbon formation tendency. The comparison is made at a pressure of 1 bar and temperature is varied between 200 and 1200 °C. The feed conditions for all these processes are kept based on stoichiometry as per equation (1)–(6) (SRM: H_2O/CH_4 = 1, POX: O_2/CH_4 = 0.5, DRM: CO_2/CH_4 = 1). Following this comparison, the proposed process of CARGEN and DRM+COSORB are described. Finally, an economic comparison between various combinations of the aforementioned processes in terms of their operational cost and carbon footprint is presented.

3.1 Comparison of syngas ratio

In this analysis, the effect of reforming technology on the H_2/CO ratio of syngas is studied at different temperatures. The relative atomistic ratio of the feed components in terms of O:C:H in each reforming technology is key to determining the H_2/CO ratio of the syngas produced. The O:C:H ratio in the feed is 1:1:6 for SRM, 1:1:4 for POX, and 1:1:2 for DRM. This shows that the hydrogen content in SRM is the highest, followed by POX and the lowest in DRM. This low quantity of hydrogen in DRM results in significantly low H_2:CO of 1:1, which is not suitable for downstream operations that require at least 2:1. A comparison of H_2:CO of the three reforming technologies as a function of temperature is provided in Table 2.

The ratio of hydrogen to carbon monoxide at different temperature conditions provided in Table 2 is seen to be highest for SRM, followed by POX and lowest for DRM. Notably, this is in the same order as that of the O:C:H ratio stated earlier. If a hydrogen rich gas, steam, for example, is added, it will change the relative ratio of hydrogen to carbon monoxide and consequently other products from the reaction system.

Table 2. Syngas ratio of the three reforming technologies as a function of temperature.

Temperature, °C	H₂/CO ratio		
	SRM	**POX**	**DRM**
200	33062.3	2649.5	940.3
300	1164.8	271.9	116.6
400	106.3	55.8	26.2
500	23.9	16.0	8.0
600	8.3	5.7	3.0
700	4.1	2.8	1.5
800	3.2	2.1	1.1
900	3.0	2.0	1.0
1000	3.0	2.0	1.0
1100	3.0	2.0	1.0
1200	3.0	2.0	1.0

3.2 Comparison of energy requirements

The energy requirement to drive the reforming reactions forms the major portion of energy requirements of reforming processes. Here, the energy requirements of the three reforming processes (SRM, POX, DRM) are shown as a function of temperature at a reaction pressure of 1 bar and depicted in Figure 3. The feed composition is kept at stoichiometric condition, as indicated earlier. It can be observed that POX technology produces energy as its energy requirements are in negative. However, with an increase in temperature, the energy requirements tend to increase; at ~ 750 °C they tend to cross the zero-energy line, indicating that it becomes net positive after this temperature. On the other hand, both SRM and DRM processes are observed to require energy for the reaction to happen under all temperature conditions. Another point to note is that energy requirements of the DRM process are more than that of SRM beyond 650 °C temperature, while below 650 °C, the energy requirements from the SRM are slightly higher than that of DRM.

Figure 3 shows the energy requirements for SRM, POX and DRM. However, this cannot be used to make comparisons since each of these processes produce syngas of a different H₂/CO ratio and their

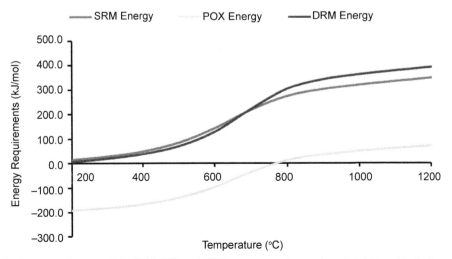

Figure 3. Energy requirements of the SRM, POX and DRM processes vs temperature at stoichiometric feed composition at 1 bar pressure.

carbon footprint of oxidant production varies with the process. Section 3.6 includes comparison results that compare the overall carbon footprint. Nevertheless, Figure 3 helps in understanding the energy requirements at different temperatures of the reformer reactor for different processes.

3.3 Comparison of carbon formation tendency

This section compares the carbon formation tendency of each reforming technology as a function of temperature. The operational conditions of feed composition, reaction pressure and temperature are identical to the previous sections.

Carbon formation is a result of a combination of two major side reactions: The Boudouard reaction and the methane decomposition reaction, which happen in all reforming processes. However, their extent varies depending upon reaction conditions, oxidant type and its concentration. Equation (7) and equation (8) illustrate the aforementioned side reactions.

From several experimental and thermodynamic modelling studies, it is proven that the extent of methane decomposition reaction at low reforming temperature of 500 °C is very high compared to Boudouard reaction. Many industrial reports also claim that methane decomposition reactions are extremely prone at the reactor entrance, wherein the concentration of methane is very high, compared to any other reaction products. Near the reactor inlets, heating is generally done in order to increase the temperature of the reaction gases to desirable reforming temperatures of 900 °C and beyond. During this transition, since the gases pass through coking temperatures, a huge quantity of carbon is deposited near the inlet. On the other hand, on the exits of the reactor, since the carbon monoxide concentrations are high, the carbon disproportionation tendency is much higher when temperatures are quenched to the target outlet temperatures of 400 °C. Formation of solid carbon at the entrance and the exit of the reactor bed is not the only concern that the industry faces, the other and the most severe challenge is related to catalyst deactivation. Since reforming reactions take place in the presence of catalyst, the formation of carbon deactivates the catalyst by clogging its active sites. In addition to this, sintering is another problem that is widely reported in the scientific literature. Due to these challenges, carbon formation in the reformer is undesirable. Much of the scientific attention in the global community at present is focused on the reduction or elimination of solid carbon from the reforming process. Extensive literature studies that report novel methods of reduction of carbon formation tendency of the reforming reaction are available; some of these studies are either process related while others are purely catalytic. Figure 4 illustrates the carbon formation tendency of the three reforming reactions at stoichiometric feed condition of each reforming process and 1 bar pressure.

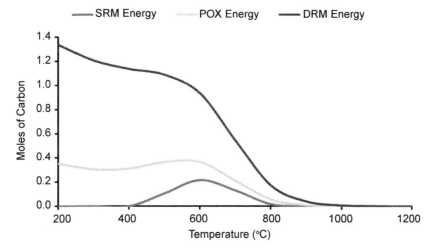

Figure 4. Carbon formation comparison between SRM, POX and DRM process at stoichiometric feed conditions and pressure of 1 bar.

From Figure 4, it can be clearly observed that the carbon formation tendency from the DRM reaction is higher than the other two reforming techniques (DRM > POX > SRM). The lowest carbon formation is observed in SRM, as it contains the highest amount of hydrogen among the reforming techniques. In conjunction to the discussion provided in the previous section of syngas ratio, it could be observed that the ratio of O:C:H also carries huge implication on carbon formation tendency. Since the DRM process has the highest ratio of C/H compared to the other reforming technologies, the carbon formation tendency of this reaction is the highest. Therefore, in order to reduce coke formation tendency, it would be advisable to either remove excessive carbon and or increase hydrogen concentration in order to reduce the C/H ratio. In addition to this, it can also be observed that low temperatures favor more carbon formation than higher temperatures for DRM and POX, while for SRM, a peculiar trend is observed, in which carbon formation is only in the 400 °C to 800 °C temperature range.

3.4 CARGEN—Co-production of syngas and carbon black

As discussed in the previous sections, the process challenges of carbon formation and low syngas ratio are the most critical hurdles in the implementation of DRM technology on a commercial scale. One approach to overcome the challenge of high carbon formation in DRM is to segregate the carbon formation reaction, and the syngas formation reaction. In this manner, it is possible to systematically improve the C/H ratio that favors desirable operational conditions for industrial operation. From thermodynamic assessments, a novel approach to segregate the two reactions (carbon formation and syngas formation) is proposed which helps in addressing the catalyst deactivation problem and syngas ratio problem. In particular, the proposed reactor system utilizes the benefit of lower operational temperatures and high C/H ratio to selectively produce high quality carbon from CO_2 and CH_4, while utilizing the benefit of high temperature and low C/H ratio to produce syngas as desirable product. These reactions are made to essentially happen in two separate reactors that produce only a single product separately. Figure 5 illustrates the process concept. As the focus of the first reactor is to produce solid carbon, it has been termed as "CARGEN" or CARbon GENerator reactor, the second reactor is the normal reforming process to produce syngas. In short, this process perceives carbon as desirable marketable product instead of a problem that should be avoided. By doing so, two separate products can be obtained from 2 sequential reactors, thereby utilizing the DRM reaction.

In Figure 5, two plots of carbon formation as a function of temperature are shown above both reactor blocks. The blue zone in the first plot referring to CARGEN, is to indicate the operational window of the first reactor in terms of temperature. This operational window defines the most favorable operational zone for a CARGEN reaction. The second plot, showing a green zone, refers to the conditions in which carbon formation is minimal, and pertains to the second reactor. Inherently, in the first system, since

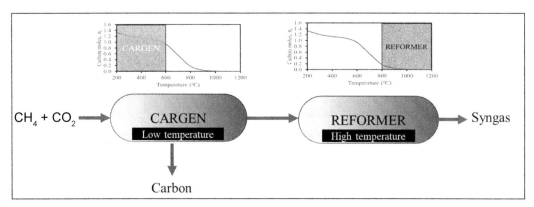

Figure 5. Illustration of novel approach to implement DRM process. CARGEN (Or Carbon Generator) indicates the first reactor, which produces solid carbon as a desirable product, while REFORMER indicates the second reactor, which produces syngas as desirable product.

carbon is the main product, the resultant syngas ratio is relatively high, however, its yield is low because the selectivity towards syngas under low temperatures is relatively low. Additionally, a big portion of the feed is expected to remain unconverted in the first reactor, as thermodynamics under these conditions do not favor 100% conversions under these conditions. Nevertheless, the portion of feed that gets converted will only form solid carbon. In conjunction with the previous section, this method could also be seen as a way to reduce the C/H ratio load on the reactants of the second reactor (Reformer unit), as all the products from the first reactor are essentially fed to the second reactor. The operational window of the second reactor pertains to high temperature conditions beyond 800 °C temperature and are expected to lead towards higher conversions of close of 100%. However, under these conditions, syngas is selectively produced. In this way, the reactor segregation based on the type of product produced helps to make a symbiotic relationship between the two reactors.

Now, will the carbon formation in the first reactor affect the catalyst and the design of the process? Wouldn't it lead to frequent shutdowns? The answer to these two important questions is that the know-how for handling "coking" reactions is already established and has been practiced extensively in the chemical industry. Chemical refineries generally employ an important unit, called a "coker" unit, that converts the residues of the Vacuum Distillation Unit (VDU) into pet-coke and light gases using catalyst and thermal cracking. This unit is known to produce around 400–500 tons per day(tpd) of solid coke in big refineries and utilize well-known fluidized bed concept for their operation. The first reactor will be operated under these conditions and, therefore, not many design challenges are expected. However, while segregating the coke formation and syngas formation reactions, an important problem of the DRM reaction is solved by alleviating its C/H ratio load on "actual" reformer unit, which operates on the product gases of the first reactor.

There are numerous opportunities to further improve the two-reactor setup discussed above. The main opportunity lies in the symbiotic relationship between the two reactors in terms of both energy and mass exchange. Mass exchange has already been discussed earlier, and it is simply the transfer of products from the first reactor to the second reactor for upgradation. As for heat exchange, the opportunity lies in the fact that the two reactors are operated at different operational temperatures. Since the product gases from the first reactor are low in temperature, relative to product gases of the second reactor, there is a possibility of preheating the product gases of the first reactor by the relatively hot gases exiting from the second. As carbon formation tendency increases with an increase in pressure, if the first reactor is operated at high-pressure condition, and second reactor at low pressure (syngas formation favors low-pressure condition), then there is also scope to generate external work. In this, the high-pressure gases from the first reactor will be passed through a gas turbine to derive external work, while low-pressure gases will be sent to the second reactor for reforming reaction. An illustration of this scheme is provided in the Figure 6 below.

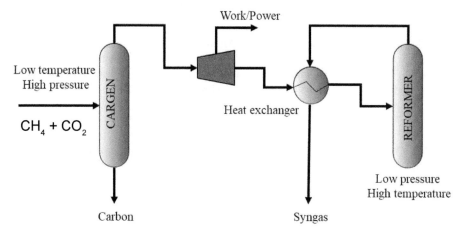

Figure 6. Illustration of the symbiotic relationship between the CARGEN and REFORMER reactor in terms of opportunities presented in terms of mass and heat exchange.

3.4.1 Variants of the two-reactor setup

In this section, two variants of the two-reactor setup process are presented. These alternative designs are "sub-sets" or alternatives to the original process and demonstrate a perspective in to the flexibility of the proposed scheme to meet the different objective functions desirable by end consumer or the downstream process plant. These case studies demonstrate the flexibility of the two processes in producing different products, while utilizing the same energy requirements.

(a) Case Study 1:

This case study demonstrates a situation in which the product gases from the first reactor are fed to the second reactor directly without any pre-treatment or mixing with external feed.

In this, a mixture of CH_4, CO_2 and O_2 are compressed and fed to the CARGEN reactor at 400 °C and 25 bar to produce solid carbon. The unreacted or partially reacted gases from the first reactor are fed to the second reactor wherein they are converted to syngas at high temperature of 820 °C and 25 bar pressure. Figure 7 provides an illustration of this case study.

As can be seen from Figure 7, the feed to the first reactor is comprised of CH_4, CO_2 and O_2. The primary reason behind the utilization of a small quantity of oxygen is to promote an internal combustion reaction that could support the endothermic energy requirements of the DRM process. *In situ* energy production in the reactor reduces the inefficiencies associated with the jacketed heat transfer, which is a general industrial practice. An energy assessment of the present case scenario indicates a total energy requirement of ~ 120 kJ/mol, which is almost 50% that of DRM. In addition to this, the total CO_2 and CH_4 conversion from the overall process is seen to be about 62% and 78%, respectively. Energy reduction at this scale could reduce the CO_2 footprints associated with fuel combustion tremendously. Additionally, the first reactor is also assessed to operate in Auto-thermal mode if the operational temperature is about 420 °C, indicating that the reactor is self-sustaining in terms of heat duty while at the same time produces a significant quantity of solid carbon. In this particular case scenario, about 0.81 moles of carbon are produced per 1.6 moles of carbon in the feed. This indicates almost 50% carbon capture in the first unit itself. The value-addition from the second reactor could be observed in the fact that it operates onto a "pretreated" and improved C/H ratio of the feed gas and, therefore, produces a higher quality syngas of ratio of 2.8, with an overall yield of 1.18 moles.

(b) Case Study 2:

This case study is slightly different from the previous case as it utilizes an additional feed of methane to the second reactor to manipulate its syngas ratio and yield. Figure 8 below illustrates the block flow diagram of the two-reactor setup under this scheme of operation.

The operational pressure of both the reactors is set to 25 bar, while the two reactors operate at different temperature conditions. The temperature of the first reactor (Or CARGEN reactor) is at 420 °C, while the second reactor is at 820 °C. Similar to the previous case study, the carbon formed in the first reactor at 420 °C and 25 bar is about 0.81, while the syngas produced from the second reactor is of a different quality due to additional methane co-feed. An energy assessment on the overall process suggests that this process would require about 118 kJ/mol of energy (47% of DRM), while producing syngas ratio of 2.7 at a yield of 1.39 moles. The implication of addition of methane is to demonstrate the flexibility of the process in producing different qualities of syngas with the addition of side reactants.

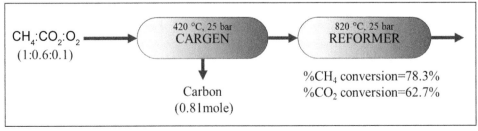

Figure 7. Case Study 1 block flow diagram of the two-reactor setup.

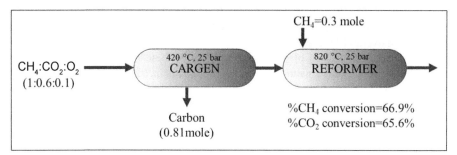

Figure 8. Case Study 2 block flow diagram of the two-reactor step.

In terms of energy requirements, both case 1 and case 2 have almost equivalent performance as both the processes require ~ 120 kJ/mol of energy. However, both the processes differ in terms of quality of syngas produced. The key benefit of the two-reactor operation is, therefore, not just limited to its tendency to overcome carbon formation issue, but also in adjusting the syngas ratio depending upon downstream process requirements. In addition to this, lower energy requirements are also realized due to breakdown of the DRM process into its two constituent reaction sets. Process optimization on different variables of the processes can further improve the performance of the CARGEN reactor system.

3.5 DRM+COSORB—Post DRM syngas ratio adjustment

The CARGEN approach produces two products, the syngas and solid carbon. However, if a carbon-free route is desired, the DRM+COSORB process offers an alternative pathway to produce syngas and an additional stream of CO while reducing the overall carbon footprint of syngas production. Figure 9 provides an illustration of the DRM+COSORB process concept. As the DRM syngas has a low syngas ratio (H_2/CO ≤ 1), the DRM+COSORB process separates a part of the CO in an absorption process. This helps boost the syngas ratio of the DRM syngas and also produces a CO stream that can be used as a feedstock for production of other petrochemicals. The advantage of this process is that it does not require the production of steam or oxygen, thereby reducing the capital costs of the syngas production unit.

The carbon footprint and operating cost comparison of DRM+COSORB process against commercial processes is discussed in the next section.

3.6 Carbon footprint and operating cost comparisons for proposed processes

Based on the approach described in section 2.3, the carbon footprints of the proposed processes of CARGEN and DRM+COSORB are shown in Figure 10. The two CARGEN cases studied have a syngas ratio of about 2.75 and, when compared to the ATR which operates in this region, there is a marked

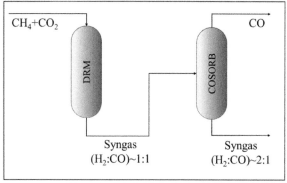

Figure 9. DRM+COSORB process.

reduction in the overall CO_2 emissions. The DRM+COSORB process can be tuned for any syngas ratio, based on the amount of CO captured in the COSORB unit. As indicated, across all syngas ratios, the process has reduced overall CO_2 emissions.

For a preliminary comparison of the operating costs of these processes, the costs of feedstock (natural gas), fuel for reformer (by natural gas firing) and oxidant production were considered. A selling price of \$75/MT was considered for the captured CO and \$200/MT was considered for the coke. Spot market prices for coke are in the range of \$300/MT and refinery sources quote a price of \$100/MT and, hence, a median price of \$200/MT was chosen for the analysis presented here. Removal of coke from the catalyst is still a technological challenge and its cost has not been included in the analysis. The cost of syngas production via different competing processes across syngas ratios is shown in Figure 11. Between syngas ratios of 2 and 3, both the studied processes (CARGEN and DRM+COSORB) have competitive operating costs.

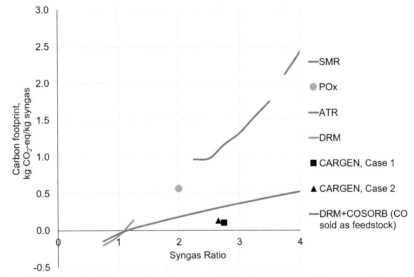

Figure 10. Carbon footprint comparison for proposed reforming processes with commercial processes.

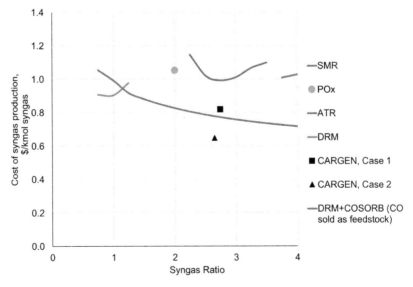

Figure 11. Operating cost comparison for proposed reforming processes.

4. Conclusions

In this study, two process modifications to the DRM reaction are proposed. These are aimed at addressing the challenges of the stand-alone DRM process. The CARGEN reactor system proposes a two-reactor sequential system, wherein the first reactor produces coke as an additional product and the second reactor produces syngas. In the DRM+COSORB process, a stand-alone DRM reformer is followed by a CO Absorption Unit, which captures CO based on the downstream H$_2$/CO ratio requirement. Both the processes produce syngas of high H$_2$/CO ratios ≥ 2 with significant reductions in overall CO$_2$ emissions. Preliminary economic assessment shows that the operating costs are lower than industrial benchmark processes. All assessments are based on thermodynamic calculations using the Gibbs free energy minimization method, wherein equilibrium compositions are assumed in product streams.

In particular, the new CARGEN process was shown to systematically handle the carbon formation challenge in a separate reactor, while producing high quality byproduct solid carbon. The novel approach of CARGEN was illustrated by using two case examples, wherein the flexibility of the new approach was shown. While the "problem" of solid carbon was solved by using a dedicated CARGEN reactor, the benefit of adjustable syngas ratio from the second reactor was also realized. The stage-wise operation of the two reactors was shown to reduce the net energy requirements to 50% of DRM, while converting at least 65% CO$_2$ per pass.

The second approach of DRM+COSORB process was also shown to mitigate the challenge of low syngas ratio of ~ 1, whereby the excess CO in the DRM syngas is removed and should be sold to an external customer.

Comparison of conventional reforming approaches with the CARGEN and DRM+COSORB processes for the same H$_2$:CO ratio using an LCA approach clearly indicates that both the processes provide a better solution for commercial implementation than the erstwhile approaches. The overall carbon footprint and operating costs for these two processes are considerably less than the ATR, which operates at syngas ratios between 2 and 3. As both CARGEN and DRM+COSORB processes produce an additional intermediate/product which enhances process economics, a detailed techno-economic analysis is underway for future scale-up and commercialization.

Acknowledgements

This work was supported by Qatar National Research Fund (QNRF), member of Qatar Foundation (Grant NPRP X-100-2-024). The statements made herein are solely the responsibility of the authors.

References

Abusrafa, A.E., Challiwala, M.S., Choudhury, H.A., Wilhite, B.A. and Elbashir, N.O. 2019. Experimental verification of 2-dimensional computational fluid dynamics modeling of supercritical fluids Fischer Tropsch reactor bed. Catalysis Today.

Afzal, S., Sengupta, D., Sarkar, A., El-Halwagi, M.M. and Elbashir, N.O. 2018. Optimization approach to the reduction of CO$_2$ emissions for syngas production involving dry reforming. ACS Sustainable Chemistry & Engineering 6(6): 7532–44. https://doi.org/10.1021/acssuschemeng.8b00235.

Alsuhaibani, A.S., Afzal, S., Challiwala, M.S., Elbashir, N.O. and El-Halwagi, M.M. 2019. The impact of the development of catalyst and reaction system of the methanol synthesis stage on the overall profitability of the entire plant: a techno-economic study. Catalysis Today.

Appl, M. and Gössling, H. 1972. Production of synthesis gas after the steam reforming process. Chemiker Zeitung 96(3).

Challiwala, M.S., Ghouri, M.M., Linke, P. and Elbashir, N.O. 2015. Kinetic and thermodynamic modelling of methane reforming technologies: comparison of conventional technologies with dry reforming. *In*: 2015 AIChE Annual Meeting.

Challiwala, M.S., Ghouri, M.M., Linke, P., El-Halwagi, M.M. and Elbashir, N.O. 2017. A combined thermo-kinetic analysis of various methane reforming technologies: comparison with dry reforming. Journal of CO$_2$ Utilization 17. https://doi.org/10.1016/j.jcou.2016.11.008.

Challiwala, M.S., Ghouri, M.M., Sengupta, D., El-Halwagi, M.M. and Elbashir, N.O. 2017. A process integration approach to the optimization of CO$_2$ Utilization via tri-reforming of methane. Computer Aided Chemical Engineering. Vol. 40. https://doi.org/10.1016/B978-0-444-63965-3.50334-2.

Choudhury, H.A., Cheng, X., Afzal, S., Prakash, A.V., Tatarchuk, B.J.T. and Elbashir, N.O. 2019. Understanding the deactivation process of a microfibrous entrapped cobalt catalyst in supercritical fluid Fischer-Tropsch synthesis. Catalysis Today, no. September 2018: 0–1. https://doi.org/10.1016/j.cattod.2019.01.031.

David, J. and Herzog, H. 2000. The cost of carbon capture. *In*: Fifth International Conference on Greenhouse Gas Control Technologies, Cairns, Australia, 13–16.

Dybkjaer, Ib. 1995. Ammonia production processes. *In*: Ammonia, 199–327. Springer.

El-Halwagi, M.M. 2017. Sustainable Design through Process Integration: Fundamentals and Applications to Industrial Pollution Prevention, Resource Conservation, and Profitability Enhancement. Butterworth-Heinemann.

Elbashir, N.O., Challiwala, M.S, Sengupta, D. and El-Halwagi, M.M. 2018. System and method for carbon and syngas production. Edited by World Intellectual Property and Organization. Production, System and Method for Carbon and Syngas, issued 2018.

Mondal Kartick et al. 2016. Dry reforming of methane to syngas: A potential alternative process for value added chemicals—a techno-economic perspective. Environmental Science and Pollution Research 23.22: 22267–22273.

Prettre, M., Eichner, C. and Perrin, M. 1946. The catalytic oxidation of methane to carbon monoxide and hydrogen. Transactions of the Faraday Society 42: 335b–339.

Rostrup-Nielsen, J.R. 1984. Sulfur-passivated nickel catalysts for carbon-free steam reforming of methane. Journal of Catalysis 85(1): 31–43. https://doi.org/10.1016/0021-9517(84)90107-6.

Rostrup-Nielsen, J.R. 1984. Catalytic steam reforming. Catalysis: Science and Technology 5: 1–117. https://www.scopus.com/inward/record.uri?eid=2-s2.0-0021158095&partnerID=40&md5=f91480d485d3760a1ba6198ccfe28060.

Rostrup-Nielsen, J.R., Christiansen, L.J. and Bak Hansen, J.H. 1988. Activity of steam reforming catalysts: role and assessment. Applied Catalysis 43(2): 287–303. https://doi.org/10.1016/S0166-9834(00)82733-5.

Rostrup-Nielsen, J.R. 2002. Syngas in perspective. Catalysis Today 71(3–4): 243–47.

Rostrup-Nielsen, J.R. and Rostrup-Nielsen, T. 2002. Large-scale hydrogen production. CATTECH 6(4): 150–59. https://doi.org/10.1023/A:1020163012266.

Rostrup-Nielsen, J. 2004. Steam reforming of hydrocarbons. a historical perspective. *In*: Studies in Surface Science and Catalysis 147: 121–26. Elsevier.

Smith, J.M. 1950. Introduction to Chemical Engineering Thermodynamics. Journal of Chemical Education. McGraw-Hill, Boston 27(10): 584. DOI: 10.1021/ed027p584.3.

Trimm, D.L. 1997. Coke formation and minimisation during steam reforming reactions. Catalysis Today 37(3): 233–38. https://doi.org/10.1016/S0920-5861(97)00014-X.

van Hook, J.P. 1980. Methane-steam reforming. Catalysis Reviews—Science and Engineering 21(1): 1–51.

Wood, D.A., Nwaoha, C. and Towler, B.F. 2012. Gas-to-Liquids (GTL): A review of an industry offering several routes for monetizing natural gas. Journal of Natural Gas Science and Engineering 9: 196–208. https://doi.org/10.1016/j.jngse.2012.07.001.

CHAPTER 13

Ranking Negative Emissions Technology Options under Uncertainty

Raymond R Tan,[1,2,]* *Elvin Michael R Almario,*[3] *Kathleen B Aviso,*[1]
Jose B Cruz, Jr[2] *and Michael Angelo B Promentilla*[1]

1. Introduction

Climate change has been recognized for some time now as a critical global environmental issue that is approaching crisis levels (Rockström et al., 2009). Atmospheric carbon dioxide (CO_2) concentration now exceeds 400 ppm, and continues to rise due to additional emissions driven by economic development and population growth. The problem is further compounded by contributions of other greenhouse gases (GHGs), such as methane (CH_4) and nitrous oxide (N_2O). Potential adverse impacts of climate change include sea level rise, weather disruptions, ocean acidification, and changes in precipitation patterns, among others. As a result of a strong scientific consensus of the gravity of the problem, public awareness of the climate issue has grown; the international community has also made significant progress towards concerted effort to mobilize solutions, especially via the Paris Agreement and the Sustainable Development Goals (SDGs) of the United Nations (UN). Nevertheless, the problem has progressed to the extent that research is now needed both for mitigation of climate change, and for adaptation to impacts that are already occurring (IPCC, 2018). For example, Tan and Foo (2018) discuss the need for integrated water management strategies as a means to reduce the vulnerability of industrial systems to reduced water availability due to climatic disruptions.

Pacala and Socolow (2004) suggested that deep cuts in GHG emissions can be achieved through massive deployment of existing low-carbon technologies, including energy efficiency enhancement, renewables, nuclear energy, and CO_2 capture and storage (CCS). They proposed different technologies to be viewed as "stabilization wedges" contributing incremental reductions whose cumulative effect would result in the required reduction of CO_2 emissions on a global scale. The concept remains applicable today, and is much more plausible than assuming that a single technology can solve the problem. Recent data suggests that even more drastic cuts in GHG emissions are needed in order to avoid catastrophic climate change with a temperature rise of about 2 °C by the year 2100. A more preferable trajectory should lead to a temperature rise of just 1.5 °C, which will result in serious but manageable impacts; however, achieving this target will require global net GHG emissions to reach zero by mid-century (IPCC, 2018). According

[1] Chemical Engineering Department, De La Salle University, 2401 Taft Avenue, 0922 Manila, Philippines.
[2] National Academy of Science and Technology, Science Heritage Building, DOST Complex, General Santos Ave, Taguig, 4044 Metro Manila, Philippines.
[3] Cruz & Associates, 19 Piko Street, Silang, Cavite 4118, Philippines.
* Corresponding author: raymond.tan@dlsu.edu.ph

to Haszeldine et al. (2018), such a trajectory will require large-scale commercial deployment of CCS and CO_2 removal (CDR) or negative emissions technologies (NETs).

CDR or NETs are different techniques that can achieve net removal of CO_2 from the atmosphere. They rely on different mechanisms for carbon fixation. For example, four decades ago, Dyson (1977) speculated on the potential of intensive cultivation of biomass (e.g., via reforestation) at a massive global scale to avert climate crisis; he computed rough estimates of the potential of such a solution, and discussed temporal, land and nutrient constraints. Today, such natural climate solutions are considered to be important stabilization wedges (Griscom et al., 2017). Other NETs include direct air capture (DAC), bioenergy with CCS (BECCS), biochar application (BA), soil carbon management (SCM), and enhanced weathering (EW). In the coming decades, these technologies will have to be deployed throughout the world to help curb GHG emissions down to net zero level by 2050. They will also have to be systematically evaluated based on various environmental and techno-economic criteria so that the best technology can be used in any given geographic region or country. Quantitative tools for multiple attribute decision making (MADM) are, therefore, essential for ensuring that the best decisions are made towards achieving sustainability (Sikdar, 2009). Such decisions also need to be made even when there are gaps in data due to corporate secrecy or due to the novelty of the technologies involved (Sikdar, 2019).

In this chapter, the classical MADM technique, known as simple additive weighting (SAW) (MacCrimmon, 1968), is used to compare selected NETs under data uncertainty. Interval numbers are used to represent the epistemic uncertainty due to the fundamental lack of knowledge about how NETs perform at commercial scales. The rest of this chapter is organized as follows: Section 2 gives a brief overview of the literature on CDR and NETs. Section 3 describes the generic interval SAW methodology. In section 4, the procedure is then applied to NETs based on data from the literature. The implications of these results are discussed and sensitivity analysis is performed. Finally, section 5 gives the conclusions and suggests prospects for future research. An appendix listing definitions of relevant acronyms is given, along with two other appendices showing the computer code used in the analysis. The latter will allow the reader to replicate or modify the results shown here.

2. Brief Overview of NETs

NETs achieve removal of CO_2 from the atmosphere through different mechanisms. This section provides a brief qualitative description of selected NETs for terrestrial deployment. McLaren (2012) and McGlashan et al. (2012) both gave early techno-economic assessments of various NETs, taking into account sequestration potential, cost and technology maturity. In McLaren (2012), maturity was quantified using the Technology Readiness Level (TRL). The TRL is a 9-point scale that was originally developed by the United States National Aeronautics and Space Administration (NASA), but which has since been adopted more widely throughout the world (Mankins, 2009). Table 1 defines the different levels in the scale. The physical and economic limits to the deployment of NETs at the global scale were estimated by Smith et al. (2016a); similar studies have scaled down the estimates to the level of single countries (Smith et al., 2016b; Alcalde et al., 2018). A more recent series of papers give comprehensive reviews of the status of NETs. The first of these papers reviews the scientific state-of-the-art of the NETs literature (Minx et al., 2018). The second paper by Fuss et al. (2018) surveyed the potential, cost, and risks associated with various NETs. The third paper discusses the role of technology innovation in the potential scale-up of NETs to commercially and environmentally significant levels (Nemet et al., 2018).

This chapter considers the terrestrial NETs considered by Smith et al. (2016b) for the United Kingdom and Alcalde et al. (2018) for Scotland. Brief descriptions of the alternatives are given here.

BECCS achieves negative emissions by combining bioenergy with CCS. Conventional bioenergy systems are nearly carbon-neutral because photosynthesis during plant growth offsets downstream CO_2 emissions from the combustion of biomass. In a system at steady state, the carbon flows to and from the atmosphere balance each other out. Application of CCS to the emissions from biomass combustion plants can dramatically reduce the CO_2 flow into the atmosphere. The carbon flows in the resulting BECCS system are imbalanced, with photosynthetic fixation becoming larger than combustion emissions. The

Table 1. TRL definitions (Mankins, 2009).

Level	TRL definition
1	Basic principles observed and reported.
2	Technology concept and/or application formulated.
3	Analytical and experimental critical function and/or characteristic proof-of-concept.
4	Component and/or breadboard validation in laboratory environment.
5	Component and/or breadboard validation in relevant environment.
6	System/subsystem model or prototype demonstration in a relevant environment.
7	System prototype demonstration in the planned operational environment.
8	Actual system completed and qualified through test and demonstration in the operational environment.
9	Actual system proven through successful system and/or mission operations.

imbalance results in a net removal of atmospheric carbon. As with conventional CCS, BECCS systems require secure geological storage of the captured CO_2. In addition, BECCS can be implemented in either dedicated biomass-fired plants, or plants that co-fire biomass with fossil fuels, such as coal. In the latter case, negative emissions only accrue from the CO_2 emissions from the biomass fraction of the fuel mix.

Afforestation and reforestation (AR) is arguably the most mature and potentially reliable NET option, and is simply the reversal of positive CO_2 emissions resulting from land use change (LUC). It is also classified as a natural climate solution (Griscom et al., 2017). Tree growth absorbs CO_2 from the atmosphere and locks up the carbon in solid form within the plant biomass. Additional carbon is also stored naturally in forest soil as a result. As pointed out by Dyson (1977), large-scale AR can result in significant climatic benefits during the period of growth. The negative emissions rate eventually declines as the carbon stock becomes saturated when forests mature over the course of several decades. Despite its simplicity, there are potential drawbacks or barriers, including competition for resources (e.g., land, fertilizer) that are also needed for agriculture to feed the world. A recent study by Bastin et al. (2019) estimated the potential for AR at 205 Gt of carbon, based on an additional 0.9×10^9 ha of forest area.

SCM entails modifying agricultural practices to minimize the loss of naturally-occurring soil carbon (Griscom et al., 2017). As crop residues also contribute to the carbon stock in soil, conditions can be optimized to encourage carbon deposition and, at the same time, minimize the decay that releases CH_4. More precise dosing of nitrogen fertilizers can also reduce N_2O emissions. Another NET option, BA, may be regarded as a special case of SCM. Carbonizing biomass via pyrolysis into biochar gives three distinct benefits. First, biochar is more chemically stable than untreated biomass, and can thus result in more permanent sequestration of carbon in soil. Secondly, under proper application conditions, biomass can enrich soil and reduce the need for agricultural inputs (e.g., irrigation water, fertilizers), thus reducing agricultural carbon footprint further. Thirdly, the gaseous and liquid co-products of biochar production can be used as carbon-neutral fuels that can displace conventional fossil fuel products, thus resulting in additional carbon credits. One major uncertainty in both SCM and BA systems is the extent and permanency of carbon storage in agricultural lands. It has been suggested that large-scale systems may require integrated techniques involving remote sensing to keep track of non-point emissions of GHGs from soil, so that the level of negative emissions can be monitored and verified (Tan, 2019).

DAC involves capture of CO_2 from air using different solvents or mass separating agents using contactors with different configurations. The underlying physical principle is similar to that of post-combustion CO_2 capture in CCS and BECCS, except that the CO_2 needs to be removed from a source with an extremely low concentration (approximately 400 ppm in air) as compared to CO_2-rich flue gas from combustion plants. The low source concentration results in thermodynamic penalties which are partially offset because there is no need to achieve high removal rates (Sanz-Perez et al., 2016). Calcium-looping systems using wet scrubbers and calcination to release a pure CO_2 stream is considered as the most mature DAC technology; other options include supported amine-based capture (McLaren, 2012).

Like CCS and BECCS, DAC requires CO_2 handling infrastructure and secure geological storage sites for its implementation.

The final NET option considered here is EW. This process takes advantage of the fact that rocks such as basalt naturally react with atmospheric CO_2 in the presence of moisture (from precipitation); the resulting reaction products dissolve in runoff, and ultimately flow into the sea which acts as the final carbon sink. Under normal conditions, this weathering process occurs at a geologically slow pace on exposed rock surfaces. EW entails artificially accelerating this process by mining the CO_2-absorbing rocks and minerals, crushing them into a powder, and applying them to agricultural or marginal lands (Smith et al., 2016b). This process creates a larger surface area for contact, and results in accelerated carbon fixation that is potentially significant for climate change mitigation purposes. The potential scale of EW is constrained more by land area and weather conditions than by mineral supply (Smith et al., 2016a). In the case of agricultural lands, application rate also needs to be calibrated in order to avoid adverse impacts on soil fertility (Smith et al., 2016b; Alcalde et al., 2018).

Each of these terrestrial NET alternatives have their own respective advantages and disadvantages. In addition to the obvious criterion of sequestration potential, the water, energy and nutrient footprints need to be considered due to the stress that they put on limited resources (Smith et al., 2016). Finally, for commercial scale-up, cost and TRL are also critical factors (McLaren, 2012). Comparison of the alternatives is also hampered by the limited historical data and knowledge, which results in uncertainties in performance or cost estimates. Thus, there is a need to develop a simple, robust methodology that allows comparison of NETs to be done based on multiple criteria, even under conditions of uncertainty.

3. Interval SAW

The basic SAW method uses the following equations:

$$\sum_j w_j d_{ij} = s_i \qquad \forall i \qquad (1)$$

$$\sum_j w_j = 1 \qquad (2)$$

where w_j is the dimensionless weight of criterion j, d_{ij} is the dimensionless score in the interval $[0, 1]$ of alternative i with respect to criterion j, and s_i is the aggregate score in the interval $[0, 1]$ of alternative i. Depending on the normalization convention applied, the optimal solution is the alternative with either the highest or lowest aggregate score, s_i. The other solutions can likewise be ranked in order of decreasing preference. From here on, it is assumed that all scores are normalized such that higher values are more desirable.

In SAW, the choice of weighting factors has a strong influence in determining the solution (Maliene et al., 2018). Since the weights reflect the decision-maker's preferences, they are inherently subjective. To manage the inherent biases that may exist in any MADM application, systematic procedures (e.g., eigenvector or geometric mean methods) for determining weights based on pairwise comparisons can be used, as in the Analytic Hierarchy Process (AHP) (Saaty, 1980). Sensitivity analysis can also be used to gauge the robustness of the rankings (Maliene et al., 2018). Cruz and Almario (2018) showed that, for any MADM problem solved via SAW, there exists a polytope in the hyperspace of the criteria weights for which the best solution retains its optimal rank; they referred to this polytope as the invariance region. Subsequently, Tan et al. (2019a) developed a procedure for tracing the invariance region.

Interval numbers offer an efficient means of representing epistemic uncertainty in mathematical models. Such uncertainty arises not from random variations (i.e., stochasticity) but from a lack of knowledge. Such uncertainties are invariably present when new and emerging technologies are characterized (Sikdar, 2019). This phenomenon is evident in the range of values reported in the literature for techno-economic characteristics of various NETs (e.g., McLaren, 2012; Smith et al., 2016; Alcalde et al., 2018). An interval number is a range of values characterized by a lower and an upper bound, which correspond to the pessimistic (i.e., conservative or risk averse) and optimistic (i.e., aggressive or risk tolerant) subjective estimates of the true (but unknown) value. In a sense, the true value is "spread out" uniformly across the interval; the width of the interval signifies the level of uncertainty of the quantity.

Arithmetic operations on interval numbers are based on operations on the limits. For example, given two interval numbers $A = [a^L, a^U]$ and $B = [b^L, b^U]$, we have:

$$[a^L, a^U] + [b^L, b^U] = [(a^L + b^L), (a^U + b^U)] \tag{3}$$

$$[a^L, a^U] - [b^L, b^U] = [(a^L - b^U), (a^U - b^L)] \tag{4}$$

$$[a^L, a^U] \times [b^L, b^U] = [(a^L \times b^L), (a^U \times b^U)] \tag{5}$$

$$[a^L, a^U]/[b^L, b^U] = [(a^L/b^U), (a^U/b^L)] \tag{6}$$

In the case of division, i.e., equation (6), the result is undefined if $0 \in [b^L, b^U]$. Crisp or ordinary numbers can be regarded as a special case of intervals with zero spread. These arithmetic operations can be used in both data pre-processing (e.g., normalization) and in the generic SAW procedure defined by equations (1) and (2). The resulting aggregate scores will thus have interval values.

Interval numbers can be ranked by pairwise comparison of their values. The trivial cases where (a) two interval numbers are exactly the same, or (b) when the worst value of one number exceeds the best value of another, need not be discussed here in detail. Some complications arise if the intervals overlap, which suggests that one number is not clearly larger than the other. In such cases, the procedure described by Sayadi et al. (2009) can be applied, wherein the intervals being compared are "collapsed" into point estimates that are weighted averages of the lower and upper bounds:

$$A > B \text{ if} \qquad (\alpha \times a^L + (1 - \alpha)a^U) > (\alpha \times b^L + (1 - \alpha)b^U) \tag{7}$$

where parameter α quantifies the degree of risk aversion of decision maker in the interval $[0, 1]$. A completely pessimistic decision-maker will use $\alpha = 1$ and will compare interval numbers based on their lower bounds (note that higher values are assumed to be more desirable here), while a completely optimistic decision-maker will use the upper bounds. This parametric control of the interval SAW procedure is a powerful and valuable feature which is particularly useful in the case of NETs, whose techno-economic parameters show wide uncertainty margins due to lack of knowledge from an extended history of large-scale use.

4. Comparison of NET Options

This case study is based on data reported by Alcalde et al. (2018) on the potential of NETs in Scotland. Their results are interesting, since they conclude that the NET potential is sufficient to offset Scotland's GHG emissions and achieve net zero emissions; the country, thus, serves as a microcosm for the kind of NET scale-up that needs to be done throughout the world. Raw data on the six NET options based on relevant criteria are shown in Table 2; with the exception of TRL, which are drawn from the paper of McLaren (2012), these data are based on the work of Alcalde et al. (2018). In this case, TRL provides a measure of the dynamic scalability of a given technology; more mature options are capable of being scaled up within the time horizon needed to achieve deep emissions cuts. Note that many of the scores show wide margins of uncertainty, in some cases spanning multiple orders of magnitude. The necessary calculations can be done in a spreadsheet, or using optimization and equation-solving software such as LINGO (Schrage, 2001). Appendices containing the LINGO code of this case study (which can be copied and pasted directly into a LINGO model file) are given at the end of the chapter in order to facilitate replication by the reader. A demonstration version of the software can be downloaded from the company website (www.lindo.com).

These raw data can then be normalized into the required dimensionless form by linear interpolation between the threshold (i.e., best and worst) values given in Table 3. This is a trivial step and, for brevity, is no longer shown here. These weights are assumed to be known *a priori*; in practice, they can be systematically determined using AHP (Saaty, 1980). The resulting dimensionless scores are given in Table 4.

Table 2. Scores of NET options with respect to techno-economic criteria (Alcalde et al., 2018; McLaren, 2012).

Alternative	Sequestration potential (Mt/y)	Water requirement (km³/y)	Energy requirement (PJ/y)	Specific cost (billion US$/y)	TRL
BECCS	[0.6, 2.4]	[1.2, 6]	[−23.16, 20.88]	[0.08, 0.32]	[4, 6]
AR	[0.34, 0.34]	[0.4, 0.8]	[0, 0]	[0.02, 0.04]	[6, 7]
SCS	[0.0588, 1.96]	[0, 0]	[0, 0]	[−0.01, 0.08]	[2, 7]
BA	[0.23, 1.5]	[0, 0]	[−30, −11.5]	[−0.19, 1.8]	[4, 6]
DAC	[0.6, 2.4]	[0.04, 0.26]	[1.56, 109.92]	[0.96, 4.99]	[4, 6]
EW	[1.4, 2.2]	[0, 0]	[4.3, 100.8]	[0.13, 12.84]	[1, 5]

Table 3. Threshold values and weights of criteria.

	Sequestration potential (Mt/y)	Water requirement (km³/y)	Energy requirement (PJ/y)	Specific cost (billion US$/y)	TRL
Best Value	2.4	0	−30	−0.19	9
Worst Value	0	6	109.92	12.84	1
Weight	0.40	0.10	0.10	0.25	0.15

Table 4. Normalized scores of NET options with respect to techno-economic criteria.

Alternative	Sequestration potential	Water requirement	Energy requirement	Specific cost	TRL
BECCS	[0.250, 1.000]	[0.000, 0.800]	[0.636, 0.951]	[0.961, 0.979]	[0.375, 0.625]
AR	[0.142, 0.142]	[0.867, 0.933]	[0.786, 0.786]	[0.982, 0.984]	[0.625, 0.750]
SCS	[0.025, 0.817]	[1.000, 1.000]	[0.786, 0.786]	[0.979, 0.986]	[0.125, 0.750]
BA	[0.096, 0.625]	[1.000, 1.000]	[0.868, 1.000]	[0.847, 1.000]	[0.375, 0.625]
DAC	[0.250, 1.000]	[0.957, 0.993]	[0.000, 0.774]	[0.602, 0.912]	[0.375, 0.625]
EW	[0.583, 0.917]	[1.000, 1.000]	[0.065, 0.755]	[0.000, 0.975]	[0.00, 0.500]

It is then possible to use SAW to determine the optimal solution based on the weights also given in Table 3. Initially, we assume a highly conservative decision-maker ($\alpha = 1$) who judges the NET alternatives based on their worst scores. In this case, the optimal solution is AR, with a dimensionless aggregate score of 0.561. Implementation of these steps in LINGO can be done using the code given in Appendix B. The code can also be modified by the reader to handle different data sets. Figure 1 shows the sensitivity analysis of aggregate scores with respect to α. For a decision-maker with moderate to low levels of risk aversion ($0 < \alpha < 0.76$), the optimal solution is BECCS. Conservative decision-makers ($0.76 < \alpha < 1$), on the other hand, will favor AR; this is a well-understood and time-tested NET and is, thus, the clear choice for the risk averse.

The decision-maker may also be interested in the stability of the optimal solution with respect to different weights. Given the baseline weights, and with $\alpha = 1$, AR was found to be the optimal alternative. The decision-maker can then determine the range of variations of the weights (i.e., the rank invariance region) for which AR remains optimal. This can be done using the procedure described by Tan et al. (2019a) with the working LINGO code in Appendix C. Table 5 gives the lower and upper limits of the criteria weights; any weight value that falls outside these bounds will definitely result in an optimal solution other than the AR option. A narrower range of weight values indicates higher sensitivity of the optimum choice to a given criterion. For example, when the weight given to sequestration potential

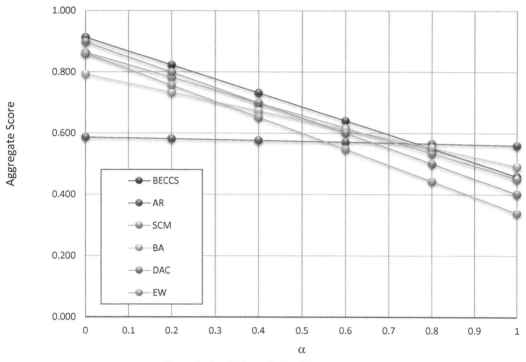

Figure 1. Sensitivity analysis with respect to α.

Table 5. Upper and lower limits of the weights of criteria.

	Sequestration potential	Water requirement	Energy requirement	Specific cost	TRL
Maximum	0.63	0.65	0.75	1	1
Baseline	0.40	0.10	0.10	0.25	0.15
Minimum	0	0	0	0	0

approaches 1, the optimal NET alternative becomes EW. On the other hand, it can also be seen that the choice of AR is completely robust to any changes in specific cost or TRL.

5. Implications for the Role of CDR/NETs in Large-Scale Carbon Management

It is clear that CDR or NETs will be needed to offset existing positive GHG emissions in order to achieve net zero emissions by mid-century, as recommended by the IPCC (2018). The different technology options being discussed in the literature need to be evaluated based on criteria such as potential, use of resources (e.g., water and energy), cost, and technological maturity. The MADM methodology described in this chapter provides a workable approach to ranking alternatives even if precise data are not available for evaluation. Nevertheless, this technique only provides a piece of the puzzle, and other aspects need to be accounted for if CDR and NETs are to become a significant carbon management strategy. First, it is notable that only AR is sufficiently mature for immediate large-scale deployment (Bastin et al., 2019). Land-based techniques (SCS, BA and EW) have been field-tested only at limited scales, well below the levels that will deliver significant benefits. BECCS and DAC have been demonstrated at the pilot plant scale, and their scale-up will be dependent on the presence of CO_2 transportation infrastructure and suitable geological storage sites.

Given the current status of technology, some key points can be identified. First, bridging the gap between research and commercial deployment of most CDR/NET options will still require significant economic investment (Nemet et al., 2018). Optimization models will be needed in order to maximize the technology maturity gains given limited financial resources (Tan et al., 2019b). There will also still be major challenges in ramping up even relatively mature CDR/NETs, such as BA, from field tests to gigaton-scale carbon management strategies; such issues include planning supply chains and developing reliable monitoring systems to verify emissions cuts (Tan, 2019). Targets for CDR/NET deployment at the level of countries (Smith et al., 2016b) or regions (Geden et al., 2019) should take into account local resource limitations, as well as integration with other carbon management strategies being used concurrently. Governance aspects (e.g., economic incentives) need to be studied for proper calibration and alignment with societal priorities (Bellamy, 2018). Finally, as with any technological solution, socio-cultural aspects need to be kept in mind, since these factors can act as barriers to their successful scale-up (Buck, 2016).

6. Conclusion

Commercial deployment of CDR or NETs in the coming decades will be critical in achieving the much-needed reductions in global GHG emissions. They will be needed to offset emissions from other human activities that are driven by economic and demographic trends. Scale-up of the best NETs for use in any given regional context requires proper evaluation of their techno-economic merits relative to stakeholder priorities that result from local conditions. They must, therefore, be evaluated, even in the presence of significant epistemic data uncertainties that are inevitable features of new and unproven technologies. Interval SAW was used here to evaluate NETs based on carbon sequestration potential, water use, energy requirement, cost, and technological maturity. The AR option is the optimal choice for conservative decision-makers, while those with moderate to high risk tolerance will favor BECCS. A rank invariance analysis procedure was then used to assess the robustness of the optimal solution to changes in criteria weights.

7. Declaration of Conflict of Interest

The authors declare no conflict of interest.

Acknowledgement

The support of the Philippine Commission on Higher Education (CHED) via the Philippine Higher Education Research Network (PHERNet) Sustainability Studies Program at De La Salle University is gratefully acknowledged. The support of a NAST Research Fellowship is acknowledged by Jose B. Cruz, Jr.

Appendix A—List of Acronyms

AHP   Analytic hierarchy process
AR   Afforestation and reforestation
BA   Biochar application
BECCS   Bioenergy with CO_2 capture and storage
CCS CO_2 capture and storage

CDR	Carbon dioxide removal
DAC	Direct air capture
EW	Enhanced weathering
GHG	Greenhouse gas
IPCC	Intergovernmental Panel on Climate Change
LUC	Land use change
MADM	Multiple-attribute decision making
NET	Negative emissions technologies
SAW	Simple additive weighting
SCM	Soil carbon management
SDG	Sustainable Development Goals
TRL	Technology readiness level
UN	United Nations

Appendix B—LINGO Code for Interval SAW

```
!OPTIONS:
```

AR	AFFORESTATION AND REFORESTATION
BA	BIOCHAR APPLICATION
BECCS	BIOENERGY WITH CCS
DAC	DIRECT AIR CAPTURE
EW	ENHANCED WEATHERING
SCM	SOIL CARBON MANAGEMENT

CRITERIA:

C	COST
ER	ENERGY REQUIREMENT
SEQ	CARBON SEQUESTRATION POTENTIAL
TRL	TECHNOLOGY READINESS LEVEL
WR	WATER REQUIREMENT

PARAMETERS:

ALPHA	RISK AVERSION INDEX OF DECISION MAKER, 1 IS RISK AVERSE
BEST (J)	IDEAL BEST VALUE FOR CRITERIA J
WORST (J)	WORST POSSIBLE VALUE FOR CRITERIA J
PERFNEG (I,J)	WORST PERFORMANCE OF TECHNOLOGY I IN CRITERIA J
PERFPOS (I,J)	BEST PERORMANCE OF TECHNOLOGY I IN CRITERIA J
W (J)	WEIGHT ASSOCIATED WITH CRITERIA J

VARIABLES:

INDEXN (I,J)	NORMALIZED VALUE FOR WORST PERFORMANCE OF TECHNOLOGY I IN CRITERIA J
INDEXP (I,J)	NORMALIZED VALUE FOR BEST PERFORMANCE OF TECHNOLOGY I IN CRITERIA J
D (I,J)	PERFORMANCE OF TECHNOLOGY I IN CRITERIA J IN CONSIDERATION OF UNCERTAINTY AND RISK AVERSION
S (I)	WEIGHTED PERFORMANCE SCORE OF TECHNOLOGY I;

```
SETS:
TECHNOLOGY: S;
CRITERIA: BEST, WORST, W;
MATRIX(TECHNOLOGY, CRITERIA): PERFPOS, PERFNEG, INDEXN, INDEXP, D;
ENDSETS

DATA:
TECHNOLOGY = BECCS   AR   SCM   BA   DAC   EW;
CRITERIA = SEQ   WR   ER   C   TRL;
PERFNEG, PERFPOS =
```

0.6	2.4	6	1.2	20.88	−23.16	0.32	0.08	4	6
0.34	0.34	0.8	0.4	0	0	0.04	0.02	6	7
0.0588	1.96	0	0	0	0	0.08	−0.01	2	7
0.23	1.5	0	0	−11.5	−30	1.8	−0.19	4	6
0.6	2.4	0.26	0.04	109.92	1.56	4.99	0.96	4	6
1.4	2.2	0	0	100.8	4.3	12.84	0.13	1	5;

```
BEST, WORST = 2.4,0, 0,6,  −30,109.92,  −0.19,12.84,  9,1;
W = 0.4   0.1   0.1   0.25   0.15;
ALPHA = 1.0;
ENDDATA

@FOR(MATRIX(I,J): INDEXN (I,J)*(BEST(J) − WORST(J)) = PERFNEG(I,J) − WORST(J);
              INDEXP (I,J)*(BEST(J) − WORST(J)) = PERFPOS(I,J) − WORST(J));

@FOR(MATRIX(I,J): D(I,J) = ALPHA*(@SMIN(INDEXN(I,J), INDEXP(I,J))) + (1 − ALPHA)*@
SMAX(INDEXN(I,J), INDEXP(I,J)));

@FOR(TECHNOLOGY(I): S(I) = @SUM(CRITERIA(J): D(I,J)*W(J)));
```

Appendix C—LINGO Code for Sensitivity Analysis with Respect to Criteria Weights

```
!OPTIONS:
AR            AFFORESTATION AND REFORESTATION
BA            BIOCHAR APPLICATION
BECCS         BIOENERGY WITH CCS
DAC           DIRECT AIR CAPTURE
EW            ENHANCED WEATHERING
SCM           SOIL CARBON MANAGEMENT

CRITERIA:
C             COST
ER            ENERGY REQUIREMENT
SEQ           CARBON SEQUESTRATION POTENTIAL
TRL           TECHNOLOGY READINESS LEVEL
WR            WATER REQUIREMENT

PARAMETERS:
ALPHA            RISK AVERSION INDEX OF DECISION MAKER, 1 IS RISK AVERSE
BEST (J)         IDEAL BEST VALUE FOR CRITERIA J
WORST (J)        WORST POSSIBLE VALUE FOR CRITERIA J
PERFNEG (I,J)    WORST PERFORMANCE OF TECHNOLOGY I IN CRITERIA J
```

PERFPOS (I,J)	BEST PERORMANCE OF TECHNOLOGY I IN CRITERIA J
W (J)	WEIGHT ASSOCIATED WITH CRITERIA J
K	INDEX OF TECHNOLOGY WITH HIGHEST SCORE S(K)

VARIABLES:

INDEXN (I,J)	NORMALIZED VALUE FOR WORST PERFORMANCE OF TECHNOLOGY I IN CRITERIA J
INDEXP (I,J)	NORMALIZED VALUE FOR BEST PERFORMANCE OF TECHNOLOGY I IN CRITERIA J
D (I,J)	PERFORMANCE OF TECHNOLOGY I IN CRITERIA J IN CONSIDERATION OF UNCERTAINTY AND RISK AVERSION
S (I)	WEIGHTED PERFORMANCE SCORE OF TECHNOLOGY I;

SETS:
TECHNOLOGY: S;
CRITERIA: BEST, WORST, W;
MATRIX(TECHNOLOGY, CRITERIA): PERFPOS, PERFNEG, INDEXN, INDEXP, D;
ENDSETS

DATA:
TECHNOLOGY = BECCS   AR   SCM   BA   DAC   EW;
CRITERIA = SEQ   WR   ER   C   TRL;
PERFNEG, PERFPOS =

0.6	2.4	6	1.2	20.88	−23.16	0.32	0.08	4	6
0.34	0.34	0.8	0.4	0	0	0.04	0.02	6	7
0.0588	1.96	0	0	0	0	0.08	−0.01	2	7
0.23	1.5	0	0	−11.5	−30	1.8	−0.19	4	6
0.6	2.4	0.26	0.04	109.92	1.56	4.99	0.96	4	6
1.4	2.2	0	0	100.8	4.3	12.84	0.13	1	5;

BEST, WORST = 2.4,0,   0,6,   −30,109.92,   −0.19,12.84,   9,1;
ALPHA = 1.0;
ENDDATA

MAX = W(1);   !The objective function can be changed depending on which weight is of interest. For example, it can be changed to "MAX = W(2);" to maximize the weight of the second criteria;

K = 2;

@SUM(CRITERIA(J): W(J)) = 1;

@FOR(TECHNOLOGY(I) | I#NE# K: @SUM(CRITERIA(J): D(I,J)*W(J)) < S(K));

@FOR(MATRIX(I,J): INDEXN (I,J)*(BEST(J) − WORST(J)) = PERFNEG(I,J) − WORST(J);
              INDEXP (I,J)*(BEST(J) − WORST(J)) = PERFPOS(I,J) − WORST(J));

@FOR(MATRIX(I,J): D(I,J) = ALPHA*(@SMIN(INDEXN(I,J), INDEXP(I,J))) + (1 − ALPHA)*@SMAX(INDEXN(I,J), INDEXP(I,J)));

@FOR(TECHNOLOGY(I): S(I) = @SUM(CRITERIA(J): D(I,J)*W(J)));

References

Alcalde, J., Smith, P., Haszeldine, R.S. and Bond, C.E. 2018. The potential for implementation of negative emission technologies in Scotland. International Journal of Greenhouse Gas Control 76: 85–91.

Bastin, J.-F., Finegold, Y., Garcia, C., Mollicone, D., Rezende, M., Routh, D., Zohner, C.M. and Crowther, T.W. 2019. The global tree restoration potential. Science 365: 76–79.

Bellamy, R. 2018. Incentivize negative emissions responsibly. Nature Energy 3: 532–534.

Buck, H.J. 2016. Rapid scale-up of negative emissions technologies: Social barriers and social implications. Climatic Change 139: 155–167.

Cruz, J.B. and Almario, E.M.R. 2018. When is the Pareto choice from a finite set invariant to variations in weight and values of multiple performance criteria? Philippine Science Letters 11: 61–70.

Dyson, F.J. 1977. Can we control the carbon dioxide in the atmosphere? Energy 2: 287–91.

Fuss, S., Lamb, W.F., Callaghan, M.W., Hilaire, J., Creutzig, F., Amann, T., Beringer, T., De Oliveira Garcia, W., Hartmann, J., Khanna, T., Luderer, G., Nemet, G.F., Rogelj, J., Smith, P., Vicente Vicente, J.L., Wilcox, J., Del Mar Zamora Dominguez, M. and Minx, J.C. 2018. Negative emissions—Part 2: Costs, potentials and side effects. Environmental Research Letters 13: Article 063002.

Geden, O., Peters, G.P. and Scott, V. 2019. Targeting carbon dioxide removal in the European Union. Climate Policy 19: 487–494.

Griscom, B.W., Adams, J., Ellis, P.W., Houghton, R.A., Lomax, G., Miteva, D.A., Schlesinger, W.H., Shoch, D., Siikamäki, J.V., Smith, P., Woodbury, P., Zganjar, C., Blackman, A., Campari, J., Conant, R.T., Delgado, C., Elias, P., Gopalakrishna, T., Hamsik, M.R., Herrero, M., Kiesecker, J., Landis, E., Laestadius, L., Leavitt, S.M., Minnemeyer, S., Polasky, S., Potapov, P., Putz, F.E., Sanderman, J., Silvius, M., Wollenberg, E. and Fargione, J. 2017. Natural climate solutions. Proceedings of the National Academy of Sciences of the United States of America 114: 11645–11650.

Haszeldine, R.S., Flude, S., Johnson, G. and Scott, V. 2018. Negative emissions technologies and carbon capture and storage to achieve the Paris Agreement commitments. Philosophical Transactions of the Royal Society A: Mathematical, Physical and Engineering Sciences, DOI: 0.1098/rsta.2016.0447.

IPCC. 2018. Summary for policymakers. *In*: Masson-Delmotte, V., Zhai, P., Pörtner, H.O., Roberts, D., Skea, J., Shukla, P.R., Pirani, A., Moufouma-Okia, W., Péan, C., Pidcock, R., Connors, S., Matthews, J.B.R., Chen, Y., Zhou, X., Gomis, M.I., Lonnoy, E., Maycock, T., Tignor, M. and Waterfield, T. (eds.). Global Warming of 1.5 °C. An IPCC Special Report on the Impacts of Global Warming of 1.5 °C above Pre-industrial Levels and Related Global Greenhouse Gas Emission Pathways, in the Context of Strengthening The Global Response to the Threat of Climate Change, Sustainable Development, and Efforts to Eradicate Poverty. World Meteorological Organization, Geneva, Switzerland.

MacCrimmon, K.R. 1968. Decision making among multiple attribute alternatives: A survey and consolidated approach. RAND Memorandum, RM-4823-ARPA.

Maliene, V., Dixon-Gough, R. and Malys, N. 2018. Dispersion of relative importance values contributes to the ranking uncertainty: Sensitivity analysis of multiple criteria decision-making methods. Applied Soft Computing 67: 286–298.

Mankins, J.C. 2009. Technology readiness assessments: A retrospective. Acta Astronomica 65: 1216–1223.

McLaren, D. 2012. A comparative global assessment of potential negative emissions technologies. Process Safety and Environmental Protection 90: 489–500.

McGlashan, N., Shah, N., Caldecott, B. and Workman, M. 2012. High-level techno-economic assessment of negative emissions technologies. Process Safety and Environmental Protection 90: 501–510.

Minx, J.C., Lamb, W.F., Callaghan, M.W., Fuss, S., Hilaire, J., Creutzig, F., Amann, T., Beringer, T., De Oliveira Garcia, W., Hartmann, J., Khanna, T., Lenzi, D., Luderer, G., Nemet, G.F., Rogelj, J., Smith, P., Vicente Vicente, J.L., Wilcox, J. and Del Mar Zamora Dominguez, M. 2018. Negative emissions—Part 1: Research landscape and synthesis. Environmental Research Letters 13: Article 063001.

Nemet, G.F., Callaghan, M.W., Creutzig, F., Fuss, S., Hartmann, J., Hilaire, J., Lamb, W.F., Minx, J.C., Rogers, S. and Smith, P. 2018. Negative emissions—Part 3: Innovation and upscaling. Environmental Research Letters 13: Article 063003.

Pacala, S. and Socolow, R. 2004. Stabilization wedges: Solving the climate problem for the next 50 years with current technologies. Science 305: 968–972.

Rockström, J., Steffen, W., Noone, K., Persson, A., Chapin, F.S., Lambin, E.F., Lenton, T.M., Scheffer, M., Folke, C., Schellnhuber, H.J., Niykvist, B., De Wit, C.A., Hughes, T., van der Leeuw, S., Rodhe, H., Sorlin, S., Snyder, P.K., Constanza, R., Svedin, U., Falkenmark, M., Karlberg, L., Corell, R.W., Fabry, V.J., Hansen, J., Walker, B., Liverman, D., Richardson, K., Crutzen, P. and Foley, J.A. 2009. A safe operating space for humanity. Nature 461: 472–475.

Saaty, T.L. 1980. The Analytic Hierarchy Process. McGraw-Hill, New York, USA.

Sanz-Perez, E.S., Murdock, C.R., Didas, S.A. and Jones, C.W. 2016. Direct capture of CO_2 from ambient air. Chemical Reviews 116: 11840–11876.

Sayadi, M.K., Heydari, M. and Shahanaghi, M. 2009. Extension of VIKOR method for decision making problem with interval numbers. Applied Mathematical Modelling 33: 2257–2262.

Schrage, L. 2001. Optimization Modeling with LINGO, 5th ed. Lindo Systems, Inc., Chicago, Illinois, USA.

Sikdar, S.K. 2009. On aggregating multiple indicators into a single metric for sustainability. Clean Technologies and Environmental Policy 11: 157–161.

Sikdar, S.K. 2019. Fractured state of decisions on sustainability: An assessment. Sustainable Production and Consumption 19: 231–237.

Smith, P., Davis, S.J., Creutzig, F., Fuss, S., Minx, J., Gabrielle, B., Kato, E., Jackson, R.B., Cowie, A., Kriegler, E., van Vuuren, D.P., Rogelj, J., Ciais, P., Milne, J., Canadell, J.G., McCollum, D., Peters, G., Andrew, R., Krey, V., Shrestha, G., Friedlingstein, P., Gasser, T., Grübler, A., Heidug, W.K., Jonas, M., Jones, C.D., Kraxner, F., Littleton, E., Lowe,

J., Moreira, J.R., Nakicenovic, N., Obersteiner, M., Patwardhan, A., Rogner, M., Rubin, E., Sharifi, A., Torvanger, A., Yamagata, Y., Edmonds, J. and Yongsung, C. 2016a. Biophysical and economic limits to negative CO_2 emissions. Nature Climate Change 6: 42–50.

Smith, P., Haszeldine, R.S. and Smith, S.M. 2016b. Preliminary assessment of the potential for, and limitations to, terrestrial negative emission technologies in the UK. Environmental Science: Processes and Impacts 18: 1400–1405.

Tan, R.R. and Foo, D.C.Y. 2018. Integrated multi-scale water management as a climate change adaptation strategy. Clean Technologies and Environmental Policy 20: 1123–1125.

Tan, R.R. 2019. Data challenges in optimizing biochar-based carbon sequestration. Renewable and Sustainable Energy Reviews 104: 174–177.

Tan, R.R., Almario, E.M.R., Aviso, K.B., Cruz, J.B., Jr. and Promentilla, M.A.B. 2019a. A methodology for tracing the rank invariance region in multi-criterion selection problems: Application to negative emission technologies. Process Integration and Optimization for Sustainability (in press, DOI: 10.1007/s41660-019-00089-4).

Tan, R.R., Aviso, K.B. and Ng, D.K.S. 2019b. Optimization models for financing innovations in green energy technologies. Renewable and Sustainable Energy Reviews 113: Article 109258.

CHAPTER 14

Carbon Management in the CO_2-Rich Natural Gas to Energy Supply-Chain

Ofélia de Queiroz Fernandes Araújo, Stefano Ferrari Interlenghi*
and *José Luiz de Medeiros*

1. Introduction

The cumulative amount of carbon dioxide (CO_2) that can be emitted is referred to as the "carbon budget". In the latest Intergovernmental Panel on Climate Change (IPCC) report, the budget for having a 50% chance of keeping the average global warming below 2 °C—the UN Paris Agreement, namely the 2D scenario—is estimated to be approximately 275 Gt of carbon (1008 Gt CO_2) (IPCC, 2014a). Clearly, the oil and gas (O&G) industry demands urgent and efficient technologies for CO_2 management, to mitigate the risk of having three quarters of the proven fossil reserves becoming unburnable (IPCC, 2014b)— the Stranded Asset Risk (SAR) scenario. Despite the 2D and SAR scenarios, the O&G industry might be building excess capacity (Musarra, 2017) and, to achieve the expected return on invested capital, production life needs to be extended beyond 2050, and a fossil fuel lock-in is expected to occur to some extent.

1.1 Carbon capture formal routes

To reduce the impacts of fossil fuels, three formal CO_2 capture routes are currently being pursued: Pre-combustion, post-combustion and oxy-combustion (Al-Mamoori et al., 2017). These concepts can be applied to all fossil fuels but are highlighted for natural gas (NG). Figure 1 sketches the three carbon capture routes associated to power generation and their main differences. Final destination of removed CO_2 in Figure 1 is considered to be enhanced oil recovery (EOR).

Pre-combustion can be summarized as a group of processes that remove all carbon (decarbonization) from a determined fuel and replace it with hydrogen before its combustion (Olajire, 2010). For example, NG can be transformed into syngas ($CO_2 + CO + H_2$) via an endothermic steam reform, where some air or O_2 is supplied in order to release the necessary heat. In a subsequent stage, CO is converted to more CO_2 and H_2 via the Water-Gas-Shift (WGS) reaction by adding steam. CO_2 is then separated from the gas stream via a post-combustion carbon capture technology, usually physical/chemical absorption (acid-gas removal, AGR), and sent to storage, or used as a feedstock, while H_2 is used as fuel for power/

Escola de Química, Federal University of Rio de Janeiro, CT, Ilha do Fundão, Rio de Janeiro, RJ, 21941-909, Brazil.
Emails: stinterlenghi@eq.ufrj.br; jlm@eq.ufrj.br
* Corresponding author: ofelia@eq.ufrj.br

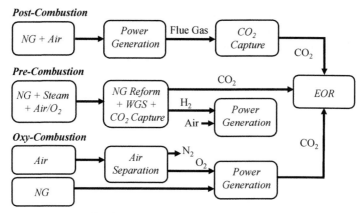

Figure 1. Carbon capture routes associated to power generation (NG ≡ Natural Gas; WGS ≡ Water-Gas-Shift reaction; EOR ≡ Enhanced Oil Recovery).

heat generation (Al-Mamoori et al., 2017). Since burning H$_2$ does not produce any carbon emissions, CO$_2$ removal before fuel combustion allows for a carbon neutral result if all carbon is captured before H$_2$ firing. A potential advantage of pre-combustion is that CO$_2$ is separated from a high-pressure stream, possibly reducing power needs and costs.

In oxy-combustion, pure O$_2$ is employed for NG firing, reducing the carbon intensity of the power generation (Al-Mamoori et al., 2017). Oxy-combustion results in a highly CO$_2$ concentrated flue-gas and steam (Araújo and de Medeiros, 2017). Water vapor can simply be condensed in order to obtain an almost pure CO$_2$ stream, avoiding capital intensive capture technologies. The main bottleneck in oxy-combustion is the cost of producing pure O$_2$ via cryogenic separation from air.

In post-combustion, CO$_2$ is separated from a flue-gas after burning NG with air (Araújo and de Medeiros, 2017). The use of air combustion generates a flue-gas with high nitrogen content, hampering CO$_2$ removal significantly. Post-combustion has gathered the most attention due to the wide array of possible technologies used and its retrofitting potential in already existing plants. As a result, post-combustion currently has the most mature technologies and application scenarios in CO$_2$ management of the NG industry and is the focus of this chapter.

1.2 Carbon capture in the oil and gas sector

Proven oil reserves are expanding as offshore exploration and production (E&P) is increasingly moving to remote areas and ultra-deepwaters (depth > 2000 m). Particularly challenging is the production of natural gas (NG) in ultra-deepwaters, beyond 250 km from the coast, with high gas-to-oil ratio (GOR) and CO$_2$-rich NG, as in the Brazilian pre-salt O&G reserves; e.g., the Libra field, whose gas has more than 40%mol CO$_2$ (Arinelli et al., 2017). These unconventional reserves are the focused scenario for being among the hardest to exploit, negatively impacting the energy return on (energy) investment (EROI) (Hall et al., 2014). Ultra-deepwater O&G processing employs Floating Production Storage and Offloading (FPSO) platforms, where, due to high GOR, ≈ 60% of the topside area is dedicated to gas processing (Araujo et al., 2017).

In parallel, to adjust investments in line with emissions reduction targets (e.g., 2D scenario), worldwide economies are enforcing mechanisms to move towards a low carbon economy; e.g., carbon taxes and emissions-trading systems, where total emissions are capped and permissions to emit CO$_2$ are traded (Energy Institute at Haas, 2016). The 2D scenario is favoring a shift towards renewable energy, demanding actions from O&G companies to mitigate SAR, extending the O&G lifetime, targeting intense carbon capture and storage (CCS).

NG is the fossil fuel with least environmental impact: In power plants it emits 50% to 60% less CO$_2$ per kW than coal, which entails other additional environmental issues, such as mercury, particulate matter

(PM) and nitrogen/sulfur oxides (NETL, 2015). NG is expected to expand at an annual rate of 6% from 2017 to 2020 (EIA, 2018), responding to $\approx$ 30% of the global power generation by 2040 (EIA, 2018). Even though NG has a clean burning, its increasing consumption has resulted in CO_2 emissions (EIA, 2016). It is also challenging that CO_2 use as an (EOR) Enhanced Oil Recovery agent is intensifying. Injection of CO_2 in oil reservoirs recovers extra barrels of oil, improving economic performance but increasing the CO_2 content in produced fluids along operation time (Zhao et al., 2016). Concomitantly, the production curve decreases (Reis et al., 2017), demanding resilient CO_2 capture technologies to maintain economic performance while mitigating environmental impacts.

This chapter is driven by the critical role of carbon capture in extending the use of fossil-based energy, with destination to EOR, which combines storage with utilization (CCU, Carbon Capture and Utilization) at a large scale in the near-term (Araújo and de Medeiros, 2017). Although expected to grow in the long-term, monetization of CO_2 through its chemical conversion to value-added products is in an earlier process-systems development, small-scaled and with market size limitations on targeted products (Araújo et al., 2014), comparatively to CCS. Nevertheless, many authors defend the idea that CCU is a more suitable short-term solution within the constraints imposed by the 2D scenario by providing a monetary incentive to close carbon loops when compared to CCS (Stuardi et al., 2019). Prospects of using CO_2 as a feedstock for chemical production include carboxylation, reduction reactions, heterogeneously catalyzed hydrogenation and photocatalytic/electrocatalytic conversion (Alper and Orhan, 2017). Despite many chemicals currently being investigated as possible CCU outputs, CO_2 to methanol or methane has the most potential application in the offshore O&G scenario under high GOR and CO_2-rich NG constraints and is briefly included among the CO_2 management technologies reviewed in this chapter.

Despite generating value and storing CO_2 in the short term, CCU is expected to contribute less than 1% to greenhouse gas mitigation targets when considering its potential scale of deployment (Stuardi et al., 2019). In addition, current trends indicate that two-thirds of the total NG consumption worldwide occurs in power/heat applications—either in furnaces of industrial processes or electricity generators (EIA, 2018). In some industries (e.g., bulk chemicals, food, metal-based durables and glass), NG supplied over 40% of their heat/power requirements in 2017 (EIA, 2018). Projections up to 2050 show that NG used for heat/power is expected to steadily grow, due to the high costs associated with fuel switch, but NG as a feedstock will stagnate (EIA, 2018). Thus, CO_2 management in the NG to energy chain is, in both the short and the long-term, more critical in its environmental and economic impacts and, as a result, the focus of this chapter.

1.3 Scope and structure

Considering CO_2-rich NG as starting point, the main CO_2 capture technologies addressed comprise chemical absorption (CA), physical absorption (PA), membrane permeation (MP), cryogenic distillation (CD), adsorption (ADS), gas liquid membrane contactors (GLMC), supersonic separators (SS) and hybrid technologies (HYB). Figure 2 shows CCS/EOR as CO_2 destination in the value-chain, composed of onshore power generation (Gas-to-Pipeline-to-Wire, GTPTW) and offshore power generation (Gas-to-Wire, GTW), both with CCS, which is the main scenario in the present approach.

In the value-chain pictured in Figure 2, CO_2-rich NG ($CO_2 \geq$ 40%mol) is decarbonated in ultra-deepwater FPSO in order to meet CO_2 specification (< 3%mol), where the thicker lines correspond to the NG path to energy. While CO_2 is injected into the source offshore reservoir for EOR, decarbonated NG is compressed and dispatched through subsea pipelines to onshore power plants, where CO_2 is captured from exhaust gases and transported back to offshore reservoirs for EOR. In the shorter offshore alternative, power is generated on a Floating Power Generation Platform (FPGP), after partial decarbonation to reduce CO_2 content to the acceptable limit of NG-fired turbines ($\cong$ 20%mol). CO_2 from NG upgrading is mixed with CO_2 from offshore post-combustion capture and injected after compression into the source reservoir for EOR. NG and CO_2 pipelines are eliminated and a HVDC cable is needed for electricity transmission to onshore facilities.

For each CO_2 capture technology, separation mechanisms, benefits and shortcomings, recent innovations, stakeholders, and technology maturity level, via the Technology Readiness Level (TRL)

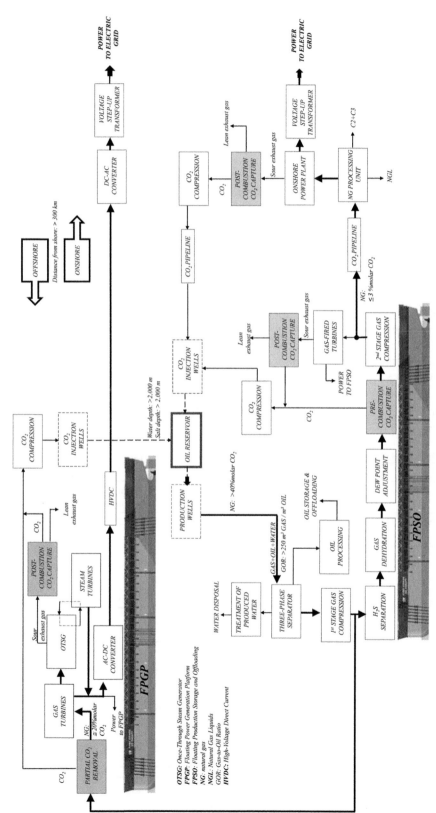

Figure 2. Value-chain of CO₂-rich natural gas to energy with decarbonation: Floating power generation vs onshore power generation.

index, are given. Still, with the focus on ultra-deepwater processing of CO_2-rich NG, the scenario of post-combustion capture is discussed, followed by a short discussion on pipeline transportation and chemical conversion of CO_2.

2. Natural Gas Decarbonation with Absorption

Absorption-based CO_2 capture is the most mature technology currently available for CO_2 removal from NG process streams. CO_2 separation from raw NG in Chemical Absorption (CA) and Physical Absorption (PA) occurs through absorption of CO_2 into a solvent, followed by solvent regeneration. CA regeneration is accomplished by stripping heat at low pressures, whilst a simple pressure reduction can recover the solvent in PA (Al-Mamoori et al., 2017). The separation mechanism in CA depends on reversible exothermic chemical reactions involving CO_2, the alkaline active component (e.g., amine or potassium carbonate) and water, creating weakly bonded intermediates at high CO_2 fugacities in absorber columns. The intermediates are dissociated at low CO_2 fugacities in heated stripper columns, which has two finalities: Vaporization of water as the CO_2 stripping agent and reversal of the exothermic absorption equilibrium to regenerate the solvent (de Medeiros et al., 2013a). In the stripper, the vapor fugacity of CO_2 is reduced, reversing the chemical absorption reaction and destroying the CO_2-amine intermediate, releasing low-pressure CO_2. The CO_2-lean solvent is recycled to the absorber (Yeo et al., 2012).

Table 1 shows the main CA solvents in use for CO_2 removal from NG. Aqueous amine and aqueous potassium carbonate solvents are presently in commercial scale and widespread in CCS applications, with large-scale demonstration plants worldwide (Araújo and de Medeiros, 2017). In NG applications, CA with aqueous monoethanolamine (MEA) has been the main CO_2 capture process, with a shift to aqueous methyl-diethanolamine (MDEA) in recent years (Olajire, 2010). Figure 3 depicts a typical CA process, where a CO_2-lean solvent (e.g., aqueous MEA, 25–30%w/w) enters the top of the absorption

Table 1. Main chemical absorption solvents. Adapted from: Polasek and Bullin (2006).

Solvent	MEA	DEA	DGA	MDEA	K_2CO_3
Solution strength (%w/w) (with water)	15–20	25–35	50–70	20–50	15–40
CO_2 loading (mol/mol)	0.3–0.35	0.3–0.8	0.3–0.4	0.4–1.0	0.1–0.25
Ability to absorb H_2S	None	Under limited conditions	None	Under most conditions	
Regeneration heat duty (GJ/t CO_2)	3.0–3.5	2.9–3.2	2.8–3.0	2.5–2.8	2–2.5

MEA = monoethanolamine; DEA = diethanolamine; DGA = diglycolamine; MDEA = methyl-diethanolamine.

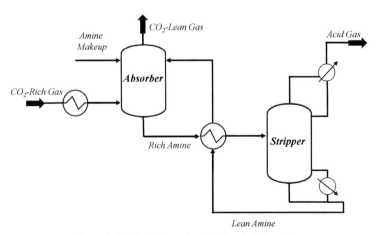

Figure 3. Typical CO_2 capture by chemical absorption.

tower operating from 40 °C to 50 °C, while the raw NG flows bottom-up in countercurrent (D'Allesandro et al., 2010). Raw NG must be cleansed of NO_x and SO_x since they irreversibly react with ethanolamines, forming stable salts that cannot be reclaimed during regeneration (Rochelle, 2009). The CO_2-rich solvent flows from the absorber bottom to the stripper, where lean aqueous ethanolamine is regenerated.

The stripper operates at high temperatures (100 °C–140 °C) and low pressures ranging from 1.5 bar to 1.8 bar absolute (Olajire, 2010) in order to reduce the stripping temperature while enhancing the stripping action of boiling water and preventing thermal degradation of the solvent. The CO_2-rich amine enters at the second stage of the stripper and flows in countercurrent to hot vapors rising from the reboiler (Songolzadeh et al., 2017). An important process parameter is the CO_2 loading of the amine solution, shown in equation (1) for MEA, defined as the ratio between the number of moles of absorbed CO_2 and the number of moles of the active component in the solvent (e.g., MEA), translating how much CO_2 is absorbed per unit of active component in the liquid. Lean solvent enters the absorber with loadings ranging from 0.2 to 0.3 mol CO_2 per mol amine and leaves with loading $\cong$ 0.5 mol CO_2/mol amine (Kothandaraman, 2010).

$$Loading = \frac{[CO_2]+[HCO_3^-]+[CO_3^=]+[MEACOO^-]}{[MEA]+[MEA^+]+[MEACOO^-]} \tag{1}$$

Low-pressure CO_2 leaves the top of the stripper and must be compressed up to 150 bar (Yu et al., 2012) or beyond (Pinto et al., 2014) for EOR applications. Over 90% CO_2 recovery can be obtained with CA in NG applications (Rochelle, 2009), requiring heat-ratios of 3–3.5 GJ/tCO_2 for solvent regeneration if MEA is used (Harkin et al., 2010). Process improvements may reduce the heat-ratio to $\approx$ 2.6 GJ/tCO_2 (Araújo and de Medeiros, 2017).

The main difference between CA and PA is the type of absorption mechanism. While CA involves reversible chemical reactions, PA occurs through a physical solubility interaction governed by the CO_2 fugacity in the vapor phase. Solubility increases as CO_2 fugacity rises and temperature decreases. While CA is better than PA at lower CO_2 fugacities (Olajire, 2010), a CO_2 fugacity threshold exists at which PA has a higher CO_2 absorption capacity then CA, which depends on the solvent. Usual physical solvents used in PA are methanol (Rectisol®), dimethyl ethers of polyethylene glycol (Selexol®), n-methyl-2-pyrrolidone (Purisol®) and propylene carbonate (Fluor Solvent™). Differently from CA, PA has low CO_2/CH_4 selectivity resulting in high hydrocarbon losses from NG, giving CA a competitive advantage.

2.1 Benefits and shortcomings

CO_2 capture through PA and CA benefits from technology maturity derived from nearly 60 years of use (Yu et al., 2012), with a great variety of solvents and additives that have been optimized for increased efficiency and high CO_2 selectivity (Olajire, 2010). Absorption schemes are easily retrofitted to older process plants and accept a wide range of CO_2 fugacities, with CA being best suited to lower fugacities and PA to higher fugacities.

The main drawback of absorption-based technologies is the high temperature and pressure swings applied in order to regenerate the CO_2-rich solvent, requiring external energy inputs despite the high heat-integration in NG processing (Al-Mamoori et al., 2017). Another aspect is the large footprint of PA and CA, normally hindering its use in CO_2-rich NG processing in offshore platforms that are constrained in area and weight (Araújo et al., 2017). One example of the large footprint is Schlumberger's amine-based process for NG decarbonation, which requires a 17 m absorber and 16 m stripper (Schlumberger, 2016).

In CA, solvent degradation occurs through salt formation with sulphur- and nitrogen-based impurities and oxidation due to the presence of small amounts of O_2. Losses via salt formation are minimized by NG pre-treatment to remove NO_x and SO_x impurities (Forsyth et al., 2017). Oxidative amine degradation is minimal as NG has low O_2 (< 0.2%mol) contents (Yu et al., 2012). A critical operational parameter is the feed CO_2 content (Rochelle, 2009), which affects solvent recirculation rate, heat consumption and equipment size. Time-varying systems—e.g., CO_2 removal from CO_2-rich NG during long-term EOR (Figure 2)—inevitably result in suboptimal operation of the designed absorption system, resulting in reduced CO_2 capture efficiency and increased heat and solvent consumption over time. Salvindera et al. (2017) used a composition controller to show that, for CA-MEA (CA with MEA 20%w/w in water), CO_2 content in the lean and CO_2-rich gases would greatly vary along operation time.

2.2 Recent focus

Research focus on CO_2 absorption consists of new process configurations to minimize heat demand (e.g., absorber intercooling, split-flow and heat integration). Moullec and Kanniche (2011) evaluated process flowsheets for CA-MEA, concluding that process modifications must be coupled with new solvents and heat-integration with the power plant. Sharma et al. (2016) used multi-objective optimization of NG decarbonation to reduce the energy penalty at high CO_2 absorption requirements, demanding energy integration/cooling schemes.

Research into new solvents aims to increase CO_2 equilibrium loading, solvent chemical resilience and decrease stripping heat duty and corrosiveness. Adeyami et al. (2016) used a hybrid PA/CA mechanism allowing for higher CO_2 loading and reducing heat requirements for stripping. Chowdbury et al. (2011, 2013) focused on how side groups in ethanolamines (e.g., methyl, ethyl, propyl) affect the CO_2 absorption rate, cyclic capacity and enthalpy of reaction. Blended amines, additives, and special amines (Bouzina et al., 2015) are another front of solvent development. Zhang et al. (2012) studied phase-changing homogeneous amine blend that exhibits phase separation upon CO_2 absorption and only the CO_2 rich-phase enters the stripper, reducing reboiler duty. Adeyami et al. (2016) synthesized three amine-based deep eutectic solvents for CO_2 capture with enhanced environmental performance. Another attempt to minimize heat demand is the use of non-aqueous solvents (NAS), aiming to reduce the heat of vaporization, partially responsible for the heat demand in the stripper (Lail et al., 2014).

Although no commercial application in CO_2 capture exists and, despite their high cost, ionic liquids are a promising alternative due to their thermal and chemical stabilities (Brennecke and Gurkan, 2010). Research is currently underway as CO_2 capture may occur through physical and/or chemical absorption, and innovative schemes explore its high degradation temperature, allowing high-pressure stripping, reducing compression costs (Barbosa et al., 2019).

Concerning sustainability aspects, Marx et al. (2011) reviewed the literature on lifecycle assessment (LCA) of CCS and noticed that most studies considering the NG industry focus on post-combustion capture from flue-gas rather than NG upgrading. Karl et al. (2009) studied the worst emissions cases using common aqueous ethanolamine solvents (MEA, MDEA and DEA) at a Norway gas-fired power plant, concluding that insecure CA processes can negative affect human health and are devastating to local vegetation and aquatic algae. Zakuciová et al. (2015) investigated CCS technologies in pre-combustion, post-combustion and oxy-combustion applications, concluding that freshwater eutrophication is the main CA environmental issue due to amine toxicity.

Kahn et al. (2017) studied hybrid solvents for CO_2 stripping at high pressure which could reduce compression costs in geological storage applications. Mazzetti et al. (2014) analyzed the economics of treating North Sea NG with CA-MEA, concluding that, under the Norwegian CO_2 taxes, the technology is adequate for EOR. Araújo et al. (2017) comparatively analyzed separation technologies for processing CO_2-rich NG in ultra-deepwater oil-gas fields. In their analysis, CA exhibited the lowest hydrocarbon losses and the lowest specific electric power consumption at the expense of the highest footprint for all investigated scenarios.

3. Natural Gas Decarbonation with Membrane Permeation

A more recent technology, relatively to CA and PA, called Membrane Permeation (MP) employs semi-permeable porous or non-porous barriers (i.e., membranes) to separate components of gas or liquid mixtures through a variety of material-dependent mechanisms (D'Allesandro et al., 2010). Its compactness finds an application niche in CO_2-rich NG decarbonation at ultra-deepwater platforms. The "game-changing" attribute of MP results from its small footprint, an appealing factor on the topside of FPSOs (Araújo et al., 2017). Additionally, MP can have high CO_2 selectivity, low energy requirements (D'Allesandro et al., 2010) and modularity, which allows easy scale-up, high flexibility and resilience to feed-changing conditions (Reis et al., 2018). MP is also a relatively simple separation system without moving parts easing control, operation and scale-up, resulting in more than 200 MP plants in the NG industry—mostly installed by UOP/Honeywell and Schlumberger in offshore rigs (Yeo et al., 2012).

Available materials for membrane production are organic (acetates, polysulfone and other polymers), inorganic (carbon, zeolite, ceramic or metallics) and mixed matrix (composite organic inorganic compounds) (Al-Mamoori et al., 2017). Separation costs are impacted by material-dependent properties (e.g., membrane porosity, permeability, selectivity, pressure range and chemical impurities resistance) and overall operational ranges. Currently, polymeric membranes are the most common type for providing reasonable separation results at reduced costs and are easier to synthesize (Adewole et al., 2013). Polymeric membranes can be cast into flat sheet, hollow-fiber or spiral-wound configurations (Zhang et al., 2013), facilitating their wide use in CO_2 separation applications. Separation of CO_2 from CH_4 and heavier hydrocarbon components in MP produces a CO_2-rich permeate and a lean-CO_2 retentate (upgraded NG), illustrated in Figure 4.

Five possible material-dependent separation mechanisms can occur in MP: (a) Knudsen diffusion, (b) molecular sieving, (c) solution-diffusion separation, (d) surface diffusion and (e) capillary condensation (Olajire, 2010). The main separation mechanisms in NG applications comprise molecular sieving—separation through molecular size—and solution-diffusion separation, in which CO_2 dissolves into the membrane material and then diffuses through it due to a concentration gradient (Ismaila et al., 2005). Independently of the mechanism, the main driving force is the trans-membrane difference of fugacities of the permeating species, resulting in flux differentiation of species. To assure a sufficiently high driving force, feed streams are usually at high pressure, requiring compression prior to MP, implying in high power demand (Yeo et al., 2012), despite the small heat requirement.

One of the most important parameters in designing MP is the permeability-selectivity tradeoff—the more permeable a membrane is, the less selective it is (Robeson, 1991). Highly permeable membrane able to process high flowrates of CO_2-rich NG would result in high methane loss to the permeate. On the other hand, a more selective membrane would require a larger membrane area (Al-Mamoori et al., 2017). A careful choice of membrane material is essential in obtaining the highest permeability while maintaining selectivity. Although the permeate (captured CO_2) is at low pressure, the retentate (treated NG) faces a small pressure loss (Araújo and de Medeiros, 2017), reducing the compression effort to pipeline dispatch pressure.

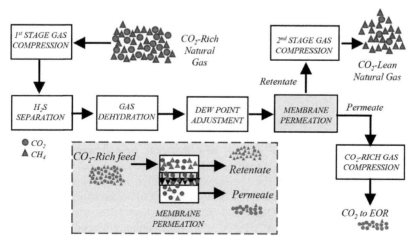

Figure 4. Upgrading CO_2-rich natural gas with membrane permeation.

3.1 Benefits and shortcomings

Offshore processing of CO_2-rich NG uses EOR, since the early production stage, to destine and monetize the high flow rate of separated CO_2. Crude oil recovery factor varies from 2 bbl/tCO_2 (close to U.S. operational data) to 4.35 bbl/tCO_2 (Azzolina et al., 2016). Early injection in the reservoir results in long-term enrichment of CO_2 in the produced gas because only approximately 60% of the injected volume is stored in the reservoir formation (Gazalpour et al., 2005) while the remainder returns to the gas processing plant. In parallel, along the operation lifetime, the production curve declines, intensifying the impact of CO_2 injection in NG composition. This scenario requires resilience and flexibility, offered by MP due to its modularity and compactness. The addition of a new MP module would suffice as a means to adapt the processing plant, while absorption-based technologies would require revamping. The small size of MP makes it the leading technology in offshore CO_2 removal applications where spatial requirements are a paramount constraint (Reis et al., 2017), although the need to recompress the captured CO_2 remains a drawback, similarly to CA and PA (Araújo et al., 2017). A second benefit of MP is that it does not require solvents, eliminating the need for handling and storage of chemicals. For applications on ultra-deepwater FPSOs, solvent makeup and waste management challenge logistics: Chemicals need to be collected, stored as waste, and transported onshore for treatment and disposal (EC, 2019). Besides, with a wide variety of materials to choose from, a high CO_2 selectivity can be achieved with very small unwanted hydrocarbon losses (Yeo et al., 2012).

A shortcoming of MP is its sensitivity to chemical impurities or membrane poisoning, mainly in polymeric NG membranes (Yeo et al., 2012). Raw NG components, such as sulfur-containing molecules, have extremely adverse effects on membrane surface with excess exposure, leading to changes in porosity or even inhibiting separation (Adewole et al., 2013). Plasticization alters initially glassy polymers by disrupting chain packing and enhancing inter-segmental mobility of polymeric chains. NG applications require adjustment of hydrocarbon dew-point (HCDP) of the gas feed (refer to Figure 4) as condensing hydrocarbons could plasticize membranes and degrade selectivity (Hao et al., 2008). In CO_2-rich NG processing, CO_2 and other hydrocarbons induce plasticization (Zhang et al., 2019), resulting in increased permeability at the expense of extreme reduction in selectivity, to the point that separating only CO_2 becomes impossible (Yeo et al., 2012). Pressure dependent plasticization occurs by dissolution of penetrants into the membrane matrix, resulting in the destruction of the membrane surface and increased chain flexibility (Suleman et al., 2016). The higher the pressure the higher the plasticization of the membrane, making this process extremely relevant in NG applications.

Membrane material selection faces tradeoffs: High resistance to plasticization, low permeability and high selectivity (Adewole et al., 2013). Depending on the material, several pre-treatments are required in order to facilitate stable separation and reduced membrane replacements. A second shortcoming of polymeric membranes is their physical aging (Olajire, 2010). Over time, polymeric membranes lose separation performance with net loss of CO_2 recovery, more pressure differential needed and lower selectivity. Consequently, continuous use of polymeric membranes results in membrane plasticization, poisoning or poor selectivity/permeability, with loss of performance before their lifespan limit is reached (Suleman et al., 2016).

3.2 Recent focus

The main development points in MP are: (a) new materials (mixed matrix membranes, MMM), (b) membrane surface treatments, (c) additives against plasticization and (d) membrane stability (chemical/thermal). MMMs are produced by embedding a filler material inside a polymeric matrix, being the most active membrane research area (Vinoba et al., 2017). Santaniello and Golemme (2018) developed an MMM from Hyflon®AD60X and SAPO-34, which takes advantage of the unusually high plasticization resistance of Hyflon®AD60X, while maintaining the good separation properties of SAPO-34. Results showed a twofold reduction in permeability but an increase in selectivity by a factor of 9 due to the tortuous paths CH_4 must follow in order to permeate (Santaniello and Golemme, 2018).

Gamali et al. (2018) and Agahei et al. (2018) added fumed silica into a Pebax (amide 6-n-ethylene oxide) matrix, achieving improved selectivity and permeability of CO_2 at high pressures ($\cong$ 3.5 MPa).

Another area of investigation is doping existing matrices to improve performance. Adewole et al. (2013) reviewed membrane modifications and reported the main changes: Crosslinking with thermal treatment, crosslinking with diamino compounds, polymer sulfonation, thermal rearrangement, polymer blending and dual-layer hollow fiber spinning. Thermal rearrangement and crosslinking are the most common modifications reported in literature. Crosslinking creates intermolecular connections through chemical bonds, resulting in membranes being more resistant to swelling and plasticization, while thermal rearrangement seeks stability at higher temperatures (Adewole et al., 2013). Wang and Hu (2018) evaluated PHA rearrangement at $\cong$ 450 °C, noticing improved permeability for CO_2.

4. Natural Gas Decarbonation with Cryogenic Distillation

When processing CO_2-rich NG ($CO_2 \geq 40\%$mol), CA and MP become less cost-effective (Maqsood et al., 2014), and cryogenic distillation (CD) finds its application niche. Low temperature distillation is a well-established process in the gas industry (e.g., O_2 and N_2 production). CO_2 can be obtained in CD as solid, liquid or cold vapor at high pressures, saving compression costs in transportation and storage. Whereas solid CO_2 may be the desired product, solid generation inside a CD column presents negative effects in process efficiency (Maqsood et al., 2014). Figure 5 shows a typical CD process and *PxT* phase envelopes for binary CO_2/CH_4 mixtures, where the dotted line represents the CO_2 freeze-out barrier—to the left of the barrier dry-ice will be present. Operational temperatures are usually below –40 °C varying from –45 °C to –80 °C, as in the Shute Creek plant of ExxonMobil (EPA, 2018), although values lower than

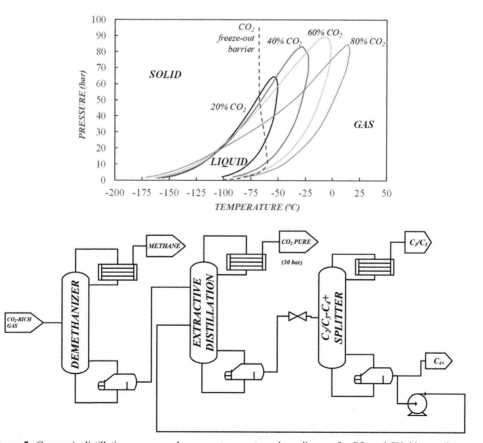

Figure 5. Cryogenic distillation process and pressure-temperature phase diagram for CO_2 and CH_4 binary mixtures.

–70 °C are not recommended. At too high temperatures, separation is not totally achieved, and, at too low temperatures, dry-ice forms in the distillation column—CO_2 freeze-out—resulting in clogging.

Higher pressures increase CH_4 losses, but reduce solid formation, avoiding freeze-out problems, and result in higher quality product (both in the lean and pure CO_2 streams). On the other hand, lower pressures reduce process efficiency, decrease quality and require higher separation time (Maqsood et al., 2014). Usual operating pressures for commercial CD range from 31 to 70 bar, as in the Total pilot plant in Southern France.

4.1 Benefits and shortcomings

CD products have high purity (> 95%), while in PA, CA and MP the CO_2 stream contains CH_4 and impurities, and traces of the solvent used (PA and CA). Additionally, CD allows CO_2 recovery at high pressure, does not need water makeup nor recurring consumable costs, and has practically no corrosion (Olajire, 2010). On the other hand, high power consumption for refrigeration is needed to promote separation (Maqsood et al., 2014). Besides its high cost and difficult control, CD has a high footprint. To avoid efficiency loss and solids in the columns, a dehydration unit is also required, and if liquified NG (LNG) is targeted, a demethanizer column is also needed (Maqsood et al., 2014).

4.2 Recent focus

Process optimization is the focus, mainly dealing with heat integration. CD has high power consumption due to the refrigeration cycles required in order to meet cooling needs, and the high operating pressure. Sun et al. (2019) proposed the addition of two inter-reboilers into a conventional three-column CD process, resulting in a total duty of 1.066 GJ/tCO_2 compared to the 1.231 GJ/tCO_2 of a conventional CD plant. Ali et al. (2018) used mixed-integer optimization to determine the best arrangement for a novel packed bed CD, concluding that power consumption and separation time could be greatly reduced with their arrangement. However, the CO_2-lean gas did not achieve methane specification, requiring further system-wide optimization.

Three commercial technologies for CD are currently available: Ryan-Holmes process (Figure 5), CryoCell® and Controlled Freeze Zone™. Hart and Gnanendran (2009) discussed field trials of the CryoCell®, identifying scenarios where CD shows superior performance relatively to traditional CA. Kelley et al. (2011) present advantages of using a single-step CD process (Controlled Freeze Zone™) for reducing the footprint and separation time. A potential niche for CD technologies applied to CO_2 separation is its possible integration into LNG (Liquefied Natural Gas) production. This integration uses the necessary refrigeration to transform NG into LNG to remove CO_2, minimizing the energy intensity of the production chain.

Xiaojun et al. (2015) studied integrating Pressurized LNG (PLNG) process with CO_2 removal to reduce footprints of both processes. Smaller equipment sizes and specific power were achieved but higher overall energy consumption for high CO_2 content was reported. The lack of toxic solvents, corrosion and waste streams could greatly improve environmental performance of NG processing and increase CD acceptance. CD economic performance is expected to improve with increasing CO_2 content in NG.

5. Natural Gas Decarbonation with Adsorption

In the adsorption process (ADS), CO_2 (the adsorbate) travels from the bulk of the CO_2-rich NG phase layering up. Through selective binding of CO_2 on the surface of a solid or condensed liquid phase (substrate), ADS separates CO_2 with minimal CH_4 loss (Al-Mamoori et al., 2017). CO_2 ADS may occur through physisorption or chemisorption links. In the former, weak Van der Waals interactions occur while, in the latter, chemical reactions like covalent bonding promote interaction (D'Allesandro et al., 2010).

Adsorbents are classified as high temperature or low temperature materials (Al-Mamoori et al., 2017). High temperature materials consist of hydrotalcites, alkaline-earth oxides (calcium and magnesium oxides), alkali silicates, zirconates and double salts and are chemisorbents. Low temperature materials consist of zeolites, carbon-based materials (e.g., carbon nanotubes, activated carbon surfaces, carbon nanofibers and graphene), molecular sieves and advanced materials, such as covalent organic frameworks (COFs), porous polymer networks (PPNs), metal-organic frameworks (MOFs). Low temperature materials for CO$_2$ ADS are mostly physisorbents. Low temperature CO$_2$ chemisorbents are rare in literature and consist of adsorbent materials with supported amines (amine-functionalized materials), in a hybrid adsorption and absorption mechanism (Al-Mamoori et al., 2017).

ADS is a cyclic process; after the adsorption phase, the active binding sites available are saturated with CO$_2$ and separation is no longer possible, characterizing CO$_2$ breakthrough (Serna-Guerrero et al., 2010). To regenerate the adsorption capacity, CO$_2$ desorption is conducted, and another adsorption cycle occurs. Adsorbent regeneration may occur by: (i) pressure reduction—Pressure-Swing Adsorption (PSA); (ii) increasing temperature—Temperature Swing Adsorption (TSA); (iii) inducing an electrical field or passing an electrical current through the adsorbent—Electrical Swing Adsorption (ESA); or (iv) hybrid regeneration—Temperature and Pressure Swing Adsorption (TPSA). The efficiency of ADS depends on the regeneration procedure, process design, operational factors, and the adsorbent material. The main criteria an adsorbent must achieve are low cost, low regeneration requirements, high surface area, high CO$_2$ selectivity, fast kinetics and long-term stability (Al-Mamoori et al., 2017). The Technology Readiness Level (TRL) (Mankins, 2009) of ADS processes applied to NG upgrading is $TRL_{TPSA} < TRL_{ESA} < TRL_{TSA} < TRL_{PSA}$.

5.1 Benefits and shortcomings

A great variety of adsorbent materials exist, such as zeolite 13X, HKUST-1 and Mg-MOF-74, making ADS one of the best technologies for trace CO$_2$ removal (D'Allesandro et al., 2010). High selectivity tradeoffs exist with total feed capacity and energy required for adsorbent regeneration—highly selective materials are costly, while commonly used adsorbents co-adsorb unwanted molecules, mainly water and hydrocarbons. To minimize co-adsorption, pre-treatment steps such as dehydration or desulfurization are required (Al-Mamoori et al., 2017). In parallel, surface treatments exist to improve selectivity—for example, the addition of supported functional groups (e.g., amines) and changing the adsorption interaction type.

The regeneration phase of the ADS consumes energy, e.g., PSA and TSA. PSA has attracted more attention than TSA for NG due to its simplicity, low cost, and lower energy requirements. TSA has higher heat requirements and long downtimes between ADS cycles for system cooling, greatly increasing operating costs, but has better CO$_2$ recovery per cycle with less adsorbent losses. PSA is recommended for NG upgrading, while TSA suites flue gas CO$_2$ removal better (Al-Mamoori et al., 2017).

A shortcoming of ADS is that no adsorbent combined to CO$_2$ desorption (PSA/TSA/ESA) is yet cost effective (D'Allesandro et al., 2010). High costs are incurred due to adsorbent material and high heat consumption in regeneration. ADS requires multiple adsorption columns to phase adsorption and regeneration, assuring continuous CO$_2$ capture.

5.2 Recent focus

ADS is not yet in full commercial scale for CO$_2$ applications in NG upgrading context, even though PSA processes are already available in other sectors (e.g., hydrogen purification) (D'Allesandro et al., 2010). The characteristic pressures and flow rates of the NG industry are much higher than in usual PSA applications (Grande et al., 2017; Riboldi and Bolland, 2017). Seeking cost effectiveness of CO$_2$ ADS, the bulk of academic research focus is in synthesizing and testing new or modified adsorbents to improve separation performance. Fu et al. (2017) modified MOF's surface with amine groups to promote stronger binding interactions between CO$_2$ and the solid. The modified UiO-66/PEI had drastically reduced surface area but increased CO$_2$ loading to a maximum of 3.13 mol/kg of solid at 25 °C. Gil et al. (2017) used wastes from the aluminium industry to synthesize CoAl-MgAl and NiAl-hydrotalcite compounds, obtaining a loading of 5.26 mol CO$_2$/kg of solid at 50 °C. Szczęśniak et al. (2017) investigated hybrid graphene and MOF adsorbents and noticed that the new hybrid surface had a greatly increased binding area, leading to a loading of 9 mol CO_2/kg of solid at 0 °C.

Although critical in the development of new materials, most of the present studies regarding new adsorbents ignore real application conditions (Al-Mamoori et al., 2017). ADS applications and its economic performance are reported through simulations of large-scale processes, employing laboratory-scale adsorption data gathered in the literature. Grande et al. (2017) simulated CO_2 separation from a gas composed of 83% methane and 10% CO_2 in a 12-tower adsorption process, concluding that huge methane losses would render ADS unfeasible for gas sweetening. Leperi et al. (2016) optimized a two-stage PSA process with varying degrees of dehydration, reporting zeolite 13X as the best performing material and concluding that the higher the water content in NG the greater the negative effect on ADS performance, with exponentially increasing costs.

New thermodynamic models to predict the behavior of ADS have been proposed, mostly of semi-empirical formulation fitted to experimental data. Elfving et al. (2017) modelled the equilibrium of CO_2 working capacity in PSA, TSA and TPSA, with good fit to experimental data. For TPSA, acceptable results occurred in high-purity CO_2 applications. Clark et al. (2013) simulated a fluidized bed ADS while Serna-Guerrero et al. (2010) used a semi-empirical model for amine-functionalized mesoporous silicas, obtaining equilibrium equations (Serna-Guerrero et al., 2010) and process kinetics (Serna-Guerrero et al., 2010).

6. Natural Gas Decarbonation with Gas Liquid Membrane Contactors

Gas Liquid Membrane Contactors (GLMC) combine absorption and membrane technologies (Norahim et al., 2017). In GLMC, a membrane provides an interfacial surface for gas-liquid contact, allowing CO_2 transfer from the NG into a solvent, where it is chemically absorbed (Figure 6), while avoiding phase dispersion (de Medeiros et al., 2013b).

CO_2 diffusion through the porous membrane interface is negligible when compared to the mass transfer rate into the absorbing solvent (de Medeiros et al., 2013b). However, if the membrane pores become filled by the liquid phase, the wetted membrane will show exponentially increased mass transfer resistance (Norahim et al., 2017). To prevent membrane wetting, the membrane should be highly hydrophobic, have high overall porosity (to minimize resistance) and exhibit high chemical resistance to withstand possible negative effects from the solvent. Therefore, non-porous (i.e., skin-dense) membranes commonly used in MP are not suitable for GLMC. Currently, the membranes developed for NG sweetening use polymeric materials due to their ease of synthesis in the desired porosity range and their hydrophobicity (Kang et al., 2017). The most common polymeric membrane materials in GLMC

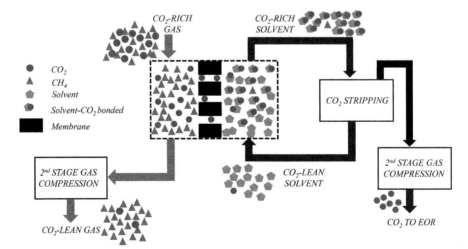

Figure 6. Typical GLMC based process for upgrading CO_2-rich NG.

applications are polypropylene, polyethylene, polyethersulfone, polysulfone, polytetrafluoroethylene and poly(vinylidene dufluoride) (PVDF). PVDF is the most promising material due to its high hydrophobicity, good chemical resistance, low cost and ease of fabrication (Norahim et al., 2017). GLMC is not a hybrid technology, as the membrane provides a physical barrier, not acting in the separation process.

6.1 Benefits and shortcomings

GLMC uses a chemical absorption mechanism spatially confined by a membrane, potentially allowing the system to share benefits of the individual technologies (MP and CA) while avoiding their major drawbacks. For instance, GLMC eliminates flooding and entrainment, common operational problems in absorption packing columns (Norahim et al., 2017). The higher packing density and surface area of GLMC results in faster separation than CA or PA. Furthermore, GLMC allows for independent control of gas and liquid flows without gravitational effects. Lastly, its modularity provides easy scaling-up (and down) and flexibility in a wide variety of scenarios (de Medeiros et al., 2013b).

The technology is released from the permeability-selectivity tradeoff found in MP technologies. Since the separation occurs through the absorption of CO_2 into a chemically selective solvent, developments of new membranes turn the focus from selectivity to permeability, aiming at increased rates of CO_2 transfer. However, selectivity is not the main operational parameter; the pressure differential through the membrane can be considerably reduced compared to the MP process, minimizing power consumption. Although GLMC has lower energy intensity, solvent stripping occurs at low-pressure, posing an energy penalty due to compression of the captured CO_2 (de Medeiros et al., 2013).

The main shortcoming in GLMC is membrane wetting over time. GLMC systems have a breakthrough pressure, i.e., the pressure at which liquid breaks the membrane barrier and diffuses into the membrane pore. The larger the pore, the lower the breakthrough pressure, so smaller pores are preferred. In the GLMC process life-span, physical aging of the membrane or effects like plasticization occur and the membrane barrier loses resistance, lowering the breakthrough pressure until membrane wetting occurs (Norahim et al., 2017). Although not yet cost effective, a thin layer of a highly hydrophobic material on the liquid side of the membrane reduces wetting (Gugliuzza and Drioli, 2007).

Another relevant aspect is that mass transfer rate is directly related to solvent viscosity (de Medeiros et al., 2013b). The higher the viscosity, the lower the mass transfer rate, highly jeopardizing the performance of GLMC. It is noteworthy that GLMC long-term cost effectiveness, when compared to other CO_2 capture technologies, has yet to be proven.

6.2 Recent focus

The main literature body in GLMC seeks enhancements in membrane materials and solvents. Rezaei et al. (2015) synthesized an MMM of PVDF and Cloisite 15A using distilled water as a solvent and, studying the effects of filler content, noting that surface area and permeation were greatly improved. Ghasem et al. (2012) added o-xylene into polyethersulfone membranes to increase hydrophobicity, obtaining increased CO_2 recovery.

A rigorous thermodynamic model was proposed by de Medeiros et al. (2013b) in order to reproduce behaviors of industrial GLMC systems, such as (i) temperature increase along GLMC length due to exothermal reactive absorption of CO_2 into aqueous MEA/MDEA solutions and (ii) equilibrium shifts in the solvent side when inert components (e.g., CH_4) penetrate through the membrane causing stripping action in the solvent. The model considers vapor-liquid reactive equilibrium (RVLE) flow in the solvent side due to CO_2-CH_4 penetration and rigorous compressible flow modelling for both gas and (two-phase) solvent sides. Quek et al. (2018) proposed a model related to membrane pore size for high pressure applications and Zolfaghari et al. (2018) modelled membrane wetting-critical points in GLMC operation. Fougerit et al. (2017) studied trans-membrane pressure in wetting, specifically in breakthrough pressure of liquids, showing that membrane wetting is independent of trans-membrane pressure. Hashemifard et al. (2015) evaluated partial wetting of membrane pores, correlating pore size with wetting potential. An overlooked aspect is GLMC's economic performance, contrasting and comparing it to more widespread technologies, and determining allowable wetting ratios.

7. Natural Gas Decarbonation with Supersonic Separators

The supersonic separator (SS) is used in the NG industry for water dew-point adjustment (WDPA) via water removal and hydrocarbon dew-point adjustment via propane and heavier hydrocarbons (C3+) removal from raw NG. New SS applications contemplate CO_2 removal from CO_2-rich NG after WDPA and HCDPA, otherwise water and C3+ would condense, hampering the fall of temperature necessary for CO_2 liquefaction, which normally occurs at colder conditions.

A supersonic separator (SS) achieves separation by expanding and accelerating a compressible fluid to supersonic speeds through a Laval nozzle, which comprises a converging section, a throat and a diverging section (Teixeira et al., 2018). Besides the Laval nozzle, SS encompasses a static swirling device in the converging section, a liquid collector after the supersonic section, and an ending diffuser, which is a continuation of the Laval diverging section. The thermodynamic transitions in SS are described in terms of the Mach number, $Ma = v/c$, where c is sound speed and v is flow velocity. The flow accelerates from subsonic to $Ma = 1$ at the throat and then becomes supersonic ($Ma > 1$) in the diverging section. The acceleration converts enthalpy into kinetic energy, reducing the fluid temperature along the flow path (de Medeiros et al., 2017), liquefying condensable components like water, C3+ and even CO_2 at special conditions. A centrifugal field is imposed by swirling vanes, pulling liquid droplets towards the walls, where collecting vanes capture them. Figure 7 sketches an SS where increasing arrows indicate axial acceleration; only the Laval and the ending diffuser are shown, where the former corresponds to the converging-diverging nozzle upstream of the liquid collector.

Downstream, the throat the flow is supersonic and temperature and pressure are low, turning the flow unstable as the discharge pressure is higher. This eventually produces a normal shock front, a flow discontinuity where axial velocity suddenly drops to subsonic ($Ma < 1$) accompanied by sudden increases of temperature, pressure and entropy. For successful SS operation, the formed liquid should be collected before the shock front, otherwise separation is lost as everything would re-vaporize through the shock. Thus, in the supersonic section, enthalpy to kinetic energy conversion changes water and C3+ to low enthalpy liquid mist, centrifugally collected by separating vanes, and removed before the shock (Arinelli et al., 2017). The shock phenomenon can be explained as follows: Assuming adiabatic SS flow, there is an analogous subsonic flow with the same mass, momentum and energy flow rates, but hotter and with greater entropy rate, which is globally stable by the 2nd Law of Thermodynamics and is accessible via an irreversible, adiabatic, sudden collapse of supersonic flow at a specific location in the diverging section. This phenomenon is the normal shock front. As any metastable collapse, normal shock is easily provoked by irreversibilities (e.g., friction) so that, as the flow accelerates beyond $Ma = 1$, the shock is more likely to occur. For NG processing, supersonic flow with high pressure recovery is unlikely to

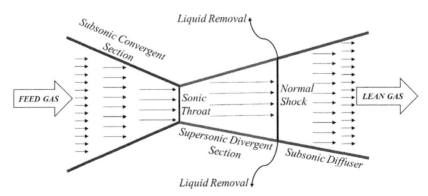

Figure 7. Sketch of a supersonic separator.

exist above $Ma = 2$ as it progressively loses stability against higher discharge pressure. At the shock, entropy is adiabatically created, at rates that increase with shock irreversibility, which increases with Ma immediately upstream of the shock, reducing the SS backpressure.

Critical parameters for SS control are SS head-loss, minimum temperature attained and gas velocity (v). Minimization of head-loss produces a pressurized lean gas (Arinelli et al., 2017), while temperature and Ma must be rigorously controlled. In the case of CO_2 removal via SS, if too low temperatures (too high Ma) are achieved, CO_2 freeze-out occurs, potentially clogging the SS (de Medeiros et al., 2019). SS geometry must be calculated in order to guarantee $Ma = 1$ at the throat; inadequate sizing would degrade SS performance with loss of lean gas and insufficient WDPA and HCDPA (de Medeiros et al., 2019).

7.1 Benefits and shortcomings

A benefit of using SS with CO_2-rich NG is that it can simultaneously perform WDPA and HCDPA (de Medeiros et al., 2019). SS can also remove CO_2 from NG, provided the NG feed has been submitted to WDPA and HCDPA in order to avoid water and HC condensations (Arinelli et al., 2017). For CO_2 removal from NG, the SS feed should be at higher pressures and lower temperatures relatively to ordinary NG WDPA and HCDPA SS applications. An advantage of SS compared to other CO_2 capture technologies is that SS retains performance when CO_2 content rises, being applicable to CO_2-rich streams (Arinelli et al., 2019).

SS is also modular, with a low footprint and no rotating parts (reducing maintenance routines). Separation occurs very rapidly, resulting in a compact equipment, especially beneficial for FPSOs (Hammer et al., 2014). A further beneficial aspect is that, despite the SS head-loss during separation, the CO_2-lean NG is obtained at high pressures as well as the condensed CO_2, reducing EOR compression costs (Arinelli et al., 2017).

The main shortcoming of SS derives from its tight pressure, temperature and velocity ranges for operability. NG must be pre-conditioned and design should be appropriate to avoid the flow crossing the solid-vapor-liquid equilibrium (SVLE) freeze-out border that traverses the VLE envelope for CO_2-rich NG feeds. In fact, current SS research shows a limited scope of applicability for CO_2 removal. For example, considering an SS for CO_2 removal from a CO_2-rich NG feed (after WDPA and HCDPA) with 45%mol CO_2, to avoid freeze-out, SS reaches a minimum CO_2 content of 21%mol in the treated NG. Thus, SS is indicated for bulk CO_2 removal from CO_2-rich NG, with a lower operating range in comparison to to CD (Arinelli et al., 2017). Lastly, SS technology is still in early stages of development and operational problems have not yet to be reported.

7.2 Recent focus

Proof of concept approaches, aiming at determining application niches, outlining benefits and shortcomings, are the bulk of SS research. Hammer et al. (2014) studied the viability of using SS for CO_2 removal from gas turbine flue-gas in FPSOs, which is the downmost step in the CO_2-rich NG to energy supply-chain, showing successful capture of CO_2 in three different systems. A more complete sequence of developments on SS process simulation is related to the Brazilian pre-salt context (Machado et al., 2012; Arinelli et al., 2017; de Medeiros et al. 2017; Teixeira et al., 2018; Brigagão et al., 2019; Arinelli et al., 2019; de Medeiros et al., 2019), bearing more than 45% of CO_2 in the gas, with two pending Brazilian patents (Teixeira et al., 2017; Brigagão et al., 2017). Arinelli et al. (2017) developed and used a Unit Operation Extension (UOE) simulating SS (SS-UOE) for the HYSYS process simulator (ASPENTECH Inc.), which simulates SS considering phase-equilibria sound speed (to precisely calculate Ma) determination based on rigorous thermodynamics for predicting phase behavior inside the equipment (de Medeiros et al. 2017). SS-UOE enables the definition of geometrical and operational parameters for NG applications. Another HYSYS UOE (PEC-UOE) was developed by de Medeiros et al. (2017) to determine phase-equilibrium c for any SS flow condition; i.e., single-phase gas, two-phase gas and liquid and three-phase gas-liquid-water. A diversity of SS configurations was tested and compared, exhibiting better results than conventional processes, not only for WDPA/HCDPA of raw NG, but also for CO_2 removal. SS outperformed traditional MP CO_2 removal because the separated CO_2 exits the SS as a high-pressure liquid, a situation especially suited for its utilization as an EOR agent, while the CO_2-rich permeate from MP is at low pressure and demands high power consumption for compression to become useful in EOR (Arinelli et al., 2017; de Medeiros et al., 2017).

Teixeira et al. (2018) modelled the utilization of SS in different scenarios, as in the recovery of thermodynamic hydrate inhibitors (THI) methanol, ethanol and monoethylene glycol (MEG) from raw NG streams, used to prevent hydrate formation in subsea pipelines. The SS operation, while aiming at THI removal from raw NG feeds, also entailed several concomitant benefits, such as WDPA and HCDPA of the NG stream, besides producing lean NG ready for exportation and crude LPG (C3+) ejected as liquid from SS. Brigagão et al. (2019) developed a process employing SS for air pre-purification in CD. This could further improve the technologies horizon by using it in a hybrid process as a CO_2 bulk removal and impurity removal unit. De Medeiros et al. (2017) simulated multiphase sound speed, both in multi-reactive and non-reactive multiphase mediums, for SS and for supersonic reactors (SR).

Another mainstream simulation of SS operation employs Computational Fluid Dynamics (CFD), as in Wen et al. (2012), where, trading off CFD and rigorous phase-equilibria calculations, thermodynamics is oversimplified, often leading to unrealistic operation conditions. For example, Wen et al. (2012) used a commercial CFD software employing SS to obtain LNG from a dehydrated NG, obtaining very low pre-shock temperatures of -116.5 °C, questioned in Arinelli et al. (2017). The rigorous thermodynamic approach in Arinelli et al. (2017) and de Medeiros et al. (2017) allows the correct determination of SS flow properties and associated phenomena, such as the prediction of the SVLE CO_2 freeze-out borders, among other special phase behaviors. Attention has been turned towards improving SS geometry so as to obtain easier removal of the solids/liquids formed and more appropriate flow characteristics. Wang et al. (2018) improved a conventional Laval nozzle to improve flow pattern inside the SS; better swirling is obtained in the divergent section, improving separation.

8. Natural Gas Decarbonation with Hybrid Technologies

None of the covered technologies has full CO_2 capture range and, to mitigate economic penalty from CCS, technologies must have their capture cost reduced, by increasing capture efficiency, while reducing power, heat consumption and environmental impacts. To meet these objectives, hybrid technologies (HYB) consist of using two or more technologies, in parallel or serial arrangements. Series HYB uses one technology for bulk CO_2 removal and another for final polishing. The bulk technology is not in charge of meeting CO_2-lean gas specifications and the polishing process faces small CO_2 contents, reducing energy consumption of both technologies (Al-Mamoori et al., 2017). In parallel arrangements, CO_2-rich NG is divided into two or more streams and each technology must meet CO_2-lean specifications. The use of parallel arrangements increases the efficiency of the technologies, but is less common than serial configurations (Al-Mamoori et al., 2017).

Song et al. (2017a) reviewed HYB processes, covering near 7400 publications up to 2017, with 60% using CA/PA, 16% MP and 4% CD. For all HYB technologies, the main research focus is on testing new combinations of technologies and materials (such as MP+CA arrangements, MP+PA arrangements, PA solvents or MP materials). Investigations into whether the use of a HYB combination is more beneficial than the use of a stand-alone technology are also critical (Song et al., 2017a). To illustrate HYB technologies, the most promising candidates are discussed.

8.1 CA/PA-MP and MP-CA/PA

MP-CA and MP-PA hybrids can be used in series and parallel arrangements. In a serial arrangement, MP acts as a bulk removal process, followed by a polishing CA step, with lower CO_2 loading when compared to traditional CA/PA (Araújo et al., 2017). Rochelle et al. (2014) also proposed a CA-MP in a serial arrangement, with CA being the bulk remover. Absorption removes $\cong 50\%$ of CO_2 from the CO_2-rich NG, and MP then removes the remaining 50%. Cost reduction derives from the lower energy required since less solvent is circulated and regenerated (Rochelle et al., 2014). In parallel arrangements, CO_2-rich NG feed is divided into two streams, one being directed to MP and another to CA/PA. The main advantage is the reduction in capital costs due to smaller equipment size—e.g., the absorber column can roughly present a twofold down-scaling (Song et al., 2017a). The main issues of CA-MP are consumption of both heat for amine regeneration and power for providing trans-membrane differential fugacity in MP and possible membrane poisoning due to solvent carry over.

Araújo et al. (2017) compared standalone and hybrid technologies, CA, PA, MP, MP-CA and PA-MP, focused on offshore processing of CO_2-rich NG under increasing CO_2 content in the associated gas (10%, 30% and 50%mol), and high NG flow (6 million Sm³/d). The authors concluded that MP-CA partially inherits MP's small footprint and is the most flexible capture in CO_2-rich NG, in ultra-deepwater offshore fields having high GOR, with early CO_2-EOR. Reis et al. (2017) used an optimization-based procedure to define the service distribution between bulk removal in MP and the CA polishing operation, subject to two constraints: Maximum CO_2 content in treated NG of 3% and minimum CO_2 content in injected gas of 75%, for raw NG with 10, 30 and 50%mol CO_2. The authors concluded that slipping CO_2 from MP to CA considerably reduces the total footprint (i.e., the required permeation area) in MP.

8.2 CA-ADS and PA-ADS (CA/PA-ADS)

CA/PA-ADS consists in the use of a slurry containing an absorbent (amines, potassium carbonate) and an adsorbent (MOF, zeolites). Both materials are regenerated with a single thermal operation at lower energy costs and present higher CO_2 selectivity. Liu et al. (2014) tested this concept in an amine/MOF HYB and showed that 1.25 mol/L CO_2 loading was achieved at 1 bar with high CO_2 selectivity and a regeneration duty of 1.44 GJ/tCO_2—standalone MEA consumes $\cong$ 3.3 GJ/tCO_2. CA-ADS can potentially reduce absorbent/adsorbent volume and requires less regeneration energy. Shortcomings are due to its high footprint and the initial cost of adsorbent materials.

8.3 CD-MP

CD-MP is the HYB scheme that has attracted the most attention. The concept is to use a CD to remove bulk CO_2, water and H_2S, resulting in a high-pressure gas. This high-pressure gas is then passed through a CO_2 selective MP for polishing. CD-MP reduces compression requirements when compared to standalone MP and refrigeration load when compared to standalone CD (Prosernat, 2017). The main advantage of CD-MP is its high flexibility, since CD is very resilient to feed changes and subsequent MP can be very CO_2 selective for optimal treatment. The disadvantages are MPs sensitivity to cold temperatures and losing part of the MP modularity.

8.4 SS-SS and SS-MP

Arinelli et al. (2019) applied the SS process in offshore NG processing with ultra-high CO_2 content (> 60%mol) for dew-point adjustments and decarbonation on a floating-hub processing 50 MMsm³/d of CO_2 ultra-rich gas, reinjecting 96% of treated CO_2-rich gas for enhanced oil recovery, while reserving 4% as fuel-gas upgraded to 20%mol CO_2 for power production. The best economic performance of SS alternative compared to SS followed by MP reflects its highest revenues from recycling condensate from the 1st SS unit, entailing 18% higher oil production.

9. Technology Readiness Level and Research Status in Natural Gas Decarbonation

Technology Readiness Level (TRL) is a method to estimate technology maturity that is widely used in literature (Bakhtiary-Davijany and Myhrvold, 2013). TRL usually varies from 1 to 9, in which lower values represent low maturity. In this section, TRLs for the addressed CO_2 capture technologies are

Table 2. Technology readiness levels and cutoffs.

	TRL	Development stage completed	Definition of the development stage
LOW	0	**Unproven Concept** Basic R&D, paper concept	Basic scientific/engineering principles observed and reported, paper concept with no completed tests or design history.
	1	**Proven Concept** Proof of concept through papers or R&D experiments	Existing studies that contain experiments or formulated applications but do not include physical models.
	2	**Validated Concept** Experimental proof of concept using physical model tests	Concept design validated with physical models. A mock-up or dummy equipment is functionally tested with reliability tests on the major components.
MEDIUM	3	**Prototype Tested** System function, performance and reliability tested	A prototype is built and passed through generic functional and performance tests with benefits and risks being experimentally demonstrated.
	4	**Environment Tested** Pre-production system environment tested	Prototype has passed all other TRL categories and is tested on a realistic environment (either simulated or real).
	5	**System Tested** Production system interface tested	A full-scale equipment is built and tested/integrated into the intended operating system with full interface and functionality being fully tested.
HIGH	6	**System Installed** Production system installed and tested	A TRL 5 (full-scale equipment with intended operational conditions) operating for over 3 years in which the first years may need additional support due to the learning curve.
	7	**Field Proven** Production system field proven	System is installed for over 3 years with proven reliability, economic and environmental results.

suggested. Table 2 presents the proposed TRL cutoff values—low, medium and high maturity—applied to CO_2 capture technologies suitable to CO_2-rich NG processing on ultra-deepwater rigs. The attributed classification is specific to the NG application and not an overall TRL.

Mature technologies for CO_2 removal from CO_2-rich NG include CA, PA, MP and CD. These four technologies have TRL = 7 since they all have commercial applications with over 3 years of operation. CA and PA are the oldest technologies, with many licensors. OP/Honeywell is one of the main players, offering both amine and carbonate-based solvents with their AmineGuard and Modified Benfield Process products. Other interested parties in CA are ExxonMobil, Basf, Schlumberger and Shell, offering full absorption packages (operation, solvent and columns), with Ineos, Evonik and Dow Chemical Company offering only solvents.

PA, depending on the solvent, has more licensors than CA, with the main contributors being UOP/ Honeywell, Lurgi and Shell with their respective Selexol, Recistol/Purisol and Sulfinol processes. PA and CA were the first carbon capture technologies to reach commercial application. For example, the Oklahoma Natural Gas Processing plant (120 MMscf/d of NG) and the Freer Texas Refinery (12 MMscf/d of a 15% CO_2 NG) in the USA use CA. More recently, a plant in Fort Nelson Canada, for geological storage of CO_2, uses CA to purify 2.2 Mt/yr of NG. As for PA, the Century Plant in the USA uses Selexol to remove CO_2 from a 5 Mt/yr NG stream and the Shenfu Dongsehn plant in China uses Rectisol to decarbonate 1,065 t/hr NG.

MP is commercially available and widespread in certain applications, such as in FPSOs, and can increase its participation in the NG chain. Most commercialized membranes in the world are produced by UOP/Honeywell (Separex/Polysep Composites), Schlumberger (Cynara, Apura and Semple) and Air Liquide (ALaS MEDAL). Acetate-based polymers are the membranes mostly applied in processing plants. The Lula oil field operated by Petrobras uses a Separex type membrane to treat 1 Mt/yr of NG having 20–40%mol CO_2, Cakerwala Gas in the Gulf of Thailand uses Cynara to treat 1,280 MMScfd of NG bearing 36%mol CO_2, demonstrating the high potential of MP for CO_2-rich NG. Castor Storage in Spain uses Separex in a 5%mol CO_2 NG, showing MP capacity to treat gases with low CO_2 content.

In contrast with PA/CA and MP, CD only has 3 commercially available technologies—Cool Energies CryoCell®, ExxonMobil's Controlled Freeze Zone™ and the Ryan/Holmes Process. Ryan/Holmes is the oldest, with uses registered as early as 1980 by the Seminole Unit (operated by Amerada Hess Company), the Willard Unit (operated by ARCO oil and gas company), the GMK South Field (operated by Mobil Oil Corporation), and the Wasson Denver Unit (operated by Shell Oil Company) (Lastari, 2009). CryoCell® has been in field test and demonstration scale plants since 2006 and Controlled Freeze Zone™ has been operating in commercial scale field tests since 2008. CD can be considered at TRL = 7, as all available technologies are field proven at large scales and have been operating for over 3 years.

Cool Energies technology consists in a cryogenic CO_2 separation in which solid CO_2 is obtained. The process requires a prior dehydration (under 5 ppm of water is allowed into cryogenic unit) and the main innovation is in the CryoCell® three phase separator of methane and CO_2. The main disadvantage is that solid CO_2 must then be liquified to be transported or stored. Controlled Freeze Zone™ attempts to simplify CD using a single-step separation in one large column. To achieve desired separation in a single stage, very low temperatures and pressures are required and CO_2 freeze-out (dry-ice formation in distillation column) will occur. Since solids are unwanted in a distillation unit—can clog tower nozzles— the ExxonMobil CD technology has two special areas inside the column in which solid CO_2 is re-liquified due to interactions between the solid and descending liquid droplets at higher temperatures. Considering that CD is already widespread in other applications (e.g., industrial gases), well-stablished actors could move into CO_2 removal from NG and a few, like Air Liquide, have already shown interest in the decarbonation market. Figure 8 summarizes the high TRL technologies providing the main stakeholders and the materials used.

For medium and low maturity technologies, a distinction in interests is made between academy and industry. ADS has medium maturity technology with TRL = 3. It must be noted that a different TRL could be given to each of the sub processses within ADS, such as PSA, TSA or ESA, since each has its own maturity level. PSA is the most advanced and represents ADS in the present analysis. Since PSA has various laboratory cycle tests reported for NG applications, it is considered TRL = 3.

SS research is mostly focused around computational simulation of possible scenarios. The main interested parties are in the academic sphere of research with low industrial interest, despite some applications of SS already existing-mainly in WDPA and HCDPA of NG and LNG processing by Twister B.V. (2017). Twister B.V. commercializes its SS model, which has been submitted to field tests, hence, its TRL = 5.

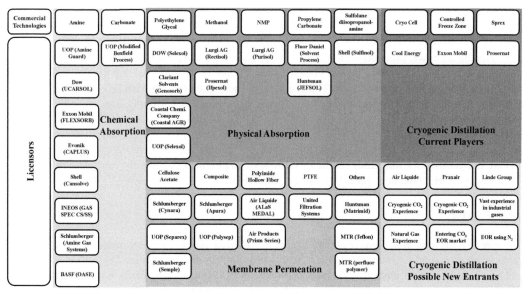

Figure 8. Summary of main players in high TRL technologies.

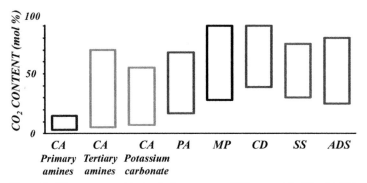

Figure 9. CO_2 management technologies vs range of CO_2 content in CO_2-rich NG.

Research in GLMC is focused on determining the best membrane/solvent combinations while reporting on basic parameters, e.g., CO_2 flux and CO_2 recovery. Tests mainly use pure CO_2 feeds, and do not integrate GLMC with upstream/downstream operations. GLMC has a TRL = 2, despite coal, flue gas and syngas applications already evolving to field testing. For CO_2 removal from NG, GLMC has been mainly confined to laboratory-scale tests. Figure 9 presents the range of operation for each technology covered in this chapter with respect to the content of CO_2 in CO_2-rich NG.

CA-based technologies are applicable to low CO_2 content but lose their edge to CD and SS that have their application niche in very high CO_2 content scenarios. CD is an excellent technology for bulk CO_2 removal but the CO_2/C_2H_6 azeotrope and high energy consumption are strong drawbacks in low CO_2 content. Since every HYB combo has a different range of action they are not represented in Figure 9.

Lastly, Table 3 summarizes the technologies for CO_2 separation from NG reviewed in this chapter, pinpointing technological gaps, main interests, TRL, availability and flexibility in variable CO_2 scenarios.

10. Post-Combustion CO_2 Capture

In contrast with from upstream decarbonation and pre-combustion, post-combustion capture (PCC) consists of removing CO_2 from flue after NG combustion. The downmost step in CO_2 management in the NG to energy value-chain, PCC is applied to fuel-fired power plants that use air as an oxidant, resulting in CO_2 diluted in nitrogen and CO_2 in the 12–15%mol range (Olajire, 2010), affecting capture performance. One of the main PCC challenges is the large energy penalty posed to power generation, representing up to 65–80% of the cost for CCS (D'Allesandro et al., 2010). Adding PCC to fossil fuel-fired power plants decreases the net power production efficiency and increases capital expenditure (Araújo and de Medeiros, 2017). Effective PCC requires CO_2 recovery with concentrations > 95.5%mol and density $\cong$ 900 kg/m³ for feasible transportation and storage. The technologies addressed for CO_2-rich NG decarbonation also apply in a PCC context. While maintaining their general characteristics, low CO_2 fugacity in the flue-gas, with high nitrogen and oxygen content, huge flowrates at low pressure creates particularities affecting technical parameters and efficacy (Araújo and de Medeiros, 2017). PCC is a context different from NG decarbonation, resulting in a distinct maturity level, i.e., certain commercially proven technologies in NG are not yet used for flue gases. In the PCC context, employing CA/PA is the most mature and cost-effective CO_2 capture method (Boot-Handford et al., 2014).

10.1 CA/PA

The low CO_2 fugacity in flue-gas results in CA being preferred over PA. CA performance stays much the same when compared to decarbonation CA, with MEA, MDEA and DEA being the most used solvents and regeneration heat requirements the main drawback. A PCC specific singularity is the increased amine degradation and equipment corrosion due to high O_2, SO_x and NO_x presence. Consequently, higher amine

Table 3. Summary of CO_2 capture technologies and the state of research.

Technology	CA	PA	ADS	MP	GLMC	CD	SS	HYB
Commercially Available	Yes	Yes	No	Yes	No	Yes	Yes	Only CD+MP
TRL	7	7	3	7	2	7	5	2–7
Energy Requirements	High	High	Medium	Low	Medium	High	Low	Low
Ease of Operation	Medium	Medium	High	High	Low	Low	Medium	Medium
Flexibility degree	Low	Low	Medium	High	High	Medium-High	Medium	High
Best CO₂ Concentration	Low	High	High	High	Medium	Medium-High	Medium-High	All
Interests	Maturity Selectivity	Maturity Effectiveness at high pressures	No Emissions Ease of Operation	Flexibility Low energy No emissions Modularity	Modularity Selectivity	High removal High product pressure	Compactness High CO₂ removal	Flexibility Potential for reduced cost
Gaps	Solvent degradation Environmental performance Pressure recovery Energy efficiency	CO₂ selectivity Slow absorption	CO₂ selectivity Cost effectiveness Cost of adsorbent Type of Regeneration (PSA, TSA, ESA)	Selectivity/ permeation Stagnation Efficiency for low pressure Pressure recovery	Long Term Performance Type of solvent Mass transfer rate Membrane wetting	Energy Consumption Size Cost	Ongoing tests Ease of operation	Real scenario tests Best technology combination

Table 4. Largest CCS projects in the power sector. Source: Koytsoumpa et al. (2018).

Project name	Boundary dam carbon CCS project–the Saskatchewan	Petra Nova carbon capture project	Sinopec Shengli power plant CCS project
Location	Saskatchewan, Canada	Texas, US	Shengli power plant, Dongying, Shangdong Province, China
Industry	Power generation (lignite/brown coal)	Power generation – pulverized coal boiler	Power generation – pulverized coal boiler
CO_2 Capacity	1 Mt/yr	1.4 Mt/yr	40,000 t CO_2/yr Scale-up to: 1 Mt/yr
Capture Process	PCC Amine Shell Global, Cansolv technology	PCC KM-CDR amine scrubbing CO_2 developed by MHI and KEPCO	PCC SINOPEC

makeup rates are required, increasing operational costs (Araújo and de Medeiros, 2017). Table 4 contains the three largest CCS projects in the power sector employing absorption PCC.

10.2 CD

CD occurs the same way in PCC as it does in NG, but separation is no longer from CH_4 but rather from N_2. Since N_2 and CO_2 liquefactions points are substantially different, separation via CD is technically possible. The main issue is the high heat consumption. Leung et al. (2014) state that flue-gas is usually obtained at high temperatures (over 600 °C in some cases) and, to reach cryogenic temperatures, would require impossibly high cooling CD. However, Leung neglected that HRSG (heat recovery and steam generation) is used to produce steam, which drives steam turbines, reducing exhaust gas temperature. A critical aspect for CD in PCC is that the operational pressure should be greater than the triple-point pressure of CO_2 (5.2 bar) in order to allow existence of liquid CO_2 for distillation, entailing compression of flue-gas. CD in PCC is limited to niche applications, in which colder flue-gas is obtained and the resulting energy penalty would be low. In addition, some of the main advantages of CD, such as high CO_2 selectivity and pressurized products, remain relevant only for the CO_2 product, while N_2 losses are irrelevant economically, opposed CH_4 losses in NG processing.

10.3 ADS

ADS has attracted great attention in PCC literature due to its heat reduction potential (Wang et al., 2011). Additionally, ADS materials are more resistant to poisoning than membranes and could minimize footprint when compared to traditional absorption. Aaron and Tsouris (2005) indicate two drawbacks that make ADS currently unfavorable to treat flue-gas. First, ADS cannot easily handle high contents of CO_2, with optimal conditions being between 0.04% and 1.5%mol. Most carbon-fired power plants have higher contents of CO_2 in flue-gas such as $\cong$ 15%v/v (Li et al., 2003). The second reason is that available sorbents are not selective enough for CO_2 separation from flue-gases.

10.4 GLMC

Compared to CA, one of the most noticeable advantages of GLMC is its high interfacial area, which can significantly reduce equipment size and, thus, lead to process intensification (Zhao et al., 2016). GLMC can offer up to 30 times more interfacial area than conventional packed tower gas absorbers, reducing absorber size tenfold. PCC GLMC is comparable to decarbonation GLMC, maintaining much of its benefits and drawbacks without major changes. Despite its great potential in PCC, GLMC is yet to prove its economic viability in the long term.

Table 5. Suggested TRL for CO_2 capture technologies in the PCC context.

TRL	Development stage completed	Technology for CO_2 capture from exhaust gas
0	Unproven Concept	–
1	Proven Concept	Supersonic Separator (SS)
2	Validated Concept	–
3	Prototype Teste	Cryogenic Distillation (CD); Adsorption (ADS)
4	Environment Tested	Membrane Permeation (MP); Gas-Liquid Membrane Contactor (GLMC)
5	System Tested	–
6	System Installed	Chemical and Physical Absorption (CA/PA)
7	Field Proven	–

10.5 SS

SS was proposed as a flue-gas treatment technology by Hammer et al. (2014). To assess the operational performance, a Laval nozzle model was implemented and successfully integrated in a steady-state process flowsheet simulator. The model includes equilibrium thermodynamics describing freeze-out of dry-ice from a gas mixture containing CO_2. A temperature/pressure-based optimization showed that supersonic expansion is a viable strategy for CO_2 capture from flue-gas of offshore gas turbines. However, SS in the PCC context is still in its incipient phase, with many proof-of-concept studies and few experimental results. Thus, more tests, including full viability and economic analysis, must be performed before SS can be considered a feasible PCC technology. Based on the literature review, Table 5 shows the suggested TRL values for each of the covered technologies in a PCC context.

11. Offshore Power Generation

When processing CO_2-rich NG at ultra-deepwater offshore rigs, in both NG to onshore facilities and CO_2 injection wells for EOR, transportation is required (refer to Figure 2). An alternative to bypass the complex subsea logistics network consists in adopting floating power-plants. Often, a simple gas cycle is preferred for power generation on platforms, due to its potentially smaller footprint and weight (Bimüller and Nord, 2015; Flatebø, 2012). On the other hand, NG combined cycle (NGCC) shows higher efficiencies and lower carbon emissions (Følgesvold et al., 2017; Song et al., 2017b; Nord and Bolland, 2013). To further reduce emissions, NGCC can be associated with PCC to directly capture CO_2. The generated power is transported through subsea cables (HVDC) to onshore facilities. The concept of transporting electricity instead of gas is known as Gas-to-Wire (GTW) (Watanabe et al., 2006).

Advances in turbo-shafts enable the firing of NG containing up to $\cong 20\%mol$ CO_2 with satisfactory performance in energy generation (Arinelli et al., 2017). GTW becomes an attractive solution to process CO_2-rich NG in ultra-deepwaters. GTW also bypasses traditional bottlenecks of high GOR by reducing logistics costs and providing additional revenue, namely power sent to grid and *in situ* CO_2-EOR. Furthermore, electricity has a more stable price than volatile oil products.

Hetland et al. (2009) evaluated the technical feasibility of a GTW-CCS system producing 580 MW of power. A traditional North Sea NG feed was fired in a four-turbine NGCC arrangement followed by an MEA CA-PCC. Results showed that offshore skids could withstand both an NGCC and a PCC plant while producing power, but further attention should be given to CO_2 transportation and water balance. Kvamsdal et al. (2010) expanded the proposed GTW by improving the systems water balance and including platform tilting effects on PCC. Optimal PCC design was able to capture 90% of the CO_2 in flue-gas, while producing 450 MW, noting that the addition of PCC incurred a 9% power loss when compared to onshore NGCCs.

More recently, Roussanaly et al. (2019) studied how rig electrification (i.e., using onshore electricity) would affect a PCC-MEA NGCC. Results showed that full electrification would attain best economic/

environmental results, reducing the electricity price from 178/258 USD/MWh to 95 USD/MWh. Nguyen et al. (2016), despite not directly studying GTW-CCS, generated a multi-objective optimization framework to analyze offshore rig performance facing increasing carbon taxes that also concluded that full electrification is the most beneficial solution. In addition, results showed aqueous amine-based CA is the best performing PCC technology for offshore rigs and that, for maximum emission reduction and profit, capture should occur with simultaneous post and pre-combustion (Nguyen et al., 2016).

In summary, with CO_2-rich NG in ultra-deepwaters and high GOR—as found in Brazilian pre-salt reservoirs—an innovative and promising technological solution consists in removing just enough CO_2 to meet NG-fired turbine standards. With turbines receiving undertreated NG, the task effort of NG upgrading is minimized, using only a CO_2 bulk-removal process, transforming the unpolished gas into power and capturing the unseparated CO_2 from the flue-gas. Hence, the floating NGCC power plant includes decarbonation and post-combustion capture, maintaining economic viability and drastically improving environmental performance. Literature favors absorption-based technologies, with a focus in CA, with high TRL and good performance in both NG and flue-gas CCS. CA is easily adapted to offshore skids and, if optimized, can have a low footprint maintaining viability though platform tilting and its effects on PCC must be considered. Additionally, since CA usually requires aqueous solvents to be used, special attention must be given to the water balance, as process water make-up is an issue in offshore conditions.

Further attention should also be given to other potential CCS technologies to determine whether performance improvements can be attained. As an example, MP could promote further footprint reduction while maintaining adequate results as high CO_2 selectivity is not necessary since NG will be burned. Furthermore, technological improvements both in CO_2 capture and NG processing will greatly affect results and potentially improve performance (Nguyen et al., 2016).

12. CO_2 Transportation

Transporting CO_2 from its capture point to its destination–offshore CO_2-EOR—is an important part of the NG to energy value-chain (Onyebuchi et al., 2018). A reliable, safe and low-cost transport system is a key feature of any CCS project. Depending on the volume of CO_2, a variety of transport modals may be utilized, ranging from road tankers to ships and pipelines (Svensson et al., 2004).

The most used modal is pipeline (Onyebuchi et al., 2018), both onshore and subsea. Continuous flow, ensured by pipelines, is essential when considering the huge flow rates of CO_2 captured from source facilities. Furthermore, pipelines have proven to be generally more cost-effective in the long-term and are currently being used worldwide (Onyebuchi et al., 2018). Pipelines require rigid specifications, as exposure to CO_2 eventually leads to corrosion and brittle fractures. Corrosion is due to the acid nature of CO_2, while brittle fractures are linked to supercritical CO_2 leaks (Rabimdran et al., 2011). Impurities in the CO_2 stream are a serious issue since they can change *PxT* envelope boundaries, within which a single-phase flow is stable. Presence of moisture above 50 ppm may lead to H_2CO_3 formation inside the pipeline, entailing corrosion potential (Forbes et al., 2008).

To optimize the mass to volume ratio, CO_2 is carried as dense phase either in liquid or supercritical conditions. Supercritical is preferred when transporting CO_2 by pipelines, implying that the pipeline operational temperature and pressure should be maintained at supercritical conditions, i.e., above 32.1 °C and 72.9 atm (Johnsen et al., 2011). The typical pressure and temperature range for a CO_2 pipeline is between 85 and 150 bar, and between 13 °C and 44 °C ensuring a stable single-phase flow throughout the pipeline (Forbes et al., 2008).

Since supercritical CO_2 is the preferred state for transportation, very specific conditions of both pipeline and fluid must exist. De Medeiros et al. (2008) developed a thermodynamic model to predict pipeline conditions while accounting for supercritical CO_2 flow and its characteristic high density and compressibility. CO_2 pipelines operate at high pressures, high densities, and high fluid compressibility, even in liquid state. That is, CO_2 can be highly compressible while reaching densities of 900–1000 kg/m^3 (de Medeiros and Araújo, 2016).

CO_2 pipelines have one more critical difference when compared to NG pipelines: CO_2 flow must occur above bubble-point pressure to avoid vaporization, which would transform the dense single-phase flow into a two-phase compressible flow with lesser total density causing acceleration and head-loss build-up (de Medeiros et al., 2008). This would generate new difficulties, like greater head-loss per km in the two-phase flow, since the presence of a low-density vapor accelerates the fluid, increasing velocity and pressure, with density loss. In summary, long distance CO_2 pipelines must operate at high pressures, high flow rates, high density and low velocity (de Medeiros and Araújo, 2016).

The pressure drop due to friction and/or gravitational effects along the pipeline is compensated by adding recompression stations. Larger diameter pipelines enable higher flow rates with smaller pressure drops and, therefore, a reduced number of recompression stations; however, larger pipelines are more expensive. Hence, a tradeoff exists in which smaller diameter pipelines have lower implementation costs but larger recompressions costs.

Currently, only a few pipelines worldwide are used to carry CO_2 and are almost all for EOR applications. The oldest one is the Canyon Reef Carriers pipeline, a 225 km pipeline built in 1972 for EOR in Texas (USA). The longest one is the 800 km Cortez pipeline which carries 20 Mt/yr of CO_2 from a natural source in Colorado to the oil fields in Denver City, Texas since 1983 (Forbes et al., 2008).

Table 6 presents the main existing long-distance CO_2, pipelines.

Table 6. Existing long-distance CO_2 pipelines. Adapted from Serpa et al., 2011.

Pipeline	Location/Start year	Capacity (Mt/yr)	Length (km)	Diameter (in)	Pressure (bar)	CO₂ source
Cortez	USA/1994	19.3	803	30"	186	McElmo Dome
Sheep Mountain	USA/1983	n/a	296	20"	n/a	Sheep Mountain
Sheep Mountain North	USA/1983	n/a	360	24"	132	Sheep Mountain
Bravo	USA/1984	7.3	350	20"	165	Bravo Dome
Central Basin	USA/1985	20	278	16"–26"	170	Denver City Hub
Bati Raman	Turkey/1983	1.1	90	n/a	170	Dodan Field
Canyon Reef Carriers	USA/1972	4.4	352	16"	140	Gasification Plant
Val Verde	USA/1998	2.5	130	10"	n/a	Gas Plant
Bairoil	USA/1986	8.3	180	n/a	n/a	Gas Plant
Weyburn	USA/2000	5	328	12"–14"	152	Gasification Plant

n/a – not available.

In Europe, long-distance pipelines for CO_2 transportation are non-existent. However, more recently, several CO_2 pipelines have started operation, with the longest ones being in the North Sea (e.g., 160 km pipeline for Snøhvit LNG project) and in the Netherlands ($\cong$ 80 km pipeline to transport CO_2 to greenhouses from Rotterdam to Amsterdam) (Serpa et al., 2011).

13. CO₂ Conversion to Methanol in FPSO

A solution for ultra-deepwater CO_2-rich NG processing, with adjustment of its CO_2 content still rich in CO_2, is to submit it to dry-reforming to produce synthesis gas (SG), which can subsequently be converted to methanol (MeOH). Any unconverted CO_2 can be further directed to enhanced oil recovery (EOR) in the oil field. Lima et al. (2016) proposed a process with PA (using propylene carbonate as a solvent) due to its easiness of regeneration (a simple expansion valve and a flash vessel). Treating high pressure CO_2-rich NG with PA gives two products: (i) a lean, high pressure, NG with low CO_2 content; and

(ii) a CO_2-rich, low pressure, gas effluent. A relevant point in the downstream processing of this CO_2 rich effluent is that PA has also affinity for hydrocarbons, yet to a lesser degree than its affinity for CO_2. This apparent drawback is explored by Lima et al. (2016), since the contaminant hydrocarbons (CH_4 and C_{2+}) react with CO_2 via dry-reforming to yield syngas accordingly to equations (2a) and (2b) (Araújo et al., 2014).

$$CH_4 + CO_2 \rightarrow 2CO + 2H_2 \qquad \Delta H = +247.3 \; kJ/mol \qquad (2a)$$

$$C_nH_{2n+2} + nCO_2 \rightarrow 2nCO + (n + 1)H_2 \qquad (2b)$$

The strongest limitation to dry-reforming is the availability of a suitable catalyst, low pressure and a high consumption of heat because the reaction is very endothermic, as seen in equation (2a). Dry-reforming also requires high temperature, which contributes to carbon deposition. In CO_2 reforming, coke deposition on the catalyst is reported to be very fast. However, when CO_2 reforming occurs simultaneously with steam reforming—i.e., bi-reforming—coke deposition is drastically reduced (Gangadharan et al., 2012). Consequently, operational conditions which, cumulatively, minimize carbon formation and maximize CO_2 conversion are employed in the reformers. In other words, an appropriate H_2O/C (steam-to-carbon) ratio is chosen so that high H_2/CO and low CO_2/CO ratios result in the syngas product. Hence, coke deposition reactions can be neglected in the analysis, so that bi-reforming encompasses only equations (3a), (3b), (4a) and (4b):

CH$_4$ Steam Reforming: $CH_4 + H_2O \rightarrow CO + 3H_2 \qquad \Delta H = +206.3 \; kJ/molCH_4 \qquad$ (3a)

HC Steam Reforming: $C_nH_{2n+2} + nH_2O \rightarrow nCO + (2n + 1)H_2 \qquad$ (3b)

CH$_4$ Dry-Reforming: $CH_4 + CO_2 \rightarrow 2CO + 2H_2 \qquad \Delta H = +247.3 \; kJ/molCO_2 \qquad$ (4a)

HC Dry-Reforming: $C_nH_{2n+2} + nCO_2 \rightarrow 2nCO + (n + 1)H_2 \qquad$ (4b)

On the one hand, in the case of reformer operating dry-reforming only, the product syngas has a low H_2/CO ratio for MeOH synthesis, as seen in equations (2a) and (2b). In this case, the water gas-shift (WGS) reaction in equation (5) can adjust this ratio at the expense of creating more CO_2 (Gangadharan et al., 2012). Nevertheless, WGS is necessary, because the reactor feed has excess of CO, which would not react to MeOH without H_2. Thus, part of the excess of CO is converted by WGS producing H_2. This extra H_2 converts the remaining excess of CO to MeOH. The reversibility of WGS is also important, since the H_2/CO ratio can be adjusted by manipulating reaction conditions.

$$CO + H_2O \leftrightarrow H_2 + CO_2 \qquad \Delta H = -51 \; kJ/mol \qquad (5)$$

$$CO + 2H_2 \leftrightarrow CH_3OH \qquad \Delta H = -90.6 \; kJ/mol \qquad (6a)$$

$$CO_2 + 3H_2 \leftrightarrow CH_3OH + H_2O \qquad (6b)$$

Lima et al. (2016) evaluated the combination of dry and steam reforming in one bi-reforming reactor (one pot reactor) and compared this to conversion segregating dry-reforming in the first reactor with WGS performed in a subsequent reactor. The second configuration uses a well-proven modular WGS converter. It is worth noting that, besides the environmental motivation, both alternatives were driven by the fact that MeOH is an important chemical commodity, used as raw material in several processes, which would bring economic advantages (i.e., a gas-to-liquids route). The main conclusion of Lima et al. (2016) is that the proposed processes are potentially amenable to a Me-FPSO. This is especially advantageous vis-à-vis the stringent weight and space limitations in FPSO plants. Figure 10 illustrates the conceptual floating process.

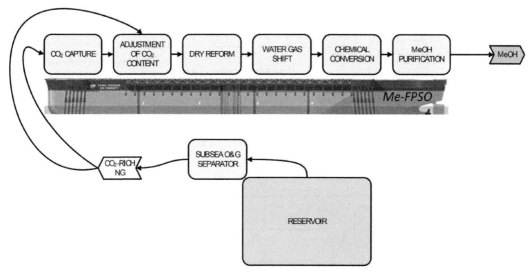

Figure 10. Floating process for CO_2 conversion to methanol.

14. Final Remarks

This chapter provides a state-of-the art review on the main technologies available for CO_2 management in the context of CO_2-rich associated NG, from ultra-deepwater O&G fields. The focus is in the gas-to-energy value-chain, where CO_2 management starts at offshore decarbonation on FPSOs. This captured CO_2 is used as an EOR agent, being monetized by the promoted enhancement in oil production (1–3 bbl/tCO_2). In the downmost side of the chain, CO_2 from flue-gas of gas-fired power plants is captured and stored through EOR in the source reservoirs. This cyclic movement of carbon occurs in a nearly closed loop, depending on the efficiency of the carbon capture technology and the distance traveled by carbon. CO_2 management, in this context, refers to the process system available to maintain carbon in closed loops, shown in Figure 11.

Carbon pipeline distances should be minimized in order to reduce costs and risk of leaks (either of the two molecules in scene). Offshore power generation in floating power generation plants (FPGP) is a technology that introduces the concept of transporting electricity instead of molecules to onshore facilities (GTW) and minimizes the distance of molecule transport.

The main challenge is the NG upgrading (the upstream edge), where the adopted technologies face weight and footprint constraints. To that end, this chapter reviews mostly mature technologies or ones with strong drivers to mature in the near-term. Maturity is approached through TRL, an indicator attributed to each reviewed technology, disregarding disruptive technologies due to their very incipient stage of development (low TRL). The mature or nearly mature technologies in the envisaged scenario are chemical and physical absorption (CA/PA), membrane permeation (MP), cryogenic distillation (CD), gas liquid membrane contactors (GLMC), supersonic separators (SS) and hybrid processes (HYB). For each technology, a brief description of the separation method is presented, contextualized in the upstream operation, while highlighting main benefits, shortcomings and recent focus.

CA and PA dominate the NG-to-energy value-chain due to their maturity. Absorption-based technologies can be applied to both pre- and post-combustion CO_2 capture with similar results. Despite their proven economic viability, CA/PA are plagued with high energy requirements and footprint, solvent toxicity and degradation—thermal (pre- and post-combustion) and oxidative (post-combustion mainly).

MP is optimal for offshore applications, such as CO_2-rich NG processing in ultra-deepwaters, due to its modularity, flexibility and lack of solvents, although for very high CO_2 content in NG, other technologies, mainly the HYB, have a competitive advantage. Various new materials and surface

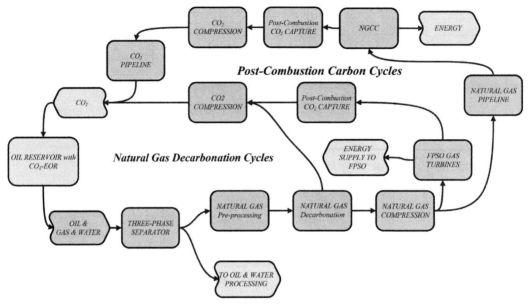

Figure 11. Carbon management cycles.

treatments are currently being tested in order to improve performance, challenged by the selectivity-permeability tradeoff—MP's main drawback of highly selective membranes being associated with low permeability and *vice versa*.

The combination of MP and CA into a single technology yields a new separation method GLMC. Its use intends to maintain CA and MP benefits while removing their major drawbacks. Nevertheless, GLMC has yet to prove its long-term economic viability in both NG upgrading and post-combustion applications, thus, research is concentrated on finding better membranes and solvents to enhance decarbonation and reduce wetting.

CD can produce pure and pressurized CO_2. CD is commercial in pre-combustion applications, but high cooling requirements have slowed its application in post-combustion. Rigid control is required in order to avoid unwanted column clogging. CD has potentially the best performance in terms of flexibility, as increased CO_2 content can be compensated for by temperature/pressure adjustments without performance loss.

SS is of great interest in NG processing in ultra-deepwaters as they promote separation in compact equipment. Additionally, SS can promote simultaneous CO_2 removal, WDPA and HCDPA. On the other hand, due to the CO_2 freeze-out border, CO_2 removal from raw NG with 45%mol CO_2 can currently limit CO_2 abatement to 21%mol—requiring a HYB configuration, by adding a CO_2 polishing process.

HYB consists of combining two or more technologies in parallel or series arrangements. The use of HYB can bypass major drawbacks, increasing flexibility and overall performance. Improvements are derived from cost reductions due to smaller equipment, lower operational costs, lower energy requirements and less stringent separation needs in a single pass. Currently CA-ADS, CA/PA-MP and CD-MP are the most promising candidates, with CD/MP already in commercial scale.

NG decarbonation and post-combustion are completely different CCS applications. While NG processing in ultra-deepwaters requires high pressure CO_2 removal, flue-gases must deal with low CO_2 fugacity and high oxygen/nitrogen content. The performance of a given separation technology is greatly affected by the context in which it is inserted. Choosing the correct CO_2 removal method becomes paramount to avoid excessive costs. A maturity analysis shows that CO_2 removal from CO_2-rich NG is more advanced than from flue-gas with more technologies reaching high TRL numbers.

Furthermore, CO_2 transportation, both onshore and offshore, has rigid safety specifications that must be met. Thus, attention should be given to the captured CO_2 properties concerning whether further costly downstream conditioning is required. To reduce overall transportation costs, NG could be directly burned to produce power which is sold to the onshore grid via subsea HVDC cables. This would reduce NG CO_2 removal requirements as new turbo-shafts achieve acceptable performance with relatively low-quality NG but increase post-combustion CO_2 removal needs.

With NG carbon content expected to worsen as CO_2-EOR becomes more widespread, stakeholders face pressing decisions as to what technology to use and in what arrangement. By providing a general view on CO_2 capture from CO_2-rich NG on ultra-deepwater rigs, this chapter contributes to design decision-making by highlighting the main applications of each technology, factors that affect their performance and what is expected to be improved in the near future.

Lastly, a conceptual processing of CO_2-rich NG on ultra-deepwater rigs involving CCU by means of chemical conversion of CO_2 to methanol is presented. Although in very early stage of development, the alternative technology for CO_2 management holds the potential to be further explored.

Acknowledgements

Authors acknowledge financial support from SHELL-Brazil (Grant 041953/2017-76). JL de Medeiros and OQF Araújo also acknowledge CNPq-Brazil (Grant 311076/2017-3).

Abbreviations and Acronyms

ADS	Adsorption
C3+	Propane and heavier paraffines
CA	Chemical Absorption
CCS	Carbon Capture and Storage
CCU	Carbon Capture and Utilization
CD	Cryogenic Distillation
CFD	Computer Fluid Dynamics
COFs	Covalent Organic Frameworks
DEC	Diethyl Carbonate
DEPG	Dimethylether polyethylene glycol
DMC	Dimethyl Carbonate
EGR	Enhanced Gas Recovery
EOR	Enhanced Oil Recovery
EOS	Equation of State
EROI	Energy Return on Investment
ESA	Electrical Swing Adsorption
FPSO	Floating Production Storage and Offloading
FPGP	Floating Power Generation Plant
GOR	Gas to Oil Ratio
GLMC	Gas Liquid Membrane Contactors
GTW	Gas to Wire
HCDP	Hydrocarbon Dew-Point
HCDPA	Hydrocarbon Dew-Point Adjustment
HYB	Hybrid
IPCC	Intergovernmental Panel on Climate Change
LNG	Liquified Natural Gas
LPG	Liquified Petroleum Gas
MDEA	Methyl-diethanolamine
MEA	Monoethanolamine
MMM	Mixed Matrix Membranes
MOFs	Metal-Organic Frameworks

MP	Membrane Permeation
NAS	Non-Aqueous Solvent
NG	Natural Gas
NGCC	Natural Gas Combined Cycle
O&G	Oil and Gas
PA	Physical Absorption
PEC-UOE	Phase-Equilibrium Sound Speed Unit Operation Extension
PCC	Post-Combustion Capture
PPNs	Porous Polymer Networks
PSA	Pressure Swing Adsorption
PTSA	Pressure Temperature Swing Adsorption
PVDF	Poly Vinylidene Difluoride
PVT	Pressure-Volume-Temperature
RVLE	Reactive Vapor-Liquid Equilibrium
SAR	Stranded Asset Risk
SR	Supersonic Reactor
SS	Supersonic Separator
SS-UOE	Supersonic Separator Unit Operation Extension
SVLE	Solid-Vapor Liquid Equilibrium
TAT	Triacetin
THI	Thermodynamic Hydrate Inhibitors
TRL	Technology Readiness Level
TSA	Temperature Swing Adsorption
UOE	Unit Operation Extension for HYSYS Simulator
WDP	Water Dew-Point
WDPA	Water Dew-Point Adjustment

References

Aaron, D. and Tsouris, C. 2005. Separation of CO_2 from flue gas: A review. Sep. Sci. and Tech. 40: 321–348.

Adewole, J.K., Ahmad, A.L., Ismail, S. and Leo, C.P. 2013. Current challenges in membrane separation of CO_2 from natural gas: A review. Int. J. of Greenh. Gas Cont. 17: 46–65. DOI: 10.1016/j.ijggc.2013.04.012.

Adeyami, I., Abu-Zahra, M.R.M. and Alnashef, I. 2016. Novel green solvents for CO_2 capture. Energy Procedia 114: 2552–2560. DOI: 10.1016/j.egypro.2017.03.1413.

Agahei, Z., Asl, V.H., Khanbabaei, G. and Dezhagah, F. 2018. The influence of fumed silica content and particle size in poly (amide 6-ethylene oxide) mixed matrix membranes for gas separation. Sep. and Pur. Tech. 199: 47–56. DOI: 10.1016/j.seppur.2018.01.035.

Al-Mamoori, A., Krishnamurthy, A., Rownaghi, A.A. and Rezaei, F. 2017. Carbon capture and utilization update. Energy Techn. 5: 769–961. DOI: 10.1002/ente.201700287.

Ali, A., Maqsood, K., Shin, L.P., Sellappah, V., Garg, S., Shariff, A.B.M. and Ganguly, S. 2018. Synthesis and mixed integer programming-based optimization of cryogenic packed bed pipeline network for purification of natural gas. J. of Clean. Prod. 171: 795–810. DOI: 10.1016/j.jclepro.2017.10.060.

Alper, E. and Orhan, O.Y. 2017. CO_2 utilization: Developments in conversion processes. Petroleum 3: 109–126. DOI: /10.1016/j.petlm.2016.11.003.

Araujo, O.Q.F., Medeiros, J.L., and Alves, R.M.B. 2014. CO_2 Utilization: A Process Systems Engineering Vision. CO_2 Sequestration and Valorization, Claudia do Rosario Vaz Morgado and Victor Paulo Peçanha Esteves, IntechOpen. DOI: 10.5772/57560. Available from: https://www.intechopen.com/books/CO2-sequestration-and-valorization/CO2-utilization-a-process-systems-engineering-vision.

Araújo, O.Q.F. and de Medeiros, J.L. 2017. Carbon capture and storage technologies: Present scenario and drivers of innovation. Curr. Opin. in Chem. Eng. 17: 22–34. DOI: 10.1016/j.coche.2017.05.004.

Araújo, O.Q.F., Reis, A.C., de Medeiros, J.L., Nascimento, J.F., Grava, W.M. and Musse, A.P.S. 2017. Comparative analysis of separation technologies for processing carbon dioxide rich natural gas in ultra-deepwater oil fields. J. of Clean. Prod. 155: 12–22. DOI: 10.1016/j.jclepro.2016.06.073.

Arinelli, L.O., Trotta, T.A.F., Teixeira, A.M., de Medeiros, J.L. and Araújo, O.Q.F. 2017. Offshore processing of rich natural gas with supersonic separator versus conventional routes. J. of Nat. Gas Sci. and Eng. 46: 199–221. DOI: 10.1016/j.jngse.2017.07.010.

Arinelli, L.O., de Medeiros, J.L., Melo, D.C., Teixeira, A.M., Brigagão, G.V., Passarelli, F.M., Grava, W.M. and Araújo, O.Q.F. 2019. Carbon capture and high-capacity supercritical fluid processing with supersonic separator: Natural gas with ultra-high CO_2 content. J. of Nat. Gas Sci. and Eng. 66: 265–283. DOI: 10.1016/j.jngse.2019.04.004.

Azzolina, N.A., Peck, W.D., Hamling, J.A., Gorecki, C.D., Ayash, S.C., Doll, T.E., Nakles, D.V. and Melze, L.S. 2016. How green is my oil? A detailed look at greenhouse gas accounting for CO_2-enhanced oil recovery (CO_2-EOR) sites. Int. J. Greenh. Gas Cont. 51: 369–79. DOI: 10.1016/j.ijggc.2016.06.008.

Bakhtiary-Davijany, H. and Myhrvold, T. 2013. On methods for maturity assessment of CO_2 capture technologies. Energy Procedia 37: 2579–2584. DOI: 10.1016/j.egypro.2013.06.141.

Barbosa, L.C., Araújo, O.Q.F. and de Medeiros, J.L. 2019. Carbon capture and adjustment of water and hydrocarbon dew-points via absorption with ionic liquid [Bmim][NTf2] in offshore processing of CO_2-rich natural gas. J. of Nat. Gas Sci. and Eng. 66: 26–41. DOI: 10.1016/j.jngse.2019.03.014.

Bimüller, J.D. and Nord, L.O. 2015. Process simulation and plant layout of a combined cycle gas turbine for offshore oil and gas installations. Journal of Power Technologies 95: 40–47.

Boot-Handford, M.E., Abanades, J.C., Anthony, E.J., Blunt, M.J., Brandani, S., Mac Dowell, N., Fernandez, J.R., Ferrari, M.C., Gross, R., Hallett, J.P., Haszeldine, R.S., Heptonstall, P., Lyngfelt, A., Makuch, Z., Mangano, E., Porter, R.T.J., Pourkashanian, M., Rochelle, G.T., Shah, N., Yao, J.G. and Fennell, P.S. 2014. Carbon capture and storage update. Energy Environ. Sci. 7: 130–189. DOI: 10.1039/ c3ee42350f.

Bouzina, Z., Dergal, F., Mokbel, I., Negadi, A., Saab, J., Jose, J. and Negadi, L. 2016. Liquid vapor equilibria of pure and aqueous solutions of diethylenetriamine or dipropylenetriamine. Fluid Phase Equilib. 414: 164–169. DOI: 10.1016/j.fluid.2016.01.022.

Brennecke, J.F. and Gurkan, B.E. 2010. Ionic liquids for CO_2 capture and emission reduction. J. Phys. Chem. Lett. (24): 3459–3464. DOI: 10.1021/jz1014828.

Brigagão, G.V., Arinelli, L.O., de Medeiros, J.L. and Araújo, O.Q.F. 2019. A new concept of air pre-purification unit for cryogenic separation: Low-pressure supersonic separator coupled with finishing adsorption. Sep. & Purf. Tech. 215: 173–189. DOI: 10.1016/j.seppur.2019.01.015.

Chowdbury, F.A., Okabe, H., Yamada, H., Onoda, M. and Fujioka, Y. 2011. Synthesis and selection of hindered new amine absorbents for CO_2 capture. Energy Procedia 4: 201–208. DOI: 10.1016/j.egypro.2011.01.042.

Chowdbury, F.A., Yamada, H. and Higashii, T. 2013. Synthesis and characterization of new absorbents for CO_2 capture. Energy Procedia 37: 265–272. DOI: 10.1016/j.egypro.2013.05.111.

Clark, S., Snider, D.M. and Spenik, J. 2013. CO_2 adsorption loop experiment with Eulerian-Lagrangian simulation. Powder Tech. 242: 100–107. DOI: 10.1016/j.powtec.2013.01.011.

D'Allesandro, D.M., Smit, B. and Long, J.R. 2010. Carbon dioxide capture: Prospects for new materials. Angewandte Chemie International 49: 6058–6082. DOI: 10.1002/anie.201000431.

de Medeiros, J.L., Versiani, B.M. and Araújo, O.Q.F. 2008. A model for pipeline transportation of supercritical CO_2 for geological storage. Journal of Pipeline Engineering 4: 253–279.

de Medeiros, J.L., Barbosa, L.C. and Araújo, O.Q.F. 2013a. Equilibrium approach for CO_2 and H_2S absorption with aqueous solutions of alkanolamines: Theory and parameter estimation. Ind. & Eng. Chem. Res. 52: 9203–9226. DOI: 10.1021/ie302558b.

de Medeiros, J.L., Nakao, A., Grava, W.M., Nascimento, J.F. and Araújo, O.Q.F. 2013b. Simulation of an offshore natural gas purification process for CO_2 removal with gas–liquid contactors employing aqueous solutions of ethanolamines. Ind. & Eng. Chem. Res. 52: 7074–7089. DOI: 10.1021/ie302507n.

de Medeiros, J.L. and Araujo, O.Q.F. 2016. CO_2 pipelines: A thermodynamic modeling with pre-salt applications. Appl. Mech. Mat. 830: 57–64. DOI: 10.4028/www.scientific.net/AMM.830.57.

de Medeiros, J.L., Arinelli, L.O. and Araújo, O.Q.F. 2017. Speed of sound multiphase and multi-reactive equilibrium streams: A numerical approach for natural gas applications. J. of Nat. Sci. & Eng. 46: 221–243. DOI: 10.1016/j.jngse.2017.08.006.

de Medeiros, J.L., Arinelli, L.O., Teixeira, A.M. and Araújo, O.Q.F. 2019. Offshore processing of CO_2-rich natural gas with supersonic separator. Multiphase Sound Speed, CO_2 Freeze-Out and HYSYS Implementation. Springer Nature Switzerland. DOI: 10.1007/978-3-030-04006-2.

[EC] European Commission. 2019. Best available techniques guidance document on upstream hydrocarbon exploration and production. ISBN: 978-92-76-01443-0. Final Guidance Document–Contract No. 070201/2015/706065/SER/ENV.F.1. Wood Environment & Infrastructure Solutions UK Limited. DOI: 10.2779/607031. Available at: ec.europa.eu/environment/integration/energy/pdf/hydrocarbons_guidance_doc.pdf. Last access date: 04/10/2019.

[EIA 2016] United States Energy Information Administration. 2016. Energy-related CO_2 emissions from natural gas surpass coal fuel use patterns change. Washington DC, USA. Available at: www.eia.gov/todayinenergy/detail.php?id=27552. Last access date: 02/25/2018.

[EIA 2018] United States Energy Information Administration. 2018. Energy Outlook 2018. Washington DC, USA. Available at: www.eia.gov/outlooks/aeo/pdf/AEO2018_FINAL_PDF.pdf. Last access date: 02/25/2018.

[Energy Institute at Haas]. 2016. Time to Unleash the Carbon Market? California, USA. energyathaas.wordpress.com/2016/06/20/time-to-unleash-the-carbon-market. Last access date: 04/23/2018.

Elfving, J., Bajamundi, C., Kauppinen, J. and Sainio, T. 2017. Modelling of equilibrium working capacity of PSA, TSA and TVSA processes for CO_2 adsorption under direct air capture conditions. Journal of CO_2 Utilization 22: 270–277. DOI: 10.1016/j.jcou.2017.10.010.

[EPA] Environmental Protection Agency. 2018. ExxonMobil Shute Creek Treating Facility Subpart RR Monitoring, Reporting and Verification Plan. Available at: https://www.epa.gov/sites/production/files/2018-06/documents/shutecreekmrvplan.pdf. Last access date: 04/10/2019.

Espinal, L., Poster, D.L., Wong-Ng, W., Allen, A.J. and Green, M.L. 2013. Measurement, standards, and data needs for CO_2 capture materials: A critical review. Environ. Sci. Technol. 47: 11960–11975.

Flatebø, Ø. 2012. Off-design Simulation of Offshore Combined Cycles. MSc. thesis submitted to the Mechanical Engineering Post-Graduation of the Norwegian University of Science and Technology.

Følgesvold, E., Skjefstad, H.S., Riboldi, L. and Nord, L.O. 2017. Combined heat and power plant on offshore oil and gas installations. Journal of Power Technologies 97: 117–126.

Forbes, S.M., Verma, P., Curry, T.E., Friedmann, S.J. and Wade, S.M. 2008. Guidelines for carbon dioxide capture, transport, and storage. Lawrence Livermore National Laboratory and Sarah M. Wade, AJW, Inc. World Res. Inst. Available at: www.wri.org/publication/guidelines-carbon-dioxide-capture-transport-and-storage#. Last access date: 04/12/2019.

Fougerit, V., Pozzobon, V., Pareau, D., Théoleyre, M.A. and Stambouli, M. 2017. Gas-liquid absorption in industrial cross-flow membrane contactors: Experimental and numerical investigation of the influence of transmembrane pressure on partial wetting. Chem. Eng. Sci. 170: 561–573. DOI: 10.1016/j.ces.2017.03.042.

Forsyth, J., Lodge, S., Consonni, S., Bona, D.D., Gatti, M., Martelli, E., Scaccabarozzi, R. and Viganò, F. 2017. Evaluation of five alternative CO_2 capture technologies with insights to inform further development. Energy Procedia 114: 2599–2610. DOI: 10.1016/j.egypro.2017.03.1419.

Fu, Q., Ding, J., Wang, W., Lu, J. and Huang, Q. 2017. Carbon dioxide adsorption over amine-functionalized MOFs. Energy Procedia 142: 2152–2157. DOI: 10.1016/j.egypro.2017.12.620.

Gangadharan, P., Kanchi, K.C. and Lou, H.H. 2012. Evaluation of the economic and environmental impact of combining dry reforming with steam reforming of methane. Chem. Eng. Res. Des. 90(11): 1956–1968. DOI: 10.1016/j.cherd.2012.04.008.

Gazalpour, F., Ren, S.R. and Tohidi, B. 2005. CO_2 EOR and storage in oil reservoirs. Oil Gas Sci. Technol. 60: 537–546. Available at: ogst.ifpenergiesnouvelles.fr/articles/ogst/abs/2005/03/gozalpour_vol60n3/gozalpour_vol60n3.html. Last access date: 04/10/2019.

Gamali, P.A., Kazemi, A., Zadmard, R., Anjareghi, M.J., Rezakhani, A., Rahighi, R. and Madani, M. 2018. Distinguished discriminatory separation of CO_2 from its methane-containing gas mixture *via* PEBAX mixed matrix membrane. Chinese J. of Chem. Eng. 26: 73–80. DOI: 10.1016/j.cjche.2017.04.002.

Ghasem, N., Al-Marzouqi, M. and Zhu, L. 2012. Preparation and properties of polyethersulfone hollow fiber membranes with o-xylene as an additive used in membrane contactors for CO_2 absorptio. Sep. & Purif. Tech. 92: 1–10. DOI: 10.1016/j.seppur.2012.03.005.

Gil, A., Arrieta, E., Vicente, M.A. and Korili, S.A. 2017. Synthesis and CO_2 adsorption properties of hydrotalcite-like compounds prepared from aluminum saline slag wastes. Chem. Eng. J. 334: 1341–1350. DOI: 10.1016/j.cej.2017.11.100.

Grande, C.A., Roussanaly, S., Anantharaman, R., Lindqvist, K., Singh, P. and Kemper, J. 2017. CO_2 capture in natural gas production by adsorption processes. Energy Procedia 114: 2259–2264. DOI: 10.1016/j.egypro.2017.03.1363.

Gugliuzza, A. and Drioli, E. 2007. PVDF and HYFLON AD membranes: Ideal interfaces for contactor applications. J. of Memb. Sci. 300: 51–62. DOI: 10.1016/j.memsci.2007.05.004.

Hall, C.A.S., Lambert, J.G. and Balogh, S.B. 2014. EROI of different fuels and the implications for society. Energy Policy 64: 141–152.

Hammer, M., Wahla, P.E., Anantharamana, R., Berstada, D. and Lervåg, K.Y. 2014. CO_2 capture from off-shore gas turbines using supersonic gas separation. Energy Procedia 63: 243–252. DOI: 10.1016/j.egypro.2014.11.026.

Hao, J., Rice, P.A. and Stern, S.A. 2008. Upgrading low-quality natural gas with H_2S and CO_2 selective polymer membranes: Part II. Process design, economics, and sensitivity study of membrane stages with recycle streams. J. Membr. Sci. 320: 108–122. DOI: 10.1016/j.memsci.2008.03.040.

Harkin, T., Hoadley, A. and Hooper, B. 2010. Reducing the energy penalty of CO_2 capture and compression using pinch analysis. J. Clean. Prod. 18: 857–866. DOI: 10.1016/j.jclepro.2010.02.011.

Hart, A. and Gnanendran, N. 2009. Cryogenic capture in natural gas. Energy Procedia 1: 697–706. DOI: 10.1016/j.egypro.2009.01.092.

Hashemifard, S.A., Matsuura, T., Ismail, A.F., Rezaei, M., Arzhandi, D., Rana, D. and Bakeri, G. 2015. Characterization of partial pore wetting in hollow fiber gas absorption membrane contactors: An EDX analysis approach. Chem. Eng. J. 281: 970–980. DOI: 10.1016/j.cej.2015.07.036.

Hetland, J., Kvamsdal, H.M., Haugen, G. and Major, F. 2009. Integrating a full carbon capture scheme onto a 450 MWe NGCC electric power generation hub for offshore operations: Presenting the Sevan GTW concept. Applied Energy 86: 2298–2307. DOI: 10.1016/j.apenergy.2009.03.019.

[IPCC 2014a]. Intergovernmental Panel on Climate Change. Climate Change. 2014. Synthesis Report, Summary for Policymakers. Paris. Available at www.ipcc.ch/report/ar5/. Last access date: 04/24/2018.

[IPCC 2014b]. Intergovernmental Panel on Climate Change. Climate Change 2014. Synthesis Report, ipcc.ch/pdf/assessment-report/ar5/syr/AR5_SYR_FINAL_All_Topics.pdf. Last access date: 04/24/2018.

Im, D., Roh, K., Kim, J., Eom, Y. and Lee, J.H. 2015. Economic assessment and optimization of the selexol process with novel additives. Int. J. of Greenh. Gas Cont. 42: 106–116. DOI: 10.1016/j.ijggc.2015.08.001.

Ismaila, A.F., Kuswaroa, T.D., Mustafaa, A. and Hasbullaha, H. 2005. Understanding the solution-diffusion mechanism in gas separation membrane for engineering students. Development in Teaching and Learning. Proceedings of the 2005 Regional Conference on Engineering Education, Johor, Malaysia. Available at: tree.utm.my/wp-content/uploads/2013/02/DTL-Ahmad-Fauzi-Ismail-Tutuk-Djoko-Kusworo-Azeman-Mustafa-Hasrinah-Hasbullah.pdf. Last access date: 17/03/2018.

Jaafari, L., Jaffray, B. and Idem, R. 2018. Screening study for selecting new activators for activating MDEA for natural gas sweetening. Sep. and Pur. Tech. 199: 320–330. DOI: 10.1016/j.seppur.2018.02.007.

Johnsen, K., Helle, K., Roneid, S. and Holt, H. 2011. DNV recommended practice: Design and operation of CO₂ pipelines. Energy Procedia 4: 3032–3039.

Kahn, S.N., Hailegiorgis, S.M., Man, Z., Garg, S. and Shariff, A.M. 2017. High-pressure absorption study of CO₂ in aqueous N-methyldiethanolamine (MDEA) and MDEA-piperazine (PZ)-1-butyl-3-methylimidazolium trifluoromethanesulfonate [bmin][OTf] hybrid solvents. J. of Mol. Liq. 249: 1236–1244. DOI: 10.1016/j.molliq.2017.11.145.

Kang, G., Chan, Z.P., Saleh, S.B.M. and Cao, Y. 2017. Removal of high concentration CO₂ from natural gas using high pressure membrane contactors. Int. J. of Greenh. Gas Con. 60: 1–9. DOI: 10.1016/j.ijggc.2017.03.003.

Karl, M., Brooks, S., Wright, R. and Knudsein, S. 2009. Amines worst case scenario, worst case studies on amine emissions from CO₂ capture plant (Task 6), NILU/NIVA.

Kelley, B.T., Valencia, J.A., Northrop, P.S. and Mart, C.J. 2011. Controlled freeze zone™ for developing sour gas reserves. Energy Procedia 4: 824–829.

Kothandaraman, A. 2010. Carbon dioxide capture by chemical absorption: A solvent comparison study, PhD. Thesis, Massachusetts Institute of Technology. Presented to the Department of Chemical Engineering, Massachusetts, USA.

Koytsoumpa, E.I., Atsonios, K., Panopoulos, K.D., Karellas, S., Kakaras, E. and Karl, J. 2015. Modelling and assessment of acid gas removal processes in coal-derived SNG production. Appl. Therm. Eng. 74: 128–135.

Kvamsdal, H.M., Hetland, J., Haugen, G., Svendsen, H.F., Major, F., Karstad, V. and Tjellander, G. 2010. Maintaining a neutral water balance in a 450 MWe NGCC–CCS power system with post-combustion carbon dioxide capture aimed at offshore operation. International Journal of Greenhouse Gas Control 4: 613–622. DOI: 10.1016/j.ijggc.2010.01.002.

Lail, M., Tanthana, J. and Coleman, L. 2014. Non-Aqueous Solvent (NAS) CO₂ capture process. Energy Procedia 63: 580–594. DOI: 10.1016/j.egypro.2014.11.063.

Lastari, F. 2009. Ryan-Holmes and Modified Ryan-Holmes Processes for LNG Production, Ph.D Thesis, Curtin University of Technology. Presented to the Department of Chemical Engineering, Australia, December 2009.

Leimbrink, M., Nikoleit, K.G., Spitzer, R., Salmon, S., Bucholz, T., Górak, A. and Skiborowski, M. 2017. Enzymatic reactive absorption of CO₂ in MDEA by means of an innovative biocatalyst delivery system. J. of Chem. Eng. 334: 1995–1205. DOI: 10.1016/j.cej.2017.11.034.

Leperi, K.T., Snurr, R.Q. and You, F. 2016. Optimization of two-stage pressure/vacuum swing adsorption with variable dehydration level for post-combustion carbon capture. Ind. and Eng. Chem. Res. 55: 3338–3350. DOI: 10.1021/acs.iecr.5b03122.

Leung, D.Y.C., Caramanna, G.C. and Maroto-Vaaler, M.M. 2014. An overview of current status of carbon dioxide capture and storage technologies. Renew. Sustain. Energy Rev. 39: 426–443. DOI: 10.1016/j.rser.2014.07.093.

Li, X., Hagaman, E., Tsouris, C. and Lee, J.W. 2003. Removal of carbon dioxide from flue gas by ammonia carbonation in the gas phase. Energy Fuels 17: 69–74. DOI: 10.1021/ef020120n.

Lima, B.C.S., Araújo, O.Q.F., de Medeiros, J.L. and Morgado, C.R.V. 2016. Technical, economic and environmental viability of offshore CO₂ reuse from natural gas by dry reforming. Applied Mechanics and Materials 830: 109–116. DOI: 10.4028/www.scientific.net/AMM.830.109.

Liu, H., Liu, B., Lin, L.C., Chen, G., Wu, Y., Wang, J., Gao, X., Lv, Y., Zhang, X., Zhang, X., Yang, L., Sun, C., Smit, B. and Wang, W. 2014. A hybrid absorption-adsorption method to efficiently capture carbon. Nat. Comm. 5: 5147. DOI: 10.1038/ncomms6147.

Machado, P.B., Monteiro, J.G.M., de Medeiros, J.L., Epsom, H.D. and Araujo, O.Q.F. 2012. Supersonic separation in onshore natural gas dew point plant. J. Nat. Gas Sci. & Eng. 6: 43–49. DOI: 10.1016/j.jngse.2012.03.001.

Maqsood, K., Mullick, A., Ali, A. and Kargupta, R. 2014. Cryogenic carbon dioxide separation from natural gas: A review based on conventional and novel emerging technologies. Rev. in Chem. Eng. 30: 453–477. DOI: 10.1515/revce-2014-0009.

Mankins, J.C. 2009. Technology readiness assessments: A retrospective. Acta Astronaut 65: 1216–1223. DOI: 10.1016/j.actaastro.2009.03.058.

Marx, J., Shcreiber, A., Zapp, P., Haines, M., Hake, J.-F. and Gale, J. 2011. Environmental evaluation of CCS using life cycle assessment—a synthesis report. Energy Procedia 4: 2448–2456. DOI: 10.1016/j.egypro.2011.02.139.

Mazzetti, M., Skagestad, R., Mathisen, A. and Eldrup, N.H. 2014. CO_2 from natural gas sweetening to kick-start EOR in the North Sea. Energy Procedia 63: 7280–7289. DOI: 10.1016/j.egypro.2014.11.764.

Moullec, Y.L. and Kanniche, M. 2011. Screening of flowsheet modifications for an efficient monoethanolamine (MEA) based post-combustion CO_2 capture. Int. J. Greenh. Gas. Control. 5(4): 727–740. DOI: 10.1016/j.ijggc.2011.03.004.

Musarra, S.P. 2017. Offshore leadership forum addresses CAPEX challenges. Offshore Magazine, 02/15/2017. Available at: www.offshore-mag.com/articles/print/volume-77/issue-2/offshore-leadership-forum/offshore-leadershipforum-addresses-capex-challenges.html. Last access date: 05/24/2018.

[NETL] National Energy Technology Laboratory. 2015. Cost and performance baseline for fossil energy plants, Volume 1: Bituminous coal (PC) and natural gas to electricity. Revision 3. November. DOE/NETL-2015/1723. United States Department of Energy. Available at: https://www.netl.doe.gov/projects/files/CostandPerformanceBaselineforFossilEnergyPlantsVolume1aBitCoalPCandNaturalGastoElectRev3_070615.pdf. Last access date: 04/15/2019.

Nguyen, T.V., Tock, L., Breuhaus, P., Maréchal, F. and Elmegaard, B. 2016. CO_2-mitigation options for the offshore oil and gas sector. Applied Energy 161: 673–694. DOI: 10.1016/j.apenergy.2015.09.088.

Norahim, N., Yaisanga, P., Faungnawakij, K., Charinpanitkul, T. and Klaysom, C. 2017. Recent membrane developments for CO_2 separation and capture. Chem. Eng. & Tech. 41: 1–14. DOI: 10.1002/ceat.201700406.

Nord, L.O. and Bolland, O. 2013. Design and off-design simulations of combined cycles for offshore oil and gas installations. Applied Thermal Engineering 5: 85–91.

Olajire, A.A. 2010. CO_2 capture and separation technologies for end-of-pipe applications: A review. Energy 35: 2610–2628. DOI: 10.1016/j.energy.2010.02.030.

Onyebuchi, V.E., Kolios, A., Hanak, D.P., Biliyok, C. and Manovic, V. 2018. A systematic review of key challenges of CO_2 transport via pipelines. Renewable and Sustainable Energy Reviews 81: 2563–2583. DOI: 10.1016/j.rser.2017.06.064.

[Prosernat]. 2017. Sprex Gas sweetening technologies. Available at: www.prosernat.com/en/solutions/upstream/gas-sweetening/sprex.html. Last access date: 30/03/2018.

Pan, R.H., Chen, Y.R., Tung, K.L. and Chang, H. 2017. Experimental and simulation study of a novel hybrid absorption and stripping membrane contactor for carbon capture. Journal of the Taiwan Institute of Chemical Engineers 81: 47–56. DOI: 10.1016/j.jtice.2017.10.009.

Pinto, A.C.C., Vaz, C.E.M., Branco, C.C.M. and Ribeiro, J. 2014. An evaluation of large capacity processing units, for ultra-deepwater and high GOR oil fields. OTC-25274-MS, Offshore Technology Conference, Houston, USA. DOI: 10.4043/25274-MS.

Polasek, J. and Bullin, J. 2006. Selecting Amines for Gas Sweetening, Bryan Research and Engineering Technical Papers. Available at: bre.com/PDF/Selecting-Amines-for-Sweetening-Units.pdf. Last access date: 24/04/2019.

Quek, C.V., Shah, N. and Chachuat, B. 2018. Modeling for design and operation of high-pressure membrane contactors in natural gas sweetening. Chem Eng. Research and Des. 132: 1005–1019. DOI: 10.1016/j.cherd.2018.01.033.

Rabimdran, P., Cote, H. and Winning, I.G. 2011. Integrity management approach to reuse of oil and gas pipelines for CO_2 transportation. pp. 04–05. *In*: Proceedings of the 6th Pipeline Technology Conference. Hannover Messe, Hannover, Germany.

Reis, A.C., de Medeiros, J.L., Nunes, G.C. and Araújo, O.Q.F. 2017. Upgrading of natural gas ultra-rich in carbon dioxide: Optimal arrangement of membrane skids and polishing with chemical absorption. J. of Clean. Prod. 165: 1013–1024. DOI: 10.1016/j.jclepro.2017.07.198.

Reis, A.C., de Medeiros, J.L., Nunes, G.C. and Araújo, O.Q.F. 2018. Lifetime oriented design of natural gas offshore processing for cleaner production and sustainability: High carbon dioxide content. J. of Clean. Prod. 200: 269–281. DOI: 10.1016/j.jclepro.2018.07.271.

Rezaei, M., Ismail, A.F., Bakeri, G., Hashemifar, S.A. and Matsuura, T. 2015. Effect of general montmorillonite and Cloisite 15A on structural parameters and performance of mixed matrix membranes contactors for CO_2 absorption. Chem. Eng. J. 260: 875–885. DOI: 10.1016/j.cej.2014.09.027.

Riboldi, L. and Bolland, O. 2017. Overview on pressure swing adsorption (PSA) as CO_2 capture technology: State-of-the-art, limits and potentials. Energy Procedia 114: 2390–2400. DOI: 10.1016/j.egypro.2017.03.1385.

Robeson, L.M. 1991. Correlation of separation factor versus permeability for polymeric membranes. J. of Mem. Sci. 62: 165–185. DOI: 10.1016/0376-7388(91)80060-J.

Rochelle, G.T. 2009. Amine scrubbing for CO_2 capture. Science. 5947: 1652–16547. DOI: 10.1126/science.1176731.

Rochelle, G.T., Chen, E., Dombrowski, K., Sexton, A. and Lani, B. 2014. Pilot plant testing of piperazine (PZ) with high temperature regeneration. Presented at NETL's CO_2 capture technology meeting, Pittsburgh, United States.

Roussanaly, S., Aasen, A., Anantharaman, R., Danielsen, B., Jakobsen, J., Heme-De-Lacotte, L., Neji, G., Sødal, A., Wahl, P.E., Vrana. T.K. and Dreux, R. 2019. Offshore power generation with carbon capture and storage to decarbonize mainland electricity and offshore oil and gas installations: A techno-economic analysis. Appl. Energy 233-234: 478–494. DOI: 10.1016/j.apenergy.2018.10.020.

Salvindera, K.M.S., Zabiria, H., Isaa, F., Aqvib, S.A., Roslana, M.A.H. and Shariffa, M. 2017. Dynamic modelling, simulation and basic control of CO_2 absorption based on high pressure pilot plant for natural gas treatment. International J. of Green. Gas Con. 70: 164–177. DOI: 10.1016/j.ijggc.2017.12.014.

Santaniello, A. and Golemme, G. 2018. Interfacial control in perfluoropolymer mixed matrix membranes for natural gas sweetening. J. of Ind. and Eng. Chem. 60: 169–176. DOI: 10.1016/j.jiec.2017.11.002.

[Schlumberger]. 2016. Amine Gas Sweetening Systems, Gas Treatments, Schlumberger Technical Specifications. Available at: www.slb.com/~/media/Files/processing-separation/product-sheets/amine-gas-sweetening-systems-ps.pdf. Last access date: 04/03/2018.

Serna-Guerrero, R., Belmabkhout, Y. and Sayari, A. 2010. Modeling CO_2 adsorption on amine functionalized mesoporous silica: 1. A semi-empirical equilibrium model. Chem. Engin. J. 161: 173–181. DOI: 10.1016/j.cej.2010.04.024.

Serpa, J., Morbee, J. and Tzimas, E. 2011. Technical and economic characteristic of a CO_2 transmission pipeline infrastructure. JRC Scientific and Technical Reports. Available at: publications.europa.eu/resource/cellar/4ab1c4e2-398e-426c-b06f-1175d3c5a403.0001.02/DOC_1. Last access date: 11/04/2018.

Sharma, I., Hoadley, A.F.A., Mahajani, S.M. and Ganesh, A. 2016. Multi-objective optimization of a Rectisol™ process for carbon capture. J. of Clean. Prod. 119: 196–206. DOI: 10.1016/j.jclepro.2016.01.078.

Song, C., Liu, Q., Ji, N., Deng, S., Zhao, J., Li, Y., Song, Y. and Li, H. 2017a. Alternative pathways for efficient CO_2 capture by hybrid processes: A review. Renew. and Sust. Energy Reviews 82: 215–231. DOI: 10.1016/j.rser.2017.09.040.

Song, M., Song, Y., Hwang, J.O., Lee, C. and Lee, W. 2017b. Samsung energy plant: All-in-one solution for floating power plants with gas-fired combined cycle gas turbines. GASTECH.

Songolzadeh, M., Soleimani, M., Ravanchi, M.T. and Songolzadeh, R. 2017. Carbon dioxide separation from flue gases: A technological review emphasizing reduction in greenhouse gas emissions. The Sci. World J. DOI: 10.1155/2014/828131.

Svensson, R., Odenberger, M., Johnsson, F. and Stromberg, L. 2004. Transportation systems for CO_2—application to carbon capture and storage. Energy Convers. Manag. 45: 2343–53. DOI: 10.1016/j.enconman.2003.11.022.

Stuardi, F.M., MacPherson, F. and Leclaire, J. 2019. Integrated CO_2 capture and utilization: A priority research direction. Current Opinion in Green and Sustainable Chemistry 46: 71–76. DOI: 10.1016/j.cogsc.2019.02.003.

Suleman, M.S., Lau, K.K. and Yeong, Y.F. 2016. Plasticization and swelling in polymeric membranes in CO_2 removal from natural gas. Chem. Eng. & Techn. 39: 1604–1616. DOI: 10.1002/ceat.201500495.

Sun, J., Rongwong, W., Liang, Z., Gao, H., Idem, R.O. and Tontiwachwuthikul, R. 2019. Simulation studies of process improvement of three-tower low-temperature distillation process to minimize energy consumption for separation of produced gas CO_2-enhanced oil recovery (EOR). The Can. J. of Chem. Eng. 93: 1266–1274. DOI: 10.1002/cjce.22214.

Szczęśniak, B., Choma, J. and Jaroniec, M. 2017. Gas adsorption properties of hybrid graphene-MOF materials. J. of Coll. and Inter. Sci. 514: 801–813. DOI: 10.1016/j.jcis.2017.11.049.

Teixeira, M.A., Arinelli, L.O., de Medeiros, J.L. and Araújo, O.Q.F. 2018. Recovery of thermodynamic hydrate inhibitors methanol, ethanol and MEG with supersonic separators in offshore natural gas processing. J. Nat. Sci. Eng. 52: 166–186. DOI: 10.1016/j.jngse.2018.01.038.

[Twister]. 2017. Bringing Safety sustainability and economic benefits to modern-day gas separation, Twister News and Events. Available at: www.twisterbv.com/praesent-at-nulla-lorem-ut-erat-nulla-laoreet-ut-est-vitae-feugiat-interdum-urna-3/. Last access date: 30/03/2018.

Vinoba, M., Bhagiyalakshmi, M., Algaheem, Y., Alomair, A.A., Pérez, A. and Rana, M.S. 2017. Recent progress of fillers in mixed matrix membranes for CO_2 separation: A review. Sep. and Pur. Tech. 188: 431–450. DOI: 10.1016/j.seppur.2017.07.051.

Wang, Q., Luo, J., Zhong, Z. and Borgna, A. 2011. CO_2 capture by solid adsorbents and their applications: Current status and new trends. Energy Environ. Sci. 4: 42–55. DOI: 10.1039/C0EE00064G.

Wang, Y. and Hu, D. 2018. Structure improvements and numerical simulation of supersonic separators with diversion cone for separation and purification. RSC Advances 8: 10228–10236. DOI: 10.1039/C7RA13198D.

Watanabe, T., Inoue, H., Horitsugi, M. and Oya, S. 2006. Gas-to-wire (GTW) system for developing small gas field and exploiting associated gas. SPE-103746-MS. International Oil & Gas Conference & Exhibition China 5–7. DOI: 10.2118/103746-MS.

Wen, C., Cao, X., Yang, Y. and Li, W. 2012. An unconventional supersonic liquefied technology for natural gas. Energy Education Science and Technology Part A: Energy Science and Research 30: 651–660.

Xiaojun, X., Lin, W. and Anzhon, G. 2016. Design and optimization of offshore natural gas liquefaction process adopting PLNG (pressurized liquified natural gas) technology. Jrn. Nat. Gas Sci. Eng. 30: 379–387. DOI: 10.1016/j.jngse.2016.02.046.

Yeo, Y.Z., Thiam, L.C., Peng, W.Z., Abdul, R.M. and Siang-Piao, C. 2012. Conventional processes and membrane technology for carbon dioxide removal from natural gas: A review. J. of Nat. Gas Chem. 21: 282–298. DOI: 10.1016/S1003-9953.

Yu, C.H., Huang, C.H. and Tan, C.S. 2012. A review of CO_2 capture by absorption and adsorption. Aerosol and Air Quality 12: 745–769. DOI: 10.4209/aaqr.2012.05.0132.

Zakuciová, K., Lapão, J. and Kočí, V. 2015. Life cycle assessment overview of carbon capture and storage technologies. Hitecarlo publications. Available at: hitecarlo.vscht.cz/files/uzel/0014795/i0rMLk3OzC9LjE8rzclRKEmtKIkPdnUNNjIwNAMA. pdf?redirected. Last access date: 10/03/2018.

Zhang, M., Deng, L., Xiang, D., Cao, B., Hosseini, S.S. and Li, P. 2019. Approaches to suppress CO_2-induced plasticization of polyimide membranes in gas separation applications. Processes 7/0051: 1–31. DOI: 10.3390/pr7010051. Available at: www.mdpi.com/2227-9717/7/1/51/pdf. Last access date: 04/10/2019.

Zhang, J., Qiao, Y. and Agar, D.W. 2012. Improvement of lipophilic-amine-based thermomorphic biphasic solvent for energy-efficient carbon capture. Energy Procedia 23: 92–101. DOI: 10.1016/j.egypro.2012.06.072.

Zhang, Y., Sunarso, J., Liu, S. and Wang, R. 2013. Current status and development of membranes for CO_2/CH_4 separation: A review. Int. J. Greenh. Gas. Control. 12: 84–107. DOI: 10.1016/j.ijggc.2012.10.009.

Zhao, H., Chang, Y. and Feng, S. 2016. Influence of produced natural gas on CO_2-crude oil systems and the cyclic CO_2 injection process. J. of Nat. Gas Sci. and Eng. 35: 144–151. DOI: 10.1016/j.jngse.2016.08.051.

Zolfaghari, A., Mousavi, S.A., Bozariomehri, R.B. and Bakhtiari, F. 2018. Gas-Liquid membrane contactors: III. Modeling study of non-uniform membrane wetting. J. of Memb. Sci. 555: 463–472. DOI: 10.1016/j.memsci.2018.03.067.

CHAPTER 15

Chemicals from Coal

A Smart Choice

Bipin Vora

1. Introduction

In the early 20th century, prior to the age of oil and gas, the coke oven industry, as it used to be called, provided ammonia, benzene, toluene, and phenol. DuPont produced a large amount of methanol at its Belle, W.V. coal mine. However, as low-cost natural gas became available around 1950, the unit was shut down. Likewise, during the 1930s and continuing through WWII, development and advances in Fischer-Tropsch technology enabled the production of straight chain hydrocarbons, waxes and fuel from coal. Production of synthesis gas also provided the route to many organic chemicals. From there to about 2000, the primary use of coal was limited to electric power generation.

The relative cost and convenience of crude oil derived fuels resulted in the displacement of coal as the premier energy source. Nonetheless, coal continues to be a staple in the energy diet of many countries, and the current prices of crude oil have triggered renewed interest in coal utilization for the production of chemicals. Table 1 shows coal utilization for the production of electricity for various countries. For some countries, like Poland and South Africa, coal accounts for more than 80% of electric power generation. As recently as 2018 in the United States, 28% of the electric power generated is from coal (IEA, 2011; IEA, 2018). As shown in Table 1, not much has changed between 2011 and 2018 regarding the utilization of coal for electricity generation.

From a carbon management perspective, the conversion of coal to chemicals is a far better option than its conversion to electricity or transportation fluids. For electricity or the transportation fluids the entire carbon content of the coal is ultimately converted to CO_2. In the case of coal to chemicals, assuming that the energy required for the several processing steps in coal to chemicals is provided by coal, about 70% of the carbon remains in the final product, polymers, plastic, fibers, whatever the final product. Furthermore, many of these products are recyclable. Thus, it is a far better option with significantly less CO_2 per unit of coal utilization.

Consultant, PE, Member NAE, AIChE Fellow, UOP/Honeywell Fellow Retired, Adjunct Professor, Chemical and Biological Engineering, IIT, Chicago; Naperville, Illinois.

Table 1. Percent of total electric power from coal in 2011 and 2018.

Country	% of electricity generated by coal—2011	% of electricity generated by coal—2018
South Africa	88	88
Poland	80	79
India	75	75
China	68	67
Australia	61	60
Germany	37	35
Japan	34	33
USA	31	28

2. Fossil Fuel Resources

Before we discuss chemicals from coal, it is important to look at resources, particularly world oil, gas and coal reserves. Figure 1 compares the geographic distribution of recoverable oil reserves in 1997 and as estimated in 2017 (BP, 2018). There has been significant consumption of crude oil between 1997 and 2017, however, the estimated recoverable crude oil reserves in 2017 are actually 45% greater than that of the 1997 estimate. In other words, we have been discovering more oil reserves as well as increasing the estimate of recoverable oil, due to technological advancements, than we are consuming. During the same period, the share of Middle East reserves declined from 59% to 48%. As shown in Figure 2, a similar effect is also true for the natural gas reserves. However, in this case, the Middle East has slightly increased its share.

Figure 2 shows the geographic distribution of natural gas reserves (BP, 2018). Total gas reserves increased from 128 trillion cubic meter in 1997 to 193 trillion cubic meter in 2017.

Figure 3 shows the geographic distribution of coal reserves in 2017 (BP, 2018). Total coal reserves in 2017 are estimated at 1035 billion tons, a net decline of 6.5% from the 1997 estimate of 1106 billion tons. However, in terms of barrels of oil equivalent (BOE) energy, this is 25% greater than the estimated combined oil and gas reserves in 2017. A further point is that there is a geographic mismatch between areas of oil and gas reserves and areas of high demand, namely, North America, Japan, Western Europe and the emerging high demand areas of Asia. On the other hand, coal reserves are advantageously located in these high energy demand areas. As a result, coal can provide security of raw materials for the energy demand of these nations.

Over the last two decades, coal prices have ranged from 30 to 110 U.S. dollars per metric ton (BP, 2018). One ton of coal generates roughly 27 million BTUs (mmBTU) of energy. In terms of energy value, coal at $50 per ton is equivalent to $1.90 per mmBTU. This is approximately ½ of the cost of natural gas in the United States during the fourth quarter of 2018. That is, in North America, the cost of coal in terms of its energy content is ½ of the cost of natural gas. However, natural gas is much more expensive than coal for the countries where natural gas is imported. For example, in India and China coal is priced at 70 to $100/MT, that is, about $2.70 to $3.80 per mmBTU, while natural gas is valued at $7 to $12 per mmBTU.

Though coal has been and continues to be a major fossil fuel source, its utilization presents significantly greater environmental challenges than the use of oil or gas. Coal combustion produces higher levels of sulfur dioxide, nitrogen oxides and particulates. The presence of mercury, arsenic, lead and other heavy metals in coal is also of concern. These are critical factors that must be considered when looking forward to future uses of coal.

The key to the utilization of coal reserves will be to use clean coal-burning technology and develop efficient processes for coal to chemicals. When these problems are solved, coal can again play a major

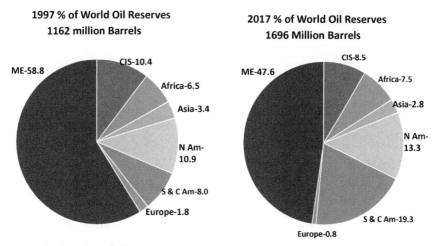

Figure 1. Distribution of world oil reserves, % (in this nomenclature, one billion denotes 10^9 and one trillion 10^{12}).

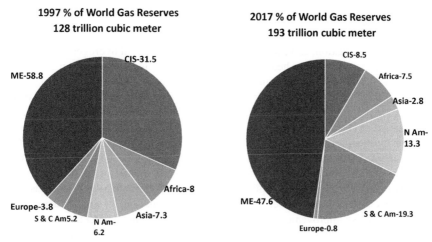

Figure 2. Estimated natural gas reserves, %.

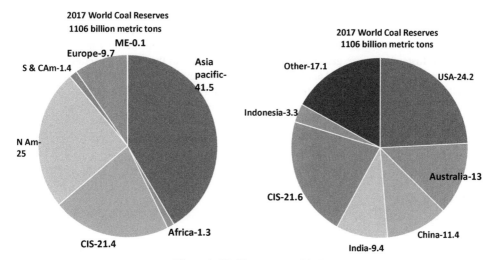

Figure 3. World gas reserves 2017.

Table 2. Coal price, $/Mt.

Year	USA	NW Europe	Japan	China
2000	30	36	36	28
2005	65	72	74	43
2010	68	92	108	110
2015	52	57	60	67
2018	73	92	113	99

role in the manufacture of chemicals. China, with its large coal reserves, made a national policy and started several projects based on coal.

Coal tar and coal oven-based chemicals, such as anthracene to carbon black and naphthalene to phthalic anhydride, have traditionally been produced, along with acetylene and its derivatives. China has exhibited continued interest and efforts for coal to chemicals, calcium carbide, VCM, PVC, ammonia and urea. More than 60% of Chinese vinyl chloride monomer (VCM) capacity is based on coal derived acetylene. All Chinese coal mine operators have long-term planning for coal to chemicals strategy.

2.1 Coal gasification and synthesis gas production

In conventional coal combustion, complete oxidation of the carbon and hydrogen content of the coal to CO_2 and H_2O is a primary goal. The heat of combustion is used to generate steam and power. On the other hand, coal gasification is a partial combustion, in which the amount of oxygen fed to the reactor is controlled in order to yield a fuel gas mixture of hydrogen and carbon monoxide $CH_4 + 0.5O_2 = CO + 2H_2$. The gasification product gas produced is called raw synthesis gas, and has considerable BTU value. The synthesis gas, after clean-up, can be more efficiently and cleanly burned in a downstream process. Alternatively, the cleaned-up synthesis gas can be used to manufacture a number of different chemicals.

One of the more significant developments in coal utilization is the cogeneration of "clean fuels", where the heat of the reaction and energy content of the waste streams are converted into electricity, with the other gasification product being liquid fuels or chemicals from fossil fuels. This is more energy-efficient than producing electricity alone. It also reduces the emissions of greenhouse gases and other pollutants. Thanks to these improvements, coal gasification is receiving greater attention. The integrated gasification combined-cycle (IGCC) system is already playing an important role in power generation.

Carapellucci et al. (2001) have investigated the performance of an IGCC power plant combining electric power generation with methanol synthesis. Figure 4 shows a schematic flow diagram of such a system. In this scheme, synthesis gas from coal gasification is used for both methanol production and power generation. In a stand-alone methanol plant, a large recycle is required in order to maximize utilization of the synthesis gas. In an integrated operation, as shown in the flow diagram, the recycle can be reduced. Instead, a purge gas stream can be fully utilized in the power generation section. This simplifies the methanol synthesis and also achieves greater energy efficiency. A similar integration can be made for DME production or for liquid fuels production with a Fischer-Tropsch unit.

The chemical reactions of gasification which uses coal are as follows:

$C + 1/2\ O_2 \rightarrow CO$

$CO + 1/2 O_2 \rightarrow CO_2$

$H_2 + \frac{1}{2} O_2 \rightarrow H_2O$

$CO + H_2O \leftrightarrow CO_2 + H_2$

$C + H_2O \rightarrow CO + H_2$

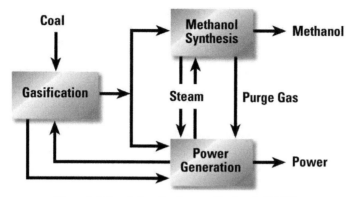

Figure 4. Integrated gasification combined cycle (IGCC).

$$C + CO_2 \rightarrow 2CO$$

$$C + 2H_2 \rightarrow CH_4$$

$$CH_4 + H_2O \rightarrow CO_2 + 3H_2$$

There are several known technologies for coal gasification, the "Texaco" gasifier (now owned by GE), Shell Global gasifier, British Gas/Lurgi gasifier, KRW gasifier and IGT U-Gas gasifier. The Texaco, Shell and Lurgi gasifiers account for the majority of gasification units worldwide. For coal to chemicals, the gasifiers are designed and operated for maximum synthesis gas production as it is a key intermediate for coal to chemicals. Lee (1997) has given a detailed review of synthesis gas technology and its uses. Synthesis gas from coal gasifiers requires significant clean-up to remove particulates, carbon dioxide and sulfur oxides. There are several well-established processes for the synthesis gas clean-up (HP, 2012), such as the UOP Benfield™ Process and the UOP Selexol™ Process. The UOP Polybed™ PSA and/or Polysep™ membrane processes may also be used for the production of hydrogen or adjusting the CO-H$_2$ ratio for downstream process applications. After clean up, the synthesis gas can either be used for power generation or for the production of chemical products, such as ammonia, methanol, DME, liquid fuels, etc.

Conversion of coal to liquids (CTL) was widely practiced in smaller capacity units in Germany during World War II. In this process, coal is converted to synthesis gas, followed by Fischer-Tropsch synthesis to liquid hydrocarbons. According to Jager (1997), several units were built in South Africa during the 1960s and 1970s, and are currently operating with a capacity of about 150,000 barrels (One barrel is equal to 42 U.S. gallons) per day.

A similar technology to CTL is the conversion of natural gas to liquid fuels ("gas to liquids," or GTL). This also involves the production of fuel from synthesis gas, though, in this case, the source of the synthesis gas is partial combustion of natural gas. Typically, the gas comes from "stranded gas" locations, where it cannot be easily utilized or transported by conventional methods. Estimates of known natural gas reserves are increasing as the rate of new discovery of unconventional gas reserves increases. Fuels and Lubes Weekly in 1993 reported that Shell began operation of a 12500 BPD gas-to-liquids plant at Bintulu in Malaysia (Fuel and Lubes, 2012). A Sasol-Chevron project of 35,000 barrels per day which began operation at the end of 2006 in Qatar was mentioned by the Catalyst Group (2004) and Shell began operation of a large GTL plant of 140,000 barrels per day in Qatar in 2011, which was reported by Independent Chemicals Information Service (ICIS, 2012).

3. Synthesis Gas Utilization

As shown in Table 3 (Khan, 2018) synthesis gas can be produced from a variety of feed sources, with coal accounting for 48.3%. Once the synthesis gas is produced, there are several downstream processes for synthesis gas to chemicals, fuels and power.

The following discussion on conversion of synthesis gas to chemicals applies to monetization of coal as well as remote natural gas. Table 4 shows the 2018 applications (Khan, 2018) of synthesis gas.

As seen in Figure 5 (Vora, 2015), once coal or natural gas is converted to synthesis gas, it opens up a number of options for making different products. Lee (1997) provides an excellent review of various processes for methane derivatives via synthesis gas. Chang (1984) provides a good review of chemicals from methanol. The question becomes: Producing which product from which feedstock is most economical? For most petrochemical processes, raw materials account for 60–70% of the cost of production. Therefore, a first analysis requires a look at the differential between the product value and the feed cost. Table 5 shows this differential using 2018 average product values at various natural gas as feed prices. One can do a similar analysis for coal. This shows the production of olefins, ethylene and propylene via methanol has the highest differential, making methanol an important intermediate. Coal at 100$/MT is equivalent to 3.80$/mmBTU.

Table 3. Feed sources for synthesis gas production.

Feed source	% of synthesis gas production
Coal	48.3
Natural Gas	46.5
Biomass/waste	3.9
Petroleum	1

Table 4. Applications for synthesis gas production in 2018.

Application	% of synthesis gas used
Chemicals	74.3
Liquid Fuels	11.6
Power	3.3
Gaseous Fuels	10.6

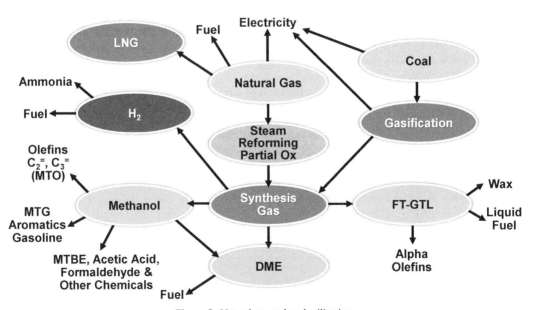

Figure 5. Natural gas and coal utilization.

Table 5. Product-feed price differential, 2018 product value at different NG price.

Product	Product value, $/MT	Approx. tons of NG per ton of product	Differential at NG 200$/MT, = $4 mmBTU	Differential at NG 400$/MT, = $8/mmBTU	Differential at NG 800$/MT, = $16/mmBTU
Gasoline, MTG	800	1.46	508	216	−368
Diesel, FT	900	1.4	620	340	−220
Ethylene, Propylene; MTO	1250	1.46	958	666	82
Propylene, MTP	1300	2	900	500	−300

Assumption: NG to Synthesis Gas efficiency 80%
Synthesis gas to Methanol-95%
MTG-90%, FT-90%, MTO-90%, MTP-65%

4. Methanol

From the previous discussion, it is clear that synthesis gas can play an important role in the utilization of coal or remote natural gas reserves. Since methanol is a key intermediate in this conversion, it is important to discuss developments in methanol markets and technology. The world methanol demand balance for its different uses is shown in Table 6. From 1995 to 2010 methanol production grew at an annual rate of 4.7%. The recent new application of methanol for the production of ethylene and propylene has given a further boost to methanol production, with 2018 total methanol production reaching 90 million MTA. Alvarado (2016) compared methanol uses in 2010 and 2015, as shown in Figure 6.

An interesting point to note is that in 2010, the use of methanol for the production of ethylene/propylene via MTO/MTP was negligible and is not represented on the chart. However, in 2015 at 18% it is the second largest application after formaldehyde.

There are several methanol technology suppliers. Lurgi GmbH, Davy and Haldor Topsøe are some of the main licensors (HP, 1993). Until 2000, a typical large methanol plant capacity was 2500 metric tons per day (MT/D). Some trends in methanol synthesis technology are particularly important for the production of light olefins from gas. First, plant capacity is increasing significantly, as exemplified by several mega-scale plants (~ 5000 MT/D) that came into operation during the 2010s, as reported by Bonarius (2005). Second, lower feedstock costs in specific geographic areas are having a major impact on methanol production economics. Technology for the production of methanol from synthesis gas is available from several licensors (Chem System, 2012), as seen in Table 7.

The early development in methanol technology is credited to Imperial Chemical Industries Ltd. (ICI). ICI first introduced the Low-Pressure Methanol (LPM) Process in 1966. In 1994, ICI Katalco introduced the Leading-Concept Methanol (LCM) Process. Later, this became part of Johnson Matthey.

The overall reaction from methane to synthesis gas to methanol can be summarized as:

$$CH_4 + \tfrac{1}{2} O_2 = CO + 2H_2 \rightarrow CH_3OH$$

Table 6. Products from methanol in 1995 and 2010.

Product, 1000 metric tons per annum (MTA)	1995	2010
Formaldehyde	7670	14880
Acetic acid	1730	4800
MTBE	8030	5280
Ethylene-Propylene	0	1920
DME, Gasoline blending	480	9600
Other	6590	12520
Total	24500	49000

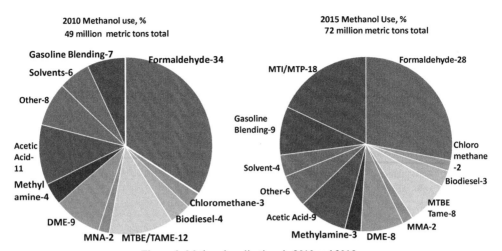

Figure 6. Methanol applications in 2010 and 2015.

Table 7. Major methanol technology licensors.

Methanol technology licensor	% Share by no. of plants	% Share by capacity
Davy Process and Johnson Matthey (JM)	24	30
JM/Uhde	5	2
JM/Jacobs	8	7
Lurgi	25	31
Mitsubishi Gas Chemical	12	18
Haldor Topsøe	15	3
JM/Toyo	3	4
Other	8	5

Synthesis gas is processed over a fixed bed of catalyst forming methanol and water. Two reactor types are most popular: An adiabatic reactor with multiple quenches of a cold stream (ICI system) or a multi-tubular reactor with internal heat exchange (Lurgi system). Both types are operated at a temperature range of 200–280 °C and low pressure of 5–7 MPa using $Cu/ZnO/Al_2O_3$ catalyst. More details are given by Lee (1990). Typical methanol properties and specifications are shown in Table 8.

4.1 Methanol production economics

Capital investment costs and the feedstock costs vary significantly for different geographic areas. In some parts of the Middle East, the natural gas price in 2018 was 1.00–1.50$/mmBTU. In the USA it ranged between $2.50 to $4 per mmBTU. It was over $8 to $12 per mmBTU in China, Japan, India and other Asian countries where LNG is imported.

During the 1980s–1990s, high feed cost units in North America and Western Europe led to significant capacity shut downs. By 1990, all production in Japan was shut down (Chem System, 2012). Subsequently, almost all new methanol units were located where natural gas was relatively low in cost, typically in the Middle East and South America. This development led to a dramatic change in the methanol industry. Since 2010, with the development of shale gas, methanol production in North America is reviving again. At the same time, China has been aggressively moving into chemicals from coal, where methanol is a key intermediate. This led to significant coal-based production of methanol in China. As was seen earlier, in Table 2, from 2000 to 2018 the coal price has ranged between 30 and 110 dollars per ton (BP, 2019). Table 9 shows the cost of methanol production for a unit producing 5000 metric tons of methanol per

Table 8. Methanol properties and specifications.

Properties	Value
Formula	CH_3OH
Molecular Weight	32.04
Specific Gravity	0.7924 g/cc
Viscosity at 20 C	0.00592 poise
Vapor Pressure at 20 C	92 mmHg
Freezing Point	−97.8 °C
Boiling point	64.7 °C
Methanol specifications	**Grade AA**
Methanol min wt%	99.85
Acetone max wt%	0.002
Aldehyde max wt%	0.001
Ethanol max wt%	0.001
Acidity (CH_3COOH) max wt%	0.003
Appearance	Free of opalescence, suspended matter and sediment
Carbonisable substances	Not darker than color standard No. 30 of ASTM D1209 Pt/Co scale
Color	Not darker than color standard No. 5 of ASTM D1209 Pt/Co scale
Permanganate Fading Time, Minutes	30
Water max wt%	0.10

Table 9. Methanol production economics in 2018.

Feedstock	Coal China	Coal USA	Natural gas China	Natural gas USA	Natural gas ME
Feedstock Cost	100 $/MT	70 $/MT	10 $/mmBTU	3.5 $/mmBTU	2 $/mmBTU
Total Capital Investment, $Million	1240	1500	930	1160	1260
Operating Cost, $/MT					
Raw Material	122.82	99.78	390	140	86
Utilities	147.94	51.53	8	5	10
Fixed Cost	38.09	41.93	26	30	33
Cash Cost of Production	173.47	193.24	424	175	79
Depreciation	60.61	74.40	43	58	63
Return on Capital	74.36	89.76	56	70	76
Total Cost of Production	443.82	357.40	523	303	267

day, based on 2018 coal prices in China and the USA and 2018 natural gas prices in China, Middle East and the USA.

4.2 Methanol derivatives

4.2.1 Acetic acid

Acetic acid is a colorless liquid with a strong and pungent smell. It is also known as ethanoic acid or methanecarboxylic acid. Acetic acid can also be produced by bacterial fermentation but this route accounts for only about 10% of world production. Production of acetic acid employing biological processes has been known for over 10,000 years, as long as wine making has been practiced. Vinegar is an aqueous solution of acetic acid. Several aerobic and anaerobic processes have been practiced (Partin, 1993). For the synthetic production of acetic acid, there are three main methods: Acetaldehyde oxidation (Fanning, 1993), hydrocarbon oxidation (Irick, 1993), and methanol carbonylation (Zoeller, 1993). Of the three, due to favorable process economics, approximately 75% of the acetic acid is produced synthetically by carbonylation of methanol.

$$CH_3OH + CO \rightarrow CH_3COOH$$

Paulik et al. (1968) at Monsanto disclosed a rhodium iodide catalyst system for methanol carbonylation that operates at a pressure of 30 to 60 atmospheres and temperature of 150–200 °C. This homogeneous liquid phase process employs an organometallic rhodium iodide complex as a catalyst. The process gives a selectivity of greater than 99% to acetic acid. The technology, known as the Monsanto Acetic Acid Process, has been the basis of all new acetic acid production worldwide. Several publications and patents describe the mechanism of the rhodium-catalyzed carbonylation reaction (Roth 1971), as well as a detailed process description, including a schematic plant design (Eby, 1983). Celanese practices a similar technology with some proprietary modifications. Monsanto granted rights to British Petroleum (BP). In 1996 BP introduced an iridium catalyst system for use in the BP Cativa Acetic Acid Process (Jones, 2000).

The main applications for acetic acid include the production of chemical compounds such as, vinyl acetate monomer (VAM), purified terephthalic acid (PTA), acetate esters, acetic anhydride, and so on. Products of commercial importance made from acetic acid are latex emulsion resins for paints, adhesives, paper coatings, textile finishing agents, cellulose acetate fibers, cigarette filter tow, and cellulosic plastics.

The significant world acetic acid market growth is due to its extensive application in the production of vinyl acetate monomer (VAM) and PTA. VAM, used in paints and adhesives, consumes approximately 40–45% of the world's acetic acid production. PTA, on the other hand, uses acetic acid as a solvent and catalyst carrying agent in the oxidation of p-xylene to produce PTA. In the process some of the acetic acid is "burned" to CO_2 and H_2O. The PTA is then used to manufacture polyethylene terephthalate (PET) fibers, resins and films, all of which have very high growth rates in consumer goods such as clothing, beverage containers and food packaging. Besides this, acetic acid is also used in the food industry and in household items as a preservative. Thus, the high demand for acetic acid in VAM and PTA production, coupled with food industry applications, would bolster the acetic acid market growth. In terms of revenue in 2014, the global value of the acetic acid Market was calculated to be 9.1 billion U.S. dollars, and is projected to reach 13.8 billion U.S. dollars by 2022. In terms of volume, the market demand in 2014 for acetic acid was 12.1 million metric tons and is projected to reach 16.8 million tons by 2022. Companies such as DuPont, (BP), Celanese, and Eastman Chemicals dominate the market.

4.2.2 Formaldehyde

Formaldehyde is an intermediate, used in the manufacture of a wide range of products. More than 60% of it is used in the production of resins, such as urea-formaldehyde, phenol-formaldehyde, and melamine-formaldehyde. Other applications include 1,4-butanediol, and polyacetal resins. In 2004, formaldehyde was classified as a carcinogen to humans by the International Agency for Cancer Research and, as such, it is a highly regulated material. In 2011, in the USA, the National Toxicology Program (NTP) reclassified formaldehyde as carcinogenic to humans (NTP, 2011). Formaldehyde is highly soluble in water and is sold as a 37% solution in water, with up to 16% methanol. For higher concentrations, solution stabilizers are required. Formaldehyde is produced by partial oxidation and dehydrogenation of methanol using either silver catalysts (Reuss, 2002) or molybdenum oxide catalysts.

$$CH_3OH \rightarrow HCHO + H_2O \qquad\qquad H = +2,400 \text{ Kcal/Kmol}$$

$$CH_3OH + \tfrac{1}{2}O_2 \rightarrow HCHO + H_2O \qquad\qquad \Delta H = -37,400 \text{ Kcal/Kmol}$$

There are several formaldehyde technology licensors. Some licensors prefer to operate at 75–85% methanol conversion with recovery and recycle of unconverted methanol. Others operate near 92–95%

conversion with no methanol recycle. BASF is one of the largest producers of formaldehyde with silver catalysts, but it does not offer the technology for license. The following companies are known to license formaldehyde technology based on silver catalysts: Dyno Industries of Norway, Karl Fischer and Josef Meissner of Germany, Mitsubishi Gas of Japan, ENI (Montedison) of Italy and DB Western and Monsanto of United States.

Another technology for formaldehyde production is based on the vapor phase oxidation of methanol using metal oxide catalysts: Iron/molybdenum oxide with small amount of cobalt, phosphorus, chromium, vanadium and copper oxides (Klissurski, 1991). These technologies are licensed by Axens (France), Dyno Industries (Norway), Haldor Topsøe (Denmark), Nippon Kasie (Japan), Lummus-CBI (United States), and Joseph Meissner and Karl Fischer (Germany) (PERP, 1996).

Despite the health concerns, the demand for formaldehyde is continuously rising due to its increasing use in the production of various resins for manufacturing purposes. Formaldehyde is also being used in the production of home building products and is known for its preservative and anti-bacterial properties. Hence, medical laboratories and some consumer products use formaldehyde as a preservative. The global consumption of formaldehyde is increasing as it is being used on a large scale for construction and remodeling activity and furniture production. Due to the excellent thermal and chemical resistance, formaldehyde-based resins are being used in manufacturing airplane and automobile parts. Formaldehyde is also being used in manufacturing anti-infective drugs, hard-gel capsules, and vaccines.

Some of the leading companies in the global formaldehyde market are Johnson Matthey Process Technologies, Foremark Performance Chemicals, Huntsman International LLC, Momentive Specialty Chemical Inc., Dynea AS, Alder S.p.A, Georgia-Pacific Chemicals LLC, Perstorp Orgnr, Celanese AG, and BASF SE. The global formaldehyde market is also expected to reach 36.6 million tons towards the end of 2026. The global formaldehyde market has been segmented into applications. Urea Formaldehyde (UF) resins and concentrates are likely to witness the highest growth in terms of volume throughout the forecast period from 2017 to 2026. Asia Pacific Excluding Japan (APEJ) is expected to dominate the global formaldehyde market. Emerging economies, such as China and India, are witnessing a rapid increase in the demand for formaldehyde for use in various industries. Formaldehyde is the most commercially important aldehyde. Production of urea, phenol-, and melamine-formaldehyde resins (UF, PF, and MF resins) accounted for nearly 70% of world consumption of formaldehyde in 2017; other large applications include polyacetal resins, pentaerythritol, methylenebis(4-phenyl isocyanate) (MDI), 1,4-butanediol (BDO), and hexamethylenetetramine (HMTA).

4.2.3 Olefins

Conversion of methanol to olefins could potentially play a large role in increasing methanol demand. Figure 7 shows China's 10-year plan for the use of coal for chemicals, presented by Yajun from National Institute of Clean-and-low-carbon Energy, Beijing in 2012 at Woodrow Wilson International Centre for Scholars. In 2010, of the total 670 million tons (MT) of coal consumed, 12% was used for the production of FT-liquids, olefins, ammonia, and methanol/DME. By 2020, coal consumption is expected to increase to 1,600 million tons, and about 28% of it will go toward the products mentioned above. The bulk of the increase is in the utilization of coal for the production of FT Liquids (5.1%), for the production of olefins (7.3%) and for DME 5.1%. In addition, 11.3% will go to the production of SNG. During the same period, the share of conventional electric power and industrial use will decline from 79.6% to 51.2%. Coal is emerging as a feedstock for new large scale methanol plants in China, and some of these plants are linked to the production of ethylene and propylene (Gregor, 2012; Hang, 2012).

Before we discuss ethylene and propylene production from methanol, we must understand the current technologies and the market. Ethylene and propylene are the two largest volume chemicals produced for the petrochemical industry, with 2018 production at 160 and 92 million metric tons, respectively. This represents an annual product value of about US$250 billion. Light olefin demand is primarily driven by polyolefin production, but other olefin derivatives, such as ethylene oxide, ethylene dichloride, propylene oxide, acrylonitrile and others, consume about 40% of the light olefins produced today. The majority of the light olefins used for petrochemical applications are produced by the steam cracking of ethane, naphtha or other gas liquids, as shown in Table 10 (Vora, 2015).

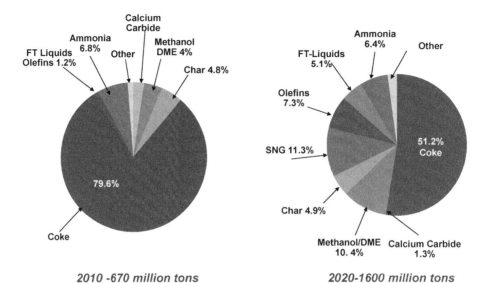

Source: Presented by Dr. Yajun Tian, National Institute of Clean-and-low-carbon Energy, Beijing, China;
at Woodrow Wilson International Center for Scholars, July 2012

Figure 7. China coal utilization planning.

Table 10. 2012 Light olefin production sources.

Production sources	**Ethylene**	**Propylene**
Ethane Cracking	35%	---
Propane Cracking	9%	58%
Butane Cracking	4%	
Naphtha Cracking	47%	
Fuel Oil Cracking	3%	
Refineries (recovered from FCC units)	---	32%
Propane Dehydrogenation (PDH)	---	5%
Others	2%	5%

The main factor in olefin production economics is the cost of feedstock, so locations for new capacity are strongly influenced by the availability of cost-advantaged feedstocks. This is evident in the large capacity build-up of ethane-based ethylene production in the Middle East since 1990. Prices for ethane in the Middle East are especially low because there are large amounts of ethane produced in association with crude oil production and countries provide incentives for ethane utilization. Availability of ethane has also increased with the discovery and production of shale gas in North America. It is seen from the data in Table 11 that the use of naphtha as a feedstock for ethylene production as a percent of total production is declining with some gains in use of LPG, ethane and the new entry of coal to olefins (CTO) or methanol to olefins (MTO). As of 2019, all CTO/MTO-based ethylene and propylene production takes place in China. The ethane growth rate is very significant.

During the 1960s, ethylene was also produced by dehydrating ethanol, but with advances in steam cracking and the availability of naphtha and light hydrocarbons, this route is economically no longer favored. Economic analyses done by consulting firms have shown that Middle East ethane crackers and the remote gas MTO have the lowest cash cost of ethylene production, followed by North American ethane crackers, based on ethane recovered from shale gas. Figure 8 shows the cash cost of ethylene production according to Chemical Market Resources, Inc (CMR, 2013). The cash cost of production is

Table 11. Ethylene production sources; mmMTA (% of total).

Feed source	2000	2010	2015	2020 estimate
Naphtha	52 (58)	58 (51)	60 (43)	68 (40)
Ethane	28 (31)	40 (34.)	50 (36)	65 (38)
LPG	10 (11)	17 (15)	21 (15)	22 (13)
MTO/CTO	0	0 (0)	9 (6)	15 (9)
Total	90 (100)	117 (100)	140 (100)	170 (100)

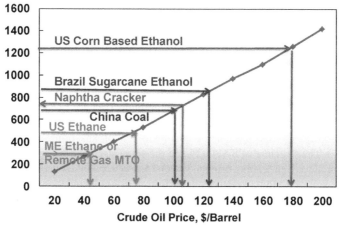

Source: Chemical Market Resources, Inc July 22nd, 2013 Volume 7 Issue 15

Figure 8. Cost of ethylene production, $/MT.

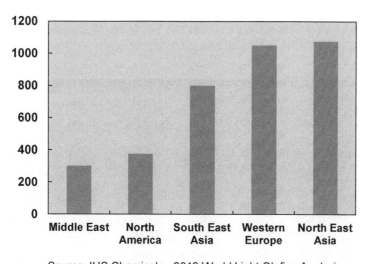

Source: IHS Chemicals - 2013 World Light Olefins Analysis

Figure 9. Cash cost of ethylene, $/MT.

similar for China's coal and naphtha cracker-based ethylene production. It is also seen that the production of ethylene from sugarcane or corn-derived ethanol in Brazil and the USA is not economical. A similar analysis done by IHS-CMAI for the cash cost of production of existing capacity on a geographic basis is shown in Figure 9. It is seen that ethylene produced from lower cost ethane gives the Middle East a significant advantage in cash costs, followed by North America, due to shale gas discoveries that have lowered the ethane and LPG prices.

When it comes to propylene, in addition to naphtha crackers, where propylene is produced in significant quantities along with ethylene, the refinery FCC units also play an important role, supplying nearly 30% of the demand in 2012. Because of the increasing use of ethane in place of naphtha for the production of ethylene, the combined production of propylene from naphtha crackers and FCC units falls short of meeting the propylene demand. Therefore, since 1990, alternate sources for propylene, such as propane dehydrogenation and metathesis, have emerged to meet the propylene supply gap (Vora, 2012).

5. Coal-to-Methanol-to-Olefins: Processes and Catalysts

Methanol-to-hydrocarbon conversion reactions were first discovered in the early 1970s using ZSM-5 (MFI) catalysts (Chang, 1977; Chang, 1994). Mobil commercialized a methanol-to-gasoline process in New Zealand in the 1980s and also developed methanol to olefins employing a ZSM-5 catalyst. With support of DOE funding, a 100-barrels per/day demonstration unit was operated in Germany (DOE, 1986). In the1980s, Edith Flanigen and her team (Wilson, 1982) at Union Carbide Corporation (UCC) discovered a new class of material, called silicoaluminumphosphate (SAPO). A particular structure with 3.8 Å pore opening known as SAPO-34, a silicon-aluminum-phosphorous-based molecular sieve, showed excellent properties for conversion of methanol to light olefins, primarily ethylene and propylene (Kaiser, 1985, 1987). The structure of SAPO-34 and the small sizes of pore openings are the keys to the high selectivity to produce light olefins using a SAPO-34 catalyst.

The small pore size of SAPO-34 restricts the diffusion of heavy and branched hydrocarbons, which leads to high selectivity to the desired olefins. On the other hand, ZSM-5 molecular sieves produce much lower light olefin yields, primarily due to larger pore openings (about 5.5 Å) in the MFI structure. Figure 9 shows a comparison of ZSM-5 and SAPO-34 structures (Vora, 2003). A further advantage of SAPO-34 is that the majority of the C_4-C_6 fraction is olefinic. This C_4-C_6 fraction can be converted to light olefins and, thus, increase the production of C_2 plus C_3 olefins to near 90%. This reduces the net purge of C_4-C_6 fraction to about 5% of the carbon yield. Reaction product distributions for methanol processed over the ZSM-5 and SAPO-34 catalysts are compared in Figure 11 (Vora, 2006).

A number of technologies based on the use of ZSM-5 or SAPO-34 as a catalyst have been developed. These are the UOP/HYDRO MTO™ ("Methanol to Olefins") Process, which employs a catalyst based on

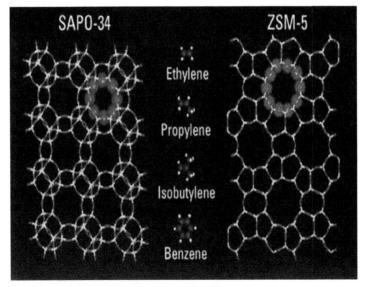

Figure 10. Framework of SAPO-34 and ZSM-5 molecular sieves.

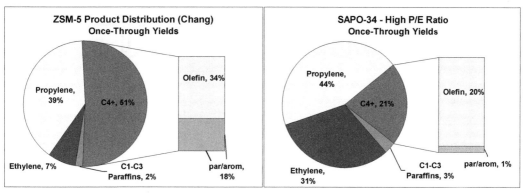

Figure 11. Once-through hydrocarbon yields for ZSM-5 and SAPO-34.

SAPO-34 material (Vora, 1998; Chen, 2004), and the Lurgi MTP™ ("Methanol to Propylene") Process, based on a ZSM-5 type catalyst (Gronemann, 2005). Similar technologies also have been developed by Dalian and Sinopec in China (Ying, 2013). In China, one commercial unit of each of these technologies came into commercial operation during 2011, and by 2018 there were 7 or more units in operation. There are more than 10 units in design and construction and these are expected to be in operation by 2022. Both Chinese technologies use catalysts containing SAPO-34 in a fluidized bed reactor with continuous catalyst circulation and regeneration. In addition, there are two methanol to propylene units licensed by Lurgi in operation in China. The MTO Process mainly produces ethylene and propylene and some C_4 olefins, while the MTP process mainly produces propylene with gasoline range C_5-plus hydrocarbon byproduct. Honeywell's UOP has announced the licensing of eleven MTO units in China. The first unit at Wison, Nanjing, successfully came onstream during the 2013. By 2019, five UOP licensed units were in operation (Funk, 2014; Senetar, 2019).

Because ZSM-5 catalysts allow larger molecular weight gasoline range materials to come out of the pores, there is less formation of coke on the catalyst relative to SAPO-34 which only allows n-butene and lower molecular weight hydrocarbons. Thus, for the ZSM-5 based catalyst system, it is feasible to design a fixed bed reactor system with cyclic regeneration, which is used in the Lurgi MTP design. On the other hand, SAPO-34-based catalyst systems employ a circulating fluidized bed reactor and regenerator similar to that used in the fluid catalytic cracking process (FCC) in petroleum refining.

5.1 Lurgi MTP process

The Lurgi MTP process uses a fixed-bed ZSM-5 catalyst manufactured by Süd-Chemie AG. It provides high propylene selectivity, low coking tendency, low propane yield and limited byproduct formation. An MTP plant with a methanol feed rate of 5000 MT/day produces 1410 MT/day propylene, 540 MT/day gasoline, and 109 MT/day LPG. Up to 60 MT/day of ethylene can optionally be recovered from the purge gas or used as fuel. Methanol, both fresh and recycled (as recovered from aqueous streams), is the feed to the MTP unit. Figure 12 (Lurgi, 2003; Wurzel, 2006) shows a schematic process flow diagram. The methanol is vaporized, superheated, and fed to a DME reactor. The DME reactor is a single-stage adiabatic reactor where most of the methanol is converted to dimethyl ether (DME) on an alumina catalyst. The reaction is exothermic and closely approaches thermodynamic equilibrium.

The product of the DME reactor is sent to three MTP reactors in parallel: Two of the reactors are in operation, while a third one is in regeneration or on stand-by. For the purposes of reaction control, each MTP reactor features six zeolite-based catalyst beds, over which the methanol/DME mixture is converted to a mixture of olefins, typically from ethylene to octenes, but such that the carbon distribution peaks at propylene. The operating temperature is about 450 °C and the operating pressure is 0.15 MPa (about 20 psia). Side products from the reaction include naphthenes, paraffins, aromatics, and light ends. The oxygen chemically bound in the methanol results in process water.

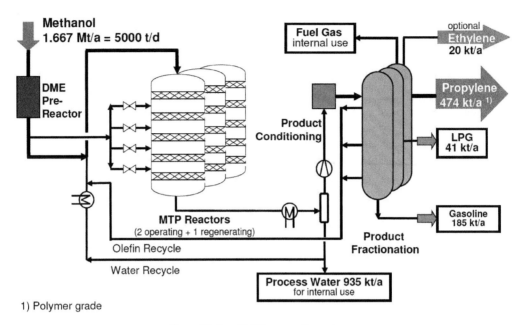

Figure 12. Lurgi MTP process flow scheme.

The regeneration of the MTP reactors is performed *in situ* by the controlled combustion of coke with an air/nitrogen mixture at temperatures similar to the normal reactor operating temperature. The MTP reactor effluent is cooled in a heat recovery system and, finally, through a quench section, in which the hydrocarbons are separated from the bulk of the water. The water is condensed and sent to the methanol and DME recovery column, from which they are recycled to the DME reactor. The water with traces of oxygenates is routed to battery limits.

The hydrocarbon vapor from the quench section is compressed to about 2.5 MPa (365 psia) by a multistage centrifugal compressor with intercoolers and partial condensers. The liquid and vapor hydrocarbons are sent to the purification section. The hydrocarbon streams are first dried by using molecular sieves before the hydrocarbon liquid is fed to a debutanizer column and the vapor is processed through a DME recovery system. The C_4+ bottom product is fed to a dehexanizer where aromatics and C_7+ are separated from the C_6-stream. The majority of the C6-fraction is sent back to the MTP reactors while the C_7+ fraction is the gasoline byproduct.

The compressed hydrocarbon vapors, including light olefins and DME, and the overhead C_4+/DME from the debutanizer are fed to a DME removal system, in which C_3-hydrocarbons are separated from C_4+ hydrocarbons and oxygenates. The methanol and DME stream are routed to the methanol recovery column for recycle to the DME reactor. The C_4 hydrocarbon fraction is recycled to the MTP reactor for further propylene production, except for a small purge that is added to the LPG byproduct stream.

The C_3-fraction is fed to the deethanizer, in which a C_2-stream is recovered as top product; one part of this stream is recycled to the MTP reactor while the rest can optionally be sent to a two-column ethylene purification unit or to fuel gas. The C_3 bottom product from the deethanizer contains about 97% propylene and 3% propane, but no methylacetylene or propadiene; it is routed through a guard bed of activated alumina and fed to the C3 splitter for the recovery of polymer-grade propylene.

5.2 UOP/HYDRO MTO process

As mentioned earlier, Edith Flanigen and her associates at UCC discovered SAPO-34 material and showed it to be a good catalyst for methanol to olefins reaction during the early 1980s. This group at UCC, then known as Catalyst, Adsorbent and Process Systems, was merged with UOP LLC, Des Plaines,

Illinois. As a result, further development for the MTO technology took place at UOP. In 1992 UOP formed a partnership with Norsk Hydro of Norway for further joint development of the technology. A fluidized bed reactor-regenerator demo for a one ton per day methanol feed was built and operated at Norsk Hydro facility in Norway for several years. At the end of 1995, a joint team of UOP and Norsk Hydro first presented their data at the World Natural Gas Symposium held in South Africa (Vora, 1997). This demonstration was not sufficient to convince potential licensees, as questions were raised regarding the quality of ethylene and propylene coming from methanol, an oxygenate feed which may result in some unknown impurities that could be detrimental to polymerization catalysts. In order to demonstrate polymer-grade propylene and ethylene from the MTO process, a partnership with Total of France was formed and a large fully integrated MTO demonstration unit was built with high purity ethylene and propylene recovery, including polymerization reactor for polyolefins. With successful demonstration by 2010, UOP licensed its first unit in China. The Total partnership was expanded to include Total's Olefins Cracking Process (OCP), a process for cracking higher C_4-C_6 olefins to propylene and ethylene. Integration of this process with MTO allowed UOP to increase ethylene-propylene yield to near 90% and the integrated process is called UOP Advanced MTO Process.

The UOP/HYDRO MTO Process can use "crude" methanol, "fuel-grade" methanol, Grade AA methanol, or even DME as feed. The choice of feedstock generally depends on project-specific situations. Figure 13 illustrates a simple flow diagram for the UOP/HYDRO MTO Process (Vora, 1998). The MTO process utilizes a circulating fluidized bed reactor that offers a number of advantages over both fixed bed reactors and other types of fluidized bed reactors. The circulating fluidized bed reactor provides better mass transfer than bubbling bed fluidized bed reactors as well as better temperature control than riser and fixed bed reactors, especially given the highly exothermic nature of the methanol-to-olefins reactions. This type of reactor has been widely used in the Fluid Catalytic Cracking (FCC) process units in petroleum refineries.

Constant catalyst activity and product composition can be maintained via continuous regeneration of a portion of used catalyst by coke burning with air. UOP's MTO catalyst has demonstrated the required selectivity, long term stability, and attrition resistance necessary for attractive economics with low operating costs.

The overall selectivity of the UOP/HYDRO MTO process is about 75–80% to ethylene and propylene on a carbon basis, and about 15% C_4 plus hydrocarbons. The balance is C_1-C_3 paraffins plus coke on catalyst. The C_4 plus material is mostly linear butenes and some pentenes. These olefins make an ideal feed to the OCP unit to further increase the yields of ethylene and propylene. Propylene to ethylene ratios

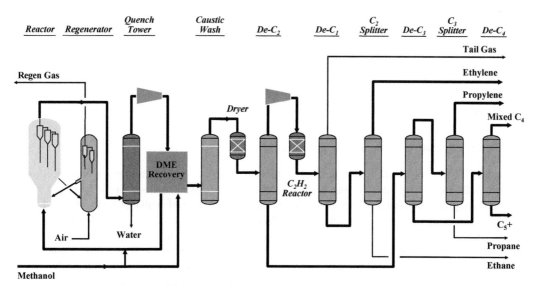

Figure 13. UOP/HYDRO MTO process flow scheme.

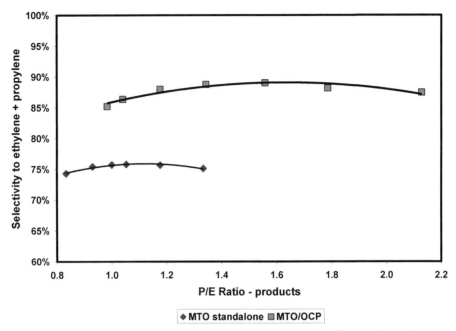

Figure 14. Olefin selectivity vs. operating severity of the UOP MTO process with and without olefin cracking process (OCP) integration.

in the product can be adjusted within the range of 0.80 to 1.33 (Figure 14) in order to reflect the relative market demand and values for ethylene and propylene. The reactor temperature is the key variable for controlling propylene to ethylene ratios, with higher temperatures leading to a higher ethylene yield. The temperature requirements have to be balanced with higher coke formation at higher temperatures.

The reactor pressure is normally dictated by mechanical considerations. Lower methanol partial pressure leads to higher selectivity to light olefins, especially ethylene. Therefore, a slight yield advantage occurs when using a crude methanol feed compared to high purity methanol. The reactor effluent is cooled and quenched to separate water from the product gas stream. The reactor provides very high conversion, so there is no need for a large recycle stream.

A small amount of unconverted oxygenates are recovered in the oxygenate recovery section, after which, the effluent is further processed in the fractionation and purification section. Conventional treating methods have been shown to be effective for removing by-products to the specification levels required for producing polymer-grade ethylene and propylene products.

As shown in Figure 14, the total ethylene plus propylene yield can be further enhanced by incorporating a cracking process to convert C_4 plus material to propylene and ethylene. Overall carbon selectivity for the integrated flow scheme approaches 90% ethylene plus propylene.

5.3 Integration of CTL/GTL with CTO/GTO

GTL and CTL processes offer large product market opportunities for natural gas and coal utilization but are challenged by high capital costs and the relatively low transportation-fuel product values. Since synthesis gas production is a common step in the manufacture of GTL and methanol, there are possibilities for integrated complexes. Figure 15 illustrates such a complex, using coal or natural gas as feedstocks, and producing both olefins and liquid fuels. Both options—coal or natural gas liquid fuels (CTL, GTL) and coal or natural gas to polymers (CTP, GTP) facilities—incorporate sizeable front-end synthesis gas units for the processing of natural gas. Over 60% of the capital cost is related to the production of synthesis gas. These units are the major contributors to the relatively high investments required for these

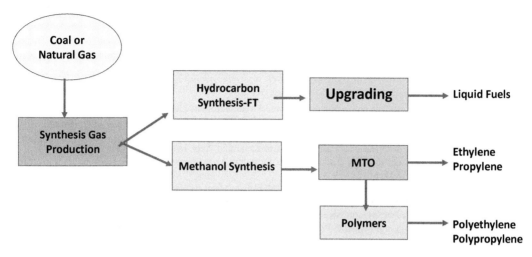

Figure 15. Integrated GTO/GTL complex for production of liquid fuels and olefins from coal or natural gas.

complexes. It follows that the integration of these facilities to co-produce fuels and chemicals could offer substantial synergistic savings.

A rough rule of thumb is that the quantity of synthesis gas required to produce 25,000 barrels per day (BPD) liquid hydrocarbons could also be used to produce in excess of one million MTA light olefins. A typical liquid fuels plant is likely to have a capacity of 100,000 BPD or more. Thus, the addition of one million MTA light olefins represents roughly 25% additional synthesis gas production capacity. The incremental production will require substantially less capital cost than a stand-alone smaller unit. The integration of methanol and liquid fuels facilities combined seen in today's oil refining sector, where there is increased focus on opportunities for petrochemicals production as some regional fuel demands change with the conversion of methanol to olefins, can provide cost saving synergies together with the production of high value-added olefins and polymer products. This would follow the current trend.

6. Conclusions

Though there are no direct routes for the conversion of coal or methane to liquid hydrocarbons or petrochemical products, these raw materials can be converted to liquid fuels or several high value-added petrochemical intermediates via synthesis gas. Examples are: Acetylene, ethylene, propylene, and methanol. These are the primary raw materials for a vast number of petrochemicals and polymer industry products, like PVC, polyethylene, polypropylene, acetic acid, acrylonitrile, formaldehyde, ammonia and many more. The vast resources of coal in areas of high demand, namely the USA, China and India, could provide long-term raw material security.

References

Alvarado, M. 2016. The Changing Face of Methanol Industry. HIS Chemical Bulletin, Issue 3.

Bonarius, J. 2005. Methanol—Too Much, Too little or Just right. Presented at CMAI World Methanol Conference, Miami, USA, December 12-14-2005.

BP Statistical Review of world Energy. 2018. https://www.bp.com/.../bp/.../energy.../statistical-review/bp-stats-review-2018-full-report....

BP Statistical Review of world Energy. 2019. https://www.bp.com/.../bp/.../energy.../statistical-review/bp-stats-review-2019-full-report....

Carapellucci, R., Cao, G. and Cocco, D. 2001. Performance of IGCC power plant integrated with methanol synthesis process. J. of Power and Energy 215(A): 347–356.

Catalyst Group Resources. 2004. GTL Technology—21st Century Advances, Spring House, PA, USA.

Chang, C.D. and Silvestri, A.J.J. 1977. The conversion of methanol and other O-compounds to hydrocarbons over zeolite catalysts. J. Catal. 47: 249–259.

Chang, C.D. 1984. Chemicals from methanol. Catal. Rev.-Sci. Eng. 26(3&4).

Chang, C.D. 1994. Chapter 4, Methanol to gasoline and olefins. *In*: Cheng, W.-H. (ed.). Methanol Production and Use. CRC Press.

Chem System PERP Report. 2012. "Methanol".

Chem Systems Formaldehyde. 1996. PERP Report No. 94/95-2, April.

CMR. 2013. Comparison of ethylene production cost. Chemical Market Resource Inc., 7(7) 22 July 2013-12-19.

DOE. 1986. Conversion of Methanol to Gasoline-Extended Project: Methanol to Olefins Demonstration Plant mile stone Report; US Department of Energy, DOE/ET/14914-H; DE86 015960, April 1986.

Eby, R.T. and Singleton, T.C. 1983. Methanol carbonylation to acetic acid. Appl. Ind. Catal. Chapter 10, 275–296.

Fanning, T. 1993. Ethylene and Acetylene based processes for Acetic Acid, Chapter 2 in Acetic Acid and its Derivatives, Agreda V., Ed. Marcel Dekker Inc.

Fuels and Lubes Weekly. 2012. Shell doubles capacity of GTL plant at Malaysia 4(43).

Gregor, J.H. 2012. Maximize Profitability and Olefin Production via UOP's Advanced MTO Technology, Presented at IHS World Methanol Conference, Madrid, Spain, Nov 27–29.

Gronemann, V. 2005. 3 in 1—Lurgi Syngas to Propylene. Presented at 2005 World Methanol Conference, Miami, FL. USA, December 12–14.

Heng, H. 2012. Focus China—Annual methanol demand to spike on MTO, MTP projects, ICIS.com, Article 9604963, 30 Oct.

HP. 1979. Hydrocarbon Processing. "Refining Processes", 71–1301.

HP. 1993. Hydrocarbon Processing. "Petrochemical Processes", March 1993 p. 70–139.

HP. 2012. Hydrocarbon Processing (HP), Gas Processes Handbook.

IAE. 2011. International Energy Agency Statistical data, https://www.iea.org/newsroom/news/2011.

ICIS. 2012. Pearl Gas to Liquid (GTL) Plant, Qatar, https://www.chemicals-technology.com/projects/pearl-gtl/.

IHS Chemicals. 2013. World Light Olefins Analysis.

Irick, G. 1993. Acetic Acid Manufacture via Hydrocarbon Oxidation, Chapter 3 Acetic Acid and its Derivatives, Marcel Dekker Inc.

Jager, B. 1997. Developments in Fischer-Tropsch technology. pp. 219–224. *In*: de Pontes, M., Espinoza, R.L., Nicolaides, C.P., Scholz, J.H. and Scurrel, M.S. (eds.). Stud. Surf. Sci. Catal. 107, Elsevier, Amsterdam.

Jones, J.H. 2000. The Cativa™ process for the manufacture of acetic acid. Platinum Metal Rev. 44(3): 94–105.

Kaiser, S.W. US Patent 4 499 327, 1985.

Kaiser, S.W. US Patent 4 524,234, 1985.

Kaiser, S.W. US Patent 4,617,242, 1987.

Khan, H.P. 2018. J. Global Synthesis Gas Overview, Stratas Advisors, A Hart Energy Company, presented at 9th China Petroleum and Chemical International Conference.

Klissurski, D. et al. 1991. Multicomponent oxide catalyst for oxidation of methanol to fomaldehyde. Appl. Catal. 77: 55–66.

Lee, S. 1990. Methanol Synthesis Technologies. CRC press.

Lee, S. 1997. Synthesis gas. Chapter 2 in Methane and Its Derivatives, Marcel Dekker Inc. New York, NY, USA.

Lee, S. 1997. Methane Derivatives via Synthesis Gas. Chapter 3 in Methane and its Derivatives, Marcel Dekker Inc, New York, NY, USA.

NTP. 2011. National Toxicology Program, Report on Carcinogens, 12th Ed. www.iaff.org/HS/PDF/12th%20Report%20on%20Carcinogens%20-%202011.pdf.

Partin, LR. and Heise, W.H. 1993. Bioderived Acetic Acid Chapter 1 in Acetic Acid and Derivatives, Agreda V., Ed. Marcel Dekker Inc.

Paulik, F.E. and Roth, J.F. 1968. Novel catalysts for the low-pressure carbonylation of methanol to acetic acid. J. Chem. Soc., Che. Commun. 1578.

Reuss, G. et al. 2002. Formaldehyde. In Ullmann's Encyclopedia of Industrial Chemistry, Wiley-VCH.

Roth, J.F., Craddock, J.H., Hershman, A. and Paulik, E. 1971. Chem. Technol. 600.

Senetar, J. 2019. Development of UOP's Advanced MTO Process—The Result of Collaborative Efforts Catalyzed by Bipin Vora; Spring 2019 AIChE Meeting, F&PD Session 105 in honor of Bipin Vora, April 1–4, New Orleans, LA.

Vora, B.V., Marker, T.L., Barger, P.T., Nilsen, H.R., Kvisle, S. and Fuglerud, T. 1997. Economic route for natural gas conversion to ethylene and propylene. pp. 87–98. *In*: de Pontes, M., Espinoza, R.L., Nicolaides, C.P., Scholz, J.H. and Scurrel, M.S. (eds.). Stud. Surf. Sci. Catal. Vol. 107, Elsevier, Amsterdam.

Vora, B.V., Eng, C. and Arnold, E. 1998. Integration of the UOP/HYDRO MTO Process into Ethylene Plants; 10th Ethylene Producers' Conference, Am. Inst. of Chemical Engineers, 8–12 March.

Vora, B.V., Pujado, P., Anderson, J. and Greer, D. 2003. Natural Gas Utilization-DME, GTL and polyolefins integration. Oil Asia Journal, September.

Vora, B.V., Bozzano, A., Foley, T. and Anderson, J. 2006. Utilize cost advantaged raw materials for light olefin production, ERTC Petrochemical Conference, February, Dusseldorf, Germany.

Vora, B.V. 2012. Development of dehydrogenation catalysts and processes. Top Catal. 55: 1297–1308.

Vora, B.V., Funk, G. and Bozzano, A. 2015. Chemicals from Natural Gas and Coal, Handbook of Petroleum Processes. 1: 883–904, Springer Reference.

Wilson, S.T., Lok, B.M. and Flenigen, E.M. 1982. US Patent 4,310,440.

Yajun, T. 2012. National Institute of Clean-and-low-carbon Energy, Beijing, Presented at Woodrow Wilson International Centre for Scholars, July.

Ying, L., Ye, M., Cheng, Y. and Li, X. 2013. A Kinetic study of methanol to Olefins Process in Fluidized Bed Reactor. The 14th International Conference on Fluidization—From Fundamentals to Products, January.

Wurzel, T. 2006. Lurgi megamethanol technology. Synthesis Gas Chemistry, DGMK Conference, October 4–6.

Zoeller, J.R. 1993. Acetic acid manufacture via Methanol carbonylation. Chapter 4 Acetic Acid, Acetic Acid and its Derivatives, Marcel Dekker Inc.

CHAPTER 16

Optimal Planning of Biomass Co-Firing Networks with Biochar-Based Carbon Sequestration

KB Aviso,[1,3,]* *JLG San Juan,*[2] *CL Sy*[2,3] *and RR Tan*[1,3]

1. Introduction

Climate change has emerged as one of the world's most critical environmental issues. According to the IPCC (2018), net global greenhouse gas (GHG) emissions need to be reduced to zero by the mid-21st Century in order to keep mean temperature rise by 2100 to a manageable level of about 1.5 °C. Furthermore, commitments made to GHG emissions under the Paris Accord result in a strong need for the deployment of low-carbon technologies in order for such cuts to be realized. Strategies include the increased use of low-carbon energy sources, such as biomass, and the large-scale deployment of carbon dioxide removal (CDR) or negative emissions technologies (NETs) which can achieve significant cuts in the release of GHGs and potentially stabilize climate in the coming decades (Haszeldine et al., 2018). Shifting to cleaner power generation will also complement the mass electrification of urban motor vehicles (Erickson, 2017). Examples of NETs are direct air capture (DAC), bioenergy with CO_2 capture and storage (BECCS), and biochar-based carbon sequestration (McGlashan et al., 2012). The potential scale and technological maturity levels of NETs were assessed by McLaren (2012). More recently, two review papers surveyed the research status (Minx et al., 2018) and techno-economics (Fuss et al., 2018) of different NETs.

Biochar-based systems offer an alternative means to achieve negative emissions. In such systems, biochar is applied to soil to achieve the net removal of carbon from the atmosphere. Thus, the carbon that was originally in atmospheric CO_2 is fixed via photosynthesis into plant biomass. The latter is subsequently converted via thermochemical processing (i.e., pyrolysis or gasification) into biochar, whose carbon content is in a predominantly chemically stable (recalcitrant) form. Application of the resulting biochar to soil thus results in permanent sequestration of this recalcitrant carbon (Woolf et al., 2010). Such systems are naturally compatible with other biomass-based energy systems, and additional benefits may accrue from modification of soil biota which further reduces the release of GHGs (He et al.,

---
[1] Chemical Engineering Department, De La Salle University, Manila, Philippines.
[2] Industrial Engineering Department, De La Salle University, Manila, Philippines.
[3] Center for Engineering and Sustainable Development Research, De La Salle University, Manila, Philippines.
* Corresponding author: kathleen.aviso@dlsu.edu.ph

2017). An advantageous feature of using biochar is the potential to produce useful energy while storing carbon (Smith, 2016). Planning of integrated, biochar-based carbon management networks (CMNs) can be facilitated through the use of computer-aided process engineering (CAPE) or process systems engineering (PSE) tools (Belmonte et al., 2017). The prospect of integrating biochar-based CMNs with biomass co-firing in power plants was recently proposed by Dang et al. (2015), who suggested a novel scheme to co-fire pyrolysis bio-oil with coal and to apply the biochar to soil; initial life-cycle analysis (LCA) showed significant potential to reduce CO_2 emissions. However, their work did not consider the optimization of such systems.

Co-firing of biomass with fossil fuels is a technologically mature approach to reducing GHG emissions, and is already widely used throughout the world (Roni et al., 2017). The feasibility of using biomass, such as agricultural waste (e.g., corn stover, rice straw, etc.), as an energy source for co-firing in modified existing coal power plants is well-established in the literature. Co-firing gives an immediate and practical mode of reducing coal usage and the associated GHG emissions. Furthermore, co-firing enables biomass to be used in existing coal power plants, instead of having to build dedicated biomass-fired plants (Madanayake et al., 2017). Co-firing biomass with coal is an attractive alternative because biomass can be integrated into existing coal-fired power plants' fuel storage and handling systems with only relatively minor retrofits, and allows for flexible operation with diverse feedstocks (Dundar et al., 2016). This flexibility leads to reduced techno-economic risk compared to the alternative of building stand-alone biomass-fired power plants that are entirely dependent on a potentially unstable supply of fuel. However, co-firing is usually limited to 10% biomass on a fuel energy basis, as higher rates of utilization may have adverse effects on plant equipment. Furthermore, it has been argued that high co-firing rates are detrimental due to the GHG emissions penalty that result from handling biomass, which has a lower energy density than coal (Miedema et al., 2017). Biomass co-firing also improves the net energy and emissions balance because the combustion of biomass residues, such as rice straw, makes use of less energy and releases less emissions when the upstream coal supply chain (i.e., mining and transportation operations) is considered in the analysis (Shafie et al., 2013). Co-firing systems in existing coal-fired power plants can be implemented using three possible configurations, namely, direct, indirect, and parallel co-firing (Agbor et al., 2014). Direct co-firing systems are characterized by the use of a single common boiler burning a blend of coal and biomass, or via separate burners for biomass and coal providing heat to a common boiler. Indirect co-firing uses gasification to convert biomass into syngas, which can then be used as the secondary fuel as in direct co-firing systems; in pyrolysis-based systems, the liquid fraction (bio-oil) can also be used as fuel along with the syngas. The residual solid biochar is available for carbon sequestration. On the other hand, parallel co-firing systems burn coal and biomass in separate boilers, which then feed into a common turbine. In such schemes, the solid biochar residue can be separated and applied to soil to achieve further GHG offsets. Thus, indirect co-firing can potentially achieve negative emissions for the biomass fraction of the power plant fuel input, even if total emissions remain positive due to the combustion of coal (Dang et al., 2015).

PSE tools can be used to facilitate planning sustainable supply chains (Cabezas et al., 2018), particularly for biomass logistics (Atashbar et al., 2018). Novel approaches to the optimization of biomass supply chains have been proposed to account for the inherently dispersed nature of biomass resources. Lim and Lam (2016) proposed a biomass element life cycle analysis (BELCA) approach that takes into account variations in biomass feedstock composition and properties to identify opportunities for blending and for matching with downstream demands. Different supply chain configurations, taking into account factors such as feedstock supply, production scale, transport distance, and the use of intermediate depots, were subjected to economic and energetic analysis by Ng and Maravelias (2017). Ng et al. (2018) then developed a framework for biomass supply chain optimization. The framework includes preprocessing to generate a simplified superstructure, and incorporates supply chain performance metrics within the resulting optimization model. It is notable that most of the literature on co-firing supply chain optimization focuses on direct co-firing (Atashbar et al., 2018). There have been studies that proposed a biomass co-firing supply chain optimization model that minimized both the cost and emissions, such as Mohd Idris et al. (2018) for oil palm biomass and Griffin et al. (2014) for mixed biomass. Pérez-Fortes et al. (2014)

developed a mixed integer linear programming (MILP) model to decide on the optimal pretreatment of biomass. The dominance of direct co-firing over indirect and parallel co-firing is attributable to the need for relatively invasive plant retrofits (Madanayake et al., 2017). This work addresses the gap in the literature by developing a two-layer supply chain model for indirect biomass co-firing coupled with biochar-based carbon sequestration.

Belmonte et al. (2017) argue that the use of PSE is essential to the proper large-scale deployment of biochar-based systems. The optimal synthesis of biochar-based CMNs was first addressed by Tan (2016), who developed a multi-period MILP to allocate biochar to different soils acting as biochar sinks. The model accounts for storage capacities and quality (contaminant) requirements. A bi-objective extension of this initial model was recently developed (Belmonte et al., 2018). Other approaches, based on pinch analysis (Tan et al., 2018) and the P-graph framework (Aviso et al., 2019), were recently developed for optimizing biochar-based CMNs. Despite recent developments in the PSE literature on mathematical models to aid in the planning of biochar-based CMNs, as well as an extensive body of literature on biomass supply chains, no spatially explicit optimization models have yet been reported that combine these two aspects in the context of low-carbon electricity generation. A natural extension to the works of Tan (2016) and Belmonte et al. (2018) is to use a supply chain perspective for biochar-based carbon sequestration.

In this chapter, a novel MILP model for optimal planning of integrated systems that combine upstream biomass allocation networks for co-firing in power plants with the biochar allocation networks for carbon sequestration is developed. The model is able to take into account biomass supply, power plant capacities and biochar application limits to enable rational system-level planning, which will be essential to enable full-scale implementation of such systems. It is assumed that the relevant data for model calibration are known *a priori*; acquisition of such data on a large scale remains a major research challenge (Tan, 2019) that is outside of the scope of this work. The rest of the chapter is organized as follows: Section 2 gives the formal problem statement that specifies model inputs and outputs. The MILP model formulation is then described in section 3. An illustrative case study is then solved in section 4 to demonstrate the model capabilities; the optimal and near-optimal solutions are analyzed to show how multiple options can be used to facilitate decision-making. Finally, conclusions and prospects for future work are given in section 5.

2. Problem Statement

The formal problem statement is given as follows:

- Given M available sources of biomass, with each source having a maximum availability S_i.
- Given N power plants that may opt to implement biomass co-firing, each with the equivalent amount of coal displaced by biomass given by b_{jh} (for each power plant j, a fixed percentage of its thermal energy requirement will be replaced with biomass if a decision to implement co-firing is made).
- Given V biomass co-firing technologies, each characterized by factors b_{jh} (the biomass requirement when technology h is implemented in power plant j) and z_h (the biochar yield per unit of biomass).
- Given P biochar sinks, with each sink having a sequestration factor F_k and a maximum biochar capacity of SEQ_k.
- Given distance d_{ij} between any given biomass source i and power plant j.
- Given distance r_{jk} between any given power plant j and biochar sink k.

The problem may be visualized as a superstructure, as shown in Figure 1, or alternatively in more compact form as an allocation matrix, as in Table 1. The objective is to determine the choice of co-firing technique to be implemented in each power plant (as denoted by variable Q_{jh}) and to determine the amount of coal replaced with biomass (C_j), the allocation of biomass (denoted by x_{ij}) and biochar (as denoted by y_{jk}), so as to minimize the total amount of carbon dioxide (CO_2) generated by the system.

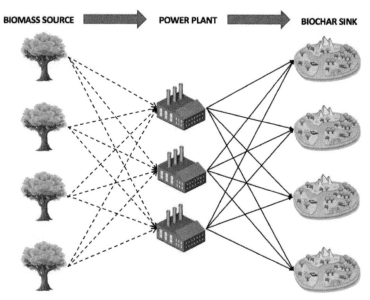

Figure 1. Representation of superstructure in schematic.

Table 1. Matrix form of the superstructure.

	Power Plant 1	Power Plant 2	Power Plant 3
Biomass Source 1			
Biomass Source 2			
Biomass Source 3			
Biomass Source 4			
Biochar Sink 1			
Biochar Sink 2			
Biochar Sink 3			
Biochar Sink 4			

3. Model Nomenclature

Sets

I	Set of biomass sources
J	Set of power plants
K	Set of biochar sinks
H	Set of available technologies for co-firing

Indices

i	Index for biomass source in set I
j	Index for power plant in set J
k	Index for biochar sinks in set K
h	Index for co-firing technologies in set H

Parameters

α	CO_2 footprint of coal combustion in Mt CO_2/Mt coal
β	CO_2 footprint of transport in Mt CO_2/Mt-km
b_{jh}	in Mt/y, represents the amount of biomass needed to replace the coal in power plant j using technology h
d_{ij}	Distance of biomass source i to power plant j in km
F_k	Sequestration factor of biochar sink k
r_{jk}	Distance of power plant j to biochar sink k in km
S_i	Available biomass from source i
SEQ_k	Maximum amount of biochar which can be sequestered by sink k
z_h	Biochar yield when using technology h

Variables

C_j	in Mt/y, refers to amount of coal replaced by biomass in power plant j
Q_{jh}	Binary variable which indicates the activation of technology h for power plant j
x_{ij}	in Mt/y, refers to the amount of biomass from source i which is used in power plant j
y_{jk}	in Mt/y, amount of biochar generated from plant j and sequestered to sink k

4. Model Formulation

The MILP model formulation is as follows:

$$\min Z = \sum_{j=1}^{N} \left(-\sum_{h=1}^{V} \alpha C_j Q_{jh} + \sum_{i=1}^{M} \beta x_{ij} d_{ij} + \sum_{k=1}^{P} \beta y_{jk} r_{jk} - \sum_{k=1}^{P} y_{jk} F_k \right) \tag{1}$$

$$\sum_{j=1}^{N} x_{ij} \leq S_i \qquad \forall i \in \{1, 2, \cdots, M\} \tag{2}$$

$$\sum_{i=1}^{M} x_{ij} = \sum_{h=1}^{V} b_{jh} Q_{jh} \qquad \forall j \in \{1, 2, \cdots, N\} \tag{3}$$

$$\sum_{h=1}^{V} Q_{jh} \leq 1 \qquad \forall j \in \{1, 2, \cdots, N\} \tag{4}$$

$$\sum_{k=1}^{P} y_{jk} = \sum_{h=1}^{V} b_{jh} Q_{jh} z_h \qquad \forall j \in \{1, 2, \cdots, N\} \tag{5}$$

$$\sum_{j=1}^{N} y_{jk} \leq SEQ_k \qquad \forall k \in \{1, 2, \cdots, P\} \tag{6}$$

$$Q_{jh} \in \{0, 1\} \qquad \forall j \in \{1, 2, \cdots, N\}, \forall h \in \{1, 2, \cdots, V\} \tag{7}$$

The objective function given by equation (1) is to minimize function Z that denotes the incremental amount of CO_2 generated by the network; it will assume a negative value if a reduction is achieved. The first term in equation (1) corresponds to the amount of CO_2 reduction resulting from reduced coal consumption (C_j) in power plant j due to displacement by biomass using technology h. Parameter

α is the CO_2 footprint per unit of coal (including both direct emissions from combustion and upstream contributions from the coal supply chain), and Q_{jh} is a binary variable which indicates the use ($Q_{jh} = 1$) or non-use ($Q_{jh} = 0$) of a co-firing technology. The second term corresponds to the CO_2 emission associated with the transport of biomass from source i to power plant j, which is proportional to the distance travelled, d_{ij}, and the amount of biomass transported, x_{ij}. The third term corresponds to the CO_2 generated in transporting the biochar from power plant j to biochar sink k which is also proportional to the distance travelled, r_{jk}, and the amount of biochar transported, y_{jk}. For the second and third terms of Equation (1), parameter β refers to the emission factor or CO_2 footprint associated with transporting biomass and biochar. Finally, the last term in Equation (1) corresponds to the amount of CO_2 sequestered from the biochar which is proportional to the amount of biochar, y_{jk}, and the sequestration factor of the sink, F_k. The latter factor accounts for direct sequestration of the recalcitrant carbon in the biochar, as well as secondary effects due to changes in GHG emissions from soil biota when biochar is applied. Equation (2) ensures that the allocation of biomass source i to the different power plants will not exceed the total amount of biomass available, S_i. The amount of biomass needed to replace the coal will depend on the type of co-firing technology selected and is given by b_{jh}. Equation (3) ensures that enough biomass is obtained from the different sources to satisfy the requirement of each power plant j. Equation (4) indicates that each power plant can only implement a maximum of one type of co-firing technology. Equation (5) on the other hand is the biochar balance which indicates that the total amount of biochar generated by a power plant as indicated by the biochar yield, z_h, should be sequestered and properly allocated to the available biochar sinks. Equation (6) ensures that the amount of biochar sequestered in sink k should not exceed the sink capacity, SEQ_k. Equation (7) indicates that the variable Q_{jh} is binary. All other variables are non-negative. Note that this MILP model can be readily solved to global optimality using the conventional branch-and-bound algorithm found in many commercial optimization software. For any given application, the model size is given by the formula in Table 2.

Table 2. Model size as function or problem scale.

Model feature	Number
Binary variables	$(N \times V)$
Continuous variables	$N + (M \times N) + (N \times P)$
Constraints	$M + N + P + (N \times V)$

5. Case Study

This representative case study considers a system with eight biomass sources, five power plants and four biochar sinks. This system size gives a cluster for which typical transportation distances for both biomass and biochar are reasonable. The representative case study gives rise to 77 continuous variables, 10 binary variables, and 27 functional constraints. The case study is implemented using the commercial software LINGO 17.0 which was run using Intel® Core™ i7-6500U processor and 8.00 GB RAM with negligible CPU time. The biomass sources are assumed to be sites for biomass collection, consolidation, and storage. The limiting data for the biomass sources is shown in Table 3. The amount of coal that must be replaced by biomass for each power plant, if co-firing is implemented, is indicated in Table 4, along with other relevant technical characteristics. It is assumed here that the co-firing rate is 10%, based on thermal energy input. The biochar sinks are tracts of agricultural or set-aside land to which biochar can be applied. The biochar sink characteristics are shown in Table 5. This includes the maximum amount of biochar that a sink can hold and the sequestration factor of the sink which corresponds to the amount of CO_2 sequestered per unit of biochar. This factor can also account for positive or negative changes in emissions of other GHGs from soil.

The amount of biochar generated from each power plant will depend on the type of co-firing technique selected, and becomes zero in the case of direct co-firing. Table 6 shows the biochar yield of each co-firing technology considered. In addition, each technology will require a different amount of biomass

Table 3. Limiting data for biomass sources.

Biomass source	Available biomass (Mt/y)
B1	0.10
B2	0.15
B3	0.10
B4	0.25
B5	0.10
B6	0.20
B7	0.12
B8	0.10

Table 4. Power plant characteristics.

Power plant	Capacity (MW)	Baseline coal consumption (Mt/y)	Amount of displaced coal (Mt/y)
P1	200	0.60	0.060
P2	250	0.75	0.075
P3	600	1.80	0.180
P4	500	1.50	0.150
P5	250	0.75	0.075

Table 5. Biochar sink characteristics.

Power plant	Maximum amount of sequestered biochar (Mt/y)	Sequestration factor (F_k) (Mt CO_2/Mt of biochar)
C1	0.04	3.20
C2	0.05	3.00
C3	1.00	2.60
C4	0.06	3.00

Table 6. Co-firing technology characteristics.

Technology		Biochar yield
T1	Direct co-firing	0
T2	Indirect co-firing	0.20

in order to supply the equivalent thermal energy of the replaced coal. The amounts of biomass required to generate the needed thermal energy are indicated in Table 7. Note that the total biomass requirement for indirect co-firing is greater than that of direct co-firing, because part of the biomass (i.e., the biochar) remains unutilized as fuel. The distances between biomass sources and power plants are shown in Table 8, while the distances between the power plants and the potential biochar sinks are shown in Table 9. It is assumed that the biomass and biochar are transported by truck, with an emission factor of 0.0001 Mt of CO_2/Mt/km (Tan, 2016). The CO_2 footprint of coal is 3.16 Mt CO_2/Mt of coal, including emissions from both the power plant and the upstream coal supply chain.

The MILP model corresponding to this case study is coded in LINGO, as shown in the Appendix. Solving the model results in an optimal CO_2 emission increment of −1.9619 Mt of CO_2/y. This result

Table 7. Biomass requirement in Mt/y.

	Co-firing technology	
	T1	**T2**
P1	0.1200	0.1500
P2	0.1500	0.1875
P3	0.3600	0.4500
P4	0.3000	0.3750
P5	0.1500	0.1875

Table 8. Distance between biomass source and power plant (d_{ij}) in km.

	P1	**P2**	**P3**	**P4**	**P5**
B1	60	120	160	220	240
B2	40	120	140	200	220
B3	30	90	140	200	220
B4	70	30	140	210	200
B5	40	40	60	140	130
B6	120	70	90	120	60
B7	80	140	80	120	160
B8	100	150	60	100	140

Table 9. Distance between power plant and biochar sink (r_{jk}) in km.

	C1	**C2**	**C3**	**C4**
P1	100	60	100	140
P2	140	100	30	60
P3	70	50	50	80
P4	100	100	110	120
P5	140	130	80	80

indicates a net reduction in CO_2 emissions, of which 77% is due to the replacement of coal with biomass and 23% to biochar sequestration. By comparison, the increment in CO_2 emission achieved when only direct co-firing is considered is -1.6967 Mt of CO_2/y, which is 13.5% less than the reduction achieved with the optimal solution. Table 10 shows the flow of biomass from source to the power plant and the flow of biochar from the power plant to the sink (shown in the shaded region). Note that direct co-firing is used in power plant P3, due to the lack of biochar sink capacity in the system.

In addition to the identification of the optimum, Voll et al. (2015) argue that the analysis of near-optimal solutions can provide valuable insights on the characteristics of good solutions to a particular problem. In addition, the actual differences in objective function values of optimal and near-optimal solutions may be insignificant in practical situations; in such cases, the near-optimal solutions may have advantages with respect to considerations that are not explicitly reflected in the optimization model formulation. Thus, an additional nine near-optimal solutions were generated to evaluate which network connections occurred most frequently in the top ten solutions of the case. These solutions were generated automatically using the MILP solver in LINGO 17.0; in the absence of such a solver feature, these solutions can be generated sequentially using additional integer-cut constraints that eliminate previously

Table 10. Optimal allocation of biomass and biochar in Mt/y.

	P1	P2	P3*	P4	P5
B1	0	0	0	0.0900	0
B2	0	0	0.0975	0.0525	0
B3	0	0	0.1000	0	0
B4	0	0.1875	0.0625	0	0
B5	0	0	0.1000	0	0
B6	0	0	0	0.0125	0.1875
B7	0	0	0	0.1200	0
B8	0	0	0	0.1000	0
C1	0	0	0	0.0400	0
C2	0	0.0150	0	0.0350	0
C3	0	0	0	0	0
C4	0	0.0225	0	0	0.0375

* Direct co-firing option is selected.

Table 11. Summary of CO_2 emissions in top ten CMNs.

Solution rank	Incremental CO_2 emission (Mt/y)
1	−1.9691
2	−1.9546
3	−1.9530
4	−1.9365
5	−1.9346
6	−1.8958
7	−1.8953
8	−1.8943
9	−1.8922
10	−1.8628

determined network topologies (Voll et al., 2015). This approach can also lead to the identification of degenerate solutions (i.e., alternative topologies with equivalent objective function values). A summary of the amount of the incremental CO_2 emissions for these different solutions are summarized in Table 11. The worst solution in this set of networks is only 5.4% worse than the optimal network in terms of CO_2 emissions reduction.

Examples of near-optimal networks which correspond to the second and fifth best solutions are also shown in Tables 12 and 13, respectively. These networks give a system-wide CO_2 incremental change of −1.9546 Mt/y and −1.9346 Mt/y, respectively. These results are just 0.74% and 1.75% worse than the optimum solution. In real life applications, such small differences may not have practical significance, so that these solutions may be interpreted as having virtually equivalent performance. The decision-maker may then select to implement a network based on other criteria not explicitly reflected in the optimization model. Two trends are also apparent in the optimal and near-optimal solutions presented here. First, due to biochar sink limitations, not all of the power plants use indirect co-firing in any given solution; some plants opt for either direct co-firing or no co-firing at all. Secondly, even if biochar sink C3 has the largest capacity, as shown in Table 5, it is utilized only sparingly due to its low sequestration factor.

Table 12. Near-optimal allocation of biomass and biochar in Mt/y (Solution 2).

	P1*	P2*	P3	P4	P5
B1	0	0	0.0750	0	0
B2	0.0200	0	0.1300	0	0
B3	0.1000	0	0	0	0
B4	0	0.1500	0.1000	0	0
B5	0	0	0.1000	0	0
B6	0	0	0	0.2000	0
B7	0	0	0.0450	0.0750	0
B8	0	0	0	0.1000	0
C1	0	0	0	0.0400	0
C2	0	0	0.0500	0	0
C3	0	0	0.0150	0	0
C4	0	0	0.0250	0.0350	0

* Direct co-firing option is selected.

Table 13. Near-optimal allocation of biomass and biochar in Mt/y (Solution 5).

	P1	P2	P3	P4*	P5
B1	0.0500	0	0	0.0175	0
B2	0	0	0.1000	0.0500	0
B3	0.1000	0	0	0	0
B4	0	0	0.2500	0	0
B5	0	0	0.1000	0	0
B6	0	0	0	0.0125	0.1875
B7	0	0	0	0.1200	0
B8	0	0	0	0.1000	0
C1	0	0	0.0400	0	0
C2	0.0300	0	0.0200	0	0
C3	0	0	0.0075	0	0
C4	0	0	0.0225	0	0.0375

* Direct co-firing option is selected.

The frequency of occurrence of network links in the top ten solutions is summarized in Table 14, and is indicated by the intensity of the shading of the cells. White indicates 0% occurrence, black indicates 100% occurrence, and intermediate shades of gray show partial occurrence in the set of solutions. Thus, it can be clearly seen which links in the network are critical, particularly for the connections between biomass sources and power plants. For example, the biomass sources B2, B4 and B5 are consistently linked to power plant P3, while B6, B7 and B8 are consistently linked to P4. These frequently occurring links represent robust features that will be relatively insensitive to deviations from modelling assumptions, such as changes in parameter values.

By comparison, it can also be seen in the bottom four rows of Table 14 that there are more variations in the biochar allocation schemes in the network. This result can be partly attributed to the selection of direct co-firing (which does not produce biochar) in many of the solutions. For example, it can be seen that in each of the solutions in Tables 10, 12 and 13, a different power plant (i.e., P1, P5 and P2,

Table 14. Frequency of connections in the top ten CMNs.

	P1	P2	P3	P4	P5
B1	White	White	Gray	Gray	White
B2	Gray	White	Black	Gray	White
B3	Gray	White	Gray	White	White
B4	White	Gray	Black	White	White
B5	White	White	Black	White	White
B6	White	White	White	Black	Gray
B7	White	White	Gray	Black	White
B8	White	White	Gray	Black	White
C1	White	White	Gray	Gray	White
C2	White	White	Gray	Gray	White
C3	White	Gray	Gray	White	White
C4	White	Gray	Gray	Gray	White

Legend: White – 0% occurrence; Black – 100% occurrence; Gray – 1–99% occurrence.

respectively) elects not to implement co-firing at all. The presence of such alternatives can potentially allow for more flexible decision-making in practical situations. These features also represent system components that are more sensitive to model assumptions; a decision-maker may seek to acquire more data before making a final selection.

6. Conclusions

In this work, a novel MILP model has been developed for the optimal planning of energy systems that integrate biomass allocation networks for co-firing in power plants with biochar distribution networks for carbon sequestration. Such integrated systems can be viewed as a special class of CMNs that allow for significant reductions of GHG emissions from electricity generation. This reduction results from both the displacement of coal and the sequestration of carbon in the biochar. Unlike previous models for optimizing biochar-based CMNs, this work explicitly accounts for the upstream biomass supply chain that precedes biochar production. Its usefulness is demonstrated using an illustrative case study. The MILP formulation also allows near-optimal networks with alternative topologies to be generated for further consideration during decision-making. The model can easily be scaled up so as to enable systematic planning of large-scale systems for commercial implementation.

Future work can focus on different extensions of the basic MILP model developed here. For instance, a multiple-objective variant can be developed in order to consider the tradeoff between emissions reduction and socio-economic aspects. Multi-period variants can be developed in order to account for seasonality of biomass feedstock supply, while multi-region formulations can include detailed geographic considerations that may restrict biomass or biochar transport. In addition, fuzzy or robust formulations will also be needed so as to account for parametric uncertainties in biomass supply and in sequestration factors. Mathematical programming can be complemented with other PSE tools, such as pinch analysis or P-graph. Finally, the model and its future variants should also be applied when planning the deployment of such systems at commercially significant scales.

Acknowledgement

This work was funded by the Comission on Higher Education (CHED) as part of the Philippine Higher Education Research Network (PHERNet) Sustainability Studies Program at De La Salle University.

Appendix—LINGO Code

LINGO is a commercial optimization software developed by LINDO Systems, Inc. (www.lindo.com). It uses an equation-based user interface and comes with a suite of optional toolboxes for solving various types of mathematical programming problems. LINGO uses the standard branch-and-bound algorithm to solve MILP problems, such as those developed here. The model corresponding to the case study is coded as follows; note that the model can be used to solve other case studies by replacing the data section with new data, which may also be cut and pasted directly from an Excel spreadsheet:

```
!SETS
I               INDEX FOR BIOMASS SOURCES
J               INDEX FOR POWER PLANT
K               INDEX FOR BIOCHAR SINKS

PARAMETERS
ALPHA       CO₂ FOOTPRINT OF COAL COMBUSTION IN MT CO₂/COAL
BETA        CO₂ FOOTPRINT OF TRANSPORT IN MT CO₂/MT-KM
Z(H)        BIOCHAR YIELD FOR TECH H
B(J,H)      IN MT/Y REPRESENTS THE AMOUNT OF BIOMASS STRAW NEEDED TO
            REPLACE THE COAL IN PLANT J USING TECH H
DIST1(I,J)  DISTANCE OF BIOMASS SOURCE I TO PLANT J
F(K)        SEQUESTRATION FACTOR OF SINK K
DIST2(J,K)  DISTANCE FROM PLANT J TO SEQUESTRATION SITE K
SUPPLY(I)   AVAILABLE BIOMASS FROM SOURCE I
SEQ(K)      MAXIMUM AMOUNT OF BIOCHAR THAT CAN BE SEQUESTERED BY SINK K
Z(H)        BIOCHAR YIELD USING TECH H

VARIABLES
C(J)        MT/Y REFERS TO REPLACED COAL IN PLANT J
Q(J,H)      BINARY VARIABLE INDICATING THAT TECH H IS ACTIVATED FOR PLANT J
X(I,J)      MT/Y REPRESENTS THE AMOUNT OF BIOMASS FROM I GOING TO PLANT J
Y(J,K)      MT/Y AMOUNT OF BIOCHAR SEQUESTERED FROM PLANT J TO LAND K;

SETS:
BSOURCE:                    SUPPLY;     !BIOMASS SOURCE I;
PSINK:                      C;          !BIOMASS POWERPLANT SINK J;
TECH:                       Z;          !BIOMASS TECH L;
CSINK:                      SEQ, F;     !BIOCHAR SINK K;
BPLINK(BSOURCE, PSINK):  X, DIST1;      !LINK BETWEEN BIOMASS SOURCE AND
                                        POWERPLANT SINK;
PTECH (PSINK, TECH):        B, Q;       !PROVIDES AVAILABLE TECHNOLOGY FOR
                                        POWERPLANT;
PCLINK (PSINK, CSINK):      Y, DIST2;   !LINK BETWEEN POWERPLANT AND BIOCHAR
                                        SINK;
ENDSETS

DATA:
BSOURCE =   B1      B2      B3      B4      B5      B6      B7      B8;
SUPPLY =    0.1     0.15    0.1     0.25    0.1     0.2     0.12    0.1;
PSINK =     P1      P2      P3      P4      P5;
C =         0.06    0.075   0.18    0.15    0.075;
```

CSINK =	C1	C2	C3	C4;
SEQ =	0.04	0.05	1	0.06;
F =	3.2	3.0	2.6	3.0;

TECH =	T1	T2;
Z =	0	0.2;
	!T1	T2;

B =	0.12	0.15			!P1;
	0.15	0.1875		!P2;	
	0.36	0.45			!P3;
	0.30	0.375			!P4;
	0.15	0.1875;		!P5;	

	!P1	P2	P3	P4	P5;	
DIST1 =	60	120	160	220	240	!B1;
	40	120	140	200	220	!B2;
	30	90	140	200	220	!B3;
	70	30	140	210	200	!B4;
	40	40	60	140	130	!B5;
	120	70	90	120	60	!B6;
	80	140	80	120	160	!B7;
	100	150	60	100	140;	!B8;

	!C1	C2	C3	C4;	
DIST2 =	100	60	100	140	!P1;
	140	100	30	60	!P2;
	70	50	50	80	!P3;
	100	100	110	120	!P4;
	140	130	80	80;	!P5;

ALPHA = 3.16;
BETA = 0.0001;

ENDDATA

MIN = @SUM(PSINK(J): -@SUM(TECH(H): ALPHA*C(J)*Q(J,H)) + @SUM(BSOURCE(I): X(I,J)*DIST1(I,J)*BETA)
    +@SUM(CSINK(K): Y(J,K)*DIST2(J,K)*BETA) – @SUM(CSINK(K): Y(J,K)*F(K)));

@FOR(BSOURCE(I): SUPPLY(I) >= @SUM(PSINK(J): X(I,J)));
@FOR(PSINK(J): @SUM(BSOURCE(I): X(I,J)) = @SUM(TECH(H): B(J,H)*Q(J,H)));
@FOR(PSINK(J): @SUM(TECH(H): Q(J,H)) <=1);
@FOR(PSINK(J): @SUM(CSINK(K): Y(J,K)) = @SUM(TECH(H): B(J,H)*Q(J,H)*Z(H)));
@FOR(CSINK(K): @SUM(PSINK(J): Y(J,K)) <= SEQ(K));
@FOR(PTECH(J,H): @BIN(Q(J,H)));

References

Agbor, E., Zhang, X. and Kumar, A. 2014. A review of biomass co-firing in North America. Renewable and Sustainable Energy Reviews 40: 930–43.

Atashbar, N.Z., Labadie, N. and Prins, C. 2018. Modelling and optimisation of biomass supply chains: A review. International Journal of Production Research 56: 3482–3506.

Aviso, K.B., Belmonte, B.A., Benjamin, M.F.D., Arogo, J.I.A., Coronel, A.L.O., Janairo, C.M.J., Foo, D.C.Y. and Tan, R.R. 2019. Synthesis of optimal and near-optimal biochar-based carbon management networks with P-graph. Journal of Cleaner Production 214: 893–901.

Belmonte, B.A., Benjamin, M.F.D. and Tan, R.R. 2017. Biochar systems in the water-energy-food nexus: The emerging role of process systems engineering. Current Opinion in Chemical Engineering 18: 32–7.

Belmonte, B.A., Benjamin, M.F.D. and Tan, R.R. 2018. Bi-objective optimization of biochar-based carbon management networks. Journal of Cleaner Production 188: 911–20.

Cabezas, H., Argoti, A., Friedler, F., Mizsey, P. and Pimentel, J. 2018. Design and engineering of sustainable process systems and supply chains by the P-graph framework. Environmental Progress and Sustainable Energy 37: 624–636.

Dang, Q., Wright, M.M. and Brown, R.C. 2015. Ultra-low carbon emissions from coal-fired power plants through bio-oil Co-firing and biochar sequestration. Environmental Science and Technology 49: 14688–95.

Dundar, B., Mcgarvey, R.G. and Aguilar, F.X. 2016. Identifying optimal multi-state collaborations for reducing CO_2 emissions by co-firing biomass in coal-burning power plants. Computers and Industrial Engineering 101: 403–15.

Erickson, L.E. 2017. Reducing greenhouse gas emissions and improving air quality: Two global challenges. Environmental Progress and Sustainable Energy 36: 982–988.

Fuss, S., Lamb, W.F., Callaghan, M.W., Hilaire, J., Creutzig, F., Amann, T., Beringer, T., De Oliveira Garcia, W., Hartmann, J., Khanna, T., Luderer, G., Nemet, G.F., Rogelj, J., Smith, P., Vicente Vicente, J.L., Wilcox, J., Del Mar Zamora Dominguez, M. and Minx, J.C. 2018. Negative emissions—Part 2: Costs, potentials and side effects. Environmental Research Letters 13: Article 063002.

Griffin, W.M., Michalek, J., Matthews, H.S. and Hassan, M.N.A. 2014. Availability of biomass residues for co-firing in peninsular Malaysia: Implications for cost and GHG emissions in the electricity sector. Energies 7: 804–23.

Haszeldine, R.S., Flude, S., Johnson, G. and Scott, V. 2018. Negative emissions technologies and carbon capture and storage to achieve the Paris Agreement commitments. Philosophical Transactions of the Royal Society A: Mathematical. Physical and Engineering Sciences 376: Article 20160447.

He, Y., Zhou, X., Jiang, L., Li, M., Du, Z., Zhou, G. et al. 2017. Effects of biochar application on soil greenhouse gas fluxes: A meta-analysis. GCB Bioenergy 9: 743–55.

IPCC. 2018. Summary for Policymakers. In: Masson-Delmotte, V., Zhai, P., Pörtner, H.O., Roberts, D., Skea, J., Shukla, P.R., Pirani, A., Moufouma-Okia, W., Péan, C., Pidcock, R., Connors, S.J., Matthews, B.R., Chen, Y., Zhou, X., Gomis, M.I., Lonnoy, E., Maycock, T., Tignor, M. and Waterfield, T. (eds.). Global warming of 1.5 °C. An IPCC Special Report on the impacts of global warming of 1.5 °C above pre-industrial levels and related global greenhouse gas emission pathways, in the context of strengthening the global response to the threat of climate change, sustainable development, and efforts to eradicate poverty. World Meteorological Organization, Geneva, Switzerland.

Lim, C.H. and Lam, H.L. 2016. Biomass supply chain optimisation via novel biomass element life cycle analysis (BELCA). Applied Energy 161: 733–45.

Madanayake, B.N., Gan, S., Eastwick, C. and Ng, H.K. 2017. Biomass as an energy source in coal co-firing and its feasibility enhancement via pre-treatment techniques. Fuel Processing Technology 159: 287–305.

McGlashan, N., Shah, N., Caldecott, B. and Workman, M. 2012. High-level techno-economic assessment of negative emissions technologies. Process Safety and Environmental Protection 90: 501–10.

McLaren, D. 2012. A comparative global assessment of potential negative emissions technologies. Process Safety and Environmental Protection 90: 489–500.

Miedema, J.H., Benders, R.M.J., Moll, H.C. and Pierie, F. 2017. Renew, reduce or become more efficient? The climate contribution of biomass co-combustion in a coal-fired power plant. Appl. Energy 187: 873–85.

Minx, J.C., Lamb, W.F., Callaghan, M.W., Fuss, S., Hilaire, J., Creutzig, F., Amann, T., Beringer, T., De Oliveira Garcia, W., Hartmann, J., Khanna, T., Lenzi, D., Luderer, G., Nemet, G.F., Rogelj, J., Smith, P., Vicente Vicente, J.L., Wilcox, J. and Del Mar Zamora Dominguez, M. 2018. Negative emissions—Part 1: Research landscape and synthesis. Environmental Research Letters 13: Article 063001.

Mohd Idris, M.N., Hashim, H. and Razak, N.H. 2018. Spatial optimisation of oil palm biomass co-firing for emissions reduction in coal-fired power plant. Journal of Cleaner Production 172: 3428–47.

Ng, R.T.L. and Maravelias, C.T. 2017. Economic and energetic analysis of biofuel supply chains. Applied Energy 205: 1571–82.

Ng, R.T.L., Kurniawan, D., Wang, H., Mariska, B., Wu, W. and Maravelias, C.T. 2018. Integrated framework for designing spatially explicit biofuel supply chains. Applied Energy 216: 116–31.

Pérez-Fortes, M., Laínez-Aguirre, J.M., Bojarski, A.D. and Puigjaner, L. 2014. Optimization of pre-treatment selection for the use of woody waste in co-combustion plants. Chemical Engineering Research and Design 92: 1539–62.

Roni, M.S., Chowdhury, S., Mamun, S., Marufuzzaman, M., Lein, W. and Johnson, S. 2017. Biomass co-firing technology with policies, challenges, and opportunities: A global review. Renewable and Sustainable Energy Reviews 78: 1089–101.

Shafie, S.M., Mahlia, T.M.I. and Masjuki, H.H. 2013. Life cycle assessment of rice straw co-firing with coal power generation in Malaysia 2013. Energy 57: 284–94.

Smith, P. 2016. Soil carbon sequestration and biochar as negative emission technologies. Global Change Biology 22: 1315–1324.

Tan, R.R. 2016. A multi-period source-sink mixed integer linear programming model for biochar-based carbon sequestration systems. Sustainable Production and Consumption 8: 57–63.

Tan, R.R. 2018. Bandyopadhyay S, Foo DCY. Graphical pinch analysis for planning biochar-based carbon management networks. Process Integration and Optimization for Sustainability 2: 159–168.

Tan, R.R. 2019. Data challenges in optimizing biochar-based carbon sequestration. Renewable and Sustainable Energy Reviews 104: 174–177.

Voll, P., Jennings, M., Hennen, M. and Shah, N. 2015. The optimum is not enough: A near-optimal solution paradigm for energy systems synthesis. Energy 82: 446–56.

Woolf, D., Amonette, J.E., Street-perrott, F.A., Lehmann, J. and Joseph, S. 2010. Sustainable biochar to mitigate global climate change. Nature Communications 1: 1–9.

CHAPTER 17

Trends in Transportation Greenhouse Gas Emissions

H Christopher Frey

1. Introduction

Transportation accounted for approximately one-quarter of global CO_2 emissions in 2016. Global transport CO_2 emissions in 2016 were 71% higher than in 1990 (IEA, 2019a). Transport emissions have increased in all major regions of the world, as indicated in Table 1, with the largest increases occurring in Asia followed by the Americas. Globally, transport CO_2 emissions are primarily from on road vehicles. In 2016, on road transport contributed 74% of total transport CO_2 emissions (IEA, 2018a). Nearly two-thirds of global transport energy consumption is for passenger mobility, of which on road light duty vehicles were the largest energy consumer, followed by aircraft, buses, 2- and 3-wheel vehicles, and rail. On road freight trucks were the major energy consumer for freight transport, followed by marine vessels, rail, and pipelines (EIA, 2016). Although accounting for only around 3% of global greenhouse gas (GHG) emissions, the GHG emissions from international aviation and shipping transport have approximately doubled from 1990 to 2016 (Olivier et al., 2017). Thus, the shares of different transportation modes with respect to total CO_2 and GHG emissions are changing with time.

Mobile emission sources include vehicles and infrastructure used for transportation of passengers and freight, and non-transportation mobile sources (USDOT, 2010; Sims et al., 2014; Frey, 2018). Transportation modes include on road vehicles, aircraft, marine vessels, rail, and pipelines. Non-transportation mobile sources include construction, farm, and industrial equipment, lawn and garden equipment, logging equipment, and recreational vehicles. These non-transportation mobile sources are sometimes included in transportation emission inventories with other non-road sources, such as aircraft, marine vessels, and rail. However, not all organizations use the same categories of transportation sources for reported data or estimates of transport greenhouse gas emissions. Thus, there can be inconsistencies in the underlying scope of transportation greenhouse gas emissions reported by different agencies.

USA GHG emissions for transportation and other mobile sources in 2016 were 2.09 Tg CO_2 equivalent, of which 96.8% was CO_2 (EPA, 2018). These sources contributed 32% of total GHG emissions in the USA. On road light duty vehicles and on road medium and heavy-duty vehicles contributed 53% and 20%, respectively, of the total mobile source emissions. Commercial aviation was the only other source category with more than 100 Tg CO_2 equivalent emissions. Marine vessels, rail, and pipelines each accounted for fewer than 50 Tg CO_2 equivalent. The total mobile emissions increased by 21.5 percent from 1990 to 2016. Most of this growth was from on road vehicles.

Glenn E. Futrell Distinguished University Professor, Department of Civil, Construction, and Environmental Engineering, North Carolina State University.

Table 1. Transport CO_2 emissions by global region from 1990 to 2016.

Year	Transport CO_2 emissions (GtCO_2) by global region				
	Americas	Asia	Europe	Africa	Oceania
1990	1.84	0.78	1.16	0.11	0.07
2000	2.27	1.24	1.15	0.16	0.09
2010	2.38	1.87	1.25	0.26	0.10
2016	2.49	2.44	1.23	0.35	0.11

Source: International Energy Agency, https://www.iea.org/statistics/co2emissions/.

In Europe, transport GHG emissions, including international aviation but excluding marine shipping, account for one quarter of total GHG emissions and were 26.1% higher in 2016 than in 1990. Increasing demand for passenger and freight transport is offsetting improvements in vehicle efficiency. Road transport accounts for 82% of transport GHG emissions. To meet the 2050 goals of the European Commission will require the cutting of transport GHG emissions by more than two-thirds (EEA, 2018).

The global stock of on road vehicles grew from approximately 850 million in 2005 to 1.3 billion in 2015, with most of the growth occurring in Asia, as indicated in Figure 1 (OICA, 2019). The potential for growth is highly dependent on many factors, including the motorization rate, which is the number of vehicles per 1,000 people. The motorization rate and the national vehicle stock in 2015 is indicated for selected countries in Figure 2. The USA has the world's largest domestic vehicle fleet and one of the highest rates of motorization, at 821 vehicles per 1,000 people. China, in contrast, has the second largest fleet but a motorization rate of only 118 vehicles per 1,000 people. With rapid economic development, the motorization rate in China will increase. The motorization rate depends on factors such as economic status but also land use patterns and access to other mobility options, such as public transportation. Approximately 10% of the global population accounts for 80 percent of motorized transportation activity. National transportation emissions typically increase as a country transitions from agricultural to industrial to service economies and as GDP per capita increases (Sims et al., 2014).

According to the reference scenario of the Energy Information Agency (EIA) in the USA, global transportation energy consumption could increase from a baseline of 104 quadrillion BTU ("quads") in 2014 to 155 quads in 2040, with the vast majority of the projected increase occurring in countries outside of the Organization for Economic Cooperation and Development (OECD), especially in Asia (EIA, 2016). The use of liquid fossil transportation fuels is expected to increase, including increases of 13 quads for diesel, 10 quads for jet fuel, and 9 quads for gasoline. While the share of natural gas for transportation use is expected to increase from 3% in 2014 to 11% in 2040, the share of vehicles powered by electricity is expected to be only 1% in 2040, even accounting for some growth in the stock of plug-in electric vehicles (PEVs). Light duty vehicle energy consumption is estimated to increase by 15 quads, followed by 13 quads for trucks, 10 quads for aircraft, and 6 quads for marine vessels. The historical baseline and projected trend for energy use for on road transport by vehicle type is illustrated in Figure 3. The key implication of this scenario is that, without strategies to the contrary, large increases in fossil fuel consumption and increases in the associated GHG emissions are likely. According to Sims et al. (2014), GHG emissions from transport could increase at a faster rate than for any other energy end-use sector through 2050.

Strategies to reduce GHG emissions from transport can typically be categorized in terms of: (a) reducing the carbon intensity of transportation fuels and energy sources; (b) improving the energy efficiency of vehicles; (c) promotion of more efficient vehicle operation (e.g., routing, speeds); (d) shifting transportation activity to less GHG-emission intensive transport modes; (e) reducing the GHG emissions intensity embodied in transportation infrastructure; and (f) reducing travel demand. The feasibility and manner in which strategies under these categories have been or could be implemented often differ substantially with regard to transportation mode. A detailed treatment of these issues is given by USDOT (2010), Sims et al. (2014) and others. This chapter provides a brief overview, based on recent

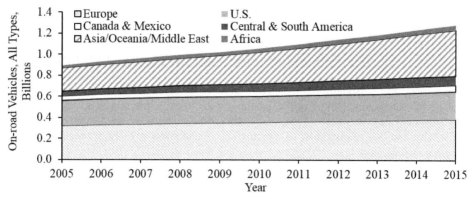

Figure 1. Number of operational on road vehicles globally, 2005 to 2015. Source: (OICA, 2019).

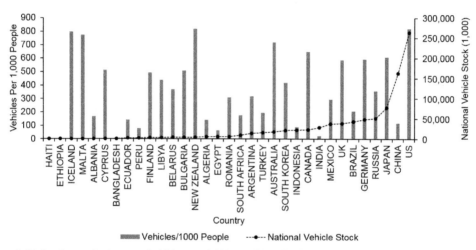

Figure 2. National motorization rate (vehicles per 1,000 people) and vehicle stock (thousands) for selected countries (OICA, 2019).

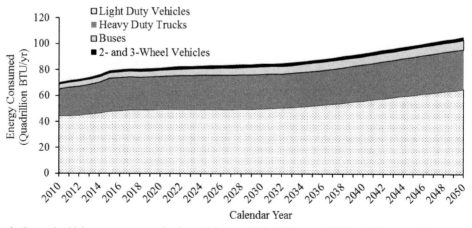

Figure 3. On-road vehicle energy consumption by vehicle type, 2010–2014 actual, 2015 to 2050 projected. Source: Energy Information Agency (2017).

literature, regarding the current status, expected trends, and possible mitigation options for transportation and GHG emissions.

2. Transportation Demand and Mode Choice for Personal Mobility

Factors that encourage personal mobility mode choice in favor of private automobile travel include decreased street connectivity, lack of mixed land use, a low local living score (i.e., not able to walk to amenities), low housing diversity score (e.g., residential and business areas are distinct and separate), low dwelling density, and lack of proximity to supermarkets. Factors that encourage more walking, cycling, or transit trips include more house diversity and greater dwelling density (Boulange et al., 2017). More compact communities are associated with less motorized travel. Car use is reduced as distance to transit is reduced (Ding et al., 2017). Policies that promote walkability and transit, such as more investment in sidewalks and transit, could reduce automobile ownership per household (Shay and Khattak, 2012). More dense population and job density would reduce auto-dependent development (Kay et al., 2014). Higher income levels are associated with higher car ownership, more use of cars for transport, and reduction in use of public transit (Shekarchian et al., 2017). Transit-oriented development (TOD) would reduce demand for transport by personal automobile and promote modal shifts to buses or rail (Kay et al., 2014). As an example, one of the wealthiest cities in the world, Hong Kong, has a very low motorization rate in large part because mass transit is ubiquitous, frequent, and inexpensive, and in part because land is expensive and, therefore, the costs of parking a car are high.

Various pricing schemes can influence behaviors that affect energy consumption and emissions. Increasing the price of driving through various mechanisms may be more effective than only providing more access to transit (Kay et al., 2014). CO_2 emissions pricing would be more effective than Vehicle Miles Traveled (VMT) pricing or a gas tax at reducing CO_2 emissions (Welch and Mishra, 2014). Other possible tax schemes focus on vehicle purchase, which would reduce the number of vehicles per household, or ownership, would which reduce the number of vehicles and VMT (Liu and Cirillo, 2016). Taxes on vehicle age (e.g., for an older SUV) might induce substitution of a newer vehicle that is lower emitting (Feng et al., 2013). Singapore restricts vehicle ownership by issuing 10 year "certificates of entitlement" for a substantial fee to purchase a vehicle (Chu, 2015).

Other price schemes that might affect travel demand or encourage modal shifts include cordon pricing, congestion pricing, and parking pricing. Cordon pricing could be applied, for example, in a central business district in order to discourage use of private automobiles for work or shopping trips. In Singapore, "electronic road pricing" tolls that vary with traffic conditions are charged in a cordoned area with the goal of maintaining target speeds on expressways and arterial roads (Chu, 2015). Such policies might discourage discretionary trips more so than work trips. Although workers could switch to a transit mode, some may simply park outside of a cordon zone. Parking pricing may be less effective at modifying behavior than cordon pricing (Azari et al., 2013).

Many schemes that might mitigate energy use and emissions may have only marginal benefits. Based on comparisons using a sketch planning tool of the estimated effect of travel demand management strategies, such as flexible workhours, rideshare programs, and incentives to use transit, transit-oriented development, reduction in transit travel time or fares, and imposition of parking and mileage based fees, combinations of these policies were not likely to achieve CO_2 emissions reductions of more than 10% by 2050 (Mahendra et al., 2012).

While built environment characteristics at both residential and job locations influence travel demand, the relative importance of these two factors may be different in China than in Western countries (Sun et al., 2017). For example, "Danwei" compounds in China are co-located work and housing areas that were centrally planned. Land use patterns in China are shifting to include housing that is market-driven, with more separation from job locations. These new housing patterns manifest the same relationships between land use and transport found in Western countries. However, Chinese cities differ from Western cities in being more populated and denser. Thus, they may not be likely to achieve the same motorization rates that are found in less dense Western cities. For example, traffic congestion in Beijing is already severe, with a car ownership level less than 50% of that in typical Western cities. Thus, perhaps mature motorization rates in China will not be as high as those in many Western countries (Wang and Zhou, 2017).

Other factors that could mitigate the need for motorized transport include sourcing localized products, prioritizing access for pedestrians and cyclists, and applying lessons learned from behavioral research regarding how and under what conditions people will choose to avoid making unnecessary motorized journals and making more use of new types of low-carbon transport (Sims et al., 2014).

3. Mode Shifts

A factor that might reduce road vehicle energy consumption is shifting to other road or to non-road transport modes. According to the Federal Transit Agency of the U.S. Department of Transportation, the CO_2 intensity of passenger transport modes ranges from 0.22 lbs CO_2/passenger-mile for both heavy rail transit and van pool to 0.96 lbs CO_2/passenger-mile for a single occupant vehicle (SOV), with bus transit having an intensity of 0.64 lbs CO_2/passenger-mile. The GHG emissions intensity of public transport modes is highly sensitive to passenger load. For example, bus transit with full seats would have GHG emissions 72% lower per passenger mile compared to average occupancy (Hodges, 2010).

However, such shifts appear likely to be modest. For example, based on policies already in place or soon to be adopted, the share of light duty VMT that might shift to buses by 2030 is estimated to be 0.2 percent in Canada, 2.7 percent in the U.S., 5.5 percent in China, and 9.5 percent in Mexico. Shifts from road transport to non-road modes, such as rail, are likely to be even more modest, ranging from 0% in the U.S. to 2.6% in China. If further policies are adopted to more aggressively promote mode shifts from LDVs to buses, the percentages increase to 0.6% for Canada, 8.2% in the USA, 16.5% percent in China, and 19.0% in Mexico (Facanha et al., 2012). Modal shifts from LDVs to buses would typically reduce gasoline consumption and increase consumption of diesel or CNG. Passenger mode shifts to rail would reduce gasoline consumption and typically increase electricity consumption. Although inter-modal shifts to non-road transport would reduce road transport energy consumption and emissions, they would lead to increases for other transport modes.

4. Light Duty Vehicle Fuels and Technology

In this section, the current status and recent developments in light duty vehicle fuels and technology are reviewed. While most of this review is based on the United States, trends in other parts of the world and internationally are also summarized.

U.S. registered on road vehicles grew from 8,000 in 1900 to 268 million in 2015, as shown in Figure 4 (FHWA, 1997; BTS, 2016). By 1950, more than half of the population of the USA lived in metropolitan areas, and by 2000 over 80% of the population were in urban areas. Most of this growth was in the suburbs. By 1960, a larger share of metropolitan populations was in suburban areas rather than in central cities. Concurrently, the population density of metropolitan areas has decreased (Hobbs and Stoops, 2002). The number of people who commute by walking or public transportation has remained at or well below 10 million each between 1960 and 2010, whereas the number of people commuting by private vehicle increased from just over 40 million in 1960 to nearly 120 million as of 2010 (AASHTO, 2015; McKenzie and Rapino, 2011).

The share of fuel consumed by U.S. LDVs in 2015 was 97% for gasoline and 3 percent for diesel. The fraction of fuel type consumed for buses was 84% diesel, 11 percent CNG, and 4 percent gasoline (Davis et al., 2017).

As shown in Figure 5, from 2011 to 2018, USA light duty advanced vehicle technology sales totaled 4 million vehicles, including 2.95 million hybrid electric vehicles (HEVs), 0.57 million battery electric vehicles (BEVs), 0.48 million plug-in hybrid electric vehicles (PHEVs), and 4,800 fuel cell electric vehicles (FCEVs). Annual sales reached a record high of 653,765 vehicles in 2018. However, the 2018 U.S. sales of electric drive vehicles were only 3.9 percent of all light duty car and truck sales, which totaled over 17 million vehicles. The three top-selling vehicles in 2018 were pickup trucks, including

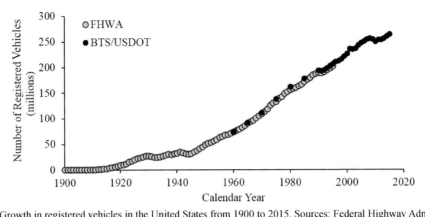

Figure 4. Growth in registered vehicles in the United States from 1900 to 2015. Sources: Federal Highway Administration (FHWA) and Bureau of Transportation Statistics of the U.S. Department of Transportation (BTS/USDOT).

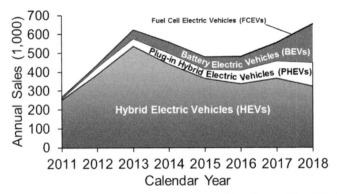

Figure 5. U.S. Annual sales from 2011 to 2018 of Hybrid Electric Vehicles (HEVs), Plug-in Hybrid Electric Vehicles (PHEVs), Battery Electric Vehicles (BEVs), and Fuel Cell Electric Vehicles (FCEVs). Source: Auto Alliance Advanced Technology Vehicle Sales Dashboard.

the Dodge Ram at 536,980, Chevrolet Silverado at 585,581, and Ford F-series at 909,330. However, consumer choices of electric drive vehicles appear to be continuously expanding. In 2019, consumers in the USA could choose from 58 plug-in vehicle models, including 24 battery electric and 34 plug-in hybrid, compared to only a few in 2012.

In the U.S., the average CO_2 emission rate of light duty vehicles has decreased by almost 50% from 1975 to 2017, from an average of 681 g/mi to 357 g CO_2/mile (EPA, 2019). Over 98% of U.S. light duty vehicles are fueled with gasoline (Frey, 2018). Light duty vehicles include passenger cars and passenger trucks. Passenger cars include sedans, couples, and wagons. Passenger trucks include pickup trucks, vans, and minivans. Sport utility vehicles (SUVs), depending on their weight and drivetrain (i.e., two-wheel, four-wheel), can be classified as either passenger cars or trucks. In 1975, 81% of new LDVs were passenger cars. However, in 2017, passenger cars accounted for only 53% of new LDVs, with the rest being passenger trucks. For some manufacturers, particularly FCA, Subaru, GM, and Ford, passenger trucks accounted for over half of new vehicle distribution in the 2017 model year (EPA, 2019).

In the U.S., the average new vehicle weight among all LDVs reached a minimum of about 3,200 lbs in 1978 and has increased to over 4,000 lbs in 2017. On average, a 2017 model year vehicle has 70% more horsepower than a 1975 model year vehicle. The average USA LDV in 2017 had 233 hp. Although vehicle CO_2 emission rates increase with vehicle weight, the increase in CO_2 emission rate in g/mile for an incremental increase in vehicle weight or vehicle horsepower is much smaller in 2018 than in 1978, largely as a result of new vehicle technologies (EPA, 2019).

A growing share of the USA new production LDV fleet incorporates emerging technologies, including but not limited to turbo-charging, gas direct injection, continuously variable transmissions or transmissions with seven or more gears, and cylinder deactivation. There is a small but growing share of HEVs, PHEVs, and BEVs, with a very small number of FCEVs available in California. For gasoline LDVs, which comprise the vast majority of USA LDVs, multi-value engines with variable valve timing are becoming more common and are found in well over 90% of new vehicles. Almost 30% of LDV gasoline engines are expected to be turbo-charged in the 2018 model year, with 80% being 4-cylinder down-sized engines. With better designed automatic transmissions, the fuel economy advantage of manual versus automatic transmissions has not only been erased but automatic transmission vehicles are, on average, more fuel efficient than manual transmission vehicles, based on the 2016 and 2017 model years (EPA, 2019).

In Europe, diesel vehicles have been promoted over gasoline for some time because of perception that their inherently more energy efficient engines, which operate at higher compression ratios than gasoline engines, would lead to lower CO_2 emissions per vehicle mile traveled. Per unit of energy, CO_2 emissions are about 5% higher for diesel fuel compared to gasoline. In the past, diesel engines have been about 30% more efficient than a comparable gasoline engine. However, with the recent advances in technologies adopted into production light duty gasoline vehicles, gasoline vehicles can have CO_2 emissions that are the same or lower than a comparable diesel version of the same vehicle, even though diesel engines retain some efficiency advantage, as has been demonstrated for a VW Golf based on chassis dynamometer and on-road measurements (Mock, 2019).

Increasing the octane rating of retail fuels, such as by increasing the blend ratio of ethanol, provides greater engine knock resistance and, thus, enables better engine performance under high load conditions. Long-term availability of higher-octane fuels would enable design of engines tailored to such fuels that would have higher compression ratios than typical of the current market. Engine efficiency can increase with increasing compression ratio. However, such reductions would likely be modest, at 9% or less for fuel octane increases of 10 Research Octane Number (RON) coupled with an increase in compression ratio of 3 (Leone et al., 2015). The life cycle implications of higher ethanol blend fuels would have to be assessed.

The emergence of autonomous vehicles could potentially disrupt light duty vehicle technology and operation. However, the implications of autonomous vehicles on energy use and GHG emissions is highly uncertain, with some studies predicting large opportunities for efficiency and emission reduction, and others identifying sources of latent demand and operational practices that could substantially increase energy consumption and emissions (Frey, 2018). For example, a shared autonomous vehicle (SAV) could provide the same travel service as ten self-owned personal vehicles, but would have 11% more travel associate with reaching the next traveler. Results are sensitive to population density, congestion, and vehicle relocation strategies (Fagnant and Kockelman, 2014).

5. Life Cycle Perspective

The impact of LDVs includes "well to wheel" (WTW) processes that include production, distribution, storage, and in-vehicle use of transportation fuels. In addition to WTW processes, there is the separate impact of vehicle manufacturing, maintenance and end-of-life processes. Thus, the total life cycle impact of LDVs includes WTW and vehicle manufacturing, maintenance and end-of-life processes. For heavy duty trucks, rail, aircraft, and ocean-going vessels, WTW comprises 93% to nearly 100% of the total GHG emissions related to the vehicle and its operation. However, for LDVs, WTW is 67% to 83% of the total GHG impacts. While in both cases WTW accounts for the majority of GHG emissions, there is more to be gained from policies aimed at lowering the GHG intensity of manufacturing, maintenance, and end-of-life for light duty on road vehicles compared to other types of vehicles (Taptich et al., 2015). For all modes, reductions in intensity of GHG emissions for WTW aspects of vehicle usage can have significant impact.

An additional consideration in life cycle assessment of GHG emissions are the emission embodied in transportation infrastructure, such as roads, railroads, airports, marine ports, and so on. For example, the GHG emissions per passenger mile can be mostly from infrastructure for some light rail systems but, in terms of total life cycle GHG emissions, passenger rail is typically much lower emitting per passenger-mile than on road passenger transport. Vehicle operation contributes the majority of the GHG emissions impact for passenger road transport (Hodges, 2010).

6. Heavy Duty Vehicles

For USA heavy duty vehicles, 90% of fuel consumed was diesel, with the balance being gasoline and a small amount of LPG (Davis et al., 2017). Diesel engines, with their higher compression ratios, have typically been more energy efficient than gasoline engines, although gasoline engines are improving in efficiency as a result of higher compression ratios, turbo-charging, and other factors, as noted earlier. GHG emission reduction strategies for diesel trucks aimed at reducing energy losses typically include improved engines and transmissions, reduction of aerodynamic drag, reduction of rolling resistance, weight reduction for the empty vehicle, reduced energy consumption for auxiliary loads, and idle reduction. Other strategies include hybridization, use of biodiesel, and powertrains for alternative fuels, such as compressed natural gas, liquefied natural gas, battery electric, and hydrogen (USDOT, 2010).

The climate impact of diesel trucks can also be reduced in part by the deployment of diesel particulate filters (DPFs) that are highly effective (typically over 99% efficient) in reducing black carbon (BC) emissions (Sims et al., 2014; Frey, 2018). BC has short-term effects on climate forcing. Therefore, reductions in BC emissions can lead to relatively rapid climate change benefits. However, DPF operation typically leads to a small reduction in overall truck energy efficiency related to engine backpressure and the process of periodically regenerating the filter to burn off deposited black carbon, thus leading to a possible trade-off with increased vehicle operational CO_2 emissions.

Based on real-world measurements using an instrumented trailer, on-road tailpipe exhaust GHG emissions were compared for diesel, diesel hybrid, and compressed natural gas long-haul freight tractors. The vehicle sample included diesel vehicle with and without selective catalytic reduction for NO_x control. The CNG vehicle had a stoichiometric burn engine with a three-way catalyst. The measured greenhouse gases included CO_2, CH_4, and N_2O. The CO_2-equivalent emissions per vehicle-mile were compared among all of the vehicle types for five driving conditions, ranging from a near-dock drayage cycle to interstate highway. There was more variability in the emission rates with respect to driving conditions than there was when comparing the vehicle technologies. The hybrid truck had the lowest average emissions rate among the vehicle types under the near-dock drayage cycle, CNG had the lowest average rate for the regional highway driving, and the diesel trucks with SCR had the lowest average rate for highway hill climbing. Thus, the emissions benefit of one technology over the other were highly route-dependent (Quiros et al., 2017).

7. Electrification and Transportation

Emissions from electric vehicles are highly dependent on how the electricity was generated. Although electric motors are substantially more efficient than gasoline or diesel engines, the overall energy efficiency of a vehicle depends on all of the losses for the energy cycle and vehicle operation, including upstream power generation efficiency, transmission losses, and battery losses. From this perspective, electric vehicles are typically not substantially more or less energy efficient than gasoline or diesel vehicles. However, a key benefit of electric vehicles is that their emissions can improve as the energy mix for power generation becomes less carbon intensive over time, without the need to replace the vehicle itself in order to achieve GHG emission reductions.

The global stock of plug-in electric vehicles, which includes battery electric vehicles and plug-in hybrid electric vehicles, was over 3 million vehicles in 2017, of which 40% were in China. Nearly all of the electric buses and electric two-wheel vehicles are in China. The global number of electric chargers reached almost 3 million in 2017 (IEA, 2018b). BEVs represented 0.6% of new car registrations in Europe in 2017. PHEVs accounted for 0.8% of European new passenger car registrations in the same year

(EEA, 2018). Globally, plug-in electric vehicle (PEV) sales, including BEVs and PHEVs, are expected to rise, reaching 8% to 26% of new vehicle sales by 2040, as indicated in Figure 6.

The International Energy Agency projects that, as a result of existing and announced policies, global coal-based power generation is expected to increase from 9.86 TWh in 2017 to 10.34 TWh in 2040, although the relative share of coal for power generation will decrease. Global natural gas-based generated power is projected to increase from 5.86 TWh in 2017 to 9.07 TWh in 2040. Under a "sustainable development" scenario, the amount of power from coal would decrease by about 80% and from natural gas would remain approximately the same. World oil demand for transport is projected to increase from 2.56 Mtoe in 2017 to 2.97 MToe in 2040.

As shown in Figure 7, EIA projects that the global electric power generation fuel mix will continue to be dominated by fossil fuels over the coming decades. Although much of the growth in power generation will be based on renewable energy sources, growth is also projected in the amount of power generated from natural gas as well as some growth in the amount of power, although not the share, from coal (EIA, 2017). Even though renewables are likely to capture a major share of power generation, it may be decades before fossil fuel drops below 50% of total power generation. The projections from IEA and EIA are based on a large number of assumptions and actual trends in the fuel mix used for power generation are unlikely to be identical to any of these projections (Frey, 2018). However, it is possible

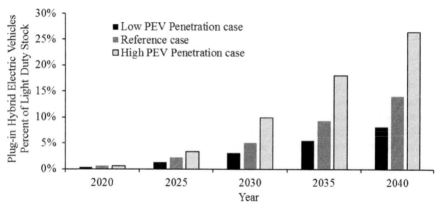

Figure 6. Projected trend from 2020 to 2040 in Worldwide Plug-in Electric Vehicles (PEVs) as percent of light-duty vehicle stock. Source: (Lynes, 2017).

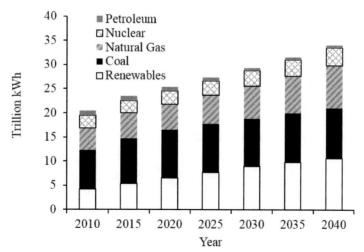

Figure 7. Trends from 2010 with projections to 2040 in the global fuel mix for electric power generation. Source: (EIA, 2017).

that electrification of vehicles has the potential to shift conventional air pollutant emissions (e.g., nitrogen oxides, carbon monoxide, hydrocarbons, fine particles) from dense high-traffic areas to the downwind regions affected by power plant plumes.

Based on an analysis of the USA's energy mix, one study concluded that simply increasing the share of electric drive vehicles (EDVs) will not, in itself, achieve GHG emission reductions, but that substantial GHG reductions require reduction in the carbon intensity of power generation. Reductions in the share of coal and increases in the share of natural gas would reduce GHG emissions for electricity used for vehicle charging. Renewable portfolio standards (RPS) require a minimum share of renewable energy in the grid mix and can lead to GHG emission reductions. Factors that affect the market penetration of EDVs include battery price and oil price. High oil costs and low battery prices would motivate a larger penetration of such vehicles into the market. However, a secondary effect of high EDV deployment is that sustained demand for electricity might lead to continued operation of existing coal plants that might otherwise be retired under a different power demand scenario (Babaee et al., 2014).

However, other studies indicate that electric vehicles might even lead to lower operational emissions of GHG, except in regions that have very high proportions (e.g., 90% or more) of coal-based power generation. For example, in an assessment of BEV GHG emissions in selected regions of China and the USA, estimated GHG emissions were lower in all cases except for the Beijing-Tianjin region. For example, GHG emissions were estimated to be 20%–40% lower in the Yangtze River and Pearl River Deltas, and were as much as 50%–60% lower in California and the U.S. northeast (Huo et al., 2015).

Based on a comparison of selected 2017 model year BEVs and conventional gasoline LDVs, as shown in Figure 8, the annual CO_2 emissions for the BEVs is highly variable, depending on the energy mix used for power generation among states in the USA, but on average would be lower than for the comparable gasoline vehicle. For example, the Bolt, Golf, i3, Clarity, Leaf, and Focus are approximately comparable in size to the Honda Civic, but would have CO_2 emissions from power generation that average 42–47% lower than the tailpipe exhaust emissions of the Honda Civic. There are some power generation energy mixes, such as for the state of West Virginia, that would lead to higher CO_2 emissions for the electric vehicles. The larger Tesla S and BYD e6 would, on average, have operational CO_2 emissions that are 38–54% lower than the Chevrolet Impala, with the Tesla being lower emitting in all states. Thus, except

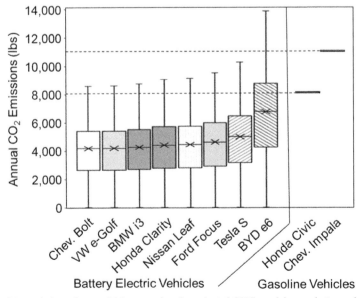

Figure 8. Annual CO_2 emissions from vehicle operation for selected 2017 model year battery electric vehicles and comparable conventional gasoline vehicles. Indirect CO_2 emissions for battery electric vehicles from power generation are estimated for each U.S. state. Values shown as box and whiskers for each BEV include the median, inter-quartile range, minimum, and maximum among USA states. Source: Alternative Fuel Data Center, U.S. Department of Energy.

for a few exceptional cases among certain states of the USA, the CO_2 emissions directly attributable to BEV operation would be lower. The actual emissions depend on the time of day of vehicle charging and the marginal emission rate associated with dispatching of power generation assets.

Battery electric trucks could reduce operational CO_2 emissions. A battery electric tractor-trailer truck with a 500-mile range would have a tare weight 6,000 kg greater than a diesel truck, resulting in loss of payload. Alternatively, a battery electric tractor-trailer truck with no additional weight would have a range of only 140 miles. Range would be reduced at cold temperatures (Sharpe, 2019).

Lithium-ion batteries are the key battery technology used in electric vehicles. Key focus areas for further development of these batteries including chemistry, energy storage capacity, manufacturing scale, and charging speed. Battery cost reductions are expected. Plug-in electric vehicle stock is expected to increase substantially by 2030, depending on policy and cost reduction, to between 125 million and 225 million vehicles. There would be concurrent growth in vehicle chargers under these scenarios. Based on current battery designs, and expected growth in the number of batteries, the demand for cobalt and lithium will increase. Research on battery chemistry includes potential reductions in cobalt demand per battery, but increases in total demand for cobalt are likely (IEA, 2018b). However, actual shortages of materials for battery production are unlikely (Le Petit, 2017).

8. Bus Transit

Based on simulation case studies for both Finland and California, the life cycle CO_2 emissions of transit buses, based on the fuel cycle and vehicle operation, per vehicle mile are lowest for battery electric buses, approximately similar for diesel hybrid and hydrogen fuel cell buses, and highest for conventional diesel and CNG buses. Electric buses may be cost-competitive with diesel buses within 10 to 20 years, whereas fuel cell buses are likely to continue to be more expensive than diesel buses. Buses that can take advantage of opportunity charging at bus stops or bus terminals could have smaller batteries, and lower costs, than buses that rely on overnight charging for a full day of driving range (Lajunen and Lipman, 2016).

A comparison of transit bus GHG emissions along a corridor found that operational emissions contribute the largest share of total life cycle emissions, and that life cycle emissions increased in order for hybrid, compressed natural gas, biodiesel, and conventional diesel transit buses (Chan et al., 2013). Comparisons of transit bus options should take into account local driving cycles, terrain, electricity generation mix, and meteorology (Xu et al., 2015).

9. Eco-Driving

So-called "eco-driving" is a technique, typically based on reducing the rate of acceleration among other strategies, for reducing vehicle energy consumption for a given trip. However, in practice, eco-driving appears to be of limited effectiveness. There have been numerous studies that have evaluated the effect of eco-driving training on bus drivers. A Calgary-based municipal fleet trained drivers to reduce idling, which was reduced by 4% to 10% per vehicle per day (Rutty et al., 2013). In Sweden, a control group, a group who received in-vehicle feedback, and a group who received feedback plus training were compared. The latter two groups reduced frequency of high acceleration and high speed and achieved 6.8% fuel savings (Strömberg and Karlsson, 2013). A study focusing on private car owners found that eco-driving training led to city and highway fuel consumption decreases of 4.6% and 2.9%, respectively, but that these reductions dropped to 2.5% and 0%, respectively, 10 months later (Barla et al., 2017).

10. Air Conditioning

Even though air conditioning (AC) is the largest individual auxiliary load in a typical passenger vehicle, few studies have examined the effect of AC operation on vehicle fuel use and emission rates. Data used to model the effect of air conditioning on exhaust emissions in the MOVES model was collected in

1997 and 1998 (EPA, 2015) and, thus, is out of date. Vehicles have typically been using HFC-134a as the refrigerant since the early 1990s. Leakage rates for refrigerants are typically estimated to be approximately 5% of the refrigerant charge. Uncertainty in the CFC-12 emissions from vehicles from Hong Kong was estimated to be ±29%, and uncertainties for HFC-134a emissions were reported to be similar (Yan et al., 2014). Under current U.S. fuel economy and GHG emission standards, manufacturers are given AC leakage credits, counted toward their corporate average GHG emissions, for improvements to AC systems based on O-rings, seals, valves, and fittings, and substitution of lower global warming potential (GWP) refrigerants. For example, HFO-1234yf, first introduced in 42,384 model year 2013 vehicles, has a GWP of 4 compared to 1,430 for HCF-134a. In 2016, 2.2 million new vehicles were produced using HFO-1234yf (EPA, 2018b).

11. Rail

Rail tends to be among the most energy efficient of transportation modes for both passenger and freight services. Improved operational practices, such as chokepoint relief and idle reduction, could reduce energy consumption and CO_2 emissions (USDOT, 2010). Infrastructure and technologies that make rail more attractive than more energy intensive transport modes, such as high-speed rail versus aircraft, could lead to inter-modal substitutions that lead to net GHG emission reductions. Methods for reducing the carbon intensity of train operations include improved aerodynamics, weight reduction, more efficient power electronics, regenerative braking, and electrification. For diesel rail, the use of drop-in biofuels can reduce carbon intensity of train operations. Hydrogen fuel cell power trains have been proposed for locomotives (Sims et al., 2014).

Electrification of existing long-distance rail services is prohibitively expensive. Possible alternatives to reducing CO_2 emissions from rail transport in such cases is the substitution of biofuels for diesel fuels, hybridization, or a transition to hydrogen-fueled locomotives (Hall et al., 2018; Graver and Frey, 2016). However, the current hydrogen fuel cycle, which is mostly based on steam reforming of natural gas, is not favorable with regard to GHG emission reductions.

12. Shipping

International shipping by marine vessels is estimated to have contributed 812 million metric tons of CO_2 emissions globally in 2015 (Olmer et al., 2017). Marine emissions are expected to increase from approximately 10% of transport related emissions in 2018 to 20% by 2060 (Hall et al., 2018). GHG emissions from maritime bunker fuels sold in Europe have increased since the 2008 global economic recession, reaching 147 Mt CO_2 in 2016. Although 10 percent below 2005 levels, these emissions would have to be reduced by 34% to meet the European 2050 target (EEA, 2018).

Over a short period from 2013 to 2015, the world fleet of ships used internationally grew by 1.5%, the amount of shipping in deadweight tons per nautical mile increased by 7%, and main engine power increased by 6% to 10%, especially for tankers, general cargo ships, and container ships. Fuel efficiency has increased slightly for some ship classes. For example, general cargo ship CO_2 intensity decreased by 5%. However, total emissions of general cargo ships increased by 9% because of increased transport supply. Although on average ship speeds did not change, the cruising speed of tankers of greater than 200,000 deadweight tons and of containers ships with more than 14,500 Twenty-foot Equivalent Units (TEU) increased by 4% and 11%, respectively. Faster cruising speed is associated with higher emission rates. In addition to emitting CO_2, approximately 21% of the total CO_2-equivalent emissions from ships includes BC (Olmer et al., 2017).

According to an estimate, CO_2 emissions from marine vessel activity centered on China is expected to increase by 8.3% per year from a baseline of 2007 to 2035. Bulk carrier, container, and tanker vessels are expected to comprise over 90% of these emissions. These increases could be partly mitigated by reduction in transport demand, reduction in energy intensity of transport, and substitution of lower carbon energy sources. A key step toward implementing policies would be regulations for monitoring, reporting, and verifying vessel activity in territorial waters. Possible policy options for reductions in marine emissions could include a maritime carbon tax or a cap-and-trade scheme (Yang et al., 2017).

Battery-electric vessels, including ferries in several European countries, have recently been introduced on a limited basis. For short-distance routes, battery-electric and hydrogen-fueled vessels may be feasible and could offer substantial GHG emissions reductions depending on the upstream energy cycle (Hall et al., 2018).

13. Aircraft

The estimated GHG emissions from aviation have approximately doubled from 1980 to 2012. There were nearly 20,000 commercial passenger aircraft in 2010. By 2050, the number of aircraft is expected to reach nearly 70,000, with GHG emissions approximately tripling compared to 2010 (Kharina et al., 2016). In Europe, GHG emissions from international aviation increased at an average rate of over 2% per year from 2013 to 2017 (EEA, 2018).

In 2016, the International Civil Aviation Organization (ICAO) completed a CO_2, or fuel efficiency, standard for aircraft that individual countries are expected to adopt by 2020. The standard is based on achieving carbon-neutral growth. While there are many potential technologies that could improve aircraft fuel efficiency, some are mutually exclusive. General categories of fuel-saving technologies include: (a) materials, such as advanced composites, that reduce weight; (b) fuselage, such as low-friction coatings and riblets; (c) modifications to control surfaces, such as wingtip devices and variable camber; (d) laminar flow; (e) changes to the external surfaces of the engine such as the nacelle and nozzles; and (f) internal changes to the engine such as higher pressure ratio, higher firing temperature, and others. Although some of these types of features exist in some form on current production aircraft, such as wingtip devices, improvements in all of these could further improve fuel efficiency compared to current production aircraft. The deployment of cost-effective packages of compatible technologies could reduce fuel consumption by as much as 25% for new aircraft in 2024 and 40% for new aircraft in 2034. These potential reductions are much greater than the more modest reductions of current new aircraft designs. The aircraft manufacturing industry typically prefers evolutionary designs rather than "clean sheet" designs that would lead to, for example, entirely new airframes. Other potential options to reduce CO_2 emissions from aircraft include biofuels, operational practices, and improved air traffic control (Kharina et al., 2016).

Perhaps the most feasible approach to decarbonizing aircraft operation is to deploy low-carbon jet fuels. The key barriers to wide-scale deployment of such fuels is the lack of feedstock supply chain and adequate capacity of an advanced fuel industry. In the short-term, low-carbon fuel policy might be better aimed at on road transportation which would facilitate longer-term adoption in the aviation sector. For example, the development of policy, infrastructure, and capacity to produce biofuels for other transportation modes, such as on road vehicles, would enable evolutionary changes to provide biofuels for aircraft. Policy options that would address other transport modes would enable decarbonization of aircraft operation including investments in sustainable biomass followed by a longer-term increasing share of advanced fuels used in aviation (Searle et al., 2019).

Aircraft are unique among transportation modes in that their emissions at 8 km to 13 km altitude can create aircraft produced condensation trail ("contrail") line-shaped ice clouds that enhance climate change. Such clouds form in part because of condensable vapors produced by aircraft engine, including water vapor, sulfuric acid, nitric acid, and low volatility hydrocarbons. Ultrafine soot particles produced in aircraft engine exhaust can also serve as condensation nuclei. However, the radiative forcing implications of these aircraft induced clouds is not well quantified, because their effect has to be disentangled from that of naturally occurring cirrus clouds. Simulation results indicate that contrails that do not retain their linear shape, known as contrail cirrus clouds, have substantially more potential climate impact than the linear contrails. Possible techniques to mitigate aircraft induced cloud formation in the short term include lean combustion technology and alternative fuels, such as biofuels with lower sulfur and lower aromatic species content. Longer term mitigation could include, for example, adoption of engine technology that would use fuels such as liquid hydrogen or liquefied natural gas, which would eliminate or reduce soot and sulfur emissions, coupled with more aerodynamic airframes, such as blended wing body technology, which would reduce the amount of energy consumed (Karcher, 2018).

Aviation operations include the use of ground support equipment at airports. Such equipment is amenable to electrification (Hall et al., 2018).

14. Policy Options

As noted earlier by Sims et al. (2014), GHG emissions from transport could increase at a faster rate than for any other energy end-use sector through 2050. Thus, substantial policy intervention will be required if these emissions are to be sufficiently mitigated.

In the USA and other countries, various policy options could be adopted at the regional, state, or local level to reduce GHG emissions from transport. Examples include incentives for purchase or operation of clean vehicles, public transit, and increase in cycling and walking, as well as efficient land-use policies. Transportation pricing policies, such as a carbon fee or mileage-based user fees, could help incentivize alternative choices with lower GHG emissions (Pacyniak et al., 2015).

Deep reductions in GHG emissions from transportation will require substantial changes to energy supply, technology, and transportation activity. Reductions in emissions from new vehicle technology are hampered by the slow rate of fleet turn-over for many transportation modes. Policy options, such as fuel taxes, vehicle efficiency standards, low carbon fuel standards, land use controls, travel demand measures and investments in new infrastructure, are among those that may be effective in achieving reduced GHG emissions, but that are also unlikely to be quickly adopted or adopted at all without significant acceptance by key stakeholders and decision makers. The co-benefits of various strategies, such as increasing energy security or reducing congestion, may be equally or more attractive to some stakeholders than GHG emissions reductions. The development and design of policy options should be informed by policy-relevant research (TRB, 2011).

An example of a policy option that appears to have broad support in many countries is standards for on road vehicles for fuel economy, GHG emissions, or both. Such standards are in place in countries that represent 80% of new vehicle sales. These standards typically focus on light duty passenger vehicles but standards are also emerging for light commercial vehicles and heavy-duty vehicles. Problems uncovered, especially in Europe, regarding gaps between official rated values and real-world CO_2 emissions are being addressed, at least partially, via improved measurement and enforcement procedures. Stringent fuel economy or GHG emissions standards are shown to drive innovation, leading to the adoption of more efficient vehicle technologies (Yang and Bandivadekar, 2017).

In general, in the absence of prioritization of effective and multipronged policies that address fuels, vehicle technology, efficient vehicle operation, mode shifts, infrastructure, and travel demand, GHG emissions from transport are likely to increase.

References

AASHTO. 2015. Commuting in America 2013: The National Report on Commuting Patterns and Trends. Washington, DC: American Association of State Highway and Transportation Official.

Auto Alliance. 2019. Advanced Technology Vehicle Sales Dashboard. https://autoalliance.org/energy-environment/advanced-technology-vehicle-sales-dashboard/, accessed 7/27/19.

Auto Alliance. 2019. Buyers Wanted! https://autoalliance.org/energy-environment/buyers-wanted/, accessed 7/27/19.

Azari, K.A. Arintono, S., Hamid, H. and Davoodi, S.R. 2013. Evaluation of demand for different trip purposes under various congestion pricing scenarios. J. Transp. Geogr. 29(May): 43–51. doi: 10.1016/j.jtrangeo.2013.01.001.

Barla, P., Gilbert-Gonthier, M., Lopez Castro, M.A. and Miranda-Moreno, L. 2017. Eco-driving training and fuel consumption: Impact, heterogeneity and sustainability. Energy Econ. 62(February): 187–194. doi: 10.1016/j.eneco.2016.12.018.

Boulange, C., Gunn, L., Giles-Corti, B., Mavoa, S., Pettit, C. and Badland, H. 2017. Examining associations between urban design attributes and transport mode choice for walking, cycling, public transport and private motor vehicle trips. J. Transp. Health 6(September): 155–166. doi: 10.1016/j.jth.2017.07.007.

BTS. 2016. Transportation Statistics Annual Report 2016. Washington, DC: Bureau of Transportation Statistics, U.S. Department of Transportation.

Capparella, J. 2019. The Best-Selling Cars, Trucks, and SUVs of 2018. Car and Driver, January 3, 2019. https://www.caranddriver.com/news/g25558401/best-selling-cars-suv-trucks-2018/, accessed 7/27/19.

Chan, S., Miranda-Moreno, L.F., Alam, A. and Hatzopoulou, M. 2013. Assessing the impact of bus technology on greenhouse gas emissions along a major corridor: A lifecycle analysis. Transportation Research Part D 20: 7–11. http://dx.doi.org/10.1016/j.trd.2013.01.004.

Chu, S. 2015. Car restraint policies and mileage in Singapore. Transp. Res. Part Policy Pract. 77(July): 404–412. doi: 10.1016/j.tra.2015.04.028.

Davis, S.C., Williams, S.E. and Boundy, R.G. 2017. Transportation Energy Data Book Edition 36. ORNL/TM-2017/513. Oak Ridge, TN: Prepared by Oak Ridge National Laboratory for U.S. Department of Energy. http://cta.ornl.gov/data/tedb36/Edition_36_Full_Doc.pdf.

Ding, C., Wang, D., Liu, C., Zhang, Y. and Yang, J. 2017. Exploring the influence of built environment on travel mode choice considering the mediating effects of car ownership and travel distance. Transp. Res. Part Policy Pract. 100(Supplement C): 65–80. doi: 10.1016/j.tra.2017.04.008.

EEA. 2018. Progress of EU transport sector towards its environment and climate objectives, Briefing No. 15/2018, European Environment Agency, Brussels, TH-AM-18-017-EN-N - ISBN 978-92-9480-020-6 - ISSN 2467-3196. Doi: 10.2800/954310.

EIA. 2016. International Energy Outlook 2016. DOE/EIA-0484, U.S. Energy Information Administration, Washington, DC.

EIA. 2017. International Energy Outlook 2017. #IEO2017. Washington, DC: U.S. Energy Information Agency. https://www.eia.gov/outlooks/ieo/.

EPA. 2015. Emission Adjustments for Temperature, Humidity, Air Conditioning, and Inspection and Maintenance for On-Road Vehicles in MOVES2014. EPA-420-R-15-020. Ann Arbor, MI: U.S. Environmental Protection Agency.

EPA. 2018. Fast Facts: U.S. Transportation Sector Greenhouse Gas Emissions 1990–2016, EPA-420-F-18-013, U.S. Environmental Protection Agency, Washington, DC.

EPA. 2018b. Greenhouse Gas Emission Standards for Light-Duty Vehicles: Manufacturer Performance Report for the 2016 Model Year. Washington, DC: U.S. Environmental Protection Agency.

EPA. 2019. The 2018 EPA Automotive Trends Report: Greenhouse Gas Emissoins, Fuel Economy, and Technology since 1975, EPA-420-R-19-002, U.S. Environmental Protection Agency, Washington, DC, March 2019.

Fagnant, D.J. and Kockelman, K.M. 2014. The travel and environmental implications of shared autonomous vehicles, using agent-based model scenarios. Transp. Res. Part C Emerg. Technol. 40(Supplement C): 1–13. doi: 10.1016/j.trc.2013.12.001.

Facanha, C., Blumberg, K. and Miller, J. 2012. Global Transportation Energy and Climate Roadmap. Washington, DC: International Council on Clean Transportation.

Feng, Y., Fullerton, D. and Gan, L. 2013. Vehicle choices, miles driven, and pollution policies. J. Regul. Econ. 44(1): 4–29. doi: 10.1007/s11149-013-9221-z.

FHWA. 1997. State Motor Vehicle Registrations by Years 1900–1995, Table MV-1. Federal Highway Administration, U.S. Department of Transportation. https://www.fhwa.dot.gov/ohim/summary95/mv200.pdf.

Frey, H.C. 2018. Trends in on road transportation energy and emissions. Journal of the Air & Waste Management Association 68(6): 514–563.

Graver, B.M., Frey, H.C. and Hu, J. 2016. Effect of biofuels on real-world emissions of passenger locomotives. Environmental Science and Technology 50(21): 12030–12039. DOI: 10.1021/acs.est.6b03567.

Hall, D., Pavlenko, N. and Lutsey, N. 2018. Beyond Road Vehicles: Survey of Zero-Emission Technology Options Across the Transport Sector, Working Paper 2018–11, International Council on Clean Transportation, Washington, DC.

Hobbs, F. and Stoops, N. 2002. Demographic Trends in the 20th Century. CENSR-4, Census 2000 Special Reports. Washington, DC: U.S. Census Bureau.

Hodges, T. 2010. Public Transportation's Role in Responding to Climate Change, Federal Transit Administration, U.S. Department of Transportation, Washington, DC.

Huo, H., Cai, H., Zhang, Q., Liu, F. and He, K. 2015. Life-cycle assessment of greenhouse gas and air emissions of electric vehicles: A comparison between China and the U.S. Atmospheric Environment 108: 107–116. http://dx.doi.org/10.1016/j.atmosenv.2015.02.073.

IEA. 2018a. CO_2 Emissions from Fuel Combustion: Highlights, 2018 Edition, International Energy Agency, 2018.

IEA. 2018b. Global EV Outlook 2018: Towards Cross-Modal Electrification, International Energy Agency, 2018.

IEA. 2019. CO_2 Emissions Statistics: An Essential Tool for Analysts and Policy Makers. https://www.iea.org/statistics/co2emissions/, accessed 7/28/19.

Karchar, B. 2018. Review article: Formation and radiative forcing of contrail cirrus. Nature Communications 9: 1824. DOI: 10.1038/s41467-018-04068-0.

Kay, A.I., Noland, R.B. and Rodier, C.J. 2014. Achieving reductions in greenhouse gases in the us road transportation sector. Energy Policy 69(June): 536–545. doi: 10.1016/j.enpol.2014.02.012.

Kharina, A., Rutherford, D. and Zeinali, M. 2016. Cost Assessment of Near and Mid-Term Technologies to Improve New Aircraft Fuel Efficiency, International Council on Clean Transportation, Washington, DC.

Lajunen, A. and Lipman, T. 2016. Lifecycle cost assessment and carbon dioxide emissions of diesel, natural gas, hybrid electric, fuel cell hybrid and electric transit buses. Energy 106: 329–342. http://dx.doi.org/10.1016/j.energy.2016.03.075.

Leone, T.G., Anderson, J.E., Davis, R.S., Iqbal, A., Reese, R.A., Shelby, M.H. and Studzinski, W.M. 2015. The effect of compression ratio, fuel octane rating, and ethanol content on spark-ignition engine efficiency. Environ. Sci. Technol. 49: 10778–10789. DOI: 10.1021/acs.est.5b01420.

Le Petit, Y. 2017. Electric vehicle life cycle analysis and raw material availability. Transport and Environment, Brussels, Belgium.

Liu, Y. and Cirillo, C. 2016. Evaluating policies to reduce greenhouse gas emissions from private transportation. Transp. Res. Part Transp. Environ. 44(May): 219–233. doi: 10.1016/j.trd.2016.02.018.

Lynes, M. 2017. Plug-in Electric Vehicles: Future Market Conditions and Adoption Rates. DOE/EIA-0484(2017). IEO2017—Issues in Focus Articles—U.S. Energy Information Administration. Washington, DC: Energy Information Agency, U.S. Department of Energy. https://www.eia.gov/outlooks/ieo/section_issues.php#pev.

Mahendra, A., Bowen, B., Simons, M. and Adler, K. 2012. Impacts of implementing transportation control measures on travel activity and emissions. Transp. Res. Rec. J. Transp. Res. Board 2287(November): 113–121. doi: 10.3141/2287-14.

McKenzie, B. and Rapino, M. 2011. Commuting in the United States: 2009. ACS-15. American Community Survey Reports. Washington, DC: U.S. Census Bureau.

Mock, P. 2018. Fact Sheet: Europe, Gasoline vs. Diesel, Comparing CO_2 Emission Levels of a Modern Medium Size Car Model Under Laboratory and On-Road Testing Conditions, International Council on Clean Transportation, Washington, DC.

OICA. 2019. Vehicles in Use. International Organization of Motor Vehicle Manufacturers. http://www.oica.net/category/vehicles-in-use/ (accessed July 28, 2019).

Olivier, J.G.J., Schure, K.M. and Peters, J.A.H.W. 2017. Trends in Global CO_2 and Total Greenhouse Gas Emissions, Publication Number 2674, PBL Netherlands Environmental Assessment Agency, The Hague, The Netherlands.

Olmer, N., Comer, B., Roy, B., Mao, X. and Rutherford, D. 2017. Greenhouse Gas Emissions from Global Shipping, 2013–2015. International Council on Clean Transportation, Washington, DC.

Pacyniak, G., Zyla, K., Arroyo, V., Goetz, M., Porter, C., Jackson, D. and Indrakanti, S. 2015. Reducing Greenhouse Gas Emissions from Transportation: Opportunities in the Northeast and Mid-Atlantic, Georgetown Climate Center, Washington, DC. https://www.georgetownclimate.org/files/report/GCC-Reducing_GHG_Emissions_from_Transportation-11.24.15.pdf.

Quiros, D.C., Smith, J., Thiruvengadam, A., Huai, T. and Hu, S. 2017. Greenhouse gas emissions from heavy-duty natural gas, hybrid, and conventional diesel on-road trucks during freight transport. Atmospheric Environment 168: 36–45. http://dx.doi.org/10.1016/j.atmosenv.2017.08.066.

Rutty, M., Matthews, L., Andrey, J. and Matto, T.D. 2013. Eco-driver training within the city of calgary's municipal fleet: Monitoring the impact. Transp. Res. Part Transp. Environ. 24(October): 44–51. doi: 10.1016/j.trd.2013.05.006.

Searle, S., Pavlenko, N., Kharina, A. and Giuntoli, J. 2019. Long-term Aviation Fuel Decarbonization: Progress, Roadblocks, and Policy Opportunities. International Council on Clean Transportation, Washington, DC.

Sharpe, B. 2019. Zero-Emission Tractor-Trailers in Canada, Working Paper 2019-04. International Council on Clean Transportation, Washington, DC.

Shay, E. and Khattak, A.J. 2012. Household travel decision chains: Residential environment, automobile ownership, trips and mode choice. Int. J. Sustain. Transp. 6(2): 88–110. doi: 10.1080/15568318.2011.560363.

Shekarchian, M., Moghavvemi, M., Zarifi, F., Moghavvemi, S., Motasemi, F. and Mahlia, T.M.I. 2017. Impact of infrastructural policies to reduce travel time expenditure of car users with significant reductions in energy consumption. Renew. Sustain. Energy Rev. 77(September): 327–335. doi: 10.1016/j.rser.2017.04.015.

Sims, R., Schaeffer, R., Creutzig, F., Cruz-Núñez, X., D'Agosto, M., Dimitriu, D., Figueroa Meza, M.J., Fulton, L., Kobayashi, S., Lah, O., McKinnon, A., Newman, P., Ouyang, M., Schauer, J.J., Sperling, D. and Tiwari, G. 2014. Transport. In: Edenhofer, O., Pichs-Madruga, R., Sokona, Y., Farahani, E., Kadner, S., Seyboth, K., Adler, A., Baum, I., Brunner, S., Eickemeier, P., Kriemann, B., Savolainen, J., Schlömer, S., von Stechow, C., Zwickel, T. and Minx, J.C. (eds.). Climate Change 2014: Mitigation of Climate Change. Contribution of Working Group III to the Fifth Assessment Report of the Intergovernmental Panel on Climate Change. Cambridge University Press, Cambridge, United Kingdom and New York, NY, USA.

Strömberg, H.K. and Karlsson, I.C.M. 2013. Comparative effects of eco-driving initiatives aimed at urban bus drivers—results from a field trial. Transp. Res. Part Transp. Environ. 22(July): 28–33. doi: 10.1016/j.trd.2013.02.011.

Sun, B., Ermagun, A. and Dan, B. 2017. Built environmental impacts on commuting mode choice and distance: Evidence from Shanghai. Transp. Res. Part Transp. Environ. 52(May). Land Use and Transportation in China 441–453. doi: 10.1016/j.trd.2016.06.001.

Taptich, M.N., Horvath, A. and Chester, M.V. 2015. Worldwide greenhouse gas reduction potentials in transportation by 2050. Journal of Industrial Ecology 20(2): 329–340. DOI: 10.1111/jiec.12391.

TRB. 2011. Policy Options for Reducing Energy Use and Greenhouse Gas Emissions from U.S. Transportation: Special Report 307. Transportation Research Board, Washington, DC: The National Academies Press. https://doi.org/10.17226/13194.

USDOT. 2010. Transportation's Role in Reducing U.S. Greenhouse Gas Emissions, Volume 1: Synthesis Report, Report to Congress, U.S. Department of Transportation, Washington, DC.

Wang, D. and Zhou, M. 2017. The built environment and travel behavior in urban china: A literature review. Transp. Res. Part Transp. Environ. 52(May). Land Use and Transportation in China 574–585. doi: 10.1016/j.trd.2016.10.031.

Welch, T.F. and Mishra, S. 2014. Envisioning an emission diet: Application of travel demand mechanisms to facilitate policy decision making. Transportation 41(3): 611–631. doi: 10.1007/s11116-013-9511-4.

Xu, Y., Gbologah, F.E., Lee, D.-Y., Liu, H., Rodgers, M.O. and Guensler, R.L. 2015. Assessment of alternative fuel and powertrain transit bus options using real-world operations data: Life-cycle fuel and emissions modeling. Applied Energy 154: 143–159. http://dx.doi.org/10.1016/j.apenergy.2015.

Yan, H.H., Guo, H. and Ou, J.M. 2014. Emissions of halocarbons from mobile vehicle air conditioning system in Hong Kong. J. Hazard. Mater. 278(August): 401–408. doi: 10.1016/j.jhazmat.2014.06.020.

Yang, H., Ma, X. and Xing, Y. 2017. Trends in CO_2 emissions from china-oriented international marine transportation activities and policy implications. Energies 10: 980, 17 pages, published 17 July 2017. doi: 10.3390/en10070980.

Yang, Z. and Bandivadekar, A. 2017. 2017 Global Update: Light-Duty Vehicle Greenhouse Gas and Fuel Economy Standards, International Council on Clean Transportation, Washington, DC.

Section 3

Wind/Solar/Hydro/Nuclear

CHAPTER 18

Nuclear Energy, the Largest Source of CO_2 Free Energy

Issues and Solutions

Anthony J Baratta

1. Introduction

Nuclear energy is the only low-carbon source of energy capable of producing 1,000s of MWs[1] day in and day out, regardless of weather or time of day. Yet many developed countries, including Germany and the U.S., are moving away from the use of nuclear energy in favor of renewables supplemented by "clean" natural gas. In the U.S., the main reason for the shift is money, secondarily influenced by public safety concerns. In Germany, it is public concern over safety in the aftermath of the Fukushima Daiichi accident.

This chapter examines these factors and explores alternatives. The fact that nuclear is currently the largest producer of low-carbon energy is discussed. The process of converting the energy in the nucleus of uranium to a useful form is explained. The risks of the process are analyzed and the three principle nuclear accidents, Three Mile Island, Chernobyl, and Fukushima Daiichi, are examined and their health consequences considered. Finally, the question of cost and competitiveness is considered for current nuclear plants; the new generation of reactors might overcome these concerns.

2. The Nuclear Process

Fission and fusion are the two types of nuclear reactions capable of producing large amounts of energy. Fission is currently the only practical process capable of commercialization. Discovered in the 1930s by Otto Hahn and Fritz Strausmann, the process of fission involves the breakup of the nuclei of heavy elements such as uranium. In the process of the breakup, energy is released in the form of radiation. This radiation is the energy associated with the movement of the fragments of the original nucleus away from the fission site.

By comparison, fusion involves combining the nuclei of light elements such as hydrogen to form heavier elements. The result is energy in the form of radiation and energy in the form of the motion of the resulting heavier nucleus. The original theory of fusion was proposed by Arthur Eddington in the 1920s

---

Emeritus Professor Nuclear Engineering, Penn State University.
[1] MW or Megawatt is a unit of power. Most modern nuclear power plants are capable of generating 1000 MW or more, enough power to run one million homes.

to explain energy generation in stars such as our sun. Fusion in the laboratory was achieved in the 1930s by Mark Oliphant.

The energy released in both fission and fusion eventually appears as heat. Current nuclear reactors use the fission of uranium to produce energy. The heat generated in the fission process is used to produce steam which drives a turbine generator, resulting in an electric current. Aside from the source of the energy, the process of generating electricity in a nuclear power plant is identical to that of many fossil fuel power plants. While no practical fusion reactor has been built, it is expected that the final result of electrical generation will rely on a similar process.

A practical system must be capable of running at high power levels for extended periods of time. In the case of fission, the process involves a chain reaction where neutrons are emitted when the uranium fissions are used to induce fissions in other uranium nuclei. As long as sufficient fuel is available, the process can continue. In fact, most commercial nuclear reactors have enough fuel to generate 1,000 MW of electricity for more than two years. The fuel is contained in the reactor core and the amount of fuel is extremely small compared to the amount of fossil fuel needed to power a conventional power plant. The reason for the difference is that a single fission generates 100 million times as much energy as the combustion of one fossil fuel atom. Nuclear processes are much more energy intensive than the chemical process of combustion due to the extremely strong nuclear forces at work.

To date, no comparable fusion reactors are able to operate for extended periods of time. However, there have been significant advances, with the latest systems achieving fusion and maintaining conditions for fusion for times measured in minutes. While this may seem small compared to fission reactors which operate for years, it does represent significant progress in the field. The U.S. has generated more energy from fusion than was consumed to cause the fusion (Herrmann, 2014). The Chinese created the conditions necessary for fusion and maintained fusion for over 100 seconds (Chinese Academy of Science, 2017), no mean achievement, since only a few years ago the times were measured in fractions of a second.

The principle safety concerns with fission reactors are the highly radioactive fragments of the uranium nucleus that are formed when the uranium fissions. These "fission products" are sufficiently radioactive that a person exposed directly to the fuel would receive a lethal dose of radiation in a matter of seconds. In normal operation, the "fission products" are contained safely in the fuel which is confined to the reactor, located inside a rugged containment structure. Under accident conditions, these barriers can and have been breached, resulting in the release of radioactive material to the environment. It is the potential release of the fission fragments to the environment during a severe accident that prompts public concern about the safety of nuclear energy.

A fusion reactor will also produce radioactive material. The difference between the two concepts is that the radioactive material from a fusion reactor will decay much more rapidly than the fission products formed in current fission reactors. It is expected that the radioactive material created in a fusion reactor will have half-lives measured in days to years compared to hundreds or thousands of years and in some cases tens of thousands of years for those from a fission reactor (Baratta, 2017). Since the radioactive waste produced in a fusion reactor is short-lived, the consequences of an accident and problems with storage would be significantly reduced.

3. Carbon Emissions from Various Energy Sources

No matter the source of energy, there are some Green House Gases (GHG) emitted. Even the generation of electricity from solar and wind have emissions of GHGs associated with them. It simply is not enough to consider just the actual production of energy from an energy source. All necessary phases of the energy production cycle, including the manufacture, deployment and maintenance of the generating system as well as the actual energy production, must be included. The upstream processes, such as the manufacture of the solar cells or the mining of uranium, must also be included in order to determine an energy source's GHG contribution.

Naturally occurring uranium is composed mostly of two isotopes, U^{235} and U^{238}. For most reactors, the amount of U^{235} in natural uranium is insufficient and must be increased through the enrichment process. The process is energy intensive, requiring 1,000s of MWs of electricity. However, where the electrical energy comes from determines the amount of GHG emitted in the upstream fuel manufacturing phase. If the electrical energy to drive the enrichment process comes from a nuclear power plant then the GHG emission is less than when the electrical energy comes from a fossil fueled electric power plant.

Similarly, the amount of GHG emitted in the production of solar cells depends on the type of cell produced. The energy required to produce crystalline solar cells is larger than for amorphous solar cells, hence, more GHG would be emitted.

Estimates of GHG emissions are also subject to where the energy is eventually generated. In the case of solar, higher efficiencies are obtained nearer the equator, resulting in fewer cells needed and, therefore, less emissions. Table 1, adapted from the International Atomic Energy Agency, shows the amount of carbon emitted in grams CO_2 equivalent per kW-hr (IAEA, 2016).

Table 1. GHG emissions for various energy sources.

Source	g CO_2-eq/kW-h
Coal	932–1132
Natural Gas	449–662
Fossil Fuel with Carbon Capture	82–232
Geothermal	62[2]
Biomass	64
Solar Photovoltaic	49
Thermal Solar	27
Wind	16
Nuclear	15
Hydropower	7

4. How Nuclear Energy Can Assist in the Reduction of GHG Emissions

From the table, the use of any of the low emission technologies would be beneficial in reducing GHG emissions more than the continued use of fossil fuels. Unfortunately, solar and wind are intermittent sources due to weather conditions, including a lack of sunshine. Typical capacity factor for wind and solar has an average of 30% worldwide (the U.S. ranges from 28% to 55% with other countries being lower) (Lazard, 2018). Capacity factor is a measure of what is generated compared to what could be generated under ideal conditions. For solar, the capacity factor for the world is in the mid 20% range (for the U.S. it ranges from 26% to 28%) (Lazard, 2018). For comparison, a typical fossil fuel or nuclear plant has a capacity factor of 94% or more. To complement an intermittent source, an additional source is necessary, one that can be operated at will and at any power level to make up the difference when the intermittent source is not available or unable to run at full rated capacity.

Hydro power is one possibility, but is very limited by the availability of sites that are acceptable. Biomass can be used, but the supply of material is limiting and costly. Nuclear can provide a low GHG source of energy to augment intermittent renewables. Nuclear does not depend on weather conditions and fuel is not of concern since sufficient uranium is available. Unfortunately, there are a number of barriers that are limiting the use of nuclear.

The first is economic. As pointed out in a previous work, the cost of natural gas (Princiotta, 2011) and the associated cost of the construction of natural gas generating stations have placed new nuclear power plants at a significant economic disadvantage. For the most recent year that data is available, the cost of electricity from a current fully depreciated nuclear power plant is $33.50 per MW-hr (NEI, 2018). While this cost is below the projected cost of new combined cycle gas turbine power plants, it is above the

---

[2] For renewables, the variation in GHG emission is small, so only the median values are presented.

cost of subsidized wind. For new nuclear, the cost is dramatically higher, in the range of $100 per MW-hr to nearly $200 per MW-hr (Lazard, 2018).

As a result, if natural gas prices remain low (compared to historical prices), there will be an increased use of natural gas across all energy sectors and an increase in liquefied natural gas exports. The electric power sector will therefore experience a continued shift to electricity generated by these fuels, driven in part by these historically low gas prices. The increase in natural gas electricity generation combined with a larger share of intermittent renewables will likely result in additional retirements of less economic coal and nuclear plants in the future (USEIA, 2019). For each MW-hr of electricity generated from a combined cycle gas-fired plant, there are 0.51 Tons of CO_2 equivalent emitted. Thus, for each MW-hr generated by a combined cycle plant that could have been generated by a nuclear plant there is 0.51 Tons of CO_2 equivalent that could have been avoided without additional carbon abatement technologies. Assuming there isn't an increase in the capacity factor of renewables (highly unlikely at this time), there will continue to be a need for some form of dispatchable electric source, such as gas or nuclear. The replacement of nuclear by gas will lead to unnecessary GHG emissions for this sector.

For maximum efficiency and lowest cost, large thermal power plants, such as nuclear, are run at or near their maximum output with capacity factors in the 90% to 95% range. When combined with intermittent sources, the nuclear plant must operate at lower power levels. Since the largest single contributor to the cost of power from a nuclear plant is the capital construction cost, the total cost needed to bring a plant to commercial operation, which is a fixed cost, the cost of the power increases significantly when the plant is run at a lower power level. There are also some technical considerations that limit the ability of the current generation of nuclear power plants to be run at very low levels without modifications. Both French and German nuclear power plants routinely adjust their output to match the needs of the electrical grid. In the U.S., this is not the practice and there are regulatory restrictions that limit the "load following" practices followed by France and Germany. Nonetheless, the designs of most of the current generation and the new generation of nuclear plants under construction are capable of load following, possibly with some modifications (Lokhov, 2019).

5. Safety Concerns

From a public perspective, the biggest concern is the possibility of an accident. To date, there have been three nuclear accidents: Three Mile Island, Chernobyl and Fukushima Daiichi. The public's perception of these events has resulted in a persuasive anti-nuclear advocacy. Unfortunately, there is a strong misunderstanding of these accidents and their impact on people and the environment.

The accident at Three Mile Island occurred in March of 1979 and resulted in near complete destruction of the fuel in the reactor. The resulting meltdown of the fuel allowed dangerous highly radioactive fission products to escape into the reactor building where almost all were contained. Very little, most notably radioactive noble gasses and some radioactive iodine, escaped to the environment. It was estimated that the release of the radionuclides could result in one extra cancer death in the population within 50 miles of Three Mile Island. Within this population of 2.2 million it was expected there would be about 540,000 cancers and 325,000 cancer deaths, not counting the accident (Battist, 1979). While some studies disagree, the principle health effect of the accident was the mental stress on the residents and not diseases associated with radiation. The stress was believed to have resulted in an increase in heart problems within the area around Three Mile Island.

In a long-term study performed by the University of Pittsburgh, it was concluded that there were health effects related to the accident. Specifically, there was an increase in mortality due to heart attacks among the group studied. At that time, they could not rule out the possibility of long latency period cancers (Talbott et al., 2000). Another series of studies found there was long term stress among the residents surrounding Three Mile Island and that such chronic stress could cause a variety of physical, emotional, and mental illnesses (Osif, 2004). More recent work suggests the low levels the residents around Three Mile Island were exposed to could explain certain anomalies observed in thyroid cancers within the group (Goldenberg, 2017). Others suggest these could be associated with the levels of radioactive iodine produced by nuclear weapons testing and suggest that additional studies be performed (Mangano and

Sherman, 2019). While one cannot say with absolute certainty there were no radiological implications from the low levels of radioactivity released, it seems there is no strong evidence to support the case that there were. Nonetheless, despite a near complete meltdown of the reactor, and a near worst case accident, the health effects appear minimal and are largely associated with the stress of events. Studies are ongoing and will likely continue for years.

By comparison, the accident at the Chernobyl reactor was much more severe, owing to the design and the lack of safety features commonly found in Western reactors. The Chernobyl reactor lacked the containment building which would have housed the nuclear reactor and associated equipment. A containment building would have held back almost all of the radioactive material in the accident.

The Chernobyl accident occurred during a physics tests that caused the reactor power to increase so rapidly that the cooling water vaporized almost instantaneously causing a sudden explosive buildup of steam. Temperatures in the reactor got so hot that the carbon used to help the nuclear reaction caught fire and the fuel melted. The result was a radioactive plume that spread throughout the surrounding area. Twenty-eight firefighters received deadly radiation doses and died. It is reasonably clear that those involved in the emergency work after the explosions suffered an increase in cancers and some circulatory issues. Similar circulatory issues were also seen at TMI among the general population near TMI. The cancers are thought to be due to the radiation dose received during the emergency. It is unclear what caused the circulatory issues. Nonetheless, the total impact on mortality was small despite what was probably the worst accident possible at a reactor, a completely uncontained release of radioactive material from the core. The World Health Organization estimates that among the 49,000 workers studied, there were about 200 deaths associated with the accident out of the nearly 5,000 deaths that occurred within this group. It should be remembered that these were the first responders and the emergency workers who fought the fire and experienced high radiation doses as a result (WHO, 2006).

Now, more than 30 years later, the general population surrounding the reactor has not seen a statistically significant increase in cancer deaths. There has been a statistically significant increase in thyroid cancers, probably as a result of the radioactive iodine released during the accident. It is expected that if the number of health effects is in proportion to the radiation dose received, there will be an additional 4,000 cancer deaths in the exposed population of 600,000. In a population of 600,000, one would expect about 200,000 cancer deaths over the lifespan of the population had the accident not occurred. To date, aside from an increase in thyroid cancers, there has not been an increase in cancer deaths in the exposed population. Studies are ongoing (WHO, 2016).

The most recent accident occurred at the Fukushima Daiichi nuclear power plant in Japan on March 11, 2011. The accident was the result of an earthquake and the resulting tsunami that occurred off the coast of Japan. The tsunami killed over 20,000 people and caused extensive damage all along the coast, including to the Fukushima Daiichi nuclear reactor's safety systems. The damage to those systems resulted in a complete loss of electrical power at the site, including the backup diesel generators. The safety systems intended to cool the reactor in the event of an unusual occurrence are run by electrical power. Because there was no electricity, four of the six reactors at the site overheated and their reactor cores melted, releasing radioactive material. When the fuel overheated, the protective fuel cladding underwent oxidation, resulting in the generation of large quantities of hydrogen. The hydrogen built up in the reactor containments detonated, damaging the containments and releasing radioactive material to the environment.

The accident was extremely severe, however, unlike Chernobyl, the population near the site and even the workers on-site received much smaller doses of radiation. Despite the low levels of radiation offsite, over 100,000 people were evacuated. The evacuation resulted in 1,000 unnecessary deaths due in part to inadequate medical care for critically ill evacuees and the elderly. Similar fatalities were noted among those evacuated because of the earthquake and tsunami. Experts concluded that the evacuation was unnecessary, since radiation levels were similar to background levels elsewhere in the world (Hasegawa et al., 2015).

The accident at Three Mile Island and the events at Fukushima Daiichi demonstrate the importance of the safety built into the design of current nuclear power plants. While the reactor and associated systems in both suffered extreme damage and will require cleanup that will last for decades, the offsite consequences were minimal. By comparison, without those built-in safety features, the accident at Chernobyl created significant off-site radioactive contamination of the surrounding area and an increase in thyroid cancer.

6. Waste Concerns

The process of uranium fission produces both short-lived and long-lived radioactive material. These materials must be disposed of in a manner such that they do not pose a risk to humans or the environment. Presently disposal methods and sites are available for all but the spent fuel. These materials consist of clothing, equipment, and chemicals that have become contaminated with radioactive materials during the operation, maintenance, and decommissioning of nuclear power plants. Such facilities are licensed by the U.S. NRC and provide reasonable assurance that there is minimal risk to the environment and human beings for at least as long as they remain a hazard.

Disposal of the spent fuel itself is still an open question in the U.S. The Nuclear Waste Policy Act designated deep geological disposal as the method for the disposal of spent fuel. Yucca Mountain was designated as the disposal site, pending evaluation and licensing. The site is located in Nevada, about 25 miles from Death Valley and 90 miles from Las Vegas. Due to opposition from the State and public opposed to having Nevada become the site for the nation's spent fuel disposal site, the process is stalled for lack of Congressional funding. Spent fuel is currently stored onsite at both operating reactors and those that have been decommissioned. For operating reactors, some of the fuel is stored in pools until it's cool enough to be placed in canisters and removed to dry storage. In the case of decommissioned reactors, the fuel is sufficiently cool so that it can be stored onsite in dry cask storage facility. As of this writing, the U.S. Congress has no plans to move forward with funding of Yucca Mountain as required by the Nuclear Waste Policy Act.

Other countries are moving forward with planning for the disposal of spent fuel. France, through a national referendum, authorized the expenditure of 25 billion Euro to construct a 500 m underground rock laboratory in eastern France situated in clays and known as the Industrial Centre for Geological Storage (Cigéo). The structure will comprise hundreds of storage tunnels covering a total area of 25 km^2 and will last for a century. The project would provide a facility not unlike the Yucca Mountain Project to store but allow for retrieval of spent fuel for possible reprocessing into new fuel. As in the U.S., though there is strong opposition despite the referendum passing to the project.

In Sweden, the Svensk Kärnbränslehantering AB or SVK is tasked with developing a spent fuel repository (Figure 1). Like the French repository, the Swedish repository uses deep geological disposal in bedrock.

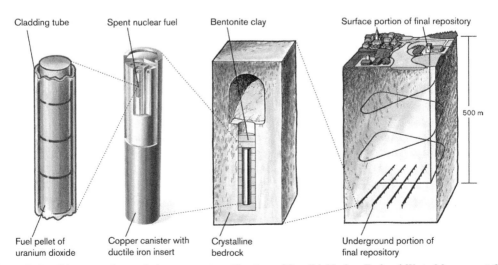

Figure 1. Swedish proposed spent fuel storage system (Courtesy of Swedish Nuclear Fuel and Waste Management Co, Illustrator: Jan Rojma).

The proposed repository is currently under licensing review by the Swedish Radiation Safety Authority. Current plans call for the facility to be operational in 2030. It is estimated the complete program will cost about SWK 141 billion (14.1 billion US$).

Both of these countries have examined the problem and have largely completed their safety and environmental reviews, concluding that deep geologic disposal is the preferred method of disposing of spent fuel. The U.S. proposal for Yucca Mountain is similar, technically, but is held up by the lack of support from the State of Nevada and its Congressional delegation. It's worth noting that Nye County, where the facility would be located, strongly supports continuing with the project design and safety assessment. The County would support construction if the final safety and environmental analysis concludes the furl could be safely stored at Yucca Mountain for generations to come.

7. Why Should Nuclear be Used?

Looking at fossil fueled power plants, the emissions from fossil fuel electric generation plants result in a variety of pollutants: Sulfur dioxide (SO_2), nitrous oxide (NO_x), particulate material and heavy metals. Sulfur dioxide causes acid rain, which is harmful to plants and to animals that live in water. SO_2 also worsens respiratory illnesses and heart diseases, particularly in children and the elderly. Nitrous oxide contributes to ground-level ozone, which irritates and damages the lungs. Particulate matter (PM) results in smog in cites and scenic areas. Coupled with ozone, PM contributes to asthma and chronic bronchitis, especially in children and the elderly. Very small, or fine PM, is also believed to cause emphysema and lung cancer. Heavy metals, such as mercury, are hazardous to human and animal health.

Estimates of the effect of burning fossil fuels suggest that they are far more damaging to health than nuclear power, even accounting for the types of accidents that have occurred. In the United Kingdom, it is estimated that deaths from air pollution from the production of electricity by burning natural gas are 100 to 1,000 times more likely than from nuclear power plants. The effects are even more dramatic compared to coal (Markandy, 2007). By comparison, the TMI, Chernobyl and Fukushima accidents had meltdowns; released radiation to the atmosphere; contaminated nearby areas with radiation yet did not produce the grim results that were postulated.

8. Advances in Nuclear Technology

Despite the inherent safety of the current generation of nuclear plants, improvements are in progress to both lower construction costs and increase safety. The latest generation of nuclear power plants are large, like the previous generation,[3] and require ten years or more to build. It is estimated that those plants currently under construction in the U.S. will cost $20 billion to complete. While they are even safer than those currently in operation, the cost and time to build them makes them economically difficult to justify. A comparable gas fired generating plant is one tenth the cost of these plants. Even when the cost of gas is included the combined cycle plant is capable of generating electricity at ½ to ¼ the cost of the latest design nuclear plant (Lazard, 2018).

Since a critical component of the cost is financing of construction, efforts are under way to reduce the time to commercial operation. Currently, the effort to reduce cost and construction time is concentrated on small modular reactors (SMR), such as the one depicted in Figure 2. The design features a self-contained steam system that can be built in a factory, eliminating the need for much of the onsite construction work typical of current and previous generation reactors. While the power output of such reactors is less than 1/3 that of the current generation of nuclear reactors, they can be combined into a multi-modular plant with 2, 3, 4 or more modules, decreasing operating cost. Their small physical size would allow them to be installed underground, making them less vulnerable to severe weather events and helping contain

---

[3] Typically for the latest generation III+ reactors are 1000 MWe to 1200 MWe.

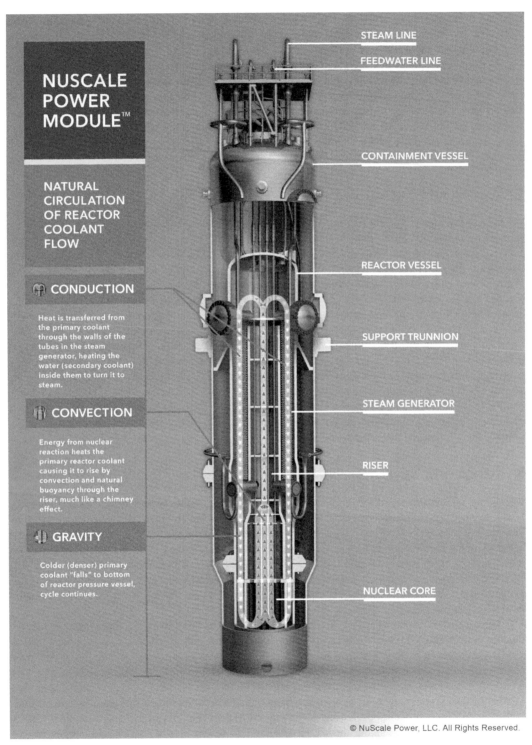

Figure 2. NuScale Power ModuleTM, image provided by NuScale Power, LLC.

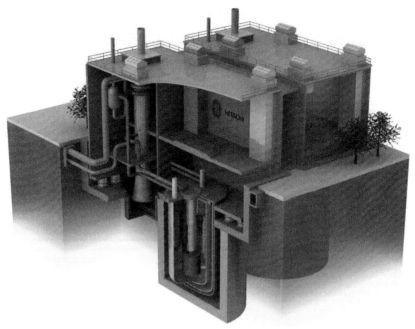

Figure 3. Cutaway of the PRISM power block. Showing two reactors and associated systems located underground (GE-Hitachi Nuclear Energy).

radioactivity in the event of a severe accident. Since all the major systems are in the module, there are no large diameter pipes that could break, causing loss of cooling, as occurred at TMI. Construction in a factory would lower cost, cutting down on construction delays associated with weather. Once installed, a module could begin operation, producing income to pay for the installation of the next and subsequent modules, further reducing costs.

Other reactor concepts are also under consideration. Fast reactors, such as the PRISM design (shown in Figure 3), use liquid metal as the coolant and produce more fuel than they consume, at the same time burning up undesirable fission products, which decreases waste production (Triplett, 2012).

Since the reactor uses liquid metal in a pool as a coolant, the system can be operated at low pressure, virtually eliminating the possibility of a loss-of-coolant accident, reducing the stress on the piping systems and increasing the efficiency. Several countries are developing similar concepts but, so far, these types of reactors have been difficult to operate reliably. Furthermore, such fast reactors, when combined with pryroprocessing, provide the option for closing the fuel cycle and the recycling of the spent nuclear fuel inventory, utilizing the valuable remaining energy content (~ 95%) while reducing the half-lives and heat loads of the remaining byproducts, thereby easing the repository requirements.

Fusion, mentioned earlier, holds promise as a source of energy. Recent advances in operation of prototype facilities have shown significant promise that a reliable operating system is possible. Assuming ITER is operational in 2025, as scheduled, the following table provides timelines developed by various countries for the operation of a commercial size fusion reactor (El-Guebaly, 2017).

However, recent advances in the use of high temperature superconducting magnets and their application in compact fusion devices may shorten this timeline (National Academy, 2019).

Unlike fission reactors, the fusion reactor would produce no long-lived fission products and would have a nearly inexhaustible fuel source that could be obtained from sea water. It is unlikely that a commercial fusion reactor will be designed and constructed in the near future. It is likely that in the next ten years we will see a demonstration fusion reactor akin to the original nuclear fission reactor, CP-1, built by Enrico Fermi and his colleagues.

Table 2. International timelines to a demonstration fusion power plant.

Timeline for International Fusion Roadmaps

9. Conclusion

Renewables can and do provide a significant source of low-GHG generated electricity. These sources must be supplemented by some form of dispatchable electric source lacking improvements in battery technology. Currently, the dominant sources of dispatchable electricity are coal, gas, and nuclear. The former two emit large quantities of GHG, offsetting some of the benefits of renewables. Nuclear is constrained by public perceptions of lack of safety and cost of construction. New designs that may help reduce costs and enhance safety are in the works. Despite these drawbacks, nuclear must be in the mix if GHG production is to be reduced over the long term.

References

Baratta, A. 2017. For a more detailed discussion of nuclear reactions and radioactive decay please see. Introduction to Nuclear Engineering, 4th ed., Pearson Education.

Battist, L., Congel, F., Buchanan, J. and Peterson, H. 1979. Population Dose and Health Impact of the Accident at the Three Mile Island Nuclear Station Preliminary Estimates for the Period March 28, 1979–April 7, 1979 (NUREG-0558), US NRC, United States.

Chinese Academy of Science. 2017. China's 'artificial sun' sets world record with 100 second steady-state high performance plasma. Phys. org. July 6, 2017, retrieved 8 July 2019 from https://phys.org/news/2017-07-china-artificial-sun-worldsteady-state.html.

El-Guebal, L. 2017. Worldwide Timelines for Fusion Energy, accessed September 11, 2019. http://sites.nationalacademies.org/cs/groups/bpasite/documents/webpage/bpa_184787.pdf.

Goldenberg, D., Russo, M., Houser, K. et al. 2017. Altered molecular profile in thyroid cancers from patients affected by Three Mile Island nuclear accident. Laryngoscope 127(suppl 3): S1–S9.

Hasegawa, A., Tanigawa, K., Ohtsuru, A., Yabe, H., Maeda, M., Shigemura, J., Ohira, T., Tominaga, T., Akashi, M., Hirohashi, N., Ishikawa, T., Kamiya, K., Shibuya, K., Yamashita, S. and Chhem, R. 2015. Health effects of radiation and other health problems in the aftermath of nuclear accidents, with an emphasis on Fukushima. The Lancet 386: 479–488.

Herrmann, M. 2014. Plasma physics: A promising advance in nuclear fusion. Nature 506(7488): 302–303.

IAEA. 2016. Climate Change and Nuclear Power 2016. International Atomic Energy Agency, Vienna.

Lokhov, A. 2011. Load-following with nuclear power plants. NEA News 2011–No. 29.2.

Lazard. 2018. Lazard's Levelized Cost of Energy Analysis—Version 12.0.

Mangano, J. and Sherman, J. 2019. In response to altered molecular profile in thyroid cancers from patients affected by Three Mile Island nuclear accident. The Laryngoscope 129: E51.

Markandya, A. and Wilkinson, P. 2007. Electricity generation and health. The Lancet 370(9591): 979–990.

National Academies of Sciences, Engineering, and Medicine. 2019. Final Report of the Committee on a Strategic Plan for U.S. Burning Plasma Research. Washington, DC: The National Academies Press. doi: https://doi.org/10.17226/25331.

NEI. 2018. Nuclear Costs in Context. Nuclear Energy Institute (Available for download at https://www.nei.org/resources/reports-briefs/nuclear-costs-in-context).

Osif, B., Baratta, B. and Conkling, T. 2004. See the discussion on health effects in TMI 25 Years Later, Penn State University Press.

Princiotta, F. 2011. See Chapter 4 in Global Climate Change—The Technology Challenge, Springer, Netherlands.

Talbott, E.O., Youk, A.O., McHugh, K.P., Shire, J.D., Zhang, A., Murphy, B.P. and Engberg, R.A. 2000. Mortality among the residents of the Three Mile Island accident area: 1979–1992. Environ Health Perspect 108: 545–552.

Triplett, B., Loewen, E. and Dooies, B. 2012. Prism: A competitive small modular sodium-cooled reactor. Nuclear Technology 178: 186.

USEIA. 2019. Annual Energy Outlook 2019, U.S. Energy Information Agency, Washington, D.C.

WHO. 2006. Health Effects of the Chernobyl Accident and Special Health Care Programs, World Health Organization, Geneva.

WHO. 2016. 1986–2016: CHERNOBYL at 30, World Health Organization, Geneva.

CHAPTER 19

Concentrated Solar Energy-Driven Multi-Generation Systems Based on the Organic Rankine Cycle Technology

*Nishith B Desai** and *Fredrik Haglind*

1. Introduction

Design of energy efficient, environmentally friendly and economically viable systems is important for sustainable development. Among the various technology options based on renewable energy sources, concentrated solar power (CSP) systems are considered to be technologies in the development stages. Many small to large-scale power plants (a few kW_e to a few MW_e) based on the CSP technology exist in different sun-rich regions worldwide. Due to the high capital cost and high levelized cost of energy (LCOE), CSP plants have not captured a large market share like those of solar photovoltaic (PV) and wind power plants. Concentrated solar power plants with cost-effective thermal energy storage can work as a base load plant with a high capacity factor. In contrast, solar PV and wind power plants with large-scale battery storage are not cost-effective. Patil et al. (2017) reported that the levelized cost of electricity (LCOE) for solar photovoltaic systems with battery storage is about 36.8% higher than that of the parabolic trough collector-powered organic Rankine cycle system with thermal energy storage. Concentrated solar power plants can also avail of the advantage of producing heat and other products, and thus work as a cogeneration, trigeneration or multi-generation unit. In contrast, solar PV and wind power plants cannot be used for heat production; therefore, the sub-systems for cooling and/or heating and/or desalination should be electricity-driven. Shalaby (2017) recommended avoiding the use solar photovoltaic systems with batteries to drive RO desalination systems because of the high capital and running costs. Commonly-used small to medium-scale, dispatchable (on demand) distributed generation systems are diesel generator-based or biomass-based systems. Biomass- and diesel-based multi-generation units can have electrical or thermal energy-driven sub-systems, depending on the resulting cost of utilities. For isolated regions and islands, the cost of electricity generation is high, as the diesel is imported from the nearby port. The use of biomass is a major concern in places with water scarcity, due to the large water footprints of biomass energy sources (Gerbens-Leenes et al., 2009).

Multi-generation systems achieve a higher efficiency and a higher energy utilization factor than plants producing only electricity (Karellas and Braimakis, 2016). Concentrated solar energy-driven

---
Department of Mechanical Engineering, Technical University of Denmark, Nils Koppels Allé, Building 403, 2800 Kongens Lyngby, Denmark.
* Corresponding author: nishithdesai17@gmail.com

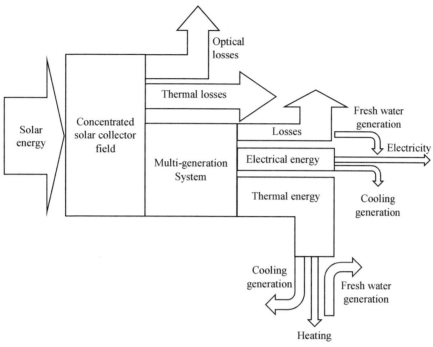

Figure 1. Representation of possible energy conversion routes of concentrated solar thermal energy-powered multi-generation systems.

multi-generation systems are also suitable for decentralized installations. Integrated systems powered by concentrated solar energy and biomass energy make up a promising option (Mathkor et al., 2015). Wu et al. (2019) proposed the integration of a concentrated solar thermal energy and power cycle system with a conventional combined cooling, heating and power system. A representation of possible energy conversion routes of concentrated solar thermal energy-powered multi-generation systems is shown in Figure 1. In the case of a typical parabolic trough collector field, the optical losses (including shading and blocking, cleanliness, shielding by bellows) are about 37% and the thermal losses (including thermal losses from piping) are about 18% (Heller, 2017). For small to medium-scale applications (a few kW_e to a few MW_e), organic Rankine cycle power systems have been demonstrated to be efficient solutions for multi-generation plants (Astolfi et al., 2017; El-Emam and Dincer, 2018). Organic Rankine cycle (ORC) power systems can be effectively used for energy sources, like concentrated solar power, biomass, waste heat, geothermal, and ocean thermal. The main advantages of organic Rankine cycle power systems employing dry and isentropic working fluids are the high isentropic efficiency of the turbine at design and part-load conditions, quick start-up, long life-time of the components, low mechanical stresses in turbine blades, automatic and unmanned operation, low operation and maintenance costs, and flexibility and ability to follow variable load profiles (Algieri and Morrone, 2012). All the mentioned characteristics make ORC units particularly suitable for supplying the electricity demand for a vapor compression refrigeration system and/or for a reverse osmosis system or the thermal energy (using high temperature working fluid vapor available at the exhaust of turbine) demand for a vapor absorption refrigeration system and/or for a water distillation system. When designed for multi-product purposes (thermal energy-driven), the system is designed with a condensation pressure higher than that of systems designed for power generation only. Hoffmann and Dall (2018) reported that the levelized cost of electricity for a solar power tower integrated Rankine cycle increases by 8.8% when used for co-generation. This is because the condensing stream leaving the turbine should be at a higher temperature in order to act as an energy source for the cogeneration application. The revenue generated from the other product (heat, fresh water, or cooling) may compensate for this low efficiency.

In this chapter, different concentrated solar energy-driven multi-generation systems based on the organic Rankine cycle technology for small to medium-scale applications are reviewed. Power generation systems are discussed in section 2. Systems generating power, fresh water and heating are presented in section 3. Section 4 describes power, cooling and heating systems. Design considerations and issues in CSP-driven multi-generation systems using ORC technology are presented in section 5. Finally, concluding remarks are given in section 6.

2. Power Generation

Parabolic trough collector (PTC)-based CSP plants, using a conventional synthetic thermal oil as a heat-transfer fluid (HTF), are the most mature CSP technology. Solar power tower (SPT) technology and linear Fresnel reflector (LFR) technology with flat mirrors and simple structure are proposed as promising alternatives to the PTC-based CSP plants. The solar power tower technology is cost-effective for large-scale applications (> 50 MW_e). The linear Fresnel reflector technology has a lower optical efficiency (Nixon and Davies, 2012; Xie et al., 2012) and requires a much higher area of installation compared to that of a PTC-based CSP plant of the same capacity (Desai and Bandyopadhyay, 2015). The paraboloid dish system is the least applied concentrated solar power technology for power generation, relative to the other technologies.

The conventional steam Rankine cycle is widely used in commercial concentrated solar power plants. Depending on the capacity of the CSP plant and steam conditions at the inlet of the turbine, the thermal efficiency of the steam Rankine cycle is in the range of 20% to 40%. Modular CSP plants with a few kW_e to a few MW_e capacity offer solutions in industrial as well as off-grid applications. For such plants, ORC power systems have been demonstrated to be an efficient solution for electricity production (Quoilin et al., 2013). Existing concentrated solar energy-powered organic Rankine cycle-based commercial/ medium-scale plants (> 500 kW_e) for different applications are listed in Table 1 (NREL, 2019; Petrollese et al., 2018; Turboden, 2019; Wendt et al., 2015). In addition, there are a few micro and small-scale CSP-ORC plants, mainly built for research and development purposes, which are not commercially viable and are, therefore, not included in the list.

A simplified schematic of a typical concentrated solar thermal energy-driven organic Rankine cycle power system is given in Figure 2. The system can be equipped with a thermal energy storage for storing the excess energy. When the stored energy is available, the ORC power system runs at full load. However, when the storage is at a minimum level and solar radiation is not sufficient, the heat transfer fluid mass flow rate is adjusted such that the solar field outlet temperature is controlled. The power system mass flow rate and turbine power output are also affected by the variations in the heat transfer fluid flow rate. Part-load efficiencies of the equipment are lower than the design condition efficiencies, and therefore, appropriate models need to be used for predicting the performance of the system. A summary of previous works on medium-scale (a few hundred kW_e to a few MW_e) concentrated solar thermal energy-powered organic Rankine cycle power systems is given in Table 2. It can be observed that the parabolic trough collector and linear Fresnel reflector are typically used for medium-scale plants. Recently, a novel nanostructured polymer foil-based concentrated solar power system, which avails the advantages of low capital cost, low operation and maintenance cost, and two-axis tracking, has been analyzed (Desai et al., 2019a; Desai et al., 2019b). This system uses a nanostructured focusing plastic film that is adhered to a glass plate.

It is important to select a proper working fluid for an organic Rankine cycle power system for cost-efficient utilization of any available heat source. For low and medium-grade heat sources, the dry and the isentropic fluids are the preferred organic working fluids, as the condition at the outlet of the turbine is always either saturated or super-heated vapor, avoiding expansion in the two-phase region (Hung, 2001; Lui et al., 2004). The promising organic working fluids for CSP-based plants are n-pentane, isopentane, hexamethyldisiloxane (MM), toluene and cyclohexane; see Table 2. In commercial, medium-scale actual plants (> 500 kW_e) n-pentane, MM or isobutene are used as working fluids in the ORC system; see Table 1. Apart from the techno-economic performance, environmental, safety, health, and legislative aspects need to be considered in the final selection of the working fluid for the ORC power system.

Table 1. List of concentrated solar energy-powered organic Rankine cycle-based commercial/medium-scale actual plants (> 500 kW$_e$) for different applications (NREL, 2019; Petrollese et al., 2018; Turboden, 2019; Wendt et al., 2015).

Name (Location)	Start year	Solar field	Solar field area (m²)	Storage	Application (net capacity)
Saguaro Power Plant (Arizona, USA)	2006	PTC	10,340	–	Electricity generation (1 MW$_e$) (currently non-operational).
Rende-CSP Plant (Calabria, Italy)	2014	LFR	9,780	–	Electricity generation (1 MW$_e$). The facility is combined with an already operating biomass-based plant (14 MW$_e$).
Airlight Energy Ait-Baha Pilot Plant (Ait Baha, Morocco)	2014	PTC	6,159	Packed-bed rock (5 h)	Electricity generation from CSP and waste heat from cement industry (hybrid plant) (2 MW$_e$).
Stillwater GeoSolar Hybrid Plant (Fallon, USA)	2015	PTC	24,778	–	Electricity generation. About 17 MW$_{th}$ from CSP combined with geothermal energy producing 33 MW$_e$. Additionally, 26.4 MW$_e$ of a solar photovoltaic plant.
Aalborg CSP-Brønderslev CSP with ORC project (Brønderslev, Denmark)	2016	PTC	26,929	–	Combined heat and electricity production from CSP (16.6 MW$_{th}$) and biomass combustion (hybrid plant) (3.8 MW$_e$).
Ottana Solar Facility (Sardinia, Italy)	2017	LFR	8,600	Two-tank direct	Power generation (0.6 MW$_e$), additionally 0.4 MW$_e$ of solar PV.

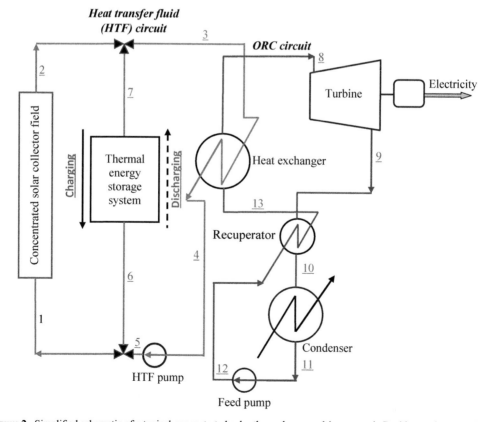

Figure 2. Simplified schematic of a typical concentrated solar thermal energy-driven organic Rankine cycle power system.

Table 2. Summary of previous works on medium-scale concentrated solar power generation based on the organic Rankine cycle technology.

Authors	Solar field	Storage	ORC capacity/working fluids	Max. temp. of ORC/HTF	Remarks
Casartelli et al. (2015)	PTC, LFR	–	2.94 MW_e and 3.57 MW_e (Toluene)	295 °C (ORC)	For cost parity, the cost of the LFR solar field should be about 50% of PTC solar field.
Cocco and Serra (2015)	LFR	Two-tank direct, thermocline	1 MW_e (Siliconic oil)	305 °C (ORC)	The cost of energy for a thermocline storage system is 420 €/MWh and for the direct two-tank system 430 €/MWh.
Cocco and Cau (2015)	PTC, LFR	Two-tank direct	1 MW_e (Siliconic oil)	305 °C (ORC)	Cost of energy (1 MW_e, 2 h storage): LFR-based plant: 380 €/MWh; PTC-based plant: 340 €/MWh.
Rodriguez et al. (2016)	LFR	Two-tank, thermocline	1 MW_e (Cyclopentane)	300 °C (ORC)	Specific cost for a thermocline storage system (€/kWh_{th}) is about 33% on average lower than that of the conventional two-tank storage system.
Desai and Bandyopadhyay (2016)	PTC, LFR	–	1 MW_e (R113, n-pentane, Cyclohexane, MDM, MM, Heptane, Toluene, R245fa, and other)	337 °C (ORC)	Cost of the LFR field to reach cost parity with a PTC-based plant: for SRC-based plants: 48% of PTC field cost; for ORC-based plants: 58% of the PTC field cost; the Steam Rankine cycle is a preferred option.
Garg et al. (2016)	PTC	Packed bed	500 kW (Isopentane, R152a, butane, isobutene, R245fa, and other)	275 °C (HTF)	Hybrid plants (5, 50 and 500 kW) powered by waste heat and solar thermal energy. Isopentane is the preferred working fluid.
Tzivanidis et al. (2016)	PTC	Single tank direct	1 MW_e (Cyclohexane, toluene, water, MDM, and other)	270 °C (HTF)	Techno-economically, Eurotrough ET-150 is a better solution compared to other PTC technologies. Cyclohexane is the preferred working fluid.
Russo et al. (2018)	LFR	Thermocline	1 MW_e (Not given)	300 °C (HTF)	For thermocline storage, forced circulation of molten salts is better compared to the natural circulation.
Javanshir et al. (2018)	SPT	–	Butane, ethanol, isobutene, R11, R141b	350 °C (ORC)	For a max cycle temperature lower than 300 °C, an ORC system (with R141b) is a better option. For high temperature, combined cycles are the better option.
Bellos and Tzivanidis (2018)	PTC	Single tank direct	238 kW to 845 kW (Toluene, cyclohexane, MDM, n-pentane)	300 °C (HTF)	Hybrid solar-waste heat-powered system. Toluene is the preferred working fluid.
Petrollese and Cocco (2019)	LFR	Two-tank direct	716 kW to 730 kW (MM, n-heptane, toluene)	222 °C (ORC)	Multi-scenario approach for the plant design. MM is the preferred working fluid.
Desai et al. (2019a)	Foil-based	Two-tank indirect	1 MW_e (n-pentane, MM)	225 °C (ORC)	A foil-based CSP plant can reduce the LCOE by up to 40% compared to the PTC-based CSP plant.

In ORC power systems, the expander is the most important component as it has the most effect on the techno-economic performance of the system. Expanders for the ORC power system can be grouped into two types: (i) turbo expanders (axial and radial turbines), and (ii) volumetric expanders (scroll expanders, screw expanders, reciprocation piston expanders, and rotary vane expanders). Turbines with an organic working fluid can reach a very high isentropic efficiency with only one or two stages. In systems with high flow rates and low pressure ratios, axial turbines (100 kW$_e$ to a few MW$_e$) are the most widely used. In contrast, radial-inflow turbines are suitable for the systems with low flow rates and high pressure ratios. However, with decreasing power output and, hence, turbine size, the rotational speed increases proportionally. Therefore, for the low power range (mainly using radial-inflow turbines, < 100 kW$_e$), it is necessary to design an adequate bearing system and to employ a high-speed generator and power electronics. Radial outflow turbine design allows a high volume flow ratio with the constant peripheral speed along the blade span (Zanellato et al., 2018). Radial-outflow turbines can be used for small to medium-scale applications with an advantage of reduced rotational speed, allowing direct coupling to a generator (Maksiuta et al., 2017). In systems with a capacity less than 50 kW$_e$, the turbines cannot be used due to high rotational speed and high cost (Imran et al., 2016). Reciprocating piston expanders (Wronski et al., 2019) and screw expanders (Bao and Zhao, 2013) can be used for small capacity plants. Scroll expanders and rotary vane expanders can be used in small or micro-scale ORC power systems (Bao and Zhao, 2013).

Apart from the expander, the heat exchangers (evaporator, recuperator, and condenser) represent a significant share of the total ORC system cost. Temperature driving force (pinch point temperature difference) and pressure drops are key performance parameters regarding heat transfers, and each heat exchanger in the power system should be sized based on these parameters. The most commonly used heat exchangers for ORC power systems are shell and tube heat exchangers (for large-scale power systems) and plate heat exchangers (for small-scale power systems, due to compactness) (Quoilin et al., 2013). Organic Rankine cycle feed pumps should meet the requirements of efficiency, controllability and low net pressure suction head. In addition, the ORC power system should be leak–proof, because the organic fluids are expensive (compared to water) and can be toxic, flammable, and have high values of global warming potential and/or ozone depletion potential. In a conventional steam Rankine cycle system, the pump electricity consumption is very low compared to the power output (low back work ratio). On the other hand, in an ORC power system, the irreversibility in the pump can reduce the overall cycle efficiency significantly (Quoilin et al., 2013).

As for the thermal energy storage technologies, the most widely-used systems for CSP-driven organic Rankine cycle systems are the conventional indirect two-tank molten salt storage technology (for large capacity) and the direct thermal oil storage technology (for small capacity). Sensible thermal energy storage using a single tank packed-bed that consists of solids (such as rocks) as the heat storage medium and a heat transfer fluid in direct contact with the solids has also been analyzed in the literature (Cocco and Serra, 2015; Russo et al., 2018). The latent heat thermal energy storage is still at the proof of concept stage because of the low thermal conductivity, resulting in slow charge and discharge processes.

2.1 *Thermodynamic analysis*

2.1.1 *Solar collector field*

The solar collector field useful heat gain, $\dot{Q}_{CL}$, can be calculated as follows:

$$\dot{Q}_{CL} = \eta_{o,CL} \cdot f_{clean} \cdot DNI \cdot IAM \cdot A_{p,CL} - U_{l,1} \cdot (T_{m,CL} - T_a) \cdot A_{p,CL} - U_{l,2} \cdot (T_{m,CL} - T_a)^2 \cdot A_{p,CL} \tag{1}$$

where $\eta_{o,CL}$ is the optical efficiency of the solar collector field, $U_{l,1}$ and $U_{l,2}$ are the heat loss coefficients based on the aperture area of the solar collector field, $A_{p,CL}$ is the aperture area of the solar collector field, $T_{m,CL}$ is the mean temperature of the solar collector field, T_a is the ambient temperature, and DNI is the direct normal irradiance. The incidence angle modifier (IAM) represents the reduction of the optical efficiency due to the incidence angle in parabolic trough collector fields and due to the incidence and the transversal angles in linear Fresnel reflector fields. The IAM for the system with two-axis tracking (paraboloid dish) is one. The cleanliness factor (f_{clean}) is the ratio of the optical efficiency in average dirty conditions to the optical efficiency with the same optical element in clean condition.

2.1.2 Organic Rankine cycle power system

The organic Rankine cycle feed pump increases the pressure of the working fluid (from state 11 to 12 in Figure 2). The power consumption of the pump, $\dot{W}_P$, is computed as follows:

$$\dot{W}_P = \frac{\dot{m}_{ORC} \cdot (h_{12s} - h_{11})}{\eta_{is,P}} \tag{2}$$

where $\dot{m}_{ORC}$ is the mass flow rate of the organic working fluid, $\eta_{is,P}$ is the isentropic efficiency of the feed pump, and h_i denotes specific enthalpy at i-th state point. The index s refers to a state achieved after an isentropic compression/expansion.

The organic working fluid in the liquid state at the maximum operating pressure (state 13) enters the heat exchanger. In the heat exchanger, heat is transferred from the high temperature heat transfer fluid, heated through the solar collector field, to the organic working fluid. Typically, this heat exchanger consists of three parts, a preheater, evaporator, and superheater. The heat transfer rate in the heat exchanger, $\dot{Q}_e$, is given as follows:

$$\dot{Q}_e = \dot{m}_{ORC} \cdot (h_8 - h_{13}) \tag{3}$$

The power output of the turbine, $\dot{W}_T$, and the gross electric output, $\dot{W}_{el,gross}$, are calculated as follows:

$$\dot{W}_T = \dot{m}_{ORC} \cdot (h_8 - h_{9s}) \cdot \eta_{is,T} \quad \text{and} \quad \dot{W}_{el,gross} = \dot{W}_T \cdot \eta_g \tag{4}$$

where $\eta_{is,T}$ is the isentropic efficiency of the turbine and η_g is the generator efficiency.

In the case of dry organic working fluids, the state point after the expansion in the turbine is superheated. The organic liquid at state 12 enters a recuperator (this component is optional) where the low-pressure organic fluid vapor from the turbine (state 9) supplies heat. Finally, the turbine exhaust is condensed in a condenser after part of its heat has been transferred in the recuperator. The heat transfer rate in the condenser, $\dot{Q}_c$, is calculated as follows:

$$\dot{Q}_c = \dot{m}_{ORC} \cdot (h_{10} - h_{11}) \tag{5}$$

3. Power, Fresh Water Generation and Heating

Reasonable water and electricity supply policies are of vital importance for the development of locations where there is inadequate water. Solar photovoltaic systems or diesel generator systems using reverse osmosis (RO) for fresh water generation are commonly used for a few kW$_e$ to a few MW$_e$ capacity plants, for simultaneous generation of electricity and fresh water. For dispatchable (on demand) electricity and fresh water generation in isolated regions and on islands, diesel generator-based systems are used. With respect to CSP-based electricity and fresh water generation systems, steam Rankine cycle power systems (Palenzuela et al., 2015), organic Rankine cycle power systems (Astolfi et al., 2017), or supercritical carbon dioxide Brayton cycle power systems (Sharan et al., 2019) can be used for power generation, and reverse osmosis systems (El-Emam and Dincer, 2018) or thermal energy-driven desalination systems (Astolfi et al., 2017) can be used for fresh water generation. Simplified schematics of a typical concentrated solar thermal energy-driven organic Rankine cycle-based electricity system with reverse osmosis-based desalination system and thermal energy-driven desalination systems are shown in Figures 3 and 4, respectively. A summary of previous works on concentrated solar power-based cogeneration systems using the ORC technology and desalination system (reverse osmosis or thermally driven) is given in Table 3.

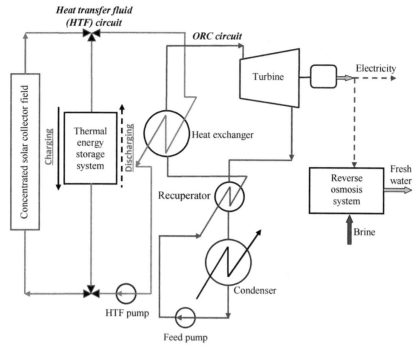

Figure 3. Simplified schematic of a typical concentrated solar thermal energy-driven organic Rankine cycle-based electricity and reverse osmosis-based desalination system.

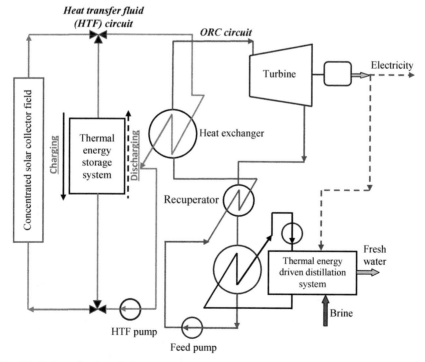

Figure 4. Simplified schematic of a typical concentrated solar thermal energy-driven organic Rankine cycle-based electricity and thermal energy-driven desalination system.

Table 3. Summary of previous works on concentrated solar energy-powered electricity and fresh water generation systems based on the organic Rankine cycle technology.

Authors	Solar field	Storage	ORC capacity/working fluids	Max. temperature of ORC/HTF	Desalination system	Remarks
Delgado and Garcia (2007)	PTC	–	100 kW$_e$ (Toluene, Octa-methylcyclotetrasiloxane, MM)	365 °C (HTF)	RO	Toluene is a promising working fluid, based on thermodynamic analysis of the system.
Bruno et al. (2008)	FPC, ETC, PTC	–	11.72 kWe and 27.82 kWe (Isopentane, n-propilbenzene, tribromomethane, dibromomethane, ethylbenzene, and other)	400 °C (ORC)	RO	Solar thermal energy-based fresh water generation cost: 4.32 €/m³ to 5.5 €/m³. PV-RO-based fresh water generation cost: 12.83 €/m³ to 14.85 €/m³. Isopentane and n-propilbenzene are promising working fluids.
Nafey and Sharaf (2010)	FPC, CPC, PTC	–	998 kW$_e$ to 1131 kW$_e$ (Toluene, dodecane, nonane, octane)	300 °C (HTF)	RO	System with direct vapor generating solar collectors. Toluene is a promising working fluid for PTC-based plants (with fresh water generation cost: 0.903 $/m³).
Nafey et al. (2010)	PTC	–	347 kW$_e$ to 662 kW$_e$ (Toluene)	340 °C (HTF)	RO	Fresh water generation cost (seawater desalination): 0.59 $/m³ to 0.89 $/m³.
Sharaf et al. (2011)	PTC	–	394 kW$_e$ to 1123 kW$_e$ (Toluene)	350 °C (HTF)	MED-TVC, MED-MVC	Plant using a multi-effect distillation system with thermal vapor compression (MED-TVC) is better than the mechanical vapor compression (MED-MVC).
Karellas et al. (2011)	PTC	–	250 kW$_e$ (R134a)	113.5 °C (ORC)	RO	Hybrid system integrating solar PV and CSP-ORC. Minimum cost of fresh water production is about 6.52 €/m³ for Chalki Island, Greece.
Sharaf (2012)	PTC	–	Not mentioned (Toluene)	350 °C (HTF)	RO, MED, MED-TVC, MED-MVC	A CSP-ORC-system with RO desalination is the best alternative (with fresh water generation cost: 0.57 $/m³).
Li et al. (2013)	PTC	–	100 kW$_e$ (MM)	400 °C (HTF)	RO	Supercritical ORC-systems have higher thermal efficiency than a subcritical ORC-system.
Mathkor et al. (2015)	PTC	–	1 MW$_e$ (Cyclopentane)	189 °C (ORC)	Single effect desalination	Hybrid system by CSP and biomass energy using an ORC-unit (1 MW$_e$), single absorption chiller (682.3 kW and a single effect desalination unit (234 m³/day). Exergy efficiency: 41.7%.
Astolfi et al. (2017)	PTC	Thermo-cline	Up to 5 MW$_e$ (n-pentane)	300 °C (HTF)	MED	Hybrid system powered by a CSP-ORC, solar PV, and DG-set with RO and MED-units. Fresh water generation cost: 1.43 $/m³ to 2.15 $/m³.
El-Emam and Dincer (2018)	PTC	Two-tank indirect	200 kW$_e$ to 500 kW$_e$ (n-octane)	340 °C (ORC)	RO	Polygeneration using an ORC-unit, vapor absorption cooling unit, desalination unit, and electrolyzer. Optimum exergy efficiency and cost rate: 30.3% and 278.9 $/h.
Desai et al. (2019b)	Foil-based	Two-tank indirect	1 MW$_e$ (n-pentane, MM, cyclopentane, isopentane, toluene)	238 °C (ORC)	MED	Lowest LCOE for cyclopentane (0.17 €/kWh$_e$) and lowest levelized cost of water for MM (0.91 €/m³).

Toluene (Delgado-Torres et al., 2007), R134a (Karellas et al., 2011), isopentane (Bruno et al., 2008), MM (Li et al., 2013), and n-octane (El-Emam and Dincer, 2018) were proposed as promising organic working fluids for CSP-driven ORC systems with RO desalination. For concentrated solar thermal energy-driven ORC-systems with a thermal energy-driven desalination system, n-pentane (Astolfi et al., 2017), toluene (Sharaf, 2012), and cyclopentane (Mathkor et al., 2015; Desai et al., 2019b) were proposed as promising organic working fluids.

The parabolic trough collector is the most widely-used CSP technology for ORC-based cogeneration systems (El-Emam and Dincer, 2018). Recently, a nanostructured polymer foil-based concentrated solar collector technology was analyzed as a promising alternative compared to a PTC-based system for ORC power systems integrated with a multi-effect distillation (MED) desalination system (Desai et al., 2019b). The assumptions related to the solar irradiation, capital cost of the sub-systems and electricity consumption significantly influence the techno-economic performance of the cogeneration system. The concentrated solar thermal energy integrated MED-system is less expensive than a RO-based desalination system (Ghobeity et al., 2011; Sharan et al., 2019). Depending on seawater salinity, membrane configuration and efficiencies of components, the specific electricity consumption for reverse osmosis systems is about 3.5 kWh$_e$/m^3 to 5 kWh$_e$/m^3 (IRENA, 2012; Sharan et al., 2019). For multi-effect seawater distillation systems, the specific electricity consumption is about 1 kWh$_e$/m^3 to 1.5 kWh$_e$/m^3 (Alfa Laval, 2018).

4. Power, Cooling and Heating

Conventional vapor compression refrigeration systems (VCRS) powered by electrical energy are widely used for cooling applications. Such systems can be powered by electrical energy produced by a concentrated solar thermal energy-based organic Rankine cycle power system. A low-grade thermal energy-driven vapor absorption refrigeration system (VARS) can also be integrated as a bottoming cycle to an organic Rankine cycle power system. A simplified schematic of a typical concentrated solar thermal energy-driven organic Rankine cycle-based vapor compression or absorption refrigeration system is given in Figure 5. A summary of previous works on CSP-driven ORC-based cooling and/or heating systems is given in Table 4.

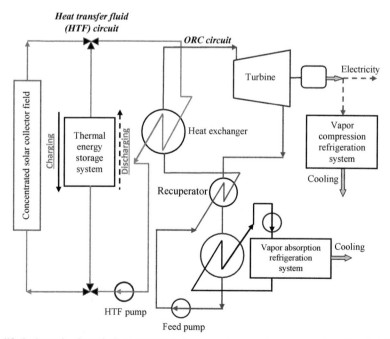

Figure 5. Simplified schematic of a typical concentrated solar thermal energy-driven organic Rankine cycle-based electricity and vapor compression or absorption refrigeration system.

Table 4. Summary of previous works on CSP-driven ORC-based cooling and/or heating systems.

Authors	Solar collector	ORC capacity/working fluids	Cooling system/refrigerants	Remarks
Al-Sulaiman et al. (2011a)	PTC	500 kW$_e$ (n-octane)	VARS (LiBr–H$_2$O)	Exergy efficiency of trigeneration system: 20% for solar mode, 7% for storage mode, and 8% for solar-storage mode.
Al-Sulaiman et al. (2011b)	SOFC–Biomass–solar thermal	500 kW$_e$ (n-octane)	VARS (LiBr–H$_2$O)	Energy efficiency of trigeneration system: 90% for solar mode, 90% for biomass mode, 76% for solid oxide fuel cell (SOFC) mode.
Buonomano et al. (2015)	Solar thermal and geothermal	6 kW$_e$ (R245fa)	VARS (LiBr–H$_2$O)	Cooling capacity: 30 kW, heating capacity: 87 kW$_{th}$. System efficiency: 69.4% for trigeneration mode, 6.4% for only power mode. Payback period range: 2.5 y to 7.6 y.
Karellas and Braimakis (2016)	PTC and biomass	1.42 kW (R134a, R152a, R245fa)	VCRS (R134a, R152a, R245fa)	Cooling capacity: 5 kW, heating capacity: 53.5 kW$_{th}$. Solar–ORC thermal, electrical and exergy efficiency: 6%, 3%, and 7%. COP of VCRS: 3.88, payback period: 7 y.
Patel et al. (2017)	PTC/LFR/dish and biomass	10 kW$_e$ (n-pentane, Toluene, R245fa)	Cascade VARS-VCRS	Fully biomass-based system better than solar-biomass system. LFR as a solar field and n-pentane as an ORC working fluid are the more appropriate choices.
Bellos and Tzivanidis (2017)	PTC	89.3 to 177.6 kW$_e$ (Toluene, n-octane, MDM, MM, etc.)	VARS (LiBr–H$_2$O)	Toluene is the more suitable working fluid with exergy efficiency of 29.42%. For toluene, net power output: 177.6 kW$_e$, heating capacity: 398.8 kW$_{th}$, cooling capacity: 947.2 kW.
Bellos et al. (2018a)	PTC and biomass	8.2 kW$_e$ (Toluene, n-octane, MDM, MM, n-heptane, cyclohexane)	VCRS (R141b, R 600, R161, R600a, and other)	Cooling capacity: 5 kW, low-temperature (50 °C) heating capacity: 7.91 kW$_{th}$, high temperature (150 °C) heating capacity: 5 kW$_{th}$. Toluene is the more suitable refrigerant. Payback period: 5.13 y.
Bellos et al. (2018b)	PTC and waste heat	146.8 kW$_e$ (Toluene, n-octane, MM, n-pentane)	VARS (LiBr–H$_2$O)	Simple payback period: 4.86 y. Cooling capacity: 413.6 kW, heating capacity: 947.1 kW$_{th}$. Toluene is the more suitable refrigerant.
Villarini et al. (2019)	LFR, CPC	25 kW$_e$ (NOVEC 649)	VARS	Cooling capacity: 17.6 kW. Energy performance of LFR-ORC-system very sensitive to location compared to CPC-ORC-system.
Bellos and Tzivanidis (2019)	PTC	10 kW$_e$ to 15 kW$_e$ (Toluene, n-octane, MM, n-pentane)	VARS (LiBr–H$_2$O)	Toluene and n-octane are the preferred refrigerants when the heat rejection temperature is about 125–135 °C.

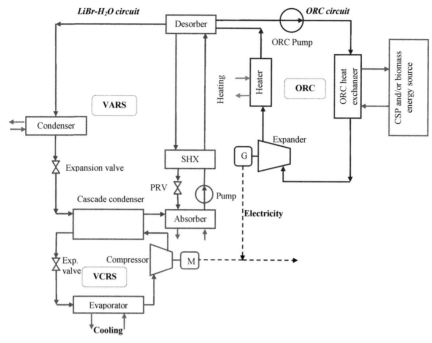

Figure 6. Simplified schematic of a typical organic Rankine cycle-based electricity and cascaded vapor compression and absorption refrigeration system (Adopted from Patel et al., 2017).

A bottoming VARS using lithium bromide-water (LiBr–H_2O) as a refrigerant is limited to space cooling at a commercial level (Tassou et al., 2010). Vapor absorption refrigeration systems with ammonia–water (NH_3–H_2O) are less advisable for food applications due to toxicity, flammability, low boiling point temperature difference of refrigerant and absorbent, low coefficient of performance and incompatibility with materials (Deng et al., 2011). Integrated systems based on adsorption cooling and liquid desiccant cooling technologies are still at the research and development phase (Jradi and Riffat, 2014).

Concentrated solar thermal energy-powered organic Rankine cycle systems integrated VARS, which works on thermal energy, are typically limited to space cooling. On the other hand, ORC-integrated VCRS, which works on electrical energy, can be used for refrigeration applications. Patel et al. (2017) proposed a concentrated solar thermal energy and biomass energy-powered ORC unit with a cascaded refrigeration system, as shown in Figure 6. In such a system, the electricity and heat duty requirements of the VCRS and VARS are fulfilled by the ORC-unit, combining the advantages of both systems. The cascaded system achieves low temperature (up to –20 °C) cooling and requires much lower electricity compared to the vapor compression refrigeration system (Patel et al., 2017).

5. Design Considerations

The key aspects of designing CSP-integrated ORC-unit-based multi-generation systems optimally are briefly covered in this section.

5.1 Solar irradiation data

The duration and intensity of the solar irradiation affect the performance, capacity factor, and economic viability of the system significantly. The sizing and configuration of the system also depend on the solar irradiation. Local factors like fog and pollution level, dust/sand storms, wind speeds and their variations also need to be considered when selecting the place of installation.

5.2 Solar collector field and thermal energy storage

Concentrated solar collector type and size and thermal energy storage type and size need to be carefully selected, as both these systems have major shares in the capital cost of the complete system. Optical efficiency and overall heat loss coefficients are crucial parameters for concentrated solar collector fields and improvements in these parameters often increase the solar field cost. The use of thermal energy storage facilitates delivery of the utilities according to the need by absorbing the variations due to fluctuations in the solar irradiation. For off-grid locations, where there are no available central grids, integration of other energy sources, fossil fuel based or biomass based, may be needed to meet the utility demands.

5.3 Organic Rankine cycle power system and other sub-systems

The selection of cycle configuration, component types and designs, and working fluid is important for efficiently converting solar thermal energy into multiple products. All these parameters are dependent on the application type and maximum capacity requirement. The other products (cooling, heating, fresh water) should be selected based on the needs of the region. The primary need (electricity, cooling, heating or fresh water) of the place is of vital importance for successful implementation. The cooling and desalination systems can be thermal energy driven or electrical energy driven, and the selection depends on the techno-economic analysis. Integration of thermal energy driven systems with an organic Rankine cycle power system enables a high energy utilization factor and high overall system efficiency. However, in such a system, the net power output is lower compared to only power generating systems with a condensing turbine. Therefore, the selection of an ORC power system and other sub-systems should be done carefully.

5.4 Load characteristics

The system design needs to be based on a detailed analysis of the part-load characteristics of the components of the multi-generation systems. Due to the mismatch between supply and demand, the major challenge is to provide a dilute and variable nature of solar energy input to the various demands. Moreover, it needs to be addressed that the actual system performance may differ from that of the design predictions due to the system inertia causing delays during the start-up and shut-down phases.

5.5 System configuration and control

For optimal system configuration, all of the aforementioned parameters need to be considered carefully. The process controls of the CSP-driven multi-generation systems based on the ORC technology should be designed as a subset of the overall plant control strategy. A proper system configuration and control provide desired products to the consumers cost-effectively and reliably.

5.6 Cost

Concentrated solar collector powered medium-scale dispatchable multi-generation energy systems with thermal energy storage are typically more costly than fossil fuel based and biomass based systems. However, factors like availability of fossil fuels and biomass as well as high carbon footprints for the former and high water footprints for the latter are the major drawbacks of these technologies.

6. Concluding Remarks

In this chapter, concentrated solar thermal energy-driven multi-generation systems based on the organic Rankine cycle technology were reviewed. Power generation, cogeneration, trigeneration, and multi-generation systems were discussed, and their possible configurations were presented. Issues related to the system design were addressed.

For solar organic Rankine cycle systems, parabolic trough collector and linear Fresnel reflector technologies are typically used. For cost parity, the cost of the linear Fresnel reflector technology (€/m²) should be about 50% to 60% lower than that of the parabolic trough collector technology. A recently analyzed nanostructured polymer foil-based concentrated solar field is a promising alternative for small to medium-scale organic Rankine cycle systems.

For fresh water generation applications, thermal energy driven multi-effect distillation is a better option than the electrical energy driven reverse osmosis system. The type of components (expander, heat exchangers and pump) and working fluid of the organic Rankine cycle power system should be decided based on the solar collector field data, type of application, and capacity of the system.

For cooling applications, depending on the temperature needed, either the electrical energy driven conventional vapor compression refrigeration systems or the thermal energy driven vapor absorption refrigeration systems can be used. The recently investigated cascaded refrigeration system is a promising alternative; however, it is currently at the research stage and no commercial plant exists as of yet.

The selection of type and size of the concentrated solar field, thermal energy storage, organic Rankine cycle power system, and other sub-systems is of vital importance for attaining a cost-effective solution. Solar irradiation data and load characteristics affect the overall system configuration and controller design.

Acknowledgement

The present work has been funded by the European Union's Horizon 2020 research and innovation programme with a Marie Skłodowska-Curie Individual Fellowship under grant agreement no. 794562 (Project: Small-scale CSP). The financial support is gratefully acknowledged.

Nomenclature

$A_{p,CL}$	aperture area of the solar collector field (m²)
DNI	direct normal irradiance (W/m²)
h	specific enthalpy (J/kg)
IAM	incidence angle modified
$\dot{m}$	mass flow rate (kg/s)
$\dot{W}$	power (W)
$\dot{Q}$	heat rate (W)
T	temperature (°C)
$U_{l,1}$	first-order heat loss coefficient based on aperture area (W/(m²·K))
$U_{l,2}$	second-order heat loss coefficient based on aperture area (W/(m²·K²))

Greek symbols

η	efficiency

Subscripts

a	ambient
c	condenser
CL	collector
e	evaporator
g	generator

is	isentropic
m	mean
o	optical
P	pump
T	turbine

Abbreviations

CSP	concentrated solar power
HMDS	hexamethyldisiloxane
HTF	heat transfer fluid
LFR	linear Fresnel reflector
ORC	organic Rankine cycle
PTC	parabolic trough collector

References

Al-Sulaiman, F.A., Dincer, I. and Hamdullahpur, F. 2011a. Exergy modeling of a new solar driven trigeneration system. Sol. Energy 85: 2228–2243.

Al-Sulaiman, F.A., Hamdullahpur, F. and Dincer, I. 2011b. Performance comparison of three trigeneration systems using organic Rankine cycles. Energy 36: 5741–5754.

Alfa Laval. 2018. <www.alfalaval.com/globalassets/documents/products/process-solutions/desalination-solutions/multi-effect-desalination/fresh-water-brochure-pee00251en-1201.pdf>, accessed 01.07.18.

Astolfi, M., Mazzola, S., Silva, P. and Macchi, E. 2017. A synergic integration of desalination and solar energy systems in stand-alone microgrids. Desalin. 419: 169–180.

Bao, J. and Zhao, L. 2013. A review of working fluid and expander selections for organic Rankine cycle. Renew. Sustain. Energy Rev. 24: 325–342.

Bellos, E. and Tzivanidis, C. 2017. Parametric analysis and optimization of a solar driven trigeneration system based on ORC and absorption heat pump. Journal of Cleaner Production 161: 493–509.

Bellos, E. and Tzivanidis, C. 2018. Investigation of a hybrid ORC driven by waste heat and solar energy. Energy Convers. Manage. 156: 427–439.

Bellos, E., Vellios, L., Theodosiou, I.C. and Tzivanidis, C. 2018a. Investigation of a solar-biomass polygeneration system. Energy Convers. Manage. 173: 283–295.

Bellos, E., Tzivanidis, C. and Torosian, K. 2018b. Energetic, exergetic and financial evaluation of a solar driven trigeneration system. Thermal Science and Engineering Progress 7: 99–106.

Bellos, E. and Tzivanidis, C. 2019. Evaluation of a solar driven trigeneration system with conventional and new criteria. International Journal of Sustainable Energy 38(3): 238–252.

Bruno, J.C., Lopez-Villada, J., Letelier, E., Romera, S. and Coronas, A. 2008. Modelling and optimisation of solar organic rankine cycle engines for reverse osmosis desalination. Appl. Therm. Eng. 28(17-18): 2212–2226.

Buonomano, A., Calise, F., Palombo, A. and Vicidomini, M. 2015. Energy and economic analysis of geothermal-solar trigeneration systems: A case study for a hotel building in Ischia. Appl. Energy 138: 224–241.

Casartelli, D., Binotti, M., Silva, P., Macchi, E., Roccaro, E. and Passera, T. 2015. Power block off-design control strategies for indirect solar ORC cycles. Energy Proc. 69: 1220–1230.

Cocco, D. and Cau, G. 2015. Energy and economic analysis of concentrating solar power plants based on parabolic trough and linear Fresnel collectors. Proc. IMechE Part A: J. Power and Energy 229(6): 677–688.

Cocco, D. and Serra, F. 2015. Performance comparison of two-tank direct and thermocline thermal energy storage systems for 1 MWe class concentrating solar power plants. Energy 81: 526–536.

Delgado-Torres, A.M. and García-Rodríguez, L. 2007. Comparison of solar technologies for driving a desalination system by means of an organic Rankine cycle. Desalin. 216(1-3): 276–291.

Deng, J., Wang, R.Z. and Han, G.Y. 2011. A review of thermally activated cooling technologies for combined cooling, heating and power systems. Prog. Energy Combust. Sci. 37: 172–203.

Desai, N.B. and Bandyopadhyay, S. 2015. Integration of parabolic trough and linear fresnel collectors for optimum design of concentrating solar thermal power plant. Clean Technologies and Environmental Policy 17(7): 1945–1961.

Desai, N.B. and Bandyopadhyay, S. 2016. Thermo-economic analysis and selection of working fluid for solar organic Rankine cycle. Appl. Therm. Eng. 95: 471–481.

Desai, N.B., Pranov, H. and Haglind, F. 2019a. Techno-economic analysis of a power generation system consisting of a foil-based concentrating solar collector and an organic Rankine cycle unit. *In*: Proceedings of the 32nd International Conference on Efficiency, Cost, Optimization, Simulation and Environmental Impact of Energy Systems (ECOS 2019).

Desai, N.B., Pranov, H. and Haglind, F. 2019b. Solar thermal energy driven organic Rankine cycle systems for electricity and fresh water generation. *In*: Proceedings of the 5th International Seminar on ORC Power Systems (ORC 2019). (Accepted for publication).

El-Emam, R.S. and Dincer, I. 2018. Investigation and assessment of a novel solar-driven integrated energy system. Energy Convers. Manage. 158: 246–255.

Garg, P., Orosz, M.S. and Kumar, P. 2016. Thermo-economic evaluation of ORCs for various working fluids. Appl. Therm. Eng. 109: 841–853.

Gerbens-Leenes, P.W., Hoekstra, A.Y. and Van der Meer, T. 2009. The water footprint of energy from biomass: A quantitative assessment and consequences of an increasing share of bio-energy in energy supply. Ecological Economics 68(4): 1052–1060.

Ghobeity, A., Noone, C.J., Papanicolas, C.N. and Mitsos, A. 2011. Optimal time-invariant operation of a power and water cogeneration solar-thermal plant. Sol. Energy 85: 2295–2320.

He, Y.-L., Mei, D.-H., Tao, W.-Q., Yang, W.-W. and Liu, H.-L. 2012. Simulation of the parabolic trough solar energy generation system with organic Rankine cycle. Appl. Energy 97: 630–641.

Heller, P. (ed.). 2017. The Performance of Concentrated Solar Power (CSP) Systems: Analysis, Measurement and Assessment. Woodhead Publishing.

Hoffmann, J.E. and Dall, E.P. 2018. Integrating desalination with concentrating solar thermal power: Namibian case study. Renew. Energy 115: 423–432.

Hung, T.C. 2001. Waste heat recovery of organic Rankine cycle using dry fluids. Energy Convers. Manage. 42: 539–553.

Imran, M., Usman, M., Park, B.S. and Lee, D.H. 2016. Volumetric expanders for low grade heat and waste heat recovery applications. Renew. Sustain. Energy Rev. 57: 1090–1109.

IRENA. 2012. Water Desalination Using Renewable Energy.

Javanshir, A., Sarunac, N. and Razzaghpanah, Z. 2018. Thermodynamic analysis and optimization of single and combined power cycles for concentrated solar power applications. Energy 157: 65–75.

Jradi, M. and Riffat, S. 2014. Tri-generation systems: Energy policies, prime movers, cooling technologies, configurations and operation strategies. Renew. Sustain. Energy Rev. 32: 396–415.

Karellas, S., Terzis, K. and Manolakos, D. 2011. Investigation of an autonomous hybrid solar thermal ORC–PV RO desalination system. The Chalki island case. Renew. Energy 36(2): 583–590.

Karellas, S. and Braimakis, K. 2016. Energy-exergy analysis and economic investigation of a cogeneration and trigeneration ORC-VCC hybrid system utilizing biomass fuel and solar power. Energy Convers. Manag. 107: 103–113.

Li, C., Kosmadakis, G., Manolakos, D., Stefanakos, E., Papadakis, G. and Goswami, D.Y. 2013. Performance investigation of concentrating solar collectors coupled with a transcritical organic Rankine cycle for power and seawater desalination co-generation. Desalin. 318: 107–117.

Liu, B.T., Chien, K.H. and Wang, C.C. 2004. Effect of working fluids on organic Rankine cycle for waste heat recovery. Energy 29(8): 1207–1217.

Maksiuta, D., Moroz, L., Burlaka, M. and Govoruschenko, Y. 2017. Study on applicability of radial-outflow turbine type for 3 MW WHR organic Rankine cycle. Energy Proc. 129: 293–300.

Mathkor, R., Agnew, B., Al-Weshahi, M. and Latrsh, F. 2015. Exergetic analysis of an integrated tri-generation organic Rankine cycle. Energies 8(8): 8835–8856.

Nafey, A.S. and Sharaf, M.A. 2010. Combined solar organic Rankine cycle with reverse osmosis desalination process: Energy, exergy, and cost evaluations. Renew. Energy 35: 2571–2580.

Nafey, A.S., Sharaf, M.A. and García-Rodríguez, L. 2010. Thermo-economic analysis of a combined solar organic Rankine cycle-reverse osmosis desalination process with different energy recovery configurations. Desalin. 261(1-2): 138–147.

Nixon, J.D. and Davies, P.A. 2012. Cost-exergy optimisation of linear Fresnel reflectors. Sol. Energy 86: 147–156.

NREL. 2019. National Renewable Energy Laboratory. <https://solarpaces.nrel.gov/by-status/operational>. Accessed on 27.05.2019.

Patel, B., Desai, N.B. and Kachhwaha, S.S. 2017. Thermo-economic analysis of solar-biomass organic Rankine cycle powered cascaded vapor compression-absorption system. Sol. Energy 157: 920–933.

Patil, V.R., Biradar, V.I., Shreyas, R., Garg, P., Orosz, M.S. and Thirumalai, N.C. 2017. Techno-economic comparison of solar organic Rankine cycle (ORC) and photovoltaic (PV) systems with energy storage. Renew. Energy 113: 1250–1260.

Petrollese, M., Cau, G. and Cocco, D. 2018. The Ottana solar facility: Dispatchable power from small-scale CSP plants based on ORC systems. Renewable Energy (in press).

Petrollese, M. and Cocco, D. 2019. Robust optimization for the preliminary design of solar organic Rankine cycle (ORC) systems. Energy Convers. Manage. 184: 338–349.

Quoilin, S., Van Den Broek, M., Declaye, S., Dewallef, P. and Lemort, V. 2013. Techno-economic survey of Organic Rankine Cycle (ORC) systems. Renew. Sustain. Energy Rev. 22: 68–186.

Rodríguez, J.M., Sánchez, D., Martínez, G.S. and Ikken, B. 2016. Techno-economic assessment of thermal energy storage solutions for a 1 MWe CSP-ORC power plant. Sol. Energy 140: 206–218.

Russo, V., Mazzei, D. and Liberatore, R. 2018. Thermal energy storage with integrated heat exchangers using stratified molten salt system for 1 MWe CSP. In AIP Conference Proceedings 2033(1): 090025.

Shalaby, S.M. 2017. Reverse osmosis desalination powered by photovoltaic and solar Rankine cycle power systems: a review. Renew. Sustain. Energy Rev. 73: 789–797.

Sharaf, M.A., Nafey, A.S. and García-Rodríguez, L. 2011. Thermo-economic analysis of solar thermal power cycles assisted MED-VC (multi effect distillation-vapor compression) desalination processes. Energy 36(5): 2753–2764.

Sharaf, M.A. 2012. Thermo-economic comparisons of different types of solar desalination processes. Journal of Solar Energy Engineering 134(3): 031001.

Sharan, P., Neises, T., McTigue, J.D. and Turchi, C. 2019. Cogeneration using multi-effect distillation and a solar-powered supercritical carbon dioxide Brayton cycle. Desalination 459: 20–33.

Tassou, S.A., Lewis, J.S., Ge, Y.T., Hadawey, A. and Chaer, I. 2010. A review of emerging technologies for food refrigeration applications. Appl. Therm. Eng. 30: 263–276.

Turboden. 2019. <https://www.turboden.com/applications/1056/concentrated-solar-power>. Accessed on 27.05.2019.

Tzivanidis, C., Bellos, E. and Antonopoulos, K.A. 2016. Energetic and financial investigation of a stand-alone solar-thermal organic Rankine cycle power plant. Energy Convers. Manage. 126: 421–433.

Villarini, M., Tascioni, R., Arteconi, A. and Cioccolanti, L. 2019. Influence of the incident radiation on the energy performance of two small-scale solar organic Rankine cycle trigenerative systems: A simulation analysis. Appl. Energy 242: 1176–1188.

Wendt, D., Mines, G., Turchi, C. and Zhu, G. 2015. Stillwater hybrid geo-solar power plant optimization analyses (No. INL/CON-15-34926). Idaho National Lab. (INL), Idaho Falls, ID (United States).

Wronski, J., Imran, M., Skovrup, M.J. and Haglind, F. 2019. Experimental and numerical analysis of a reciprocating piston expander with variable valve timing for small-scale organic Rankine cycle power systems. Appl. Energy 247: 403–416.

Wu, D., Zuo, J., Liu, Z., Han, Z., Zhang, Y., Wang, Q. and Li, P. 2019. Thermodynamic analyses and optimization of a novel CCHP system integrated organic Rankine cycle and solar thermal utilization. Energy Conversion and Management 196: 453–466.

Xie, W.T., Dai, Y.J. and Wang, R.Z. 2012. Theoretical and experimental analysis on efficiency factors and heat removal factors of Fresnel lens solar collector using different cavity receivers. Sol. Energy 86: 2458–2471.

Zanellato, L., Astolfi, M., Serafino, A., Rizzi, D. and Macchi, E. 2018. Field performance evaluation of geothermal ORC power plants with a focus on radial outflow turbines. Renew, Energy (in press).

CHAPTER 20

Solar Photovoltaic Technologies and Systems

JN Roy

1. Introduction

Global warming is a major concern as the demand for energy is ever increasing. Conventional energy resources, such as coal, natural gases and oil, are known to have high carbon emissions. For example, the greenhouse effect caused by CO_2 emission during coal plant operation is significant and a major concern. A typical coal-based thermal power plant can emit as much as 7000 Metric tons/MW CO_2 per annum. Notional damage to the environment for this is about 55 million US$ per annum. Besides this, the transport of coal is also responsible for additional GHG emissions. Disposal of end products, such as fly ash, is not environmentally friendly. The consumption of clean water, which is an ever-depleting resource, is also a major concern for coal-based power plants. A coal-based power plant typically requires about 2000 L/MWh of water. This requirement is much lower for alternative (renewable) energy generation, such as Solar Photovoltaic (Roy and Bose, 2018). Although continuous technological innovations, such as clean coal, are happening in these conventional technologies, concern remains as the demand for energy is also increasing. Particularly for the developing countries, the GDP growth rate can only be sustained by increasing the energy generation. The non-polluting conventional energy resources, such as nuclear, are no longer a preferred choice, due to some recent accidents associated with such plants. The unwanted environment damage due to radiation leaks in the atmosphere and the ocean have disastrous consequences.

The industry revolution originated due to the abundance of coal in western Europe. This was followed by the discovery of oil and gas in 1959. However, these energy resources are limited and are predicted to be exhausted by 2100, if not earlier (Roy and Bose, 2018; Gilbert, 2013). Global temperature rise is a major concern. Till 1960, the temperature rise was not noticeable, however it has risen by about +1 °C since then. Also the rapid global increase in CO_2 emission started from 1970 (280 ppm pa) and has gone up to 400 ppm pa at present (Boyle, 2004; Nelson, 2011). In the United Nations Framework Convention of Climate Change (UNFCCC, 2015), a roadmap for controlling GHG emissions has been developed. Goals have been set for several countries. India have pledged that by 2033, 40% of the electricity

FNAE, SMIEEE, FIMS, Advance Technology Development Centre and School of Energy Science & Engineering, IIT-Kharagpur-721302, India.
Email: jatinroy2000@gmail.com

generation will be from renewable sources, including bio-oil (bio ethanol and bio diesel) and large hydro. Also, the total GHG emission divided by GDP would be 33%–35% less in 2033 as compared to 2005. All these are possible by widespread development of renewable energy resources. Renewable energy is therefore getting increasing attention all over the world. Apart from the reduction of GHG emissions, these technologies help energy access and energy security (Foster et al., 2010).

2. Renewable Energy Resources

A list of major renewable energy resources is given below:

1. Solar Thermal: Thermal Energy of the sun can be converted to mechanical energy and then to electrical energy.
2. Solar Photovoltaic: Solar irradiance can be converted directly into electrical energy.
3. Wind Energy: Wind power can be used to rotate blades attached to a turbine in order to generate electricity.
4. Ocean Thermal Energy Conversion (OTEC): The temperature differential between deep water (cold) and surface water (warm) of the ocean can be used to produce electricity.
5. Wave Energy: Ocean waves carry lot of energy which can be harvested to rotate turbines and generate electricity.
6. Tidal Energy: Tidal energy can be harvested to rotate turbines and generate electricity.
7. Hydroelectric Energy: Water falling from a height has kinetic energy which can be used to rotate turbines and generate electricity. Small hydro (< 2 MW) plants fall under the renewable energy category. However large hydro (> 2 MW) is considered as a conventional energy resource.
8. Bio Energy: Bio mass can be processed to produce heat and eventually electricity.
9. Geothermal Energy: Earth's core has much higher temperature, mainly due to radioactive decay. Available heat energy at a certain depth of Earth's crust can be harvested to generate electricity.

Global energy scenario until 2017 is shown in Table 1. In 2018, 178 GW renewable power was added globally. In India, the present capacity of renewable energy is 70 GW, with a Cumulative Average Growth Rate (CAGR) of 18%. The break-up is as follows:

Solar PV: 22 GW (21%); Bio: 9 GW (13%); Wind: 34 GW (49%); Small Hydro (< 2 MW): 5 GW (7%).

The projected capacity of renewable energy in India by 2022 is 200 GW, having 100 GW Solar PV, 60 GW Wind, 10 GW Bio and 5 GW Small Hydro. The investment required would be about 125 billion US$.

It is clear from the preceding discussion that Solar Photovoltaic is getting major attention. Although the total present capacity of wind energy is highest among renewable sources, the growth is slowing down. The highest growth, reflected by additional capacity being created, is happening for Solar PV.

Table 1. Global clean energy scenario until 2017.

Energy resourse	Total capacity till 2017	Addition in 2017
Hydro	1114 GW	19 GW
Ocean	529 MW	No Addition
Solar PV	402 GW	101 GW
Bio	122 GW	8 GW
Geo Thermal	12.8 GW	0.7 GW
Solar Thermal	4.9 GW	0.1 GW
Wind	539 GW	52 GW

3. Classification of Solar Photovoltaic (SPV) Technologies

Energy conversion of solar photovoltaic (Markvart and Castaner, 2005) is a direct process. The solar energy available in the form of irradiance is directly converted into electricity using solar cells. Mono- and multi-crystalline silicon solar cells (Green, 1995), which fall under first-generation technologies, have the highest market share (about 90%). High efficiency III–V compound based Multi Junction (MJ) solar cells (Geisza et al., 2008; Stan et al., 2003) also fall under first-generation technologies. These are mainly used in space applications. Almost all the satellites/spaceships use high-efficiency MJ solar cells. Limited use of such solar cells in terrestrial applications are found in Concentrated Photovoltaic (CPV) systems (Min et al., 2009). Thin film solar cells, which are comparatively new, fall under second-generation technologies. This category of solar cells has the rest of the terrestrial market share (about 10%). Amorphous silicon (a-Si) (Street, 2005; Ahmed et al., 2017; Ahmed et al., 2019), Cadmium Telluride (CdTe) (Cusano, 1963; Hamid and Fatima, 2013; Ikegami, 1988) and Copper Indium Gallium Selenide (CIGS) (Ohlsen et al., 1993; Kronik et al., 1998) are the main second-generation technologies and are used commercially. Some solar cells, such as Dye Sensitized Solar Cells (DSSC) (Gratzel, 2003), Organic and Polymer-based solar cells (Sam-Shajing and Serdar, 2005; Bose, 2012), which are yet to be commercially exploited, fall under third-generation solar cell technologies. Although the concepts have been proven and prototypes are built, some problems, such as stability, are yet to be solved so that these cells can be used commercially. Recently, a new type of solar cell, called perovskite solar cell (Nazeerusddin and Snaith, 2015), has been drawing a lot of attention. Technologies, mainly based on quantum mechanical principle, are under fundamental research (Chuang et al., 2014). These are classified as fourth-generation technologies.

4. Solar Resources

The sun emits about 3.9×10^{26} W of power and a small fraction of this, about 1370 W/m^2, is intercepted by the earth, just outside its atmosphere (Gilbert, 2013). This is known as solar constant and its value is not the same during the entire year. As the distance between the sun and the earth varies throughout the year, the solar constant value also changes. However, the variation is small. The power at the surface of the earth is always less than the solar constant because of the attenuation of the power as the sunlight travels through the atmosphere, including clouds. This is known as irradiance and has different values at different locations depending on altitude and weather patterns. The irradiance value at a particular location varies during the day, lowest in the morning/evening and highest during noon. This also experiences seasonal variation. The irradiance variation during the day occurs as the distance travelled by sunrays from outside the atmosphere to a particular location varies during the day. This has been depicted in Figure 1. The distance is lowest when the sun is at zenith, i.e., directly overhead, as sunrays have to travel a minimum distance (h_1) or a minimum Atmospheric Mass (AM). This is referred to as AM1. At any other time of the day, the distance traveled by the sunrays is more than h_1 and, therefore, has higher atmospheric mass. At a particular time (see Figure 1), if the distance is h_2, then the atmospheric mass is defined as AM (h_2/h_1). AM1.5 is approximated as the average irradiance received throughout the day for a particular location. Solar constant is applicable to outside Earth's atmosphere and the atmospheric mass is zero (AM0).

The irradiance received by a collector at the earth surface has three main components. The major one is direct radiation, which manifests as "beam" or Direct Normal Incident (DNI). During clear sky, the perpendicular component of the irradiance received by the collector is denoted as beam or direct radiation. The incident sun rays are not perpendicular to the collector during the entire day, except when the sun is directly overhead. The power received by the collector, therefore, varies from the morning to the evening, depending on the altitude of the sun. This is given by "cosine law", see equation (1).

$$I_{DC} = I_{DG} \cos\theta \tag{1}$$

where I_{DC} is the effective direct irradiation receive by the collector. I_{DG} is the irradiance received if the sunrays are falling perpendicular to the collector plane. θ is the angle between the sunrays and the normal to the collector plane.

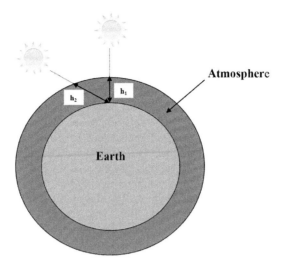

Figure 1. During two different times of a particular day, sun rays travelling different distance through the atmosphere to reach a particular point on Earth's surface.

Another important component is the diffused radiation. Scattering of sunlight in the atmosphere, including clouds, causes some part of the irradiance reaching the collector as diffused radiation. During a cloudy day, the DNI component is zero and the entire radiation received by the collector is diffused radiation. Even in clear sky conditions, there is some diffused radiation. The diffused radiations are assumed to be coming from the sky uniformly in all directions.

Another minor component of the irradiance collected by the collector is the reflected radiations. The reflected radiation from the surface adjacent to the collector and the surrounding structures, such as snow-clad mountains. The amount of the reflected radiations is decided by the ground materials and the surrounding structures/landscape. This component is very small and generally neglected.

At a particular location, the total solar energy received by a collector during a year is made up of direct and diffused radiations. The relative contribution depends on the weather pattern, such as the number of cloudy days per year at that particular location. In a tropical country like India, where there are about 300 clear sky days, the share can be 75% direct and 25% diffused in a total year. The share of diffused radiation increases in areas where there are a greater number of cloudy days.

The energy emitted by the sun can be approximated by assuming that the sun is a black body with temperature of 5800K (Roy and Bose, 2018). The spectral distribution of the energy received at AM0 has a peak at $\lambda = 0.5$ μm and the total energy is distributed in Ultraviolet (UV: $\lambda < 0.4$ μm) (7%), Visible (0.4 μm $< \lambda < 0.8$ μm) (47%) and Infra-Red (IR: $\lambda > 0.8$ μm) (46%). Some alteration of this spectral response happens due to the travel of sunrays through the atmosphere. The spectral distribution for AM1.5 can be estimated as 2.0% UV, 54% visible and 44% IR. The peak shifts to the right marginally and remains close to 0.5 μm.

5. Photovoltaic Materials and Device

The first- and second-generation solar cells, which are primarily based on p-n junction diodes, are made of semiconductor materials. A part of the solar spectrum falling on the semiconductor is absorbed and electron-hole pairs are generated. They are separated due to the in-built potential of p-n junction diode and are then allowed to flow across a load, generating current and voltage. The number of electron hole pairs generated depends on how great a portion of the solar spectrum is absorbed by the semiconductor. There is a cut-off wavelength ($\lambda_{cut-off}$), which is related to the band gap ($E_g = 1.24/\lambda_{cut-off}$) of the semiconductor material. The part of the spectrum having $\lambda < \lambda_{cut-off}$ is absorbed by the semiconductor and the rest ($\lambda > \lambda_{cut-off}$) is not absorbed. The current (I_{SC}) produced by a photovoltaic device is directly proportional to

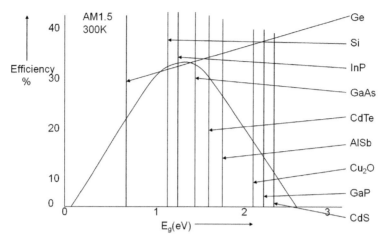

Figure 2. Ideal (theoretical) efficiency of a photovoltaic device as a function of the band gap of the material. Maximum efficiency is at a band gap of about 1.45 eV.

the number of holes and electrons generated by the incident light. The semiconductor materials having a higher band gap (lower $\lambda_{cut-off}$) intercept only a small part of the spectrum, resulting in a smaller current. On the other hand, the voltage (V_{OC}) generated by the photovoltaic devices depends on the potential energy of the carriers generated. The greater the band gap, the greater the potential energy and, thus, the voltage. It may be noted that the higher energy photons ($\lambda > \lambda_{cut-off}$) in the solar spectrum generate higher energy electron-hole pairs. However, they ultimately settle to the band gap energy by losing part of the energy in terms of heat. In the limiting case; $E_g \to 0$: $I_{SC} \to \infty$, $V_{OC} \to 0$ and Power (P) $\to 0$ and $E_g \to \infty$: $I_{SC} \to 0$, $V_{OC} \to \infty$ and Power (P) $\to 0$. The efficiency (P_{out}/P_{in}) depends on the total power falling on the material (P_{in}) and the amount of power converted to electrical energy (P_{out}). The theoretical efficiency, therefore, has strong dependency on the band gap, as shown in Figure 2.

The ideal efficiency of a solar cell occurs at about 1.4 eV (see Figure 2). CdTe has a band gap close to this value and is, therefore, an ideal choice as a solar cell material. Silicon is an elemental semiconductor having a band gap of about 1.1 eV. Although it is not ideal, this material is preferred due to its other advantages. The band gaps of ternary and quaternary compounds, such as CIGS, GaInAs, GaAlInAs, etc., can be tuned by varying the composition of the elements. It is possible to have the band gap adjusted to the ideal value for such materials and to enhance the theoretical efficiency by tandem or Multi Junction (MJ) solar cells. In this, the p-n junctions made with materials with different band gaps are stacked; the highest band gap material being at the top. The majority of the solar spectrum can then be absorbed and the energy loss due to photons having energy greater than the band gap is minimized.

6. Electrical Characteristics of Solar Cells

A solar cell is essentially a p-n junction diode (Figure 3). Electron hole pairs are generated due to incident photons. These are separated and travel in the opposite directions aided by in-built potential present in the depletion region and by concentration gradient. They eventually recombine across the load. The flow of charge carriers across the load results in current and associated voltage drop, depending on the load resistance. The equivalent circuit of the solar cell is shown in Figure 4. Photocurrent I_L generated due to light partly flows through the diode (I_D) and the rest through the load (I), resulting a voltage drop across the load (V). There are some unwanted resistances; series (R_S) and shunt (R_{Sh}). Resistances of the neutral regions of the semiconductor (see Figure 3) and the contact resistances are responsible for R_S, which is small but finite. There is an additional voltage drop across this resistance, therefore, the effective voltage drop across the load is reduced. The shunt resistance appears due to the resistance of the depletion regions (see Figure 3), and even for a perfectly made p-n junction, the resistance is finite due to the presence of minority carriers. Imperfection occurring during fabrication of the p-n junction may reduce the shunt

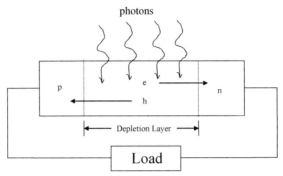

Figure 3. p-n junction diode as solar cell.

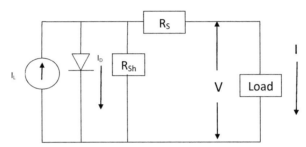

Figure 4. Equivalent circuit of solar cell.

resistance further. There is a current across the shunt resistance reducing the effective current across the load. The ideal R_{Sh} is infinite. It is clear that the current and the voltage across the load change as the load resistance (R_L) varies. In the limiting cases; $R_L = 0$ (short circuit): $I = I_{SC}$ and $V = 0$ and $R_L = \infty$ (open circuit): $I = 0$ and $V = V_{OC}$. I_{SC} and V_{OC} are known as short circuit current and open circuit voltage, respectively. In both the extreme cases, the power ($P = V \times I$) is zero. It can be seen that $I_{SC} = I_L$ as the entire photo-generated current I_L flows across the load in short circuit condition. The I-V characteristics of a solar cell at a particular irradiance can be obtained by varying the load resistance from zero (short circuit) to infinity (open circuit). This characteristic is shown in Figure 5.

The current appears in the fourth quadrant, positive V and negative I, indicating negative power. This means that the power is extracted from the system. An ideal I-V characteristic ($R_S = 0$ and $R_{Sh} = \infty$) is shown in Figure 5. For convenience, the I-V characteristic of the solar cells is drawn (Figure 6) in the first quadrant, with an understanding that the current is negative. The P-V characteristic is superimposed in this figure. Various parameters of the solar cells are defined (see Figure 6) as follows:

a) V_{OC}: Open circuit voltage when the solar cell is not connected to load ($R_L \to \infty$) and the solar cell is open circuited. The current across the load is, therefore, zero.

b) I_{SC}: Short circuit current when there is no load and the solar cell is short circuited ($R_L \to 0$). The voltage drop across the load is zero.

c) MPP: Maximum Power Point at which the power is maximum (P_m).

d) V_m: The voltage at MPP. The power extracted from the solar cell is maximum at this voltage.

e) I_m: The current at MPP. This corresponds to current delivered to the load by the solar cell when the voltage across the load is V_m.

f) Fill factor = $(V_m \times I_m)/(V_{OC} \times I_{SC})$. This signifies the actual power extracted ($V_m \times I_m$) as against the absolute maximum ($V_{OC} \times I_{SC}$).

g) η: Efficiency defined by $(P_{out}/P_{in}) = (V_m \times I_m)/(\text{Irradiance} \times \text{Area})$. The irradiance is defined as power per unit area (W/m^2), with the area being measured in m^2.

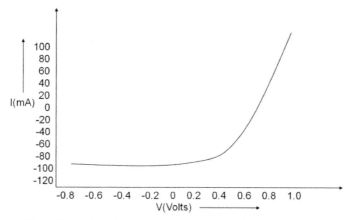

Figure 5. I-V characteristics of an ideal ($R_S = 0$ and $R_{Sh} = \infty$) solar cell.

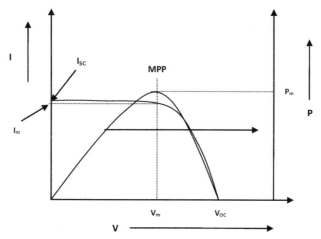

Figure 6. I-V and P-V characteristics of a solar cell.

The governing equations can be derived from the equivalent circuit (Figure 4).

$$I = I_D - I_L + (V + IR_S)/R_{Sh} \tag{2}$$

$$= I_S (e^{q(V-IR_S)/kT} - 1) - I_L + (V + IR_S)/R_{Sh} \tag{3}$$

where I_S is the reverse saturation current of the diode, k is the Boltzmann constant (8.6×10^{-5} eV/K) and T is the temperature in K. In case an ideal solar cell is assumed ($R_S = 0$ and $R_{Sh} = \infty$), the above equation can be simplified as:

$$I = I_S [e^{(qV)/(kT)} - 1)] - I_L \tag{4}$$

The open circuit voltage (V_{OC}) can be obtained by putting $I = 0$ and $V = V_{OC}$ in equation (4)

$$V_{OC} = (kT/q) \ln [(I_L/I_S + 1)] \tag{5}$$

For $I_L \gg I_S$,

$$V_{OC} = (kT/q) \ln (I_L/I_S) \tag{6}$$

The short circuit current (I_{SC}) can be obtained by putting $V = 0$ and $I = I_{SC}$ in equation (4).

$$I_{SC} = -I_L \tag{7}$$

It is evident that the power (P_m) obtained by a solar cell depends on the intensity of the light which is known as irradiance (W/m^2). As the irradiance increases, the number of photons available to generate hole-electron pairs also increases. This essentially increases the current as a greater number of charge carriers are available. The short circuit current (I_{SC}) has linear relation with the irradiance. This means that, if the intensity is doubled, the short circuit current also doubles. The I-V and P-V characteristics of a solar cell at various irradiance levels is shown in Figure 7. Figure 8(b) also shows P-V characteristics of a solar cell at different irradiances. The open circuit voltage (V_{OC}) primarily depends on the band gap of the semiconductor material used to make the solar cell. Therefore, open circuit voltage does not have a strong dependence on the irradiance. Marginal increase of the open circuit voltage occurs due to low or moderate increase of irradiance as the photo-generated population of the electrons and holes increase in the conduction and the valence bands, respectively; higher energy levels are now being occupied by these carriers, resulting in an overall increase of their potential energy. In Concentrated Photovoltaic (CPV), where the light intensity is very high (> 100X), this effect is prominent. Significantly more V_{OC} can be obtained at concentrated light as compared to normal (1X) irradiance. The efficiency of CPV is, therefore, higher. As the irradiance is much higher for CPV, the heating of the solar cell becomes a major issue. Additional cooling arrangements should be made to keep the temperature of the solar cell under control.

The temperature effect is very prominent in solar cells. However, in this case, the current is almost immune to temperature change. At a fixed irradiance, the current generated by a solar cell mainly depends on the band gap of the semiconductor material used. The effect of the temperature on the band gap is not significant. In fact, there is a small decrease of the band gap as the temperature increases. Therefore, there is a small increase in the current as temperature increases. The open circuit voltage has strong dependency on the temperature. There is a significant reduction of the open circuit voltage as the temperature increases, see equation (4). This happens due to an increase in the current (I_D) through the diode as the I_S increases with the temperature. An additional voltage drop occurs in the diode due to this increased current, resulting in an overall voltage drop across the load. The I-V and P-V characteristics at various temperatures are shown in Figure 8. Temperature co-efficient value (T.C. = $\Delta X/\Delta T$) of a particular parameter (X) captures the change (ΔX) of that parameter due to a certain change in temperature (ΔT). The temperature co-efficient of voltage (V_{OC}) is denoted as β. Typical value of β for c-Si technology is –0.36%/°C. The negative sign indicates that the V_{OC} decreases as the temperature increases. The temperature coefficients of current (I_{SC}) and power (P_m) are denoted as α and γ, respectively. For c-Si technology, typical values for α and γ are +0.05%/°C and –0.4%/°C, respectively. a-Si and CdTe technologies have lower temperature coefficients. CIGS technology, on the other hand, has comparable temperature coefficients to c-Si technology.

As the output of a solar cell varies due to the change of the irradiance and the temperature, it is important to define a Standard Test Condition (STC) for generating specifications. As described earlier,

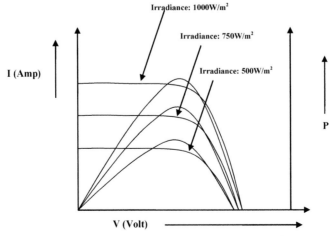

Figure 7. I-V and P-V characteristics of a solar cell at different irradiance.

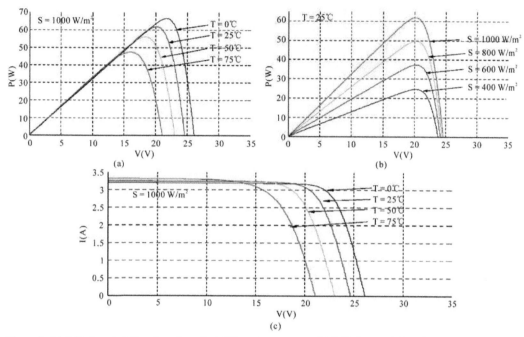

Figure 8. (a) P-V characteristics of a solar cell at different temperature, (b) P-V characteristics of a solar cell at different irradiance and (c) I-V characteristics of a solar cell at different temperature.

the solar spectrum and the angle of incidence of the incoming light also changes the output. The solar spectrum is somewhat different at AM0, AM1 and AM1.5 conditions. The STC for solar cell testing is defined as 1000 W/m² irradiance in AM1.5 spectrum falling normal to the cells having a temperature of 25 °C.

7. Solar Photovoltaic Technology

Solar cells are essentially made of p-n junction diodes. Electrical connections are taken by using metal. A detailed description of the fabrication technology is out of the scope of the present chapter. Interested readers may refer to the book (Roy and Bose, 2018) co-authored by the author of this chapter. c-Si solar cell is typically made using a 156 mm × 156 mm p-type silicon wafer, by converting the top portion of it to n-type using doping. This forms the n-p junction. Metal contacts are taken from top (n-type) and bottom (p-type) for electrical connections. Front and back view of a typical c-Si solar cell is shown in Figure 9. The metal connection is made up of fingers appearing as thin horizontal lines and bus bars appearing as thick vertical lines. The fingers collect the photo-generated carriers (electrons) from the n-side of the diode. The bus bars accumulate these carriers so that these can be taken out from the cells (see Figure 9). The finger configuration is required, as the complete front surface cannot be covered by the metal. As the metal is opaque, no light will then reach the silicon. The back surface has no such restrictions. Therefore, the entire back surface is covered by metal, which also includes bus bars for external connections.

Several advancements in c-Si technology are made to improve the efficiency. Some of the important advancements are (a) double side printing (Roy and Bose, 2018), (b) selective emitter (Roder et al., 2010), (c) Back Surface Field (BSF) (Roy and Bose, 2018), (d) buried contact (Antoniadis et al., 2010), (e) Light Induced Plating (LIP) (Roy and Bose, 2018), (f) Metal Wrap Through (MWT) (Magnone et al., 2014), (g) Emitter Wrap Through (Kiefer et al., 2011), (h) Passivated Emitter and Rear Contact (PERC) (Wang and Green, 1994), (i) Passivated Emitter Rear Locally Diffused (PERL) (Wang et al., 1995),

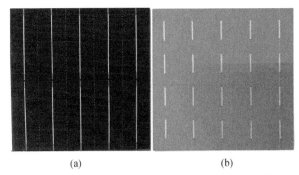

(a)	(b)

Figure 9. (a) Front and (b) Back view of a c-Si solar cell.

(j) Passivated Emitter Rear Totally Diffused (PERT) (Zaho et al., 1999) and (k) Inter-digitated Back Contact (IBC) (Zanuecoli et al., 2015).

High efficiency III-V cell technology (Roy and Bose, 2018) uses thin films of high-quality single crystal III-V compound semiconductor materials. Stacks of n-type and p-type thin films are typically deposited on Ge substrate to form solar cells. Multi-Junction (MJ) solar cells are made by using stacks of p-n junctions using ternary and quaternary III–V compound semiconductor materials. The band gaps of such materials are adjusted by varying the composition of the constituent elements. Photolithography steps are also used to define various regions. The cost of such solar cells is very high due to costly materials (in gas form) as well as processes, such as deposition and photolithography. These cells are used in space applications, such as satellites, where the overall cost implication of the solar cell is not significant. These cells are also used in Concentrated Photovoltaic (CPV) systems where the cost implication of using solar cells is significantly reduced as the area covered by the cells is much smaller.

a-Si, CdTe and CIGS solar cell technologies also use thin film deposition techniques (Roy and Bose, 2018). However, the materials (in gas form) and processes used in these technologies are not that costly. Amorphous or micro-crystalline thin film layers of p and n are deposited on a TCO (Transparent Conducting Oxide) coated glass substrate to form the solar cells. Instead of photolithography, LASER are used to define various regions and reduce the processing cost. Although these solar cells are cheaper, the efficiencies are lower due to lower mobilities of the deposited materials.

The fabrication of Dye Sensitized Solar Cell (DSSC) and organic solar cell is even simpler. Light sensitive dye can be coated on cheap plastic or ceramic substrates to form the solar cells. Some problems, such as stability, are yet to be solved before they can be commercially used.

A representative specification of a commercially available c-Si solar cell is given in Table 2. It may be noted that the parameters are measured at STC. The power of the individual c-Si solar cell is very small and many cells need to be electrically connected in order to enhance the available power. This is done by making modules (Roy and Bose, 2018), sometime called panels, generally consisting of series connected cells. A 60 cells module, which is common, has 60 cells connected in series, generally consisting of a 10 × 6 matrix. Modules with cells other than 60, e.g., 12, 24, 36, 48, 72, or 96, are also available. The top bus bars of the first cell are connected to the second cell's bottom bus bars in order to connect the two cells in series. One string consisting of 10 cells are made using this process. Six such strings are made for a 60 cells module. The strings are made by an automated process called Tabbing and Stringing (T&S). These strings are then connected in series by a process called bussing. In this, the strings are electrically connected using copper strips. The entire cell matrix is then laminated using an arrangement called "layup", consisting of layers of glass, Ethyl Vinyl Acetate (EVA)-I, cell matrix, Ethyl Vinyl Acetate (EVA)-II and tedlar sheet from top to bottom in this sequence.

The lamination is done to protect the solar cells from the external environment after installation. Galvanized aluminum frames are then attached on all four sides to provide more support and make it compatible for easy installation. A Junction Box (JB) consisting of two external cables is attached at the back side of the module. Electrical connections from output of the cell matrix to the cables are provided by the JB. The electrical connection between modules are established using these cables. Both series and

Table 2. Representative specifications of a commercially available c-Si solar cell.

Parameter	Value
V_{OC}	0.625 V
V_m	0.508 V
I_{SC}	9.08 A
I_m	8.71 A
F.F.	0.78
P_m	4.425 W_P*
Size	156 mm × 156 mm
Efficiency	18.2%

* Subscript P in W_P indicate power extracted at peak (MPP) of the P-V curve.

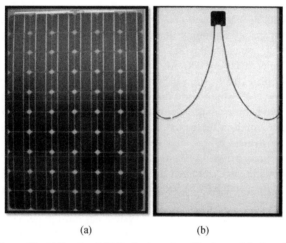

(a)   (b)

Figure 10. (a) Front and (b) Back view of a c-Si solar module (panel).

Table 3. Representative specifications of commercially available 60 cells c-Si solar module.

Parameter	Value
V_{OC}	37.45 V
V_m	30.46 V
I_{SC}	9.08 A
I_m	8.71 A
F.F.	0.78
P_m	265 W_P*
Size	1639 mm × 982 mm
Efficiency	16.46%

* Subscript P in W_P indicate power extracted at peak (MPP) of the P-V curve.

parallel connections of modules are possible. Figure 10 shows the front and back side views of a 60 cells c-Si module.

A representative specifications of a commercially available 60 cells c-Si module measured at STC are given in Table 3. III–V compound semiconductor based solar cells technology also has similar approach. The cells, typically 100 mm in diameter, are made first and then several of them are assembled to make the solar panels for the satellite. There are no standard sizes of the panels as it is tailor made as per the requirement of a particular satellite. a-Si, CdTe and CIGS thin film solar cell are made using a large TCO coated glasses. Therefore there are no separate cell manufacturing and module assembly steps; the modules are made directly.

8. Solar Photovoltaic Systems

SPV systems fall broadly in two categories. Smaller systems used for home/office are known as off-grid/grid interactive systems. These systems are installed typically on roof tops and have varying installed capacity, from < 1 kW_p to few hundreds of kW_p. Off-grid/grid interactive systems can have provisions for storage using a battery. However, the cost of SPV systems with storage increases depending on the amount of storage required. Apart from applications of providing electricity to homes/offices, smaller SPV systems are also used for water pumps, street lights, cell phone towers and many other such applications. A representative Single Line Diagram (SLD) of off-grid/grid interactive SPV system is shown in Figure 11.

An SPV array consists of numbers of solar modules connected in series and parallel, decided on the basis of the installed capacity and array design. The details of array design are described later. A Power Controller Unit (PCU) consists of a DC-DC converter, Maximum Power Point Tracking (MPPT) controller, Charge Controller (CC) and DC-AC inverter. The DC-DC converter and MPPT controller ensure that the DC power provided by the SPV array is always at MPP. A charge controller is required for management of the power flowing in and out of the battery. During a charging cycle, the power is stored in the battery. Although this primarily comes from the SPV, it is possible to charge the battery from the grid power. The power stored in the battery can be supplied to the load (through DC-AC inverter) when the SPV power is not available, e.g., after sunset. DC-AC converts the DC power to AC power for supplying to the load. In entirely off-grid systems, also known as standalone SPV systems, the grid power is not available. The load requirements are met from the SPV during the day and from the battery during the night. Part of the power generated by SPV during the day has to be stored in a battery in addition to the power supplied to the load.

The charge controller is not required if the SPV system has no storage. The PCU consisting of only DC-DC converter along with MPPT and DC-AC inverter is known as "solar inverter". Efficiency of solar inverters are significantly higher than PCUs with charge controllers. As the battery has a high cost, SPV systems are designed with no storage or minimum storage in order to cater to essential loads at night or during power outages. The amount of power produced by an SPV array having a specific installed capacity depends on irradiance level and temperature, which varies throughout the day and the year. It is possible that even during daytime, the power required by the load is more than the power produced by an SPV array. In such situations, the excess required power is provided by the grid. Such systems are known as grid-interactive SPV systems.

The estimation of energy output of smaller SPV systems with a specific installed capacity is done using calculations based on "peak sun hour", which is defined next. Such calculations are not very accurate and a long-term average is taken. It is to be noted that accurate instantaneous power calculations are not a requirement for smaller systems. For example, a household with a connected load of 4 kW, for example, does not require 4 kW all the time, as all the appliances are not switched on simultaneously. The average use of the load, which is also not precisely defined, is important. The SPV generation is also not fixed throughout the day as the irradiance and the ambient temperature vary. The estimation of peak sun hour of a particular place is done by plotting the irradiance versus the time of day, as shown in Figure 12. The irradiance increases from zero after the sun rises from the horizon and reaches its maximum when the sun is at the zenith position. During the afternoon, the intensity decreases and goes to zero at sunset. This is shown by the solid curve of Figure 12. The power produced by a solar panel varies as the irradiance changes. For example, a 265 W_p SPV module will produce 265 W only when the irradiance is 1000 W/m^2, if the temperature effect is not considered. The temperature effect will be added later.

The area (A_1 in Figure 12) under the Irradiance-Time (between sunset and sunrise) curve is the energy received per day. An equivalent time has to be estimated for a normalized power of

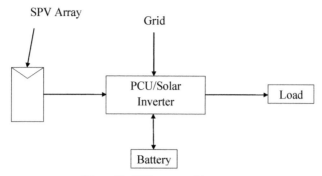

Figure 11. SPV system with storage.

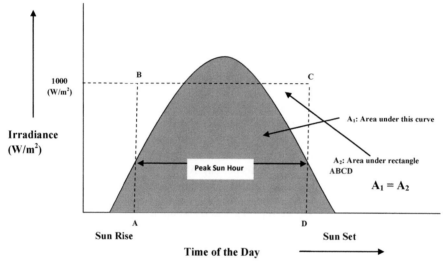

Figure 12. Estimation of peak sun hour.

1000 W/m². This is done by finding out an equivalent area corresponding to the power of 1000 W/m² (see Figure 12). The area (A_2) of the rectangle ABCD is equal to A_1. The corresponding time (AD) is known as "peak sun hour". This essentially means that the total energy received per day is (1000 W/m² × Peak sun hour). For example, if peak sun hour per day value at a particular place is 4 hrs, a 265 W_p solar module will produce (265 W × 4) = 1060 Wh of energy per day. As the peak sun hour per day value is not same throughout the year, an average value is taken for energy generation calculations. For example, a 265 W_p SPV module may produce less or more than the average value of 1060 Wh, or 1.06 kWh, depending upon the irradiance at that particular day. A long-term average is more accurate. For example, the 265 W_p SPV module is estimated to produce (1.06 × 365) kWh = 386.9 kWh energy per year. This estimation is more accurate. The peak sun hour per day values for different locations are published by various agencies. For tracking-based systems, the peak sun hour value at a particular location is higher than that of fixed systems at that location. In fixed SPV systems, solar modules are kept at fixed orientation with an angle equal to the latitude of the place of installation. The modules face south in the northern hemisphere and face north in the southern hemisphere. In single axis tracking-based SPV systems, the solar modules are oriented in the east-west direction and rotate as per the sun movement during the day; from morning to evening. Dual axes tracking-based SPV systems also take seasonal sun movement into account. Duel axes tracking is not generally used for conventional SPV systems as the complexity overrides the benefits achieved. However, in Concentrated Photovoltaic (CPV) systems, the use of duel axes tracking is a must.

The temperature effect also has to be considered for the estimation of energy. A 265 W_p SPV module will produce 265 W at 1000 W/m^2 only when the cell temperature is 25 °C. The ambient temperature varies during the day and throughout the year, depending on the weather conditions. Here also, a long-term average is considered in order to estimate the ambient temperature. As mentioned earlier, the cells of a module heat up during irradiation. Therefore, the cell temperature is more than the ambient temperature during operation. This heating due to irradiance is also not constant, as the irradiance itself varies. The estimation is typically done by assuming the average cell temperature is about 20 °C more than the ambient. Therefore, if the ambient (average) temperature at a particular location is 35 °C, then the cell temperature (average) is about 55 °C (35 °C + 20 °C). Temperature co-efficient (see section 6) is used to incorporate temperature effect into the energy estimation. If the cell temperature is 55 °C, which is 30 °C (55 °C–25 °C) more than STC, the power of the module changes by –12% (–0.4% × 30). This indicates 12% less power than STC. A 265 W_p module will produce 31.8 W (12% × 265 W_p) less power. The power produced by the module at an effective cell temperature of 55 °C is, therefore, 233.2 W_p (265 W–31.8 W). Considering both irradiance and temperature effect, the 265 W_p module produces 932.8 Wh (233.2 × 4) per day and about 340.5 kWh per year, if the peak sun hour and the ambient temperature of that location are 4 hr and 35 °C, respectively, and with the assumption that the cell temperature is 20 °C higher than that of the ambient. The difference between the cell temperature and the ambient temperature is more for tracking systems than fixed systems.

The design of an SPV system starts with the determination of required installed capacity based on the load profile of the customer. The energy generation estimation of a specific installed capacity SPV plant can be done using the method described above. For example, based on the load profile of a household, it has been estimated that the household would require a total of 20 kWh energy per day. Out of that, 15 kWh is used during the day and the rest (5 kWh) during the night. This can be estimated by listing all the electrical appliances (lights, fans, air conditioning units, laptops, TVs, fridges, etc.) being used, along with their wattage rating and the operating hours per day. A 4 kW_p solar installation would be required in case the peak sun hour per day value at the location of installation is 5 hrs. The energy produced for this 5 kW_p system is 20 kWh (4 kW × 5 Hrs). Out of this, 5 kWh is stored in the battery and the rest (15 kWh) is used to cater to the load during the daytime. During the night, the requirement is met from the stored energy in the battery. A simplistic view is presented here and several other issues, such as efficiency of PCU/Solar Inverter, efficiency of the battery during both charging and discharging cycles, provision for cloudy day (redundancy), etc., are not included in these design calculations.

In case 265 W_p modules are used, a 4 kW_p SPV system would require 15 such modules. Few options are available for array design. It can have one string consisting of 15 modules (1 × 15) or 3 strings of 5 modules (3 × 5) or 5 strings of 3 modules (5 × 3). As discussed earlier, a string consists of series-connected modules and strings connected in parallel to complete the array. The string design is done by considering the voltage specifications, mainly the maximum voltage and operating voltage range of MPPT of the PCU/Solar Inverter. The maximum current is also seen, but not very important as the PCU/Solar Inverters have very high current ratings. The maximum voltage of a 1 × 15 array is 561.75 V (V_{oc} × 15 = 37.45 V × 15), see Table 3 for V_{oc} value. The operating voltage is 456.9 V (V_m × 15 = 30.46V × 15). The operating current is 8.7 A (I_m × 1 = 8.71 A × 1). The maximum current (I_{sc}) is not considered here as the inverter does not come into the picture during a short circuit condition. A (3 × 5) array has a maximum voltage of 187.25 V, operating voltage of 152.3 V and operating current of 26.13 A. Similarly, a (5 × 3) array has a maximum voltage of 112.35 V, operating voltage of 91.38 V and operating current of 43.55 A. The array design must be made while keeping the PCU/Inverter specifications in mind. Apart from maximum voltage, both minimum and maximum of the MPPT voltage range must considered. For more accurate calculations, the temperature effects on V_{oc} and V_m must be considered. In case more than one of the array configurations are found suitable, the final choice can be made to optimize the utilization of the available space.

The bigger SPV systems, starting from 1 MW and going up to hundreds of MW, are used for generating electricity which is directly exported to the grid. Such SPV systems are called "on grid" or SPV power plants. A representative Single Line Diagram (SLD) of such systems is shown in Figure 13.

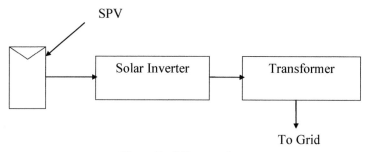

Figure 13. SPV power plant.

In addition to the SPV array and the solar inverter, a transformer is required in order to convert the output of the inverter, which is typically at 415 V, to high voltage (11 KV, 33 KV, 66 KV, etc.), depending on the requirement of the electricity grid. The power is then evacuated to the grid at high voltage.

The energy estimation of SPV power plants has to be done accurately. The cost of energy pumped to the grid is paid to the installer at a predecided price per unit (kWh). This is typically paid per year basis. Incorrect estimation of total energy to be produced per year can upset the business model. Apart from total energy generated per year, the accurate estimation of energy generated during a shorter time-scale is also important. The power generated by SPV plants is combined with the conventional power and then supplied to the consumer. It is important, therefore, to know how much power is available from the SPV plant at different times, so that proper planning can be done. The data related to energy generation per hour or even per minute are important for such planning. The simple calculations which are used for off-grid/grid interactive SPV systems, described earlier, cannot be used for energy estimation for on grid SPV systems. The inaccuracies, particularly for the shorter duration energy generation associated with such calculations, are high.

Simulation tools are used for estimation of energy generation in SPV power plants. Apart from the energy estimation, such tools are capable of configuring string design, array design and simulating performance-related parameters, such as Performance Ratio (PR), which is defined later. There are some "free to use" simulation tools, which are widely used for academia and beginners. PVWATT (www.pvwatts.nrel.gov) from NREL is one such example. Some reputed solar inverter companies also provide free-to-use simulation tools. The most widely-used simulator is PVSyst (www.pvsyst.com). A brief description on PVSyst simulation methodology has been given below.

One of the main inputs required for PVSyst is the irradiance and the ambient data for the entire year. The accuracy of the energy estimation strongly depends on how accurate this data is. PVSyst gets default irradiance and ambient data, which is freely available, from NASA. Data available from METEONORM (www.meteonorm.com) is known to have better accuracy. This is available at a price and can be seamlessly imported to PVSyst for simulation. "Paid" and "free" data are available from other sources as well. Location-related input is provided by specifying the latitude of the place. Other important inputs are SPV module parameters, inverter specifications, soiling loss, DC and AC cable-related losses, transformer loss, transmission loss, etc. The simulator can provide design parameters, such as optimum tilt angle for SPV module, string/array design, etc. The loss diagram, the energy output and the PR are the primary output parameters.

The energy output per month, as shown in Figure 14, is most commonly used. In this, actual output of the plant is also shown for comparison. It is possible to obtain the energy output for an hour or for the entire day. However, the long-term average, such as for an entire year, is more accurate. For example, the energy output estimation done for an hour is less accurate than the energy output estimation done for a day. Similarly, energy output estimation done for a year is more accurate than the energy output estimation done for a month. A closer look at Figure 14 reveals that there are some inaccuracies between the simulated energy output and actual energy obtained for each month. The overall inaccuracy reduces significantly when the energy estimation is done for a year. The inaccuracy is as high as +4.5% for the

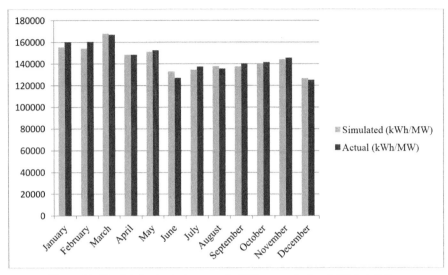

Figure 14. Simulated and actual energy of a SPV power plant per MW_p installation.

month of April. However, the inaccuracy of the total energy estimated for the entire year is only +0.6%. This is a tracker-based SPV plant and the total actual energy output per year is 1,741,451 kWh per MW_p installation. The corresponding simulated value of 1,730,277 kWh per MW_p installation.

The Performance Ratio (PR), which is also estimated by PVSyst, is another important parameter. This is defined as given in equation (8).

$$PR = [(E)/(Ir \times A \times \eta)] \times 100\% \tag{8}$$

where E is the energy obtained in kWh, Ir is the irradiance in kWh/m^2, A is the total module area in m^2 and η is the efficiency of the module.

PR essentially provides the energy output, normalized with respect to irradiance. The quality of SPV plants is judged on their PR values. Exactly identical plants installed in two different geographical locations produce different energy outputs, as the irradiance and the ambient conditions are different. However, the PR values are expected to be similar, as the difference of irradiances between these two locations are nullified for calculation of PR. It should be noted that the temperature effect is not normalized. Therefore, ideal PR of two identical SPV plants may be somewhat different. Good quality plants have PR values of 85% or more. In case the PR value of a SPV plant is less than 80%, it falls under the category of poor quality. The PR estimation is typically done on an annual basis.

9. Conclusion

Due to the widespread development of solar PV, the cost has come down drastically and there are ample examples of economic viability of solar PV plants. In most places, it has crossed the grid parity milestone, which means that the cost of electricity generated from solar PV is less than the cost of electricity generated from conventional resources. As the subsidy is no longer required for such scenarios, accelerated growth is now expected. However, the major hindrance is still the requirement of a very large initial investment. The economic viability depends on the reliability of the components used, particularly the solar modules. The modules are expected to last 25 years or more and the Levelized Cost of Energy (LCOE) is calculated accordingly. However, if the modules last only 5 years, for example, due to poor reliability, the economic viability is no longer achieved. Although warranty commitment from the module manufacturers are intended to take care of this, it is not fool proof. Under the warranty, the manufacturers commit that the modules will generate power for at least 25 years with a typical degradation of 10% in first 10 years and another 10% in next 15 years. However, 25 years is a long time, and a particular manufacturer may not last that long. Moreover, due to the advancement of technology, the module specifications are continuously been upgraded, so if a module has to be replaced 10 years down the line, it may not be possible to get one with the same specifications. Additional safeguards, in terms of insurance coverage, are being introduced gradually. It is expected that there will be a continuous improvement of trust due to maturing of the technology and the introduction of more such safeguard mechanisms.

Exponential growth of solar PV may create another major issue related to coal-based power plants. As solar electricity is becoming cheaper than coal-based electricity, the former will be given preference. However, the entire supply during the night has to be provided by coal-based electricity. The Capacity Utilization Factor (CUF) of coal-based power plants is then compromised, putting stress on their economic viability. It is, therefore, prudent to have a planned growth strategy, keeping both solar and coal-based power plants in mind.

References

Ahmed, G., Mandal, S., Barua, A.K., Bhattacharya, T.K. and Roy, J.N. 2017. Band offset reduction at defect-rich p/i interface through a wide bandgap a-SiO:H buffer layer. IEEE Journal of Photovoltaics 414.

Ahmed, G., Mandal, S., Barua, A.K., Bhattacharya, T.K. and Roy, J.N. 2019. Mixed-phase nc-SiO:H interlayer to improve light trapping and shunt quenching in a-Si:H solar cell. IEEE Journal of Photovoltaics 18.

Antoniadis, H., Jiang, F., Shan, W. and Liui, Y. 2010. All screen-printed mass-produced silicon ink selective emitter solar cells. Proc. of Photovol. Specialist Conf. (PVSC) 1193.

Bose, D.N. 2012. Organic Photovoltaics in Semiconductor Materials and Devices: New Age, Delhi.

Boyle, G. 2004. Renewable Energy: Oxford University Press.

Chuang, C.H.M., Brown, P., Bulovic, V. and Bawendi, M.G. 2014. Improved Performance and stability in quantum dot solar cells through band alignment engineering. Nature Materials 796.

Cusano, D.A. 1963. CdTe solar cells and photovoltaic heterojunctions in II–VI compounds: Sol. St. Electron 217.

Foster, R., Ghassemi, M. and Cota, A. 2010. Solar Energy, Renewable Energy and Environment. CRC Press.

Geisza, J.F., Friedmann, D.J., Ward, J.S., Duda, A., Olavarria, W.J., Moriarty, T.E., Kiehl, J.T., Romero, M.J., Norman, A.G. and Jones, K.M. 2008. 40.8% efficiency inverted triple-junction solar cell with two independent metamorphic junctions. Applied Physics Letters 123505-1-3.

Gilbert, M.M. 2013. Renewable and Efficient Electric Power Systems. IEEE Press John Wiley & Sons.

Gratzel, M. 2003. Dye sensitized solar cells. Journal of Photochemistry and Photobiology C. Photochemistry Reviews 145.

Green, M.A. 1995. Silicon Solar Cells-Advanced Principles and Practice. Bridge Printery, Sydney.

Hamid, F. and Fatima, B. 2013. Characterization and modeling of CdS/CdTe heterojunction thin film solar cell for high efficiency performance. International Journal of Photoenergy 13.

Ikegami, S. 1988. CdS/CdTe solar cells by the screen-printing-sintering technique: Fabrication, photovoltaic properties and applications. Solar Cells 89.

Kiefer, K., Ulzhofer, C., Brendemuhl, T., Harder, N.P., Brendel, R., Mertens, V., Bordihn, S., Peters, C. and Muller, J.W. 2011. High efficiency N-type emitter wrap through silicon solar cells. IEEE Journal of Photovoltaic 53.

Kronik, L., Cahen, D. and Schock, H.W. 1998. Effects of sodium on polycrystalline Cu (In,Ga)Se$_2$ and its solar cell performance. Advanced Materials 31.

Markvart, T. and Castaner, L. 2005. Practical Handbook of Photovoltaics: Fundamentals and Applications. Elsevier.

Magnone, P., Rose, R.D., Tonini, D., Frei, M., Zanuccoli, M., Belli, A., Galiazzo, M., Sangiorgi, E. and Fiegna, C. 2014. Numerical simulation on the influence of via and rear emitter in MWT solar cell. IEEE Journal of Photovolt 1032.

Min, C., Nuofu, C., Xiaoli, Y., Wang, Y., Yiming, B. and Xingwang, Z. 2009. Thermal analysis and test for single concentrator solar cells. J. of Semiconductors 4.

Nazeerusddin, M.K. and Snaith, H.K. 2015. Perovskite Photovoltaics: MRS Bulletin 40.

Nelson, V. 2011. Introduction to Renewable Energy. CRC Press.

Ohlsen, H.J., Bodegard, M., Kylner, A., Stolt, L., Hariskos, D., Ruckh, M. and Schock, H.W. 1993. ZnO/CdS/Cu(In,Ga)Se$_2$ thin film solar cells with improved performance. Proc. 23rd IEEE Photovoltaic Specialists Conference 364.

Roder, T.C., Eisele, S.J., Grabitz, P., Wagner, C., Kulushish, G., Kohler, J.R. and Werner, J.H. 2010. Add-on laser tailored selective emitter solar cells. Progr. Photovolt. Res. Appl. 505.

Roy, J.N. and Bose, D.N. 2018. Photovoltaic Science and Technology. Cambridge University Press.

Sam-Shajing, S. and Niyazi Serdar, S. 2005. Organic Photovoltaics: Mechanisms, Materials, and Devices (Optical Engineering). CRC Press, Taylor & Francis Group, Boca Raton.

Sharps, P.R., Stan, M.A., Aiken, D.I., Clevenger, B., Hills, J.S. and Fatemi, N.S. 2003. Multi-junction cells with monolithic bypass diodes. World Conference on Photovoltaic Energy Conversion 626.

Street, R.A. 2005. Hydrogenated Amorphous Silicon. Cambridge University Press.

Wang, Z.A. and Green, M.A. 1994. Series resistance caused by the localized rear contact in high efficiency silicon solar cells. Solar Energy Materials and Solar Cells 89.

Wang, Z.A., Altermatt, A.P. and Green, M. 1995. Twenty four percent efficiency silicon solar cells with double layer antireflection coating and reduced resistance loss. Applied Physics Letter 3636.

Zhao, J., Wang, A. and Green, M.A. 1999. 24.5% Efficiency Silicon PERT Cells on MCZ Substrates and 24.7% Efficiency PERL Cells on FZ Substrate: Progress in Photovoltaic: Research and Applications 471.

Zanuecoli, M., Magnone, P., Sangiorgi, E. and Fiegna, C. 2015. Analysis of the impact of geometrical and technological parameters on recombination losses in interdigitated back contact solar cells. Solar Energy 37.

CHAPTER 21

Reducing the Carbon Footprint of Wind Energy
What Can Be Learned from Life-Cycle Studies?

Melanie Sattler

1. Introduction

A renewable energy source that has recently seen rapid growth is wind energy. U.S. wind capacity grew by 13% in 2017, resulting in 6.3% of U.S. power being produced by wind energy (American Wind Energy Association, 2017). Wind turbines can convert wind to electric power without generating greenhouse gas emissions; however, over their entire life cycle (raw material acquisition, manufacturing, operation, and end-of-life), turbines do contribute pollutants that cause climate change, including carbon dioxide and methane (Alsaleh, 2016). To compare the effects of wind energy generation with other energy options, carbon dioxide equivalent (CO_2-e) emissions must be estimated over the complete life cycle.

A variety of life cycle assessment (LCA) and carbon footprint studies have been conducted regarding greenhouse gas (GHG) emissions and other environmental impacts of wind turbines (e.g., impacts on wildlife, noise, resource consumption). Wang and Wang (2015) reviewed 64 studies in terms of GHG emissions, as well as wind turbine impacts on noise pollution, bird and bat fatalities, and land surfaces. Saidur et al. (2011) reviewed 54 studies of wind turbine carbon dioxide emissions, as well as impacts on wildlife, water consumption, noise, and visual aesthetics. Davidsson et al. (2012) compared 12 life cycle assessments of wind energy systems in terms of methods used, energy use accounting, quantification of energy production, energy performance and primary energy, natural resources, and recycling. Leung and Yang (2012) reviewed 80 studies in terms of wind turbine impacts on noise and visual impacts, impacts on animals and birds, and local climate; however, impacts on global climate through greenhouse gas emissions were not included. Arvesen and Hertwich (2012) provided a comprehensive review of the literature up to 2012, categorizing 59 studies by on-shore vs. off-shore turbines, turbine size, turbine lifetime, geographic scope, and impact categories included. From the existing literature, they found a mean emission value of 19 (± 13) g CO_2-e/kWh of electricity generated from wind turbines.

Previous studies can provide useful insights regarding the phase of the wind turbine life cycle that produces the most GHG emissions, and sensitivity of emissions to changes in materials, manufacturing processes, and other parameters. This chapter will, thus, summarize insights from the literature concerning effective approaches for reducing the life-cycle carbon footprint of wind energy.

University of Texas at Arlington, Dept. of Civil Engineering, 416 Yates St., Suite 416, Arlington, TX 76010.
Email: Sattler@uta.edu

2. Which Phase of a Wind Turbine's Life Cycle Produces the Greatest CO_2-e Emissions?

Figure 1 shows a flow chart of wind turbine life-cycle phases. Figure 2 shows the contribution of turbine life-cycle phases to overall CO_2-e emissions for selected studies for large (> 1 MW) on-shore wind turbines. Every study shown in Figure 2 found the raw material/manufacturing phase (sometimes combined with the installation phase) to contribute the most to greenhouse gas emissions. Arvesen and Hertwich (2012), in their review of 59 studies that included smaller turbines < 1 MW, as well as off-shore turbines, also found that the manufacturing phase dominates energy consumption and climate change indicators, sometimes up to 90%. Hence, the manufacturing phase should be targeted for reducing greenhouse gas contributions of wind energy. It will be discussed in more detail in the next section.

In Figure 2, the installation phase ranks 2nd or 3rd in importance, if it was included as a separate phase. Hence, after raw material acquisition/manufacturing, the installation phase should be targeted to reduce GHG emissions. The installation phase consists of the use of heavy-duty diesel equipment for digging/moving soil, pouring the foundation, and installing the turbine tower, rotor, and nacelle (Figure 3). Methods of reducing emissions from the installation phase would include the use of heavy-duty equipment which is more fuel-efficient or operates on alternative fuels.

In Figure 3, O&M ranks 2nd, 3rd, or 4th in importance, depending on the study. The operation and maintenance phase includes inspection trips, change of oil, lubrication of gears and the generator, replacement of turbine parts, and repair of the turbines when they break down. Interestingly, the parts targeted for replacement varied among the studies (brake system, generator, turbine blade, part of the nacelle), which may account for differences in rankings. Differences in emissions from the electricity mix used for operation and maintenance (global, USA, and Brazil) may also explain variations in the rankings.

Similar to O&M, transportation ranks 2nd, 3rd, or 4th, depending on the study. Tremeac and Meunier (2009), who found transportation to rank 2nd, included a concrete tower, which was heavier to transport compared to steel, used in most other studies. Interestingly, both Alsaleh and Sattler (2019) and Rajaei and Tinjum (2013) included overseas transport of turbine parts, but still found transportation to rank 4th. Rajaei and Tinjum (2013) also found that eliminating transportation of components from overseas, via local manufacture, could reduce transportation GHG emissions by 22%.

In Figure 2, end-of-life consistently ranks last, regardless of whether the parts not recycled were combusted (Martinez et al., 2009) or landfilled (Tremeac and Meunier, 2009; Alsaleh and Sattler, 2019). In a number of studies, CO_2-e emissions from the end-of-life phase were considered negative because benefits of recycling the turbine parts were included.

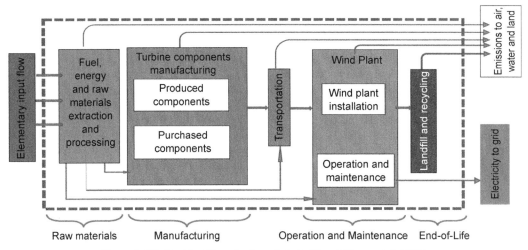

Figure 1. Wind turbine life cycle phases (adapted from D'Souza et al., 2011).

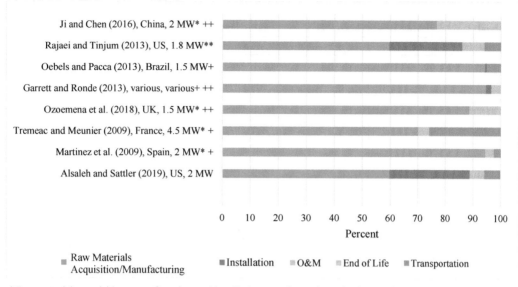

* Raw materials acquisition, manufacturing, and installation were lumped together in one phase.
+ End-of-life CO_2-e were negative due to recycling of turbine parts.
** End-of-life impacts not quantified.
++ Transport considered as part of other phases.

Figure 2. Contributions of wind turbine life cycle phases to CO_2-e emissions.

Figure 3. Major turbine parts docera.com/wiki.

In summary, according to Figure 2, to reduce life cycle carbon emissions, the raw material acquisition/manufacturing phase should be targeted, followed by installation. Impacts of raw material acquisition/manufacturing for specific materials will be discussed in the following section.

3. Which Turbine Material Generates the Most CO_2-e Emissions during Manufacturing?

Table 1 shows an example of materials used for a turbine for a Gamesa G83 2-MW turbine. The mass percentages are of the entire turbine mass, including the nacelle, rotor, tower, wiring, and foundation. The

Table 1. CO_2-e emissions from turbine material manufacturing for Gamesa G83 2 MW turbines (Gamesa, 2013).

Material	Mass (kg)	Mass (%)	SimaPro material category	kg CO_2-e	CO_2-e (%)
Concrete	1,116,000	75.7%	Concrete block	105,750	11.0%
Low alloy steel	227,866	15.5%	Steel, low-alloyed	374,096	38.9%
Corrugated steel	44,000	3.0%	Steel, low-alloyed, hot rolled	83,523	8.7%
Casting	33,084	2.2%	Cast iron	57,974	6.0%
High alloy steel	22,356	1.5%	Steel, chromium steel 18/8	100,897	10.5%
Fiberglass	15,875	1.1%	Glass fiber reinforced plastic, polyamide, injection molded	141,144	14.7%
Aluminium	4,036	0.3%	Aluminum, primary, ingot	40,811	4.2%
Polymer	3,807	0.3%	Polyethylene, high density, granulate	6,177	0.6%
Copper	2392	0.2%	Copper	15,825	1.6%
Adhesive	1,361	0.1%	Adhesive for metal	6,352	0.7%
Painting	1283	0.1%	Acrylic varnish, without water, in 87.5% solution state	3,272	0.3%
Components electric/electronic	905	0.1%	Electronics for control units	24,131	2.5%
Lubricant	628	0.04%	Lubricating oil	690	0.1%
Brass	38	0.003%	Brass	237	0.02%
TOTAL	1,473,631	100.0%		960,880	100.0%

substation is excluded. CO_2-e emissions estimates were generated using SimaPro Software version 8.3.2, with all databases selected, including Ecoinvent 3, ELCD, EU & DK Input Output Database, Industry data 2.0, Methods, Swiss Input Output Database, US-EI 2.2, and USLCI.

According to Table 1, the material with the highest mass is concrete (76%), but manufacturing it produces only 11% of the CO_2-e emissions of the materials. This is because the average emissions intensity (kg of CO_2-e emitted per kg of material manufactured) for concrete is low, at 0.15 (the value for cement is 0.89; concrete contains water and aggregate mixed with cement). Carbon emission intensity values for other major turbine components are higher than that for concrete: 6.1 for stainless steel, 0.88 to 3.29 for other steel, 1.51 for cast iron, 2.6 for fiberglass, and 8.14 for aluminium (Winnepeg, Canada, 2012). Manufacture of low alloy steel, for example, which makes up 15.5% of the materials by mass, generates 39% of the CO_2-e emissions. According to Table 1, manufacture of low alloy steel, corrugated steel, and high alloy steel together produces 58.1% of CO_2-e emissions.

Most (83%) of the low alloy steel shown in Table 1 is used in the tower of the turbine. Accordingly, Alsaleh and Sattler (2019) found that the manufacture of the tower contributed > 40% of greenhouse gas emissions. Similarly, for a 1.5 MW turbine in France, Oebels and Pacca (2013) found the steel tower to contribute > 50% of the CO_2 emissions from the manufacturing phase. Guezuraga et al. (2011) and Ardente et al. (2008) also found the steel tower to contribute over 50% and around 50%, respectively, of energy consumption over the turbine's lifetime.

Hence, in order to reduce climate change contributions of turbines, steel used in the tower should be targeted for emission reductions. Methods of accomplishing this will be discussed in detail in the next section.

A few wind turbine studies (Martinez et al., 2009; Rajaei and Tinjum, 2013) have found the foundation to produce the greatest CO_2 emissions, primarily due to the cement. Berndt (2015) examined the influence of concrete mix design on turbine life cycle GHG emissions. Replacing 40 MPa class concrete with 32 MPa class concrete reduced CO_2 emissions by 11%. Replacing 65% of the cement with blast furnace slag reduced CO_2 emissions by around 44%. Use of recycled concrete aggregate reduced emissions moderately.

4. How can CO_2 Emissions from Steel Manufacturing for Turbine Parts be Reduced?

Using recycled steel would substantially reduce CO_2 emissions, as would the use of renewable energy in steel manufacturing. In addition, a number of changes and innovations in the steel-making process can lower steel's carbon footprint. Basic oxygen furnace (BOF) steel-making emits 4 times more CO_2 than electric furnace steel-making; hence, replacing BOF processes with electric furnace processes would reduce GHG emissions by 75% from one of the most-energy intensive parts of the steel-making process (Turner, 2011). Using charcoal to replace part of the coal and coke used in steelmaking can reduce CO_2 emissions by about 50% without requiring substantial modification of the steelworks (CSIRO, 2015).

Oebels and Pacca (2013) found that replacing the steel tower with a cement tower, with reinforcing steel, decreased CO_2 emissions by 6.4% overall. Although manufacturing emissions decreased by more than 6.4%, transportation emissions increased, due to the greater weight of concrete.

Ardente et al. (2008) found that replacing 40% of the steel in the turbine transformer with copper reduced CO_2 emissions. Substituting carbon fibers for glass fibers in reinforced plastics increased energy consumption by 12%, but substituting flax fibers decreased energy consumption by 1%.

5. What Other Factors can Reduce Turbine CO_2-e Emissions?

5.1 Turbine design

Changes in design of wind turbines can affect GHG emissions over their lifetimes. Several studies have described such changes, as described in this section.

Guezuraga et al. (2011) compared GHG emissions from a 1.8 MW gearless turbine with a 2.0 MW turbine with gearbox. The gearless turbine generated 8.8 g CO_2-e/kWh compared with 9.7 for the geared turbine. Ozoemena et al. (2018) found that a turbine design with advanced rotors and reduced tower mass increased CO_2-e emissions by 6%; however, a design with drivetrain improvements using permanent magnet generators decreased CO_2-e emissions by 13%.

Vertical axis wind turbines have lower power ratings compared to horizontal axis turbines (Uddin and Kumar, 2014). However, Rashedi et al. (2013) found vertical axis wind turbines to have lower CO_2-e emissions per kWh compared to horizontal axis turbines.

Simons and Cheung (2016) developed a method to quantify wind farm carbon dioxide emissions and energy production in order to enable the design of more efficient wind farms by varying turbine material, hub height, and blade length.

5.2 Turbine size

Generally, turbines in the MW size range emit less GHG emissions per kWh of electricity generated than turbines in the kW size range. Crawford (2007) found an 11% increase in energy yield for a 3 MW turbine compared to an 850 kW turbine. Tremeac and Meunier (2009) found that a smaller wind turbine (250 W) produced greater CO_2-e emissions per kWh electricity produced than a 4.5 MW turbine. Jungblutz et al. (2005) found the environmental impact per kWh to decrease for larger wind turbines, an effect of scaling. Raadal et al. (2011) surveyed 63 life cycle studies conducted between 1990 and 2019, and found that GHG emissions from wind power varied from 4.6 to 55.4 g CO_2-e/kWh. The low value was for a 3 MW turbine; GHG emissions were found to decrease with increasing turbine size, reflecting economies of scale.

5.3 Turbine lifetime

Currently, horizontal axis wind turbines are generally assumed to have a lifespan of 20 years (Simons and Cheong, 2016). However, in practice, lifetimes of 30 years and over have been achieved (Garrett and Ronde, 2013). When the turbine lifespan is extended via proactive maintenance, GHG emissions from maintenance increase; however, this can be outweighed by the additional energy production, so that the overall GHG emissions per kWh of electricity generated decreases, because the pollutants from manufacturing are distributed over more years.

Alsaleh and Sattler (2019) examined changes in CO_2-e/kWh power produced and net energy production when the turbine life span was extended via increased maintenance. They found that CO_2-e/kWh power produced decreased from 0.053 to 0.042 to 0.035 when the turbine life span was extended from 20 to 25 and 30 years, respectively. They also found that for the 20, 25, and 30-year life spans, the turbines produced 3.2, 4.2, and 5.1 times more energy than they consumed over their life cycle, respectively.

Turbine lifetimes could also be increased via the use of newly developed materials designed to be stronger and/or require less maintenance. For example, nickel-catalyzed growth of iron-aluminium microparticles and nanoparticles at defect sites in a low-density, aluminium-rich steel were found to strengthen the steel without compromising its pliability (Kim et al., 2015).

5.4 Wind farm location and layout

The greater the power production from a given wind turbine per time, the lower the CO_2-e emissions per kWh (Garrett and Ronde, 2013). Wind farm location and layout can, thus, be chosen to maximize energy production and decrease GHG emissions per kWh power produced.

Kusiak and Song (2010) developed algorithms to predict wind farm power production based on weather forecasting data. The use of these algorithms would enable selection of the windfarm location to maximize power production based on higher wind speeds. Larsen and Rethore (2013) developed the TOPFARM system to optimize wind farm layout based on power production, as well as cost and fatigue loads. Barthelmie and Jensen's (2010) research findings proposed that placing horizontal wind turbines close together reduces their efficiency; specifically, for each distance equivalent to a rotor diameter that the turbines are placed closer together, there is a 1.3% loss in efficiency.

Arvesen and Hertwich (2012), from their review of 59 studies, report that off-shore turbines show comparable or slightly higher emissions than on-shore systems comprised of large turbines, despite higher wind capacity factors due to higher resource requirements.

5.5 Manufacturing location

Manufacturing location can also be chosen in order to take advantage of renewable energy for manufacturing, and to minimize transport distances and, thus, minimize GHG emissions due to transport. Guezuraga et al. (2011) found that manufacture of turbine components in China produced more CO_2-e/kWh than manufacture of components in Germany or Denmark, due to longer required transport distances and greater use of black coal for component production in China. Oebels and Pacca (2013) found 7.1 g/kWh for a 1.5 MW wind turbine, which is relatively low, based on the clean energy mix of Brazil (87% renewables).

5.6 End-of-life recycling

Compared to recycled materials, acquisition and processing of virgin materials requires greater energy and, thus, produces more GHG emissions when the energy is supplied by fossil fuels. Turbine materials, including metals (steel, iron, aluminium), plastics, and concrete, are excellent candidates for recycling. The greater the amount of material recycled, the lower the end-of-life CO_2-e emissions (Martinez et al., 2010).

Table 2 summarizes disposal information included in several large (> 1 MW) wind turbine studies. All 3 studies assume a 90% or greater recycling percentage for metals, two assume 90% recycling of plastics, and one assumes 90% recycling of concrete. Martinez et al. (2009) and Tremeac and Meunier

Table 2. Disposal information included in various large wind turbine studies.

Study	Turbine part disposal					
	Metals	**Plastics**	**Electrical/ Electronics**	**Fiber-glass**	**Lubricants/ Grease/Oils**	**Other**
Alsaleh and Sattler (2019)	98% R 2% L	90% R 10% L	50% R 50% L	100% L	100% L	Paints/adhesives: 100% L Cables: 99% R, 1% L
Martinez et al. (2009)	90–95% R	PVC 100% L Other 100% C	N/A	100% L	100% C	Rubber: 100% C
Tremeac and Meunier (2009)	90% R, 10% L (nacelle)	90% R, 10% L (polyester from nacelle)	N/A	98% R, 2% L (blades)	N/A	Concrete tower: 90% R, 10% L

Note: R: Recycling, L: Landfilling, C: Combustion.

(2009) found CO_2-e emissions from the end-of-life phase to be negative because the benefits of recycling the turbine parts were allocated to that phase, rather than to the product, which is subsequently made from the recycled materials, as was done by Alsaleh and Sattler (2019).

6. Conclusions and Recommendations

Among the phases of a turbine's life cycle, studies consistently report that manufacturing produces the greatest CO_2-e emissions. Among turbine materials, steel causes the most GHG emissions during the manufacturing process, according to most studies. Using recycled steel, as well as changes in the steel production process—use of renewable energy, use of an electric furnace rather than a basic oxygen furnace, and substitution of charcoal for a portion of coal and coke—can reduce steel's carbon footprint. Further research to reduce CO_2 emissions from steel production is recommended.

Additional strategies for reducing GHG emissions from wind energy include changes in turbine design, use of turbines in the MW range rather than the kW range, extending turbine lifetimes via proactive maintenance or use of new materials, selection of wind farm location and layout to maximize energy production, selection of manufacturing location to take advantage of renewable energy resources and minimize transport distances, and end-of-life recycling of turbine materials.

References

Alsaleh, A. 2016. Life Cycle Assessment of Greenhouse Gas Emissions, Traditional Air Pollutants, Water Depletion, and Cumulative Energy Demand from 2-MW Wind Turbines in Texas. Ph.D. dissertation and defense. University of Texas at Arlington.

Alsaleh, A. and Sattler, M. 2019. Life cycle assessment of large wind turbines in the US. Clean Technologies and Environmental Policy 21(4): 887–903.

American Wind Energy Association. 2017. US Wind Industry Annual Market Report: Executive Summary. Available at: https://www.awea.org/AnnualMarketReport.aspx?ItemNumber=11563&RDtoken=34167&userID= (Accessed 5/18).

Ardente, F., Beccali, M., Cellura, M. and Lo Brano, V. 2008. Energy performances and life cycle assessment of an Italian wind farm. Renewable and Sustainable Energy Reviews 12: 200–217.

Arvesen, A. and Hertwich, E. 2012. Assessing the life cycle environmental impacts of wind power: A review of present knowledge and research needs. Renewable and Sustainable Energy Reviews 16(8): 5994–6006.

Barthelmie, R.J. and Jensen, L.E. 2010. Evaluation of wind farm efficiency and wind turbine wakes at the Nysted offshore wind farm. Wind Energy 13(6): 573–586. DOI:10.1002/we.408.

Berndt, M.L. 2015. Influence of concrete mix design on CO_2 emissions for large wind turbine foundations. Renewable Energy 83: 608–614.

Crawford, R.H. 2007. Life-cycle energy analysis of wind turbines—an assessment of the effect of size on energy yield. WIT Transactions on Ecology and the Environment 105: 155–164. WIT Press, ISSN 1743-3541.

CSIRO. 2015. Environmentally-friendly steelmaking. https://www.csiro.au/en/Research/MRF/Areas/Community-and-environment/Responsible-resource-development/Green-steelmaking, accessed 4/19.

Davidsson, S., Höök, M. and Wall, G. 2012. A review of life cycle assessments on wind energy systems. Int. J. Life Cycle Assess. 17: 729–742. DOI 10.1007/s11367-012-0397-8.

D'Souza, N., Gbegbaje-Das, E. and Shonfield, P. 2011. Life cycle assessment of electricity production from a Vestas V112 turbine wind plant. Copenhagen, Denmark.

Gamesa Corp. 2013. The wind turbine manufacturer in Spain. Available at: http://www.gamesacorp.com/en/cargarAplicacionPresenciaGlobal.do?tipo=P. Accessed on 09/17/2016.

Garrett, P. and Rønde, K. 2013. Life cycle assessment of wind power: comprehensive results from a state-of-the-art approach. International Journal of Life Cycle Assessment 18: 37–48. DOI 10.1007/s11367-012-0445-4.

Guezuraga, B., Zauner, R. and Pölz, W. 2012. Life cycle assessment of two different 2 MW class wind turbines. Renewable Energy 37: 37–44.

Ji, S. and Chen, B. 2016. LCA-based carbon footprint of a typical wind farm in China. Energy Procedia 88: 250–256.

Jungbluth, N., Bauer, C., Dones, R. and Frischknecht, R. 2005. Life cycle assessment for emerging technologies: Case studies for photovoltaic and wind power. International Journal of Life Cycle Assessment 10(1): 24–34.

Kim, S.H., Kim, H. and Kim, N.J. 2015. Brittle intermetallic compound makes ultra-strong low-density steel with large ductility. Nature 518: 7–79. DOI: 10.1038/nature14144.

Kumar, I., Tyner, W.E. and Sinha, K.C. 2016. Input–output life cycle environmental assessment of greenhouse gas emissions from utility scale wind energy in the United States. Energy Policy 89: 294–301.

Kusiak, A. and Song, Z. 2010. Design of wind farm layout for maximum wind energy capture. Renewable Energy 35: 685–694.

Larsen, G.C. and Réthoré, P.E. 2013. TOPFARM a tool for wind farm optimization. 10th Deep Sea Offshore Wind R&D Conference, DeepWind'2013. Energy Procedia 35: 317–324.

Leung, D.Y.C. and Yang, Y. 2012. Wind energy development and its environmental impact: A review. Renewable and Sustainable Energy Reviews 16: 1031–1039.

Martínez, E., Sanz, F., Pellegrini, S., Jiménez, E. and Blanco, J. 2009. Life cycle assessment of a multi-megawatt wind turbine. Renewable Energy 34: 667–673.

Martinez, E., Jiménez, E., Blanco, J. and Sanz, F. 2010. LCA sensitivity analysis of a multi-megawatt wind turbine. Applied Energy 87: 2293–2303.

Oebels, K.B. and Pacca, S. 2013. Life cycle assessment of a non-shore wind farm located at the north eastern coast of Brazil. Renewable Energy 53: 60–70.

Ozoemena, M., Cheung, W.M. and Hasan, R. 2018. Comparative LCA of technology improvement opportunities for a 1.5-MW wind turbine in the context of an onshore wind farm. Clean Technologies and Environmental Policy 20: 173–190.

Raadal, H.L., Gagnonb, L., Modahla, I.S. and Hanssenaet, O.J. 2011. Life cycle greenhouse gas (GHG) emissions from the generation of wind and hydro power. Renewable and Sustainable Energy Reviews 15: 3417–3422.

Rajaei, M. and Tinjum, J.M. 2013. Life cycle assessment of energy balance and emissions of a wind energy plant. Geotech. Geol. Eng. 31: 1663–1670. DOI 10.1007/s10706-013-9637-3.

Rashedi, A., Sridhar, I. and Tsend, K.J. 2013. Life cycle assessment of 50 MW wind firms and strategies for impact reduction. Renewable and Sustainable Energy Reviews 21: 89–101.

Saidur, R., Rahim, N.A., Islam, M.R. and Solangi, K.H. 2011. Environmental impact of wind energy. Renewable and Sustainable Energy Reviews 15: 2423–2430.

Simons, P.J. and Cheung, W.M. 2016. Development of a quantitative analysis system for greener and economically sustainable wind farms. Journal of Cleaner Production 133: 886–898. http://dx.doi.org/10.1016/j.jclepro.2016.06.030 0959-6526.

Tremeac, B. and Meunier, F. 2009. Life cycle analysis of 4.5 MW and 250 W wind turbines. Renewable and Sustainable Energy Reviews 13: 2104–2110.

Turner, M. 2011. Mitigating Iron and Steel Emissions. The Global Network for Climate Solutions Fact Sheets, Nico Tyabji, ed. Columbia Climate Center, Earth Institute, Columbia University, www.thegncs.org/, accessed 4/19.

Uddin, M.S. and Kumar, S. 2014. Energy, emissions and environmental impact analysis of wind turbine using life cycle assessment technique. Journal of Cleaner Production 69: 153–164.

Wang, S. and Wang, S. 2015. Impacts of wind energy on environment: A review. Renewable and Sustainable Energy Reviews 49: 437–443.

Winnipeg, Canada. 2012. Corporate Finance, Materials Management, Documents, Policies and Information, WSTP South End Plant Process Section Report, Appendix H. Emission factors in kg CO_2-equivalent per unit. https://www.winnipeg.ca/finance/findata/matmgt/documents/2012/682-2012/682-2012_Appendix_H-WSTP_South_End_Plant_Process_Selection_Report/Appendix%207.pdf, Accessed 5/19.

CHAPTER 22

Hydropower

A Low-Carbon Power Source

Ånund Killingtveit

1. Introduction and Summary

Hydropower is a renewable energy source where electrical energy is derived from the potential energy of water moving from higher to lower elevations. Hydropower is probably the most efficient technology for production of renewable energy, with an efficiency of 90% or more, "water-to-wire". Hydropower projects can be grouped into four main types: Run-of-River (ROR), Storage (Reservoir-based), Pumped-Storage (PSH) and In-stream (Hydrokinetic). Hydropower projects are always site-specific, tailored for an optimal use of available head and water resources. Hydropower stations can be found in the range of less than 1 kW up to more than 20 GW (20000 MW). In fact, 9 out of the 10 largest power plants in the world are hydropower plants!

Energy from hydropower today fulfils about 16% of the global electricity demand, and the potential is still not fully utilized. It has been estimated that only around 25–30% of the global hydropower resources have yet been developed, and that most of the remaining economic potential probably could be fully developed by 2050. This could increase capacity and energy production to more than two times the current level. Hydropower is the largest source of renewable energy in the electricity sector, with a share of 62% of the renewables, nearly twice all other renewables combined.

Hydropower can be very cost competitive and has traditionally been the only renewable technology that could produce electricity at equal or lower cost, compared to thermal energy sources, like coal, oil or gas, typically in the range of 2–10 U.S. cent/kWh.

Hydropower offers significant potential for carbon emission reductions due to generally low greenhouse gas (GHG) emissions compared both to other renewables and thermal power plants.

In addition to its role in energy generation, hydropower often has a vital role by supporting grid stability and security of supply, energy storage and frequency and voltage control. Hydropower reservoirs can store large quantities of energy and, in the future, is expected to have an increasingly important role in balancing other more variable renewable energy sources, like wind and solar power, which cannot store energy.

Professor Emeritus, Department of Civil and Environmental Engineering, Norwegian University of Science and Technology (NTNU), Trondheim, Norway.
Email: aanund.killingtveit@ntnu.no

Hydropower systems can also provide important water management services, like flood control, navigation, water supply and irrigation. Many reservoirs are designed for multi-purpose use, where income from energy generation pays most of the investment and operation costs, while other uses may benefit from water storage and regulated release of water.

Hydropower projects can have a substantial impact on flow regime in rivers, and may lead to negative and positive impact on ecosystem and humans. Therefore, both for existing and future projects, environmental and social impacts need to be carefully assessed, evaluated and mitigated as far as possible.

On a global basis, the potential for further hydropower development is significant. The remaining economic resource potential is large, > 5000 TWh, and in the near and medium future, resource depletion is not likely to restrict hydropower development. Since only 25% of the technical potential and less than 50% of the economic potential has been developed, the prospects for further development are good. If these TWh's are replacing electricity from coal fired plants, the total reduction of GHG emissions could amount to 5000 Million tons of CO_2 each year.

2. Hydropower History and Status Today

2.1 Historical development

The first hydroelectric power plant was installed in a country house in Cragside, Rothbury, England, in 1870. Industrial use of hydropower started a few years later, for example in Grand Rapids, Michigan (1880) and at Niagara Falls, New York (1881). The world's first hydroelectric power station with a capacity of 12.5 kW was commissioned on 30 September 1882 on Fox River at the Vulcan Street Plant, Appleton, Wisconsin, USA, lighting two paper mills and a residence (IPCC, 2011). Early hydropower plants were much more reliable and efficient than the fossil fuel-fired plants and this resulted in a rapid deployment of small- to medium-scale hydropower stations distributed where there was an adequate supply of water and a need for electricity. By 1886, there were 45 hydroelectric power plants in operation in the USA. Many other countries followed suit, such as Norway, where the first hydropower plant began operation in 1885 in the town of Skien. Only a few years later, in 1891, the town of Hammerfest, far north of the Arctic circle, had electric streetlights supplied from a municipal hydropower system and in 1897 India saw its first hydroelectric power station of 130 kW, supplying a small town.

The share of hydropower in the global energy mix has been around 16–17% for many decades, and is expected to remain high even though strong growth in other renewables, in particular wind and solar PV, may lead to a somewhat lower share in the coming decades. The technical potential for increased hydropower generation is large enough to meet substantial further deployment both in the medium (2030) and long term (2050).

World hydropower generation has increased by more than a factor of 6 over the years from 1960 (690 TWh) to 2018 (4239 TWh), see Figure 1. The first 35 years, from 1960 to 1994, there was a steady increase of 50 TWh/year. Then, during the next 10 years, the growth slowed down to less than 20 TWh/year.

The slowdown in the decade around 2000 has been explained to be a result of an increasingly negative view on hydropower due to its environmental impacts, and to changes in funding policy for large infrastructure and energy projects. Then, from 2004 onwards, the growth increased again, now at around 100 TWh/year, mostly as a result of a very rapid expansion of hydropower in China and some other developing countries.

The strong upswing from 2004 can be explained by the increasing concern about climate change and the political determination to move from carbon-based technologies to renewable sources. Another important driver may have been the efforts led by the International Hydropower Association (IHA) and a multi-stakeholder range of partners in promoting greater sustainability through the development and use of "sustainability guidelines" (IHA, 2019).

The fast growth in hydropower seen since 2004 also continued during the last years, up to 4239 TWh in 2018 (IEA, 2018). Last year, from 2017 to 2018, the growth in hydropower (129 TWh) was nearly the

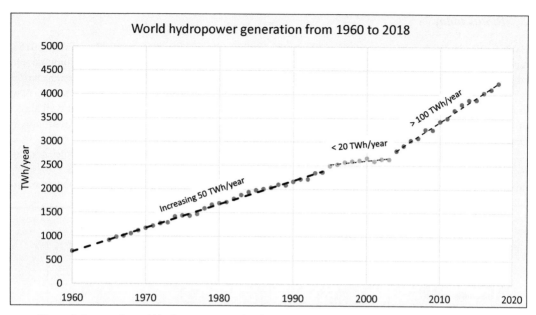

Figure 1. Increase in world hydropower generation from 1960 to 2018 (Ritchie and Roser, 2019; IEA, 2018).

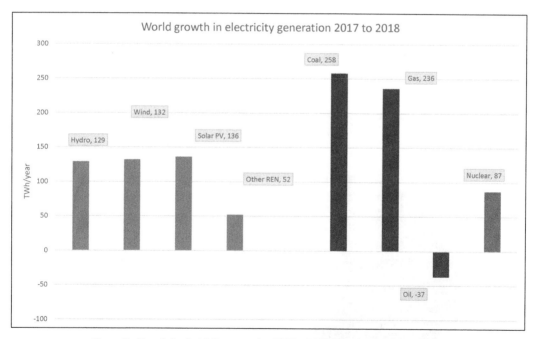

Figure 2. Growth in electricity generation 2017 to 2018 by technology (IEA, 2018).

same as that of wind (132 TWh) and solar PV (136 TWh), see Figure 2. Unfortunately, the strong growth in renewables was still not enough to meet total demand increase (900 TWh) and generation from fossil fuel powered plants increased even more than for renewables, by 258 TWh for coal and 236 TWh for gas, according to IEA (IEA, 2018).

2.2 Status today

Hydropower is today a major source of electricity, with an annual generation of 4239 TWh (2018), representing almost 16% of the global electricity generation. This means that more than 1 billion people covered their electricity consumption from hydropower. Hydropower generation in six different world regions in 2016 can be seen in Table 1. Most recent regional data are still from 2016. Hydropower is the third largest source of electricity generation, behind coal (37.9%) and natural gas (22.8%) but well ahead of nuclear (10.2%), wind (4.6%) and solar (2.1%), see Figure 3. It is worth noticing that hydropower generation is still more than twice that of wind and solar power combined.

Today, hydropower is produced in 159 countries, the top 10 hydropower producing countries in the world in 2016 are listed in Table 2. Four leading countries: China, Canada, Brazil and the USA, together produced 2253 TWh, more than half of the world total. China alone produced almost twice as much

Table 1. Hydropower generation and capacity in six different regions (IHA, 2017; IRENA, 2018A).

Region	Hydropower generation 2016 (TWh/year)	Hydropower capacity *) 2016 (GW)	Capacity factor (Average)
North America	702	177	0.45
South America	709	170	0.48
Europe	595	190	0.36
Africa	106	28	0.43
Asia/Eurasia	1950	537	0.41
Australasia/Oceania	40	13	0.35
World	**4102**	**1115**	**0.42**

*) Not including Pumped Storage.

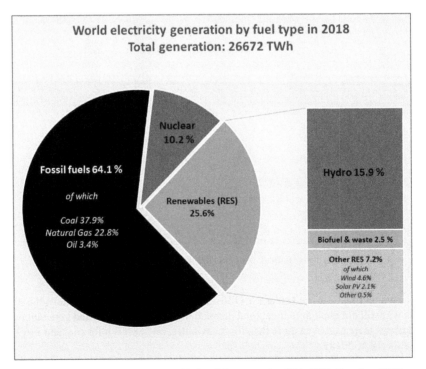

Figure 3. Share of fuel source for world electricity generation (IEA, 2018; Enerdata, 2018).

Table 2. Generation in the 10 top hydropower countries in 2016 (Enerdata, 2018; IRENA, 2018A).

Country	Total generation (2016)	Renewable (RES)	Hydro	Hydro % share of	
	TWh	TWh	TWh	RES	Total
China	6165	1523	1193	78.4	19.4
Canada	677	434	387	89.3	57.2
Brazil	579	466	381	81.8	65.8
USA	4316	637	292	45.9	6.8
Russia	1090	186	186	99.8	17.0
Norway	150	145	144	99.1	96.0
India	1463	189	130	68.8	8.9
Japan	1025	159	85	53.6	8.3
Venezuela	106	76	76	99.9	71.7
Turkey	274	90	67	74.5	24.5
Sum/Average	**15845**	**3903**	**2942**	**75**	**18.6**

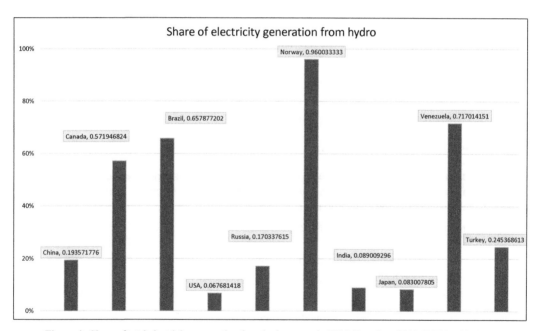

Figure 4. Share of total electricity generation from hydropower in 2016 (Enerdata, 2018; IRENA, 2018A).

hydropower as the whole of Europe, and is still increasing its hydropower generation capacity rapidly. Nearly 20% of all electricity generation in China now comes from hydropower. The table illustrates the high share of hydropower among renewables in these countries, from 46% to nearly 100% and 75% on average. The hydropower share of total generation varies from only 6.8% (USA) to 96% (Norway), on average 18.6%, see Figure 4.

It may come as a surprise, but almost all the largest power plants in the world are based on hydropower. Figure 5 shows the capacity and typical annual generation in the 10 world's largest power plants (by capacity). Nine out of these are hydropower plants. The only non-hydro is the nuclear plant Kashiwazaki-Kariwa in Japan. This station was, however, taken out of operation after the Fukushima disaster and has not been operational since, so generation is not shown.

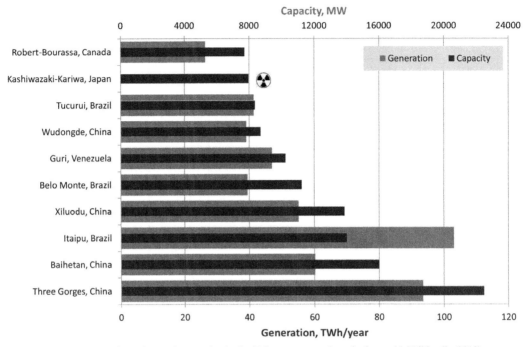

Figure 5. Capacity and annual generation in the 10 largest power plants in the world (Wikipedia, 2019).

3. Main Components of Hydropower System

3.1 Structure of a hydropower plant

A hydropower plant includes civil structures to store, divert and transport water, and mechanical and electrical components in order to convert energy to mechanical and electrical energy. A hydropower system typically consists of the following main components:

– Storage reservoir(s) (Not for Run-of-River plants)
– Intake in the reservoir or in the river
– Waterways to collect and transport water to the powerhouse
– Powerhouse with mechanical and electrical equipment
– Waterways to transport water back from powerhouse to outlet in downstream water body.

The main technical components are illustrated on Figure 6:

– Dam creating a reservoir or intake pond
– Intake with gates for control of water release
– Waterway ("head race") to transport water to the power plant (tunnel, canal, pipe…)
– Surge tank for control of pressure variations and to eliminate water hammer
– Penstock (Except instream and ultra-low head projects)
– Powerhouse housing electro mechanical equipment
– Turbine ("Prime mover")
– Generator
– Control equipment for monitoring and operation
– Transformer
– Power lines
– Waterway ("tail race") to transport water from the powerplant to downstream water body.

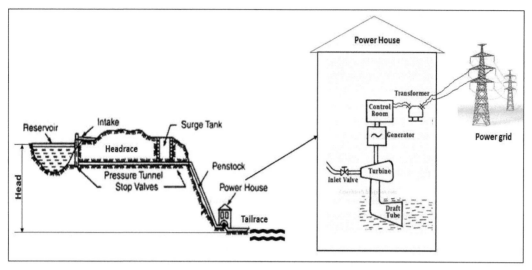

Figure 6. Structure and main components of a high-head storage hydropower plant.

In addition, there will usually be other important components, such as gates and valves, trash racks, ventilation system, drainage system, governor, power cables, switchyards, etc.

A hydropower plant is normally tailored to optimize the utilization of available water and head. All the components shown in Figure 6 may not be needed for all types of hydro projects, for example reservoir, surge tank, tunnel or penstock.

There are two basic types of turbine, reaction and impulse, that can be used in order to maximize efficiency and reliability, depending on head and water flow. The most common types are Pelton (impulse) and Francis turbine (reaction) for high and medium head situations, and Kaplan and Propeller turbines (reaction) for lower head systems. The efficiency of a turbine varies very much depending on relative discharge (% of full load), therefore it is very important to be able to run at or as close as possible to the "best-point" for most of the time. For a run-of-river plant with large variation in inflow, this may be difficult if only one turbine is installed. Two or more turbines will allow operation at better efficiency over a wider range of flow, but also increase the cost of construction.

3.2 Classification of hydropower projects

Hydropower projects can be found over a continuum in scale, from very small units < 1 kW up to megaprojects like China's Three Gorges with 22.5 GW installed capacity (22500000 kW) and an annual generation of close to 100 TWh/year. Hydropower projects are usually classified into four main types, depending on their purpose and technical solutions:

- Storage hydro—with a reservoir
- Run-of-river (RoR) hydro—without a reservoir
- Pumped storage hydro (PSH)
- Instream/hydrokinetic hydro

Hydropower projects are also often classified according to size (pico, micro, mini, small, medium, large) or by head (low, medium, high). Classification according to size is administratively simple, but, to some degree, is arbitrary and used for simplifying the procedure of fiscal support and taxation. Concepts like 'small hydro' or 'large hydro' are not very useful as indicators of impacts, economics or other characteristics. There is yet no consensus on how to classify by size, different definitions are used in different countries (IPCC, 2011).

3.3 Run-of-river (RoR) hydropower plants

An RoR hydropower plant is a plant where little or no water storage is provided, it generates electricity from the available flow of the river at any given time. Such plants may sometimes include a short-term storage or "pondage", giving from a few hours up to daily flexibility in adapting generation to the demand profile. The generation profile will, to a large degree, be determined by the natural river flow conditions, or by the release profile from upstream storage if it is part of a cascade. In the absence of any pondage or upstream reservoirs, generation depends entirely on flow and typically may have substantial daily, seasonal and year-to-year variations.

3.4 Storage hydropower plants

Storage hydropower plants include reservoir(s) to impound water, which can be stored and released later when needed. Water stored in reservoirs provides flexibility to generate electricity on demand, reducing dependency on the variability of inflow. Large reservoirs can store inflow for months or even years, but are usually designed for seasonal storage, to store water during wet seasons and supply water during dry seasons. With the ability to control water flows, storage reservoirs are often also used in multi-purpose projects, providing additional benefits like flood control, water supply, irrigation, navigation and recreation.

3.5 Pumped-Storage Hydropower plants (PSH)

In a PSH, water can be pumped from a lower reservoir into an upper reservoir when energy demand is low and released back from the upper reservoir through turbines to generate electricity later, when needed. This cycle can happen several times a day. The round-trip efficiency (pumping generation) is high, from 75% to 80%. In this way, excess electricity, for example from wind power, can be stored in a battery. Pumped storage currently represents 99% of all on-grid electricity storage in the world. PSH projects will usually not be a net producer, but may have some natural inflow to the upper reservoir, which will increase the generation. Energy stored in a PSH is directly proportional to the water volume stored in the upper reservoir and the elevation difference between reservoirs. A major advantage of PSHs is their ability to interact with other variable renewables, such as wind and solar power. PSH installations can store excess energy during periods of high wind or high insolation, and provide backup reserve which is immediately dispatchable during periods when the other variable power sources are unavailable.

3.6 In-stream (hydrokinetic) hydropower plants

In-stream energy can be derived from the movement (kinetic energy) of water flowing in rivers and canals, or from tidal flow and ocean currents. This technology differs from traditional hydropower plants, which rely on the elevation and pressure difference (head) between the intake and outlet. Hydrokinetic devices are placed directly in the stream of flowing water and energy is extracted from the kinetic energy in the water, like wind turbines in air. Kinetic energy in the flowing water is converted to mechanical energy by a propeller that drives a generator which produces electricity.

Because it is powered by kinetic energy instead of potential energy, it is also known as a 'zero-head' turbine. As such, no dams and/or head differential are necessary for the operation of this device; the course of a river remains almost in its natural state. However, the efficiency and amount of energy that can be extracted is low, and the cost high. So far, very few of these power plants have been put in operational use in rivers and contribution to electricity generation is still insignificant.

4. Hydropower Resources—Overview

Hydropower is a renewable energy source utilizing water moving in the hydrological cycle (water cycle), which is "powered" by radiation energy from the sun. Close to 50% of all solar energy reaching the surface of the earth is used for evaporating water and thereby converted to latent energy in water vapour. Air currents bring some of this water vapour in over land, where it condenses into clouds and precipitation. Precipitation falling on land generates runoff which moves back to the ocean under the influence of gravity. This runoff is the basis for hydropower generation and, since the water cycle is powered by solar energy, water will continue to flow as long as the sun shines, ensuring a renewable energy supply. The distribution of precipitation and elevation on land area determine the runoff volume and timing and thereby the potential for hydropower generation.

4.1 How to compute hydropower generation potential

Hydropower (except hydrokinetic plants) is generated by converting potential energy in water as it moves from high to a lower elevation, into mechanical and electrical energy. The output (Power, P) depends on three factors, flow (Q), head (H) and efficiency (η) and can be computed using equation 1:

$$P = \rho * g * Q * H * \eta \ (W) \tag{1}$$

P   Electrical power output (J/s = W)

ρ Density of water (1000 kg/m^3)

g   Acceleration of gravity (9.81 m/s^2)

Q   Water flow per unit time (m^3/s)

H   Elevation drop or Head (m)

η Efficiency in the conversion process

Assuming values for density and acceleration of gravity as given above, the equation can be simplified to give output directly in kW, equation 2:

$$P = 9.81 * Q * H * \eta \ (kW) \tag{2}$$

The amount of energy (E) produced depends on the duration of the flow. If duration Δt is given in hours (h) the amount of energy produced (kilowatt hour, kWh) can be computed using equation 3:

$$E = P * \Delta t \ (kWh) \tag{3}$$

Another useful equation gives the energy output in kWh per m^3 of water as a function of H and η, often called "Energy equivalent of water" (EEKV), equation 4:

$$EEKV = P * 1h/Q*3600s \ \ = \ \ 9.81*H*\eta/3600 \ \ (kWh/m^3) \tag{4}$$

In addition to kWh, other commonly-used units of electrical energy are megawatt hour (MWh), Gigawatt hour (GWh) and Terawatt hour (TWh), where:

1 MWh = 1000 kWh   (10^6 Wh)

1 GWh = 1000 MWh   (10^9 Wh)

1 TWh = 1000 GWh   (10^{12} Wh)

For comparison to other renewable or thermal energy sources, the unit Exajoule (EJ) is often used. One EJ (10^{18} J) equals 277.78 TWh, and one TWh equals 0.0036 EJ. The global hydropower production in 2018 was 4239 TWh, corresponding to 15.26 EJ.

The capacity factor (Cf) of a power plant is the ratio of the total energy output from the power plant over a period of time (typically one year) to its potential output if it had operated at full rated (nameplate) capacity over the same period. This is also called plant load factor (plf) or capacity utilization factor (CUF). Typical capacity factors for hydropower plants are in the range of 0.35 to 0.5, world average is 0.42 for hydropower, see Table 1. Due to maintenance or other limitation, sometimes equipment is not available for power generation even if water and demand are there. This is known as "plant availability factor" (PAF).

4.2 Definition of hydropower potential

In order to compute how much hydropower that can be generated (the potential) within an area (a catchment, a region, a country, etc.), the usual procedure is to identify all feasible sites with a suitable combination of flow (Q) and Head (H), where hydropower plants can be located. The potential at each site is computed by applying equations (1–3), considering the topography, volume of water and its variability in time. Results are usually given as potential annual energy generation (GWh/year), for each site. Hydropower potential within the area is then computed by adding up the potential production from all feasible sites, omitting sites with environmental restrictions, high cost or social constraints.

It can be argued that this definition of potential is not precise because the selection of 'feasible sites' will depend on technology, economic parameters and social and environmental preferences, and all these can change with time. This has led to the use of other definitions like: 'Theoretical potential', 'Technical potential' and 'Economic potential'. Some also argue for the use of definition 'Sustainable potential'.

4.3 Global and regional hydropower potential

The International Journal on Hydropower & Dams World Atlas & Industry Guide (IJHD), provides the most comprehensive inventory of current hydropower installed capacity and annual generation, and hydropower resource potential, updated every year. The Atlas provides three measures of hydropower resource potential, all in terms of annual generation, but gives no detailed explanation of how the potentials were computed or limits for what is defined as technically or economically feasible. A possible explanation for each, based on other sources, is given in parenthesis:

Gross theoretical potential	(Gross potential at all known sites)
Technically feasible potential	(Portion of potential with cost low enough to justify site assessment)
Economically feasible potential	(Potential with cost less or equal to large thermal power plants)

The total global technical potential for hydropower is estimated at 15778 TWh/year (IJHD, 2017), this is nearly four times the current generation. Economically feasible potential was estimated at 9623 TWh/year, more than twice the existing generation in 2018 (4239 TWh/year). The different potential estimates are summarized in Table 3 and visualized in Figure 7.

Table 3. Global hydropower potential (IJHD, 2017).

Region	Gross theoretical potential		Technically feasible		Economically feasible	
	TWh/year	%	TWh/year	%	TWh/year	%
North America	7601	18	1891	12	1045	11
South America	7848	19	2859	18	1728	18
Europe	3136	7	1195	8	852	9
Africa	4423	11	1647	10	1124	12
Asia	18248	44	8000	51	4786	50
Australasia/Oceania	658	2	186	1	89	1
World	**41914**	100	**15778**	100	**9624**	100

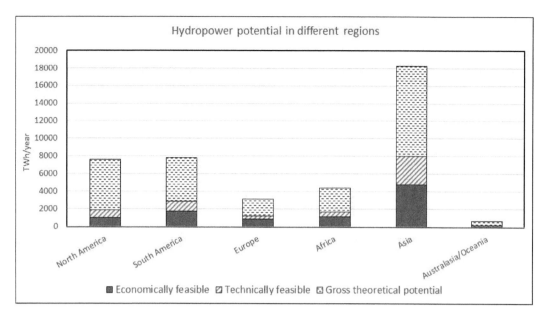

Figure 7. Different estimates of hydropower potential in six regions (IJHD, 2017).

5. Cost Issues

Cost of hydropower is site specific and can, therefore, vary considerably from one project to another. The main cost components are:

(1) Upfront investment (capital cost)
(2) Operation and maintenance (O&M) cost
(3) Decommissioning cost

The Levelized Cost Of Energy (LCOE) includes all these cost elements for the entire lifetime of the project, and is usually given in units of US c/kWh or US $/MWh. The LCOE for hydropower depends on these cost components, but also on several others:

(4) Capacity factor
(5) Lifetime of project
(6) Cost of capital (discountrate).

Capital cost includes the cost of civil structures (dams, tunnels, powerhouse, etc.), electro-mechanical equipment (turbine, gates, governor, generator, transformer, control systems, etc.), access roads, powerlines, cost of planning and cost of mitigation measures (mitigation, resettlement, fish ladders, etc.). The cost of civil structures is usually the dominating share in large projects (60–70%), but for small projects, the electro-mechanical cost can be 50% or more. Typical capital cost for hydropower varies from < 1000 $/kW up to 3000 $/kW capacity for large hydro and from 1500 to 6000 $/kW or even higher for small hydro (IPCC, 2011; IRENA, 2015; NREL, 2012; IHA, 2018). However, there are also many examples of projects with costs as low as 500 $/kW, for especially good sites.

Typical lifetime for hydropower varies from 40 to 80 years or more (IPCC, 2011). Electro-mechanical equipment usually has a shorter lifetime and civil structures a longer lifetime (50–100%). During an average life span of 80 years, it will be necessary to replace much of the electro-mechanical equipment at least once; the cost of this is included in the typical 2.5% per year assumed O&M cost.

This can be compared to LCOE for other renewable energy sources (RES), see Figure 8 which is based on data from a large number of recent projects, compiled and published by The International

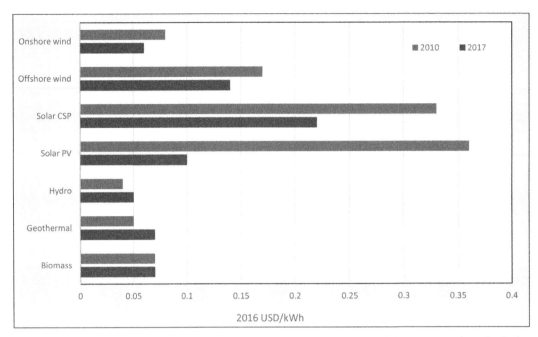

Figure 8. Average global levelized cost (USD/kWh) of electricity from utility-scale renewable power generation technologies, 2010 and 2017. Based on data from (IRENA, 2019).

Renewable Energy Agency (IRENA, 2015; IRENA, 2017). The figure shows the cost trend for the seven most important renewable technologies during recent years (2010–2016). Hydropower is, on average, still the technology with the lowest average cost, followed by onshore wind, geothermal and bioenergy. A closer look into IRENAs cost database reveals that small hydro generally has a higher cost than large hydro, and that there are large regional differences. Regions like Europe and North America, where a larger share of available resources have already been developed, show a higher cost than regions with large untapped potential, like Asia and South America.

The following is a summary of comments from the IRENA report on power cost of renewables (IRENA, 2015):

"Hydropower produces some of the lowest-cost electricity of any generation technology. The LCOE of large-scale hydro projects at excellent sites can be as low as USD 0.02/kWh, while average costs are around USD 0.05/kWh where untapped economic resources remain. Small-scale hydropower can also be very economic, although typically it has higher costs and is sometimes more suitable as an option for electrification that can provide low-cost electricity to remote communities or for the local grid.

There is a clear cost dichotomy for hydropower between regions with remaining economic resources to exploit and those where most of the economic resources have been exploited already. Asia, Africa and South America all experience LCOEs for hydropower projects of on average USD 0.04 to USD 0.05/kWh. In contrast, in regions which have exploited their most economic resources, weighted average LCOE ranges are around USD 0.09 to USD 0.10/kWh (e.g., in Europe, Eurasia, North America and Oceania). In addition to the higher costs, these regions are also constrained in the amount of economic capacity that still remains to be added." (IRENA, 2015).

An overview of typical cost components in a large (500 MW) hydropower plant is given in Figure 9. Civil cost components (dam, tunnels, powerhouse, etc.) will typically add up to over 50% of the total cost, while electro-mechanical components are typically < 20%. For small hydro and projects without dams and long tunnels, the share of electro-mechanical cost is usually larger, up to 50% or more.

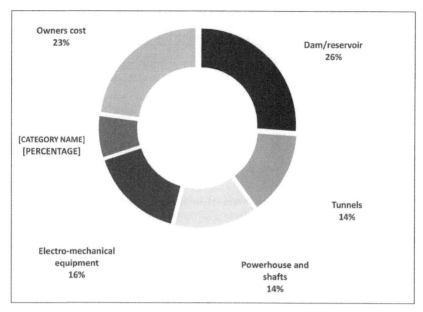

Figure 9. Capital cost breakdown for a 500 MW hydropower plant. Based on data from (NREL, 2012).

6. Sustainability Issues

Sustainable development has been defined as ".. development that meets the needs of the present without compromising the ability of future generations to meet their own needs" (WCED, 1987). In the process of transition from a fossil fuel to a renewable energy system, it is very important to ensure that the future renewable energy system will be sustainable, considering both economic, social and environmental sustainability. The sustainability issue is of particular importance for hydropower, since it depends on and can have significant impact on an equally important resource: Water.

Besides economic sustainability, which always will have high priority, other important aspects to be assessed, evaluated and documented in a sustainability analysis are as follows:

– Environmental and Social impacts
– Greenhouse gas (GHG) emissions and 'carbon footprint'
– Energy Payback Ratio/Energy Return On Investment (EPR/EROI)
– Water consumption and 'water footprint'
– Sediment issues - Reservoir sedimentation
– Climate change impacts and climate resilience

Hydropower has many advantages compared to other sources of electrical energy, especially compared to those based on fossil fuels: it is renewable, clean, largely carbon-free and well suited for integration into the grid, often supporting the stability of the grid. But hydropower can also have negative impacts, especially on the local level, where dams and reservoirs, weirs, diversions and changes in river flow may disturb ecosystems and create problems.

Hydropower schemes will often modify the flow regime in rivers and the water levels regime in lakes and reservoirs and can, therefore, have negative consequences for ecosystems and biodiversity. RoR projects usually do not change the flow regime, while storage hydro projects typically lead to changes in the seasonal flow regime, for example by decreasing spring and summer flows and increasing winter flows in Nordic rivers.

The natural transport of sediments may also be altered, creating problems both in the reservoir and intake pond where sediments are deposited, and downstream of dams where reduced sediment concentrations may lead to increased bank and riverbed erosion, and decreased stability.

Table 4. Median life-cycle carbon equivalent intensity (IHA, 2018B).

Technology	Carbon emission (g CO_2–eq/kWh)
Coal	820
Gas	490
Solar PV	48
Hydropower	18.5
Wind Offshore	12
Nuclear	12
Wind Onshore	11

The energy payback ratio (EPR) is the ratio of total energy produced during a system's normal lifespan to the energy required to build, maintain and fuel that system. Other metrics that refer to the same basic calculation include the energy returned on energy invested (EROI). A high EPR indicates good performance. Lifecycle energy payback ratios for hydropower plants reach the highest values of all energy technologies, ranging from 170 to 267 for run-of-river, and from 205 to 280 for reservoirs (IPCC, 2011).

A few observations of high greenhouse gas (GHG) emissions from tropical hydropower reservoirs has been raising doubt about the low GHG emission levels for hydropower, but it is still not clear if these observations represent a net emission or emissions that occur naturally, for example from a wetland (Prairie et al., 2018).

The comparison of typical carbon emission caused by various electricity production sources is shown in Table 4. One can see that the life-cycle emission of hydropower is very low compared to fossil fuel-based sources, lower than for Solar PV and in the same range as nuclear and wind power.

In order to assess the long-term sustainability of hydropower projects, it is necessary to assess all relevant social, economic and environmental consequences of the project. Sustainability issues are becoming increasingly important for hydropower developers, and the International Hydropower Association (IHA) has, therefore, developed the Hydropower Sustainability Assessment Protocol, containing tools to guide in planning, implementation and operation of hydropower projects (IHA, 2019). The protocol includes between 19 and 23 relevant sustainability topics, depending on the development stage of the project. It is the result of a long process involving many different stakeholders: Social and environmental NGO's, governments, commercial organizations, development banks and the hydropower sector. Some of the topics included are biodiversity, indigenous people, infrastructure safety, resettlement, water quality, erosion and sedimentation and downstream flow regimes. The tools can be downloaded from IHA web pages (IHA, 2019).

7. Integration into Water Management System

Water, energy and food production are closely linked. On the one hand, water availability is crucial for many energy technologies, including hydropower and cooling water for thermal power plants, and on the other hand, energy is needed to secure water supply for agriculture, industries and households, particularly in water-scarce areas. This mutual dependence, the 'water-energy-food nexus', has led to the understanding that water, energy and food production must be addressed in a holistic way, also considering impacts of climate change and project sustainability. The challenge to provide energy, water and food for an increasing population in a sustainable way, will require improved regional and global water governance.

Since hydropower projects often include large water storage facilities (reservoirs), hydropower can play an important role in providing both energy, water and food security (Killingtveit, 2014; IEA Hydropower, 2018). Out of 45000 large dams and water reservoirs in the world, about 11000 (25%) have hydropower capacity. Therefore, hydropower development is often also part of water management systems, as much as energy management systems, both of which are increasingly becoming climate driven.

Water management services consists of both qualitative and quantitative functions, such as:

– Flood control
– Drought control
– Ground water stabilization
– Water supply (increased water availability for other uses especially during lean period and areas where water is not available)
– Sediment management
– Habitat management (via flow management during the year)
– Barrier to saline water intrusion
– Navigation (transport)
– Irrigation (agriculture, food production)
– Leisure and tourism
– Aquaculture (fisheries and food)
– Cooling water (Industry, thermal power)

8. Integration into Broader Power System

In addition to providing capacity to meet electricity demand, hydropower has several characteristics that enable it to provide other important services to make power systems operation more reliable:

– Start generating electricity on very short notice and with low start-upcosts.
– Provide rapid changes in generation and have a high part-load efficiency (fast ramping).
– Restore a power station to operation without relying on the electric power transmission network (black start capability during grid failure).

The ability to rapidly change output in response to system needs without suffering large decreases in efficiency, makes hydropower plants well suited to providing the balancing services called regulation and load-following. The almost instantaneous regulating of production of electricity from hydro generators make hydropower very cost-attractive for balancing purposes (IEA Hydropower, 2018).

The need for balancing arises from the time lag between planning the production and the actual consumption. Demand may be stochastic due to the influence of varying weather, and there may also be stochastic events on the producer and transmission side, like accidents within generators and transformers, mishaps on the transmission lines and other disturbances.

Because the physical balance between supply and demand must be continuous, access to regulating supply up or down is essential if short-term physical restriction of demand or blackout is to be avoided. It is believed that the introduction of more intermittent energy will increase the profitability of hydro in the balancing market (EASE/EERA, 2018; IEA, 2014).

Today's electricity system is changing rapidly, creating new opportunities for hydropower to contribute to system resilience, reliability, and affordability. Therefore, interest in PSH is increasing, particularly in regions where solar PV and wind power are reaching high levels of penetration or are growing rapidly. PSH is among the most efficient and flexible large-scale means of storing energy available today. It is highly cost-effective when compared with other sources of energy storage, like a battery, on life cycle basis (Krueger et al., 2018). It allows not only the production of electric energy, as hydropower plants do, but also the storage of energy in the form of the gravitational potential energy of the water. During periods with high demand or high energy prices, the water, stored in an upper reservoir, is released through turbines to a lower reservoir in order to produce electricity. During periods with low demand or energy prices, the water is pumped back from the lower reservoir to the upper reservoir in order to store it. PSH technology can ramp up to full production capacity within minutes, providing a quick response for peak-load energy supply and making it a useful tool for many grid services (EASE/ EERA, 2018; IEA, 2014):

- Provision of contingency reserve to restore the balance of supply and demand
- Provision of regulation reserve
- Load following
- Load shifting (energy arbitrage)
- Black start
- Voltage support

The majority of existing pumped storage capacity, 176 GW, is found in Europe, Japan and the United States of America. According to data from IHA, this is expected to increase rapidly in the future. There are now more than 100 projects in the pipeline around the world, totaling 75 GW of new capacity. This is expected to increase the world total PSH capacity to 250 GW. The majority of the new projects will be operational by 2030 (IHA, 2018A).

9. Future Deployment of Hydropower

The transition from a fossil (carbon-based) electricity system to a renewable system has high political priority, and all possible renewable sources needs to be developed. So far, hydropower, wind power and solar PV have given the largest contribution to a "greener" power system. Also, in the future, these three seem to have to take the lion's share in the process. Hydropower will have a special role, as the only renewable technology (beside bioenergy) that can provide energy storage and dispatchable power. Increasing the share of wind and solar power is expected to increase the demand for such services from hydropower.

Based on information about hydropower potential (Table 3) and what has already been developed (Table 1), it is possible to get an estimate of remaining resources. This information is summarized in Table 5 for six main regions and globally. As seen in the table, the remaining economically feasible potential is still well above 5000 TWh. Not all of this will be possible to develop, mainly due to social and environmental restrictions, but the figure still shows a large potential for further development, much more than what has already been developed. Furthermore, if the cost limit is increased, we can see that the technically feasible resources are very large, up to nearly 12000 TWh or 2.5 times the existing development.

Looking at the different regions, one can see that, for economically feasible projects, there are three distinct classes: Europe and North America both have about 30% left to develop, Africa over 90% and the rest of the world around 55–60%.

Considering hydropower's moderate cost, long lifetime, high energy payback ratio, low GHG emissions and important energy management services, combined with manageable social and

Table 5. Remaining (undeveloped) hydropower resources by 2016.

Region	Remaining (undeveloped) share of technically feasible hydropower		Remaining (undeveloped) share of economically feasible hydropower	
	TWh/year	%	TWh/year	%
North America	1189	63	343	33
South America	2150	75	1019	59
Europe	600	50	257	30
Africa	1541	94	1018	91
Asia	6050	76	2836	59
Australasia/Oceania	146	78	49	55
World	**11676**	**74**	**5522**	**57**

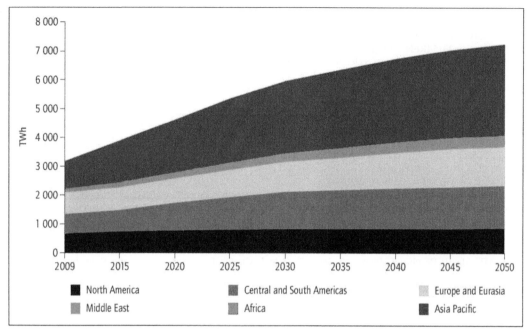

Figure 10. Annual hydroelectricity generation till 2050 in the Hydropower Roadmap vision (IEA, 2012).

environmental impacts, it seems safe to assume that hydropower will continue to be developed up to a level of at least twice the current generation by 2050, possibly before.

One example of projections for future deployment is given in Figure 10. This projection was based on data from 2009, and has been surprisingly good up to 2018, though slightly too low.

References

EASE/EERA. 2018. Energy Storage Technology Development Roadmap. https://www.eeraset.eu/ease-eera-energy-storage-technology-development-roadmap-download-your-copy/.

Enerdata. 2018. Global Energy Statistical Yearbook. https://yearbook.enerdata.net/electricity/world-electricity-production-statistics.html.

IEA. 2012. Technology Roadmap–Hydropower OECD/IEA. https://www.iea.org/publications/freepublications/publication/2012_Hydropower_Roadmap.pdf.

IEA. 2014. Technology Roadmap Energy Storage. https://www.iea.org/publications/freepublications/publication/TechnologyRoadmapEnergystorage.pdf.

IEA. 2017. Key World Energy Statistics. https://www.iea.org/publications/freepublications/publication/KeyWorld2017.pdf.

IEA. 2018. Global Energy & CO$_2$ Status Report. https://www.iea.org/geco/data/.

IEA Hydropower. 2018. Valuing Hydropower Services: The Economic Value of Energy and Water Management Services provided by Hydropower Projects with Storage. https://www.ieahydro.org/news/2018/2/iea-hydros-report-on-valuing-hydropower-servicesavailable-for-download.

IHA. 2017. Hydropower Status Report. https://www.hydropower.org/publications/2017-hydropower-status-report.

IHA. 2018A. The world's water battery: Pumped hydropower storage and the clean energy transition. IHA working paper December 2018. https://www.hydropower.org/publications/the-world's-water-battery-pumped-hydropowerstorage-and-the-clean-energy-transition.

IHA. 2018B. Hydropower status report: sector trends and insights. International Hydropower Association, London, 2018, p 104. https://www.hydropower.org/publications/2018-hydropower-status-report.

IHA. 2019. Hydropower Sustainability Guidelines. http://www.hydrosustainability.org/Hydropower-Sustainability-Guidelines.aspx.

IJHD. 2017. World Atlas & Industry Guide. International Journal of Hydropower and Dams, Wallington, Surrey, UK.

IPCC. 2011. IPCC Special Report on Renewable Energy Sources and Climate Change Mitigation. Prepared by Working Group III of the Intergovernmental Panel on Climate Change. *In*: Edenhofer, O., Pichs-Madruga, R., Sokona, Y.,

Seyboth, K., Matschoss, P., Kadner, S., Zwickel, T., Eickemeier, P., Hansen, G., Schlömer, S. and von Stechow, C. (eds.). Cambridge University Press, Cambridge, United Kingdom and New York, NY, USA, 1075 pp.

IRENA. 2015. Renewable Power Generation Costs in 2014. https://www.irena.org/-/media/Files/IRENA/Agency/Publication/2015/IRENA_RE_Power_Costs_2014_report.pdf.

IRENA. 2017. Rethinking Energy. Accelerating the global energy transformation. ISBN 978-92-96111-06-6. https://www.irena.org/publications/2014/Sep/REthinking-EnergyTowards-a-new-power-system.

IRENA. 2018A. Database on Renewable Electricity Capacity and Generation Statistics. https://www.irena.org/Statistics/View-Data-by-Topic/Capacity-and-Generation/Query-Tool.

IRENA. 2018B. Renewable Energy Statistics. https://www.irena.org/media/Files/IRENA/Agency/Publication/2018/Jul/IRENA_Renewable_Energy_Statistics_2018.pdf.

IRENA. 2019. Renewable Power Generation Costs in 2018. ISBN: 978-92-9260-126-3, May 2019. https://www.irena.org/publications/2019/May/Renewable-power-generation-costs-in-2018.

Killingtveit, Å. 2014. Hydroelectric Power. *In*: Letcher, T. Future Energy. Improved, Sustainable and Clean Options for Our Planet. Elsevier ISBN 978-0-08-099424-6.

Krueger, K. et al. 2018. Li-Ion Battery versus Pumped Storage—A Comparison of Raw Material, Investment Costs and CO_2-Footprints. Seventh International Conference and Exhibition on Water Resources and Renewable Energy Development in Asia. Danang, Vietnam, March 2018, p 10.

NREL. 2012. National Renewable Energy Laboratory. Cost and Performance Data for Power Generation Technologies. Prepared by Black & Veatch. https://www.bv.com/docs/reports-studies/nrel-cost-report.pdf.

Prairie, Y.T., Alm, J., Beaulieu, J., Barros, N., Battin, T., Cole, J., Giorgio, P., DelSontro, T., Guerin, F., Harby, A., Harrison, J., Mercier-Blais, S., Serca, D., Sobek, S. and Vachon, D. 2018. Greenhouse gas emissions from freshwater reservoirs: What does the atmosphere see? Ecosystems 21: 1058–1071.

Ritchie, H. and Roser, M. 2019. Renewable Energy. Published online at OurWorldInData.org. https://ourworldindata.org/renewable-energy.

WCED. 1987. Our common future. Report from United Nations World Commission on Environment and Development (WCED). Oxford, University Press. https://en.wikipedia.org/wiki/Our_Common_Future.

Wikipedia. 2017. List of Largest Power Stations. https://en.wikipedia.org/wiki/List_of_largest_power_stations.

Index

Printed and bound by CPI Group (UK) Ltd, Croydon, CR0 4YY

24/10/2024

01778288-0014